Energy Systems

Series Editor:

Panos M. Pardalos, University of Florida, USA

For further volumes:
http://www.springer.com/series/8368

Steffen Rebennack • Panos M. Pardalos
Mario V.F. Pereira • Niko A. Iliadis
Editors

Handbook of Power Systems II

Springer

Editors

Dr. Steffen Rebennack
Colorado School of Mines
Division of Economics and Business
Engineering Hall
816 15th Street
Golden, Colorado 80401
USA
steffen.rebennack@gmail.com

Prof. Panos M. Pardalos
University of Florida
Department of Industrial and Systems Engineering
303 Weil Hall, P.O. Box 116595
Gainesville FL 32611-6595
USA
pardalos@ufl.edu

Dr. Mario V. F. Pereira
Centro Empresarial
Rio Praia de Botafogo
228/1701-A-Botafogo
CEP: 22250-040
Rio de Janeiro, RJ
Brazil
mario@psr-inc.com

Dr. Niko A. Iliadis
EnerCoRD
Plastira Street 4
Nea Smyrni 171
21, Athens
Greece
niko.iliadis@enercord.com

ISBN: 978-3-642-12685-7 e-ISBN: 978-3-642-12686-4
DOI 10.1007/978-3-642-12686-4
Springer Heidelberg Dordrecht London New York

Library of Congress Control Number: 2010921798

Cover illustration: Cover art is designed by Elias Tyligadas

Cover design: WMXDesign GmbH, Heidelberg

Printed on acid-free paper

Springer is part of Springer Science+Business Media (www.springer.com)

To our families.

Preface of Volume II

Power systems are undeniably considered as one of the most important infrastructures of a country. Their importance arises from a multitude of reasons of technical, social and economical natures. Technical, as the commodity involved requires continuous balancing and cannot be stored in an efficient way. Social, because power has become an essential commodity to the life of every person in the greatest part of our planet. Economical, as every industry relates not only its operations but also its financial viability in most cases with the availability and the prices of the power.

The reasons mentioned above have made power systems a subject of great interest for the scientific community. Moreover, given the nature and the specificities of the subject, sciences such as mathematics, engineering, economics, law and social sciences have joined forces to propose solutions.

In addition to the specificities and inherent difficulties of the power systems problems, this industry has gone through significant changes. We could refer to these changes from an engineering and economical perspective. In the last 40 years, important advances have been made in the efficiency and emissions of power generation, and in the transmission systems of it along with a series of domains that assist in the operation of these systems. Nevertheless, the engineering perspective changes had a small effect comparing to these that were made in the field of economics where an entire industry shifted from a long-standing monopoly to a competitive deregulated market.

The study of such complex systems can be realized through appropriate modelling and application of advance optimization algorithms that consider simultaneously the technical, economical, financial, legal and social characteristics of the power system considered. The term technical refers to the specificities of each asset that shall be modelled in order for the latter to be adequately represented for the purpose of the problem. Economical characteristics reflect the structure and operation of the market along with the price of power and the sources, conventional or renewable, used to be generated. Economical characteristics are strongly related with the financial objectives of each entity operating a power system, and consist in the adequate description and fulfillment of the financial targets and risk profile. Legal specificities consist in the laws and regulations that are used for the operation of the power system. Social characteristics are described through a series of parameters that have to be considered in the operation of the power system and reflect the issues related to the population within this system.

The authors of this handbook are from a mathematical and engineering background with an in-depth understanding of economics and financial engineering to apply their knowledge in what is know as modelling and optimization. The focus of this handbook is to propose a selection of articles that outline the modelling and optimization techniques in the field of power systems when applied to solve the large spectrum of problems that arise in the power system industry. The above mentioned spectrum of problems is divided in the following chapters according to its nature: Operation Planning, Expansion Planning, Transmission and Distribution Modelling, Forecasting, Energy Auctions and Markets, and Risk Management.

Operation planning is the process of operating the generation assets under the technical, economical, financial, social and legal criteria that are imposed within a certain area. Operation is divided according to the technical characteristics required and the operation of the markets in real time, short term and medium term. Within these categories the main differences in modelling vary in technical details, time step and time horizon. Nevertheless, in all three categories the objective is the optimal operation, by either minimizing costs or maximizing net profits, while considering the criteria referred above.

Expansion planning is the process of optimizing the evolution and development of a power system within a certain area. The objective is to minimize the costs or maximize the net profit for the sum of building and operation of assets within a system. According to the focus on the problem, an emphasis might be given in the generation or the transmission assets while taking into consideration technical, economical, financial, social and legal criteria. The time-step used can vary between 1 month and 1 quarter, and the time horizon can be up to 25 years.

Transmission modelling is the process of describing adequately the network of a power system to apply certain optimization algorithms. The objective is to define the optimal operation under technical, economical, financial, social and legal criteria. In the last 10 years and because of the increasing importance of natural gas in power generation, electricity and gas networks are modelled jointly.

Forecasting in energy is applied for electricity and fuel price, renewable energy sources availability and weather. Although complex models and algorithms have been developed, forecasting also uses historical measured data, which require important infrastructure. Hence, the measurement of the value of information also enters into the equation where an optimal decision has to be made between the extent of the forecasting and its impact to the optimization result.

The creation of the markets and the competitive environment in power systems have created the energy auctions. The commodity can be power, transmission capacity, balancing services, secondary reserve and other components of the system. The participation of the auction might be cooperative or non-cooperative, where players focus on the maximization of their results. Therefore, the market participant focus on improving their bidding strategies, forecast the behavior of their competitors and measure their influence on the market.

Risk management in the financial field has emerged in the power systems in the last two decades and plays actually an important role. In this field the entities that participate in the market while looking to maximize their net profits are heavily concerned with their exposure to financial risk. The latter is directly related to the operation of the assets and also with a variety of external factors. Hence, risk mangers model their portfolios and look to combine optimally the operation of their assets by using the financial instruments that are available in the market.

We take this opportunity to thank all contributors and the anonymous referees for their valuable comments and suggestions, and the publisher for helping to produce this volume.

February 2010

Steffen Rebennack
Panos M. Pardalos
Mario V.F. Pereira
Niko A. Iliadis

Contents of Volume II

Contents of Volume I

Contributors

Laetitia Andrieu EDF R&D, OSIRIS, 1 avenue du Général de Gaulle, 92140 Clamart, France, laetitia.andrieu@edf.fr

Julián Barquín Institute for Research in Technology (IIT), Advanced Technical Engineering School (ICAI), Pontifical Comillas University, Alberto Aguilera 23, 28015 Madrid, Spain, julian.barquin@iit.upcomillas.es

Mette Bjørndal Department of Finance and Management Science, Norwegian School of Economics and Business Administration (NHH), Helleveien 30, 5045 Bergen, Norway, mette.bjorndal@nhh.no

Simon Blake Department of Engineering and Computing, Durham University, Durham, UK, s.r.blake@durham.ac.uk

Miguel Carrión Department of Electrical Engineering, EUITI, Universidad de Castilla – La Mancha, Edificio Sabatini, Campus Antigua Fábrica de Armas, 45071 Toledo, Spain, Miguel.Carrion@uclm.es

Efraim Centeno Institute for Research in Technology (IIT), Advanced Technical Engineering School (ICAI), Pontifical Comillas University, Alberto Aguilera 23, 28015 Madrid, Spain

Santiago Cerisola Universidad Pontificia Comillas, Alberto Aguilera 23, 28015 Madrid, Spain, santiago.cerisola@upcomillas.es

Silvano Cincotti Department of Biophysical and Electronic Engineering, University of Genoa, Via Opera Pia 11a, 16146 Genoa, Italy, cincottig@dibe.unige.it

Antonio J. Conejo Department of Electrical Engineering, Universidad de Castilla – La Mancha, Campus Universitario, s/n, 13071 Ciudad Real, Spain

Alberto Cruz Instituto de Investigación Tecnológica, Escuela Técnica Superior de Ingeniería – ICAI, Universidad Pontificia Comillas, C/Alberto Aguilera 23, 28015 Madrid, Spain, Alberto.Cruz@iit.upcomillas.es

Michel De Lara ENPC Paris Tech, 6–8 avenue Blaise Pascal, Cité Descartes – Champs sur Marne, 77455 Marne la Vallée Cedex 2, France, delara@cermics.enpc.fr

Pablo Dueñas Institute for Research in Technology (IIT), Advanced Technical Engineering School (ICAI), Pontifical Comillas University, Alberto Aguilera 23, 28015 Madrid, Spain

Andreas Eichhorn Humboldt University, 10099 Berlin, Germany, eichhorn@math.hu-berlin.de

Félix Fernández Institute for Research in Technology (IIT), Advanced Technical Engineering School (ICAI), Pontifical Comillas University, Alberto Aguilera 23, 28015 Madrid, Spain

Stein-Erik Fleten Department of Industrial Economics and Technology Management, Norwegian University of Science and Technology, 7491 Trondheim, Norway, stein-erik-fleten@iot.ntnu.no

Maria Fonoberova Aimdyn, Inc., 1919 State St., Santa Barbara, CA 93101, USA, mfonoberova@aimdyn.com

Raquel García-Bertrand Department of Electrical Engineering, Universidad de Castilla – La Mancha, Campus Universitario, s/n, 13071 Ciudad Real, Spain

Robert Golob Faculty of Electrical Engineering, University of Ljubljana, Tržaška 25, 1000 Ljubljana, Slovenia, robert.golob@fe.uni-lj.si

Eric Guerci GREQAM, Université d'Aix-Marseille, 2 rue de la Charité, 13002 Marseille, France, eric.guerci@univmed.fr

Holger Heitsch Humboldt University, 10099 Berlin, Germany, heitsch@math.hu-berlin.de

Ronald Hochreiter Department of Finance, Accounting and Statistics, WU Vienna University of Economics and Business, Augasse 2-6, 1090 Vienna, Austria, ronald.hochreiter@wu.ac.at

Rabih A. Jabr Department of Electrical and Computer Engineering, American University of Beirut, P.O. Box 11-0236, Riad El-Solh, Beirut 1107 2020, Lebanon, rabih.jabr@aub.edu.lb

Kurt Jörnsten Department of Finance and Management Science, Norwegian School of Economics and Business Administration (NHH), Helleveien 30, 5045 Bergen, Norway, kurt.jornsten@nhh.no

Loi Lei Lai City University London, UK, l.l.lai@city.ac.uk

Jesus M. Latorre Universidad Pontificia Comillas, Alberto Aguilera 23, 28015 Madrid, Spain, jesus.latorre@upcomillas.es

Jacques Lawarree Department of Economics, University of Washington, Box 353330, Seattle, WA 98195, USA, lawarree@u.washington.edu

Guang Li Market Operations Support, Electric Reliability Council of Texas, 2705 West Lake Drive, Taylor, TX 76574, USA, gli@ercot.com

Chen-Ching Liu School of Electrical, Electronic and Mechanical Engineering, University College Dublin, National University of Ireland, Belfield, Dublin 4, Ireland, liu@ucd.ie

Juan Marín Instituto de Investigación Tecnológica, Escuela Técnica Superior de Ingeniería – ICAI, Universidad Pontificia Comillas, C/Alberto Aguilera 23, 28015 Madrid, Spain, Juan.Marin@iit.upcomillas.es

Antonio Muñoz Instituto de Investigación Tecnológica, Escuela Técnica Superior de Ingeniería – ICAI, Universidad Pontificia Comillas, C/Alberto Aguilera 23, 28015 Madrid, Spain, Antonio.Munoz@iit.upcomillas.es

Alex D. Papalexopoulos ECCO International, Inc., 268 Bush Street, Suite 3633, San Francisco, CA 94104, USA, alexp@eccointl.com

Salvador Pineda Department of Electrical Engineering, Universidad de Castilla – La Mancha, Campus Universitario, s/n, 13071 Ciudad Real, Spain

Martin Povh Faculty of Electrical Engineering, University of Ljubljana, Tržaška 25, 1000 Ljubljana, Slovenia, martin.povh@fe.uni-lj.si

Andres Ramos Universidad Pontificia Comillas, Alberto Aguilera 23, 28015 Madrid, Spain, andres.ramos@upcomillas.es

MohammadAli Rastegar Department of Biophysical and Electronic Engineering, University of Genoa, Via Opera Pia 11a, 16146 Genoa, Italy, rastegar@dibe.unige.it

E. Grant Read Energy Modelling Research Group, University of Canterbury, Private Bag 4800, Christchurch 8140, New Zealand, grant.read@canterbury.ac.nz

Javier Reneses Institute for Research in Technology (IIT), Advanced Technical Engineering School (ICAI), Pontifical Comillas University, Alberto Aguilera 23, 28015 Madrid, Spain

Werner Römisch Humboldt University, 10099 Berlin, Germany, romisch@math.hu-berlin.de

Eugenio F. Sánchez-Úbeda Instituto de Investigación Tecnológica, Escuela Técnica Superior de Ingeniería – ICAI, Universidad Pontificia Comillas, C/Alberto Aguilera 23, 28015 Madrid, Spain, Eugenio.Sanchez@iit.upcomillas.es

Babacar Seck ENPC Paris Tech, 6–8 avenue Blaise Pascal, Cité Descartes – Champs sur Marne, 77455 Marne la Vallée Cedex 2, France, seck@cermics.enpc.fr

Philip Taylor Department of Engineering and Computing, Durham University, Durham, UK, p.c.taylor@durham.ac.uk

James Tipping Energy Modelling Research Group, University of Canterbury, Private Bag 4800, Christchurch 8140, New Zealand, james.tipping@gmail.com

Miguel Vázquez Institute for Research in Technology (IIT), Advanced Technical Engineering School (ICAI), Pontifical Comillas University, Alberto Aguilera 23, 28015 Madrid, Spain

David Wozabal Department of Statistics and Decision Support Systems, University of Vienna, Universitätsstraße 5, 1010 Vienna, Austria, david.wozabal@univie.ac.at

Long Zhou City University London, London, UK, long.zhou.1@city.ac.uk

Part I
Transmission and Distribution Modeling

Recent Developments in Optimal Power Flow Modeling Techniques

Rabih A. Jabr

Abstract This article discusses recent advances in mathematical modeling techniques of transmission networks and control devices within the scope of optimal power flow (OPF) implementations. Emphasis is on the newly proposed concept of representing meshed power networks using an extended conic quadratic (ECQ) model and its amenability to solution by using interior-point codes. Modeling of both classical power control devices and modern unified power flow controller (UPFC) technology is described in relation to the ECQ network format. Applications of OPF including economic dispatching, loss minimization, constrained power flow solutions, and transfer capability computation are presented. Numerical examples that can serve as testing benchmarks for future software developments are reported on a sample test network.

Keywords Economic dispatching · Interior-point methods · Load flow control · Loss minimization · Nonlinear programming · Optimization methods · Transfer capability

1 Introduction

The optimal power flow (OPF) is an optimization problem that seeks to minimize or maximize an objective function while satisfying physical and technical constraints on the power network. The versatility of the OPF has kept it amongst the active research problems since it was first discussed by Carpentier (1962). The OPF has numerous applications, which include the following (Wang et al. 2007; Wood and Wollenberg 1996):

1. Coordinating generation patterns and other control variables for achieving minimum cost operation

R.A. Jabr
Department of Electrical and Computer Engineering, American University of Beirut, P.O. Box 11-0236, Riad El-Solh, Beirut 1107 2020, Lebanon
e-mail: rabih.jabr@aub.edu.lb

S. Rebennack et al. (eds.), *Handbook of Power Systems II*, Energy Systems,
DOI 10.1007/978-3-642-12686-4_1, © Springer-Verlag Berlin Heidelberg 2010

2. Setting of generator voltages, transformer taps, and VAR sources for loss minimization
3. Implementing both preventive and corrective control strategies to satisfy system security constraints
4. Stress testing of a planned transmission network, for instance, computation of the operating security limit in the context of maximum transfer capability
5. Providing a core pricing mechanism for trading in electricity markets, for instance, the computation of the locational marginal cost at any bus in the network
6. Providing a technique for congestion management in restructured power networks
7. Controlling of flexible AC transmission systems (FACTS) for better utilization of existing power capacities

This article centers on recent developments that have occurred in OPF modeling approaches. In particular, it covers aspects related to the physical network representation, operational constraints, classical power flow control and FACTS devices, OPF objective functions and formulations, and optimization techniques. It also includes numerical examples, which could serve as benchmarks for future OPF software research and development.

A recent advancement that has appeared in the power systems literature is the extended conic quadratic (ECQ) format (Jabr 2007; 2008) for OPF modeling and solution via primal-dual interior-point methods. In retrospect, previous research on interior-point OPF programs reported the use of the voltage polar coordinates model (Granville 1994; Jabr 2003; Rider et al. 2004; Wang et al. 2007; Wu et al. 1994), the voltage rectangular model (Capitanescu et al. 2007; Torres and Quintana 1998; Wei et al. 1998), and the current mismatch formulation (Zhang et al. 2005). The advantages of the ECQ format include the simple and efficient computation of the Jacobian and Lagrangian Hessian matrices in interior-point methods and the use of linear programming scaling techniques for improving the numerical conditioning of the problem.

2 Physical Network Representation

Consider a power system operating in steady-state under normal conditions. The system is assumed to be balanced and is represented by a single-phase network formed of N buses. Denote the complex rectangular representation of an element in the $N \times N$ bus admittance matrix by $\hat{Y}_{in} = G_{in} + jB_{in}$. In OPF formulations, the network model is accounted for via the enforcement of the real and reactive power injection constraints:

$$P_i = P_{gi} - P_{di}, \tag{1}$$

$$Q_i = Q_{gi} - Q_{di} + Q_{ci}, \tag{2}$$

where

- P_{gi}/Q_{gi} is the real/reactive power generated at bus $i (i = 1, \ldots, N)$.
- P_{di}/Q_{di} is the real/reactive power demand of the load at bus $i (i = 1, \ldots, N)$.
- Q_{ci} is the reactive power injected by a capacitor at bus $i (i = 1, \ldots, N)$.

There are different representations of the power injections in terms of the elements of the bus admittance matrix and the bus voltages, namely the classical model with voltages in polar or rectangular coordinates and the more recent extended conic quadratic model.

2.1 Classical Model

Assume that bus voltages are expressed in polar form ($\tilde{U}_i = U_i \angle \theta_i$). The real and reactive injected power at an arbitrary bus i is given by (Grainger and Stevenson 1994)

$$P_i = U_i^2 G_{ii} + \sum_{\substack{n=1 \\ n \neq i}}^{N} [U_i U_n G_{in} \cos(\theta_i - \theta_n) + U_i U_n B_{in} \sin(\theta_i - \theta_n)], \tag{3}$$

$$Q_i = -U_i^2 B_{ii} - \sum_{\substack{n=1 \\ n \neq i}}^{N} [U_i U_n B_{in} \cos(\theta_i - \theta_n) - U_i U_n G_{in} \sin(\theta_i - \theta_n)]. \tag{4}$$

2.2 Extended Conic Quadratic Model

The ECQ network model is obtained by defining (Jabr 2007)

$$R_{in} = U_i U_n \cos(\theta_i - \theta_n) \text{ for every branch } i - n, \tag{5}$$

$$T_{in} = U_i U_n \sin(\theta_i - \theta_n) \text{ for every branch } i - n, \tag{6}$$

$$u_i = U_i^2/\sqrt{2} \text{ for every bus } i. \tag{7}$$

The substitution of (5)–(7) in the nonlinear power flow equations (3)–(4) yields the following linear equations:

$$P_i = \sqrt{2} G_{ii} u_i + \sum_{\substack{n=1 \\ n \neq i}}^{N} [G_{in} R_{in} + B_{in} T_{in}], \tag{8}$$

$$Q_i = -\sqrt{2} B_{ii} u_i - \sum_{\substack{n=1 \\ n \neq i}}^{N} [B_{in} R_{in} - G_{in} T_{in}]. \tag{9}$$

From the definitions (5)–(7), it follows that

$$R_{in}^2 + T_{in}^2 - 2u_i u_n = 0, \tag{10}$$
$$\theta_i - \theta_n - \tan^{-1}(T_{in}/R_{in}) = 0. \tag{11}$$

Equations (8)–(11) are referred to as the extended conic quadratic format of the load flow equations (Jabr 2008). In the ECQ format, it is understood that the angle at the slack bus is zero and both u_i and R_{in} take only non-negative values. The bus voltage magnitudes can be deduced from the ECQ variables by solving (7)

$$U_i = (\sqrt{2}u_i)^{1/2}. \tag{12}$$

3 Operational Constraints

Several operational restrictions must be taken into account in the OPF formulation, including generator capability constraints, voltage constraints, and branch flow limits.

3.1 Generator Capability Constraints

A generator must be operated such that it stays within the limits of its stability and power rating. The power rating usually depends on thermal restrictions. If the rating is exceeded for a long time, it may cause damage; unless it is exceeded by a large amount, the machine will continue to function. On the other hand, the stability limit if exceeded even for a short period of time may cause the machine to lose synchronism (Sterling 1978).

The generator capability constraints are most accurately accounted for using the capability chart, which shows the normal loading and operation limits of the generator (Sterling 1978). In OPF, it is possible to model the capability chart using a trapezoidal approximation (Chebbo and Irving 1995); however, a further simplification is obtained by using box constraints:

$$P_{gi}^{\min} \le P_{gi} \le P_{gi}^{\max}, \tag{13}$$
$$Q_{gi}^{\min} \le Q_{gi} \le Q_{gi}^{\max}, \tag{14}$$

where

- $P_{gi}^{\min}/Q_{gi}^{\min}$ is the minimum real/reactive power generated at bus i
- $P_{gi}^{\max}/Q_{gi}^{\max}$ is the maximum real/reactive power generated at bus i

3.2 Voltage Constraints

In OPF, the generator voltage refers to the voltage maintained at the high voltage side of the generator transformer. The voltage limits are usually a few points off the rated terminal voltage (Debs 1988):

$$U_i^{\min} \leq U_i \leq U_i^{\max}, \tag{15}$$

where $U_i^{\min}$ and $U_i^{\max}$ are the minimum and maximum allowable limits of the generator bus voltage magnitude. This enables the stator terminal voltage to be maintained at a constant value, normally the design figure, and allows the necessary reactive power contribution to be achieved by manually tapping the generator transformer against the constant stator volts (British Electricity International 1991).

It is also common to consider the minimum and maximum voltage limits (15) at load buses. The load voltage limits are chosen such that they do not cause damage to the electric system or customer facilities. In terms of the ECQ model variables, the voltage constraints reduce to

$$(U_i^{\min})^2 \Big/ \sqrt{2} \leq u_i \leq (U_i^{\max})^2 \Big/ \sqrt{2}. \tag{16}$$

3.3 Branch Flow Constraints

Thermal limits establish the maximum amount of current that a transmission facility can conduct for a specified period of time without sustaining permanent damage or violating public safety (North American Electric Reliability Council 1995). By using the ECQ format, it is possible to limit the (squared) magnitude of the line current in line $i - n$ and leaving bus i using a linear equation (Ruíz Muñoz and Gómez Expósito 1992):

$$I_{in}^2 = \sqrt{2}A_{in}u_i + \sqrt{2}B_{in}u_n - 2C_{in}R_{in} + 2D_{in}T_{in} \leq (I_{in}^{\max})^2, \tag{17}$$

where

$$\begin{aligned} A_{in} &= g_{in}^2 + (b_{in} + b_{sh}/2)^2, \\ B_{in} &= g_{in}^2 + b_{in}^2, \\ C_{in} &= g_{in}^2 + b_{in}(b_{in} + b_{sh}/2), \\ D_{in} &= g_{in}b_{sh}/2. \end{aligned}$$

In the above equations, g_{in} and b_{in} are the series conductance and susceptance in the π equivalent model, $b_{sh}/2$ is the 1/2 charging susceptance, and $I_{in}^{\max}$ is the line current rating. It is a common practice to limit the line current magnitude at both the sending and receiving ends of each branch.

For short lines, the thermal limit dictates the maximum power transfer. However, practical stability considerations set the transfer limit for longer lines (in excess of 150 miles) (Saadat 1999). The transient stability limits can be roughly approximated by constraints on active power flow:

$$P_{in} = \sqrt{2} g_{in} u_i - g_{in} R_{in} - b_{in} T_{in} \leq P_{in}^{\max}, \tag{18}$$

where $P_{in}^{\max}$ is the stability limit expressed through the real power in line $i - n$ and leaving bus i. This limit is obtained from stability studies.

4 Tap-Changing and Regulating Transformers

Almost all transformers are equipped with taps on windings to adjust the ratio of transformation. Regulating transformers are also used to provide small adjustments of voltage magnitudes or phase angles. Tap-changers are mainly employed to control bus voltage magnitudes, whereas phase-shifters are limited to the control of active power flows. Some transformers regulate both the magnitude and phase angle (Grainger and Stevenson 1994). Previous researchers studied methods for accommodating tap-changers and phase-shifters in Newton's method. These techniques are well documented in Acha et al. (2004).

The tap-changing/voltage regulating and phase-shifting transformers can be accounted for using the regulating transformer model in Fig. 1, where the admittance $\hat{y}_{t(ij)}$ is in series with an ideal transformer representing the complex tap ratio $1 : \hat{a}_{(ij)}$. The subscript (ij) is dropped to simplify the notation. Because the complex power on either side of the ideal transformer is the same, the equivalent power injection model of the regulating transformer can be represented as in Fig. 2 in which the quantities at the fictitious bus x are constrained as follows (Jabr 2003):

$$a^{\min} \leq \frac{U_j}{U_x} \leq a^{\max}, \tag{19}$$

$$\varphi^{\min} \leq \theta_j - \theta_x \leq \varphi^{\max}. \tag{20}$$

In the above equations, $[a^{\min}, a^{\max}]$ and $[\phi^{\min}, \phi^{\max}]$ are the intervals for the magnitude and angle of the complex tap ratio $\hat{a} = a\angle\phi$. The tap-changing (or voltage

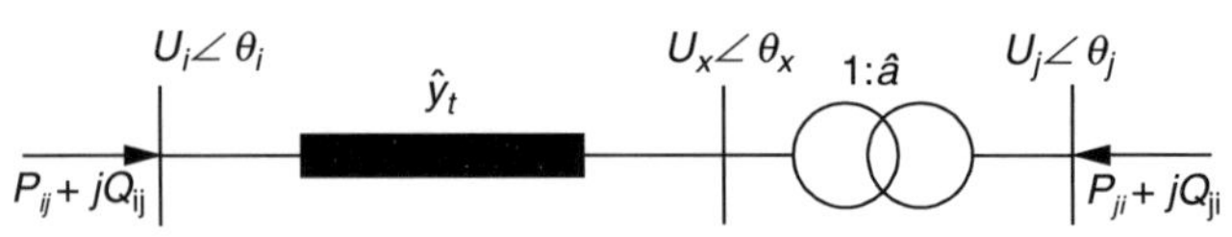

Fig. 1 Regulating transformer equivalent circuit

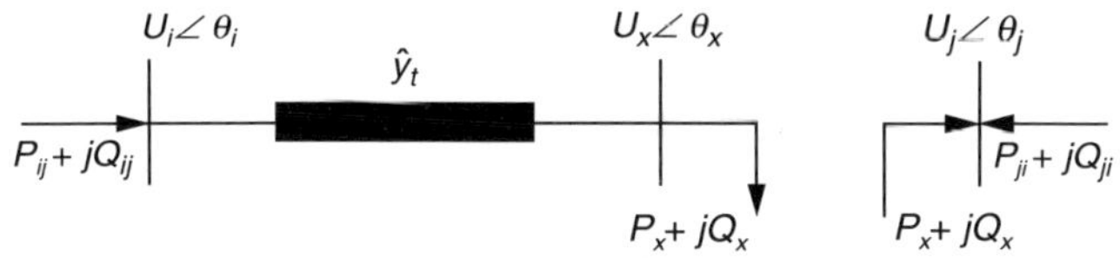

Fig. 2 Regulating transformer power injection model

regulating) transformer model can be obtained by setting

$$\varphi^{\min} = \varphi^{\max} = 0. \tag{21}$$

Similarly,

$$a^{\min} = a^{\max} = 1 \tag{22}$$

results in the phase-shifter model.

Equation (20) can be used with the ECQ format. Equation (19) can be easily placed in a form compatible with this format by using the substitutions in Sect. 2.2 for U_j and U_x, so that (19) becomes

$$(a^{\min})^2 u_x \leq u_j \leq (a^{\max})^2 u_x. \tag{23}$$

Figure 2 shows that the lossless ideal transformer model requires that the active/reactive power extracted from bus x is injected into bus j. It is possible to account for this constraint without introducing the additional variables P_x and Q_x by adding the active/reactive injection equation at bus x to the active/reactive injection equation at bus j. The result is equivalent to combining buses x and j into one super-node and writing the active/reactive injection (8)/(9) at this node. The super-node equation sets the summation of power flows leaving buses x and j to zero.

5 FACTS Devices

FACTS provide additional degrees of freedom in the power network operation by allowing control of bus voltages and power flows. The unified power flow controller (UPFC) is one of the most comprehensive FACTS devices. When installed at one end of a line, it can provide full controllability of the bus voltage magnitude, the active power line flow, and the reactive power line flow (Acha et al. 2004; Zhang et al. 2006).

The principle of operation of the UPFC has been previously reported in the power systems literature (Gyugyi 1992). Figure 3 shows its equivalent circuit under steady-state operating conditions (Acha et al. 2004; Zhang et al. 2006). The equivalent circuit includes two voltage sources operating at the fundamental frequency and

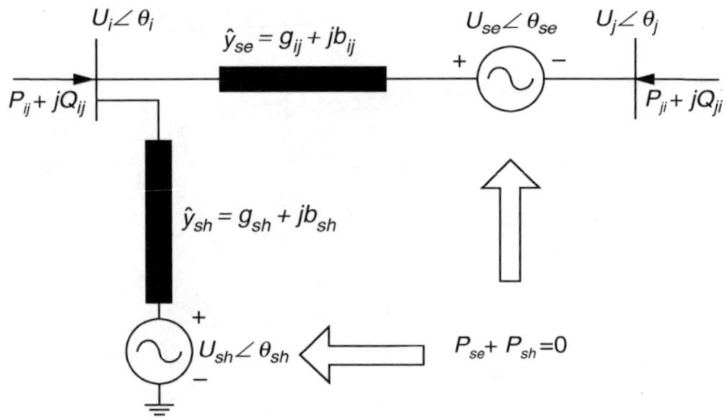

Fig. 3 UPFC equivalent circuit

two impedances. The voltage sources represent the fundamental Fourier series component of the AC converter output voltage waveforms and the impedances model the resistances and leakage inductances of the coupling transformers. To simplify the presentation, the resistances of the coupling transformers are assumed to be negligible and the losses in the converter valves are neglected.

Acha et al. (2004), Ambriz-Pérez et al. (1998), Zhang and Handschin (2001), and Zhang et al. (2006) present Newton and interior-point methods for including a detailed model of the UPFC in OPF studies, that is, the UPFC control parameters (voltage magnitude and angle in the series and shunt converters) are treated as independent variables in the optimization process. A downside of this comprehensive modeling is that the success of the iterative solution becomes sensitive to the choice of the initial UPFC control parameters. Another representation is the power injection model (PIM) proposed in Handschin and Lehmköster (1999). Because the PIM is a strict linear representation of the UPFC, it does not contribute to the nonconvexity of the power flow equations (Lehmköster 2002). Moreover, it does not suffer from problems related to initial point selection.

For a UPFC connected between buses i and j, let the series and shunt voltage sources be represented as phasors in polar form: $\tilde{U}_{se(ij)} = U_{se(ij)}\angle\theta_{se(ij)}$ and $\tilde{U}_{sh(ij)} = U_{sh(ij)}\angle\theta_{sh(ij)}$. To avoid clutter, the subscript (ij) is dropped below. Based on the equivalent circuit in Fig. 3, the active power injection at bus i is

$$P_i = -U_iU_jb_{ij}\sin(\theta_i - \theta_j) - U_iU_{se}b_{ij}\sin(\theta_i - \theta_{se}) - U_iU_{sh}b_{sh}\sin(\theta_i - \theta_{sh}). \quad (24)$$

The first term in (24) is identical to the conventional load flow equation of a transmission device with series susceptance b_{ij} and shunt susceptance b_{sh}. The last two terms can be used to define $-P_i^{FD}$, an active power injection at bus i attributed to the FACTS device's series and shunt voltage sources. Equation (24) can be written as

$$P_i - P_i^{FD} = -U_iU_jb_{ij}\sin(\theta_i - \theta_j), \quad (25)$$

where

$$P_i^{FD} = -U_i U_{se} b_{ij} \sin(\theta_i - \theta_{se}) - U_i U_{sh} b_{sh} \sin(\theta_i - \theta_{sh}). \tag{26}$$

Similarly, the FACTS device's active injection at bus j and reactive injections at buses i and j are

$$P_j^{FD} = U_j U_{se} b_{ij} \sin(\theta_j - \theta_{se}), \tag{27}$$

$$Q_i^{FD} = U_i U_{se} b_{ij} \cos(\theta_i - \theta_{se}) + U_i U_{sh} b_{sh} \cos(\theta_i - \theta_{sh}), \tag{28}$$

$$Q_j^{FD} = -U_j U_{se} b_{ij} \cos(\theta_j - \theta_{se}). \tag{29}$$

By assuming lossless converter valves, the active power exchange among converters via the DC link is zero (Acha et al. 2004; Handschin and Lehmköster 1999; Lehmköster 2002; Zhang et al. 2006), that is,

$$P_i + P_j = P_i^{FD} + P_j^{FD} = 0. \tag{30}$$

Therefore, the UPFC power injection model can be represented as in Fig. 4, where the UPFC active power injection into bus j, $-P_j^{FD}$, is equal to P_i^{FD}.

The UPFC PIM can be directly integrated into the ECQ OPF program by the following:

1. Including the series and shunt coupling transformers into the bus admittance matrix computation.
2. Treating the UPFC injection quantities as additional variables.

Equations (26)–(29) can be used to define upper and lower bounds on each of the UPFC injections. Moreover, the voltage magnitude and angle of the series and shunt voltage sources in Fig. 3 can be deduced from the UPFC PIM by solving (26)–(29) to yield the following closed-form solution (Jabr 2008):

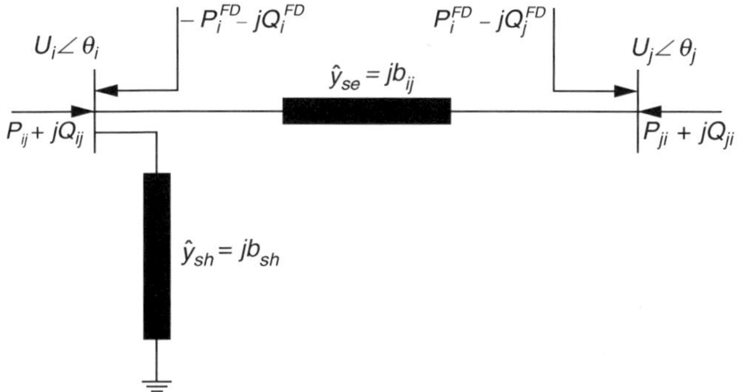

Fig. 4 UPFC power injection model

$$U_{se} = \frac{-\sqrt{(P_j^{FD})^2 + (Q_j^{FD})^2}}{U_j b_{ij}}, \quad (31)$$

$$\theta_{se} = \theta_j - \text{atan2}(-P_j^{FD}, Q_j^{FD}), \quad (32)$$

$$U_{sh} = \frac{-\sqrt{(P_i^{FD1})^2 + (Q_i^{FD1})^2}}{U_i b_{sh}}, \quad (33)$$

$$\theta_{sh} = \theta_i - \text{atan2}(P_i^{FD1}, Q_i^{FD1}). \quad (34)$$

In the above equations, atan2 is the four-quadrant arctangent function and

$$P_i^{FD1} = P_i^{FD} + U_i U_{se} b_{ij} \sin(\theta_i - \theta_{se}), \quad (35)$$

$$Q_i^{FD1} = -Q_i^{FD} + U_i U_{se} b_{ij} \cos(\theta_i - \theta_{se}). \quad (36)$$

6 OPF Objective Functions and Formulations

The OPF can provide solutions to different operational and planning problems depending on the choice of the objective function and constraints. Herein, four types of problems are considered: economic dispatching, loss minimization, constrained power flow solution, and operating security limit determination.

6.1 Economic Dispatching

Economic dispatching is one of the energy control center functions that has great influence on power system economics. Given a set of network conditions and a forecast load, generator ordering optimally determines, within the plant technical limitations, the set of units that should be brought in or taken out of service such that there is sufficient generation to meet the load and reserve requirement. The generation ordering phase, also commonly known as unit commitment, has economic dispatching as a subproblem (Wood and Wollenberg 1996). Generation ordering must be performed several hours in advance so that plant can be run-up and synchronized prior to loading (Sterling 1978). As real time approaches, a better forecast of the load is obtained and economic dispatching is executed again to satisfy the operational constraints and to minimize the operational costs. This is done by reallocating the generation amongst the committed units. The OPF can be used to accurately model the requirements of the economic dispatching problem.

Traditionally, the objective dictating the operation of power systems has been economy. The objective is the minimum generation cost

$$\text{minimize} \sum_i c_{0i} + c_{1i} P_{gi} + c_{2i} P_{gi}^2, \quad (37)$$

where P_{gi} is the active power supplied by a generator connected to bus i; c_{0i}, c_{1i}, and c_{2i} are the corresponding cost curve coefficients. It is also straightforward to model convex piecewise-linear cost curves that appear in market operation by using interpolatory variables (Jabr 2003). The constraints of the OPF dispatching problem are the following:

1. ECQ network constraints given by (1), (2), (8), and (9) for all buses and by (10) and (11) for all branches
2. Generator capability constraints given by (13)–(14) for all generating units
3. Voltage constraints at the slack, generator, and load buses given by (16)
4. Branch flow constraints given by (17) or (18) for all lines
5. Tap-changing and regulating transformer PIM constraints as described in Sect. 4
6. UPFC PIM constraints as described in Sect. 5

6.2 Loss Minimization

Loss minimization is a reactive optimization problem in which the real power generation, except at the slack bus, is assumed to be held at prespecified values. The problem is formulated to minimize the real power loss in the network, or equivalently the power injection into the slack bus, by optimally setting the generation voltages, VAR sources, transformer taps, and the relevant parameters of other power flow controllers.

The objective of the loss minimization problem is the minimum active power loss:

$$\text{minimize } P_s, \tag{38}$$

where P_s is the real power injected into the slack bus as given by (8). The feasible region is commonly formed of the following constraints:

1. ECQ network constraints given by (1), (2), (8), and (9) for all buses and by (10) and (11) for all branches
2. Real power generation values except at the slack bus
3. Reactive power constraints for voltage-controlled buses given by (14)
4. Voltage constraints at the slack, generator, and load buses given by (16)
5. Tap-changing and regulating transformer PIM constraints as described in Sect. 4
6. UPFC PIM constraints as described in Sect. 5

6.3 Constrained Power Flow Solution

The Newton–Raphson algorithm (Fuerte-Esquivel and Acha 1997; Fuerte-Esquivel et al. 2000) is currently the industry standard for power flow because of its quadratic convergence properties. In the presence of tap-changers, regulating transformers, or

FACTS devices, the control targets (nodal voltage magnitudes and active/reactive power flows) are accounted for as equality constraints. In practice, safeguards have to be implemented in case any of the targets is not attainable because of the network's physical constraints or technical limits. A nonattainable target translates into an empty feasible region and consequently leads to divergence of the numerical method. To circumvent such cases in the Newton–Raphson power flow, Acha et al. (2004) and Zhang et al. (2006) employ a limit checking technique combined with control equality relaxation. An alternative approach proposed in Xiao et al. (2002) is to formulate the load flow control problem as a nonlinear program whose objective is to minimize deviations from prespecified control target values. This yields a nearest available solution for cases in which control targets are not achievable.

An OPF framework for load flow control allows multiple constraint enforcement without resort to specifically tailored strategies for fixing multi-violated constraints at their offending limits. The objective is to minimize the L_1-norm of deviations between the dependent controlled quantities and the corresponding target values:

$$\text{minimize} \sum_{k=1}^{N_c} \left| h_k^T x - C_k \right|, \tag{39}$$

where

- $h_k^T x$ is a linear function representing the value of the control quantity, which can be active power flow, a reactive power flow, or u_i (the equivalent of a voltage magnitude).
- h_k^T is a row vector derived from the ECQ power flow model in Sect. 2.2.
- x is a column vector of state variables.
- C_k is the value of the kth control target ($k = 1, \cdots, N_c$).

The constraints of the load-flow control problem are the following:

1. ECQ network constraints given by (1), (2), (8), and (9) for all buses and by (10) and (11) for all branches
2. Real power generation values except at the slack bus
3. Reactive power constraints for voltage-controlled buses given by (14)
4. Slack and generator bus voltage values
5. Tap-changing and regulating transformer PIM constraints as described in Sect. 4
6. UPFC PIM constraints as described in Sect. 5

In case reactive power generation limits are to be enforced at the expense of voltage magnitudes, the voltage magnitudes at the slack and generator buses are also modeled as control quantities in (39). This would allow the voltage magnitude at the slack or generator bus to deviate from the specified value to satisfy reactive power generation limits.

Therefore, the load flow control method requires the solution of a direct least absolute value nonlinear programming problem. There are two mathematically equivalent nonlinear programming representations of the above problem. Both representations are computationally tractable using interior-point methods. It has been shown in Jabr (2005) that, for least absolute value state estimation, one of the two formulations results in a more numerically robust interior-point implementation. The superior representation substitutes (39) with a linear objective function, functional equality constraints, and positively bounded variables (Jabr and Pal 2008):

$$\text{minimize} \sum_{k=1}^{N_c} (r_k + s_k) \text{ subject to} \tag{40}$$

$$h_k^T x - C_k + r_k - s_k = 0, \tag{41}$$

$$r_k \geq 0, \quad s_k \geq 0; \quad k = 1, \cdots, N_c. \tag{42}$$

6.4 Operating Security Limit Determination

The operating security limit determination requires computing the limit for either the system loadablity or transfer capability based on computer simulations of the power network under a specific set of operating conditions and constraints. These limits may be based on nonlinear alternating current (AC) simulations or linear direct current (DC) simulations of the network (North American Electric Reliability Council 1995). Although the DC simulation techniques (Hamoud 2000) are both computationally efficient and numerically stable, they are less accurate than their AC counterparts. Distribution factors (Ejebe et al. 2000) that are also based on the DC power flow are very fast but are valid only under small parameter perturbations. The most common AC simulation techniques include repeated power flow (Gao et al. 1996), continuation power flow (Ajjarapu and Christy 1992; Cañizares and Alvarado 1993), and optimization methods. Recent optimization-based transfer capability computation techniques employ the primal-dual interior-point method (Dai et al. 2000; Irisarri et al. 1997) and have been tested on networks that include different models of FACTS devices (Xiao et al. 2003; Zhang 2005; Zhang and Handschin 2002; Zhang et al. 2006).

The computation of the system loadability limit simulates total system load increase, whereas the computation of the transfer capability limit assumes load increase at a specified region or buses. Any of the two limits can be computed from an OPF formulation whose objective is

$$\text{maximize}\, \lambda \tag{43}$$

subject to the load change constraints

$$P_{di} = \lambda P_{di}^0, \quad Q_{di} = \lambda Q_{di}^0; \quad i \in \Omega_d \tag{44}$$

where λ is the load increase parameter, P_{di} and Q_{di} are the real and reactive power demand of the load at bus i, the superscript 0 denotes the load base case, and Ω_d is the set of buses with variable loads having constant power factor. The other problem constraints are the same as in the economic dispatching problem in Sect. 6.1.

7 OPF Solution Techniques

This section reviews various optimization algorithms that have been applied to the OPF problem. It also includes the implementation details of one of the most promising algorithms for OPF, the primal-dual interior-point method (Jabr 2003; 2005).

7.1 Classification of Solution Methods

Solution methods applied to OPF can be broadly classified as either calculus-based methods or search-based methods. The most popular calculus-based methods are the active set methods and the interior-point methods. Active set methods include the Lagrange–Newton method (Sun et al. 1984), sequential quadratic programming (SQP) (Burchett et al. 1984), simplex-based sequential linear programming (SLP) (Alsac et al. 1990; Chebbo and Irving 1995), and the method of feasible directions (Salgado et al. 1990). The main obstacle in the active set methods lies in the identification of the binding inequality constraints at optimality. Essentially, these methods have to generate a guess of the active set at every iteration and test it for feasibility or optimality. This is different from path-following interior-point methods (Wright 1997) where the active set is determined asymptotically as the solution is approached. The power systems literature reports applications of several improvements over the pure path-following primal-dual interior-point method (Granville 1994): (a) Mehrotra's predictor–corrector technique (Jabr 2003; Torres and Quintana 1998; Wei et al. 1998; Wu et al. 1994), (b) Gondizio's multiple-centrality corrections (Capitanescu et al. 2007; Torres and Quintana 2001), (c) trust region technique (Min and Shengsong 2005), and (d) optimal step length control (Rider et al. 2004; Wang et al. 2007).

The search-based methods comprise various artificial intelligence (AI) techniques such as genetic algorithms. Although genetic algorithms are significantly less efficient than their calculus-based counterparts, they are capable of handling non-convex cost curves (Lai and Sinha 2008) and discrete control variables (Bakirtzis et al. 2002), for instance discrete transformer tap positions. Recent research has also shown that calculus-based optimization methods are capable of handling discrete controls through their use with ordinal optimization search procedures (Lin et al. 2004). A detailed survey of the application of heuristic and randomized-based strategies in power systems is given in Lai and Sinha (2008).

7.2 A Primal-Dual Interior-Point Approach

For notational simplicity, consider the following nonlinear programming problem representing the OPF formulation:

$$\text{minimize } f(\mathbf{x}) \text{ subject to} \tag{45}$$
$$\boldsymbol{c}(\boldsymbol{x}) = 0 \tag{46}$$
$$\boldsymbol{d}(\boldsymbol{x}) - \boldsymbol{w} = 0 \tag{47}$$
$$\boldsymbol{w} \geq 0 \tag{48}$$

where $f : \Re^n \rightarrow \Re$, $\boldsymbol{c} : \Re^n \rightarrow \Re^p$, and $\boldsymbol{d} : \Re^n \rightarrow \Re^q$. The functions f, $\boldsymbol{c}$, and $\boldsymbol{d}$ are assumed to be twice continuously differentiable. The details of the primal-dual interior-point (PDIP) method applied to (45)–(48) are found in Jabr (2005). Section 7.2.1 includes an outline of the algorithm. Issues related to scaling and problem-specific implementation details are discussed in Sects. 7.2.2 and 7.2.3, respectively.

7.2.1 PDIP Algorithm

Given a starting point $(\boldsymbol{x}, \boldsymbol{w}, \boldsymbol{y}, \boldsymbol{z})$ with $(\boldsymbol{w}, \boldsymbol{z}) > 0$

repeat

$$\text{Set } \mu = \frac{\boldsymbol{w}^T \boldsymbol{z}}{p}; \tag{49}$$

Solve for $(\Delta \boldsymbol{x}^{\text{aff}}, \Delta \boldsymbol{w}^{\text{aff}}, \Delta \boldsymbol{y}^{\text{aff}}, \Delta \boldsymbol{z}^{\text{aff}})$:

$$\begin{bmatrix} \boldsymbol{M} & 0 & -\boldsymbol{A}^T & -\boldsymbol{B}^T \\ 0 & \boldsymbol{Z} & 0 & \boldsymbol{W} \\ \boldsymbol{A} & 0 & 0 & 0 \\ \boldsymbol{B} & -\boldsymbol{I} & 0 & 0 \end{bmatrix} \begin{bmatrix} \Delta \boldsymbol{x}^{\text{aff}} \\ \Delta \boldsymbol{w}^{\text{aff}} \\ \Delta \boldsymbol{y}^{\text{aff}} \\ \Delta \boldsymbol{z}^{\text{aff}} \end{bmatrix} = \begin{bmatrix} -\nabla f + \boldsymbol{A}^T \boldsymbol{y} + \boldsymbol{B}^T \boldsymbol{z} \\ -\boldsymbol{W}\boldsymbol{Z}\boldsymbol{e} \\ -\boldsymbol{c} \\ -\boldsymbol{d} + \boldsymbol{w} \end{bmatrix}, \tag{50}$$

where

$$\boldsymbol{W} = diag(w_1, w_2, \ldots, w_p), \tag{51}$$
$$\boldsymbol{Z} = diag(z_1, z_2, \ldots, z_p), \tag{52}$$
$$\boldsymbol{e} = (1, 1, \ldots, 1)^T, \tag{53}$$
$$\boldsymbol{A} = \nabla \boldsymbol{c}, \tag{54}$$
$$\boldsymbol{B} = \nabla \boldsymbol{d}, \tag{55}$$
$$\boldsymbol{M} = \nabla^2 f - \sum_{j=1}^{q} y_j \nabla^2 c_j - \sum_{i=1}^{p} z_i \nabla^2 d_i; \tag{56}$$

$$\text{Compute}\,\alpha_{\text{aff}} = \arg\max_{\alpha\in(0,1]} \{(\boldsymbol{w}, \boldsymbol{z}) + \alpha(\Delta\boldsymbol{w}^{\text{aff}}, \Delta\boldsymbol{z}^{\text{aff}}) \geq 0\}; \tag{57}$$

$$\text{Set}\,\mu_{\text{aff}} = \frac{(\boldsymbol{w} + \alpha_{\text{aff}}\Delta\boldsymbol{w}^{\text{aff}})^T(\boldsymbol{z} + \alpha_{\text{aff}}\Delta\boldsymbol{z}^{\text{aff}})}{p}; \tag{58}$$

$$\text{Set}\,\sigma = (\mu_{\text{aff}}/\mu)^2; \tag{59}$$

Solve for $(\Delta\boldsymbol{x}, \Delta\boldsymbol{w}, \Delta\boldsymbol{y}, \Delta\boldsymbol{z})$:

$$\begin{bmatrix} \boldsymbol{M} & 0 & -\boldsymbol{A}^T & -\boldsymbol{B}^T \\ 0 & \boldsymbol{Z} & 0 & \boldsymbol{W} \\ \boldsymbol{A} & 0 & 0 & 0 \\ \boldsymbol{B} & -\boldsymbol{I} & 0 & 0 \end{bmatrix} \begin{bmatrix} \Delta\boldsymbol{x} \\ \Delta\boldsymbol{w} \\ \Delta\boldsymbol{y} \\ \Delta\boldsymbol{z} \end{bmatrix} = \begin{bmatrix} -\nabla f + \boldsymbol{A}^T\boldsymbol{y} + \boldsymbol{B}^T\boldsymbol{z} \\ \sigma\mu\boldsymbol{e} - \boldsymbol{W}\boldsymbol{Z}\boldsymbol{e} - \Delta\boldsymbol{W}^{\text{aff}}\Delta\boldsymbol{Z}^{\text{aff}}\boldsymbol{e} \\ -\boldsymbol{c} \\ -\boldsymbol{d} + \boldsymbol{w} \end{bmatrix}, \tag{60}$$

where $\Delta\boldsymbol{W}^{\text{aff}}$ and $\Delta\boldsymbol{Z}^{\text{aff}}$ are defined using the *diag* operator:

$$\text{Compute}\,\alpha = \arg\max_{\alpha\in(0,1]} \{(\boldsymbol{w}, \boldsymbol{z}) + \alpha(\Delta\boldsymbol{w}, \Delta\boldsymbol{z}) \geq 0\}; \tag{61}$$

$$\text{Set}\,(\boldsymbol{x}, \boldsymbol{w}, \boldsymbol{y}, \boldsymbol{z}) \leftarrow (\boldsymbol{x}, \boldsymbol{w}, \boldsymbol{y}, \boldsymbol{z}) + \alpha(\Delta\boldsymbol{x}, \Delta\boldsymbol{w}, \Delta\boldsymbol{y}, \Delta\boldsymbol{z}); \tag{62}$$

until convergence test is satisfied.

The convergence test requires satisfying the relative primal infeasibility given by

$$pfeas = \frac{\|(\boldsymbol{c}; \boldsymbol{d} - \boldsymbol{w})\|_1}{1 + \|\boldsymbol{w}\|_1} \leq \varepsilon, \tag{63}$$

the relative dual infeasibility given by

$$dfeas = \frac{\left\|\nabla f - \boldsymbol{A}^T\boldsymbol{y} - \boldsymbol{B}^T\boldsymbol{z}\right\|_1}{1 + \|\boldsymbol{z}\|_1} \leq \varepsilon, \tag{64}$$

and the relative duality gap given by

$$opt = \frac{\left|\boldsymbol{y}^T\boldsymbol{c} + \boldsymbol{z}^T\boldsymbol{d} + \boldsymbol{x}^T\boldsymbol{r}_D\right|}{1 + \left|f - \boldsymbol{y}^T\boldsymbol{c} - \boldsymbol{z}^T\boldsymbol{d} - \boldsymbol{x}^T\boldsymbol{r}_D\right|} \leq \varepsilon. \tag{65}$$

The precision parameter ε is set to 10^{-8}.

7.2.2 Scaling

Scaling of the objective and constraint functions is achieved by multiplying them with suitable constants so that the typical numerical values of the scaled functions in the optimization program are roughly of the same order of magnitude. In linear programming, the most widely used method is scaling rows and columns to have unit norms (Tosovic 1973). In SQP algorithms, it is recommended to scale the problem so that the cost function gradient is of similar magnitude to each of the constraint function gradients (Belegundu and Chandrupatla 1999).

The OPF using the ECQ format includes conic constraints (10), arctangent constraints (11), and a set of linear constraints. Because only the coefficients of the linear constraints are dependent on the parameters of the network, the coefficient matrix of linear equations is scaled such that its rows have unit norm. This procedure is similar to what is done in linear programming. The objective function is scaled so that for typical values of the variables, it is of the same order of magnitude as the constraint functions. The above choices are based on numerical experimentation and have shown to result in convergence with a reduced number of interior-point iterations.

7.2.3 Implementation Details

The PDIP solver requires, at each iteration, the computation of the Jacobian (54), (55) and the Lagrangian Hessian (56) matrices. The Jacobian matrix corresponding to all constraints except (10) and (11) is constant throughout all iterations. For the feasibility constraints (10) and (11), the formulae for computing the Jacobian and Hessian elements are given by (g' and g'' denote the left-hand sides of (10) and (11), respectively):

$$\frac{\partial g'}{\partial R_{in}} = 2R_{in}, \quad \frac{\partial g'}{\partial T_{in}} = 2T_{in}, \tag{66}$$

$$\frac{\partial g'}{\partial u_i} = -2u_n, \quad \frac{\partial g'}{\partial u_n} = -2u_i, \tag{67}$$

$$\frac{\partial^2 g'}{\partial R_{in}^2} = \frac{\partial^2 g'}{\partial T_{in}^2} = 2, \quad \frac{\partial^2 g'}{\partial u_n \partial u_i} = \frac{\partial^2 g'}{\partial u_i \partial u_n} = -2, \tag{68}$$

$$\frac{\partial g''}{\partial \theta_i} = 1, \quad \frac{\partial g''}{\partial \theta_n} = -1, \tag{69}$$

$$\frac{\partial g''}{\partial R_{in}} = \frac{T_{in}}{R_{in}^2 + T_{in}^2}, \quad \frac{\partial g''}{\partial T_{in}} = \frac{-R_{in}}{R_{in}^2 + T_{in}^2}, \tag{70}$$

$$\frac{\partial^2 g''}{\partial R_{in}^2} = \frac{-2R_{in}T_{in}}{(R_{in}^2 + T_{in}^2)^2}, \quad \frac{\partial^2 g''}{\partial T_{in}^2} = \frac{2R_{in}T_{in}}{(R_{in}^2 + T_{in}^2)^2}, \tag{71}$$

$$\frac{\partial^2 g''}{\partial R_{in}\partial T_{in}} = \frac{\partial^2 g''}{\partial T_{in}\partial R_{in}} = \frac{R_{in}^2 - T_{in}^2}{(R_{in}^2 + T_{in}^2)^2}. \tag{72}$$

The initial vector $\boldsymbol{x}$ for the primal variables is chosen to correspond to a flat start:

$$u_i = 1\big/\sqrt{2} \text{ and } \theta_i = 0 \text{ for all buses,} \tag{73}$$

$$R_{in} = 1, \quad T_{in} = 0 \text{ for all branches,} \tag{74}$$

$$P_i^{FD} = Q_i^{FD} = 0 \text{ for all UPFC injections,} \tag{75}$$
$$P_{gi} = (P_{gi}^{\min} + P_{gi}^{\max})/2 \text{ for all generators,} \tag{76}$$
$$Q_{gi} = (Q_{gi}^{\min} + Q_{gi}^{\max})/2 \text{ for all generators.} \tag{77}$$
$$\lambda = 1 \text{ (initial load increase parameter)} \tag{78}$$

The starting solution for the PDIP method should also specify values for $\boldsymbol{w}$ and for the Lagrange multipliers $\boldsymbol{y}$ and $\boldsymbol{z}$ (see Sect. 7.2.1):

$$w_i = \max(|d_i(\mathbf{x})|, \rho), \quad \rho \in [0.01, 0.1] \tag{79}$$
$$z_i = \rho / w_i, \tag{80}$$
$$y_i = 0. \tag{81}$$

8 Numerical Examples

The OPF program was implemented in MATLAB. The program has been run successfully on an Intel Core 2 Duo Processor T5300 (1.73 GHz) PC with 1 GB RAM. The numerical results presented herein are obtained from simulations on the IEEE 30-bus test system. A total of 11 cases are presented according to the following schedule:

1. Economic dispatching in cases 1–4 (simulation results are in Table 1)
2. Loss minimization in cases 5–6 (simulation results are in Table 2)
3. Constrained power flow solution in cases 7–8 (simulation results are in Table 2)
4. Operating security limit determination in cases 9–11 (simulation results are in Table 3)

The simulation results in Tables 1–3 are limited to the control variables such that it is possible to recover the complete system state from a load-flow solution. The objective function together with the PDIP iteration counts and execution times are also included. The conditions under which the testing was performed in each of the cases are detailed below.

Case 1: Economic dispatching of the base case. The OPF finds a slightly lower dispatch cost of 802.360 £/h as compared to the one in Alsac and Stott (1974).

Case 2: Economic dispatching with three UPFC devices installed in lines 3–4, 12–15, and 25–27 with the shunt branches at buses 3, 12, and 25. Each UPFC was initially simulated under normal operation, that is, for the control of bus voltage and active and reactive power flows:

$$U_3 = U_{12} = U_{25} = 1 \text{ pu},$$
$$P_{3-4} = 65 \text{ MW}, \quad P_{12-15} = 25 \text{ MW}, \quad P_{25-27} = 6 \text{ MW},$$
$$Q_{3-4} = -5 \text{ MVAR}, \quad Q_{12-15} = 10 \text{ MVAR}, \quad Q_{25-27} = -5 \text{ MVAR},$$

Table 1 Simulation results for cases 1–4

Control variables		Case 1	Case 2	Case 3	Case 4
Bus voltage (pu)	U_1	1.050	1.028	1.028	1.050
	U_2	1.038	1.018	1.018	1.040
	U_5	1.011	0.992	0.992	1.014
	U_8	1.019	1.001	1.001	1.022
	U_{11}	1.085	1.100	1.074	1.071
	U_{13}	1.087	1.024	1.031	1.066
Generated power (MW)	P_{G2}	48.840	48.992	48.918	48.828
	P_{G5}	21.513	21.626	21.598	21.504
	P_{G8}	22.140	22.441	22.296	22.003
	P_{G11}	12.236	12.453	12.212	12.211
	P_{G13}	12.000	12.000	12.000	12.000
Tap (pu)	$a_{(6-9)}$	1.034	0.990	0.962	1.039
	$a_{(6-10)}$	0.909	0.900	1.100	0.907
	$a_{(4-12)}$	0.998	1.033	1.015	0.990
	$a_{(28-27)}$	0.941	0.945	0.939	0.959
UPFC series voltage (pu)	$U_{se(3-4)}$	–	0.100	0.101	0.086
	$U_{se(12-15)}$	–	0.059	0.057	0.016
	$U_{se(25-27)}$	–	0.153	0.046	0.012
UPFC series angle (deg.)	$\theta_{se(3-4)}$	–	−105.250	−105.907	−102.059
	$\theta_{se(12-15)}$	–	−162.732	−131.075	−106.832
	$\theta_{se(25-27)}$	–	−121.497	162.244	108.596
UPFC shunt voltage (pu)	$U_{sh(3-4)}$	–	1.003	1.004	1.033
	$U_{sh(12-15)}$	–	0.997	0.997	1.063
	$U_{sh(25-27)}$	–	1.004	1.005	1.043
UPFC shunt angle (deg.)	$\theta_{sh(3-4)}$	–	−7.190	−7.186	−6.566
	$\theta_{sh(12-15)}$	–	−11.432	−11.286	−10.551
	$\theta_{sh(25-27)}$	–	−16.569	−12.579	−12.192
Dispatch cost (£/h)		802.360	806.810	804.124	802.093
PDIP	iterations	12	9	10	11
	time (s)	0.14	0.11	0.12	0.13

The dispatch cost in this case (806.810 £/h) is higher as compared to case 1 due to the above operational constraints on voltage and power flow.

Case 3: This case is similar to case 2, except that the UPFC in line 3–4 is used to control the bus voltage and active and reactive power flows, the UPFC in line 12–15 is used to control the bus voltage and the real power flows, and the UPFC in line 25–27 is used to control the bus voltage and the reactive power flows:

$$U_3 = U_{12} = U_{25} = 1 \text{ pu},$$
$$P_{3-4} = 65 \text{ MW}, \quad P_{12-15} = 25 \text{ MW},$$
$$Q_{3-4} = -5 \text{ MVAR}, \quad Q_{25-27} = -5 \text{ MVAR}.$$

The corresponding dispatch cost (804.124 £/h) is expectedly lower as compared to case 2 due to relaxing two of the operational constraints on P_{25-27} and Q_{12-15}.

Table 2 Simulation results for cases 5–8

Control variables		Case 5	Case 6	Case 7	Case 8
Bus voltage (pu)	U_1	1.100	1.100	1.060	1.067
	U_2	1.085	1.085	1.043	1.045
	U_5	1.053	1.052	1.010	1.010
	U_8	1.059	1.054	1.010	1.005
	U_{11}	1.100	1.100	1.082	1.082
	U_{13}	1.100	1.063	1.071	1.071
Tap (pu)	$a_{(6-9)}$	1.070	1.024	0.978	0.978
	$a_{(6-10)}$	0.911	0.900	0.969	0.969
	$a_{(4-12)}$	1.007	1.100	0.932	0.932
	$a_{(28-27)}$	0.964	0.947	0.968	0.968
UPFC series voltage (pu)	$U_{se(12-15)}$	–	0.234	–	0.255
UPFC series angle (deg.)	$\theta_{se(12-15)}$	–	38.606	–	32.804
UPFC shunt voltage (pu)	$U_{sh(12-15)}$	–	1.069	–	1.030
UPFC shunt angle (deg.)	$\theta_{sh(12-15)}$	–	−13.085	–	−13.868
Real power loss (MW)		16.042	21.888	17.562	23.827
PDIP	iterations	10	11	8	9
	time (s)	0.12	0.13	0.10	0.11

Table 3 Simulation results for cases 9–11

Control variables		Case 9	Case 10	Case 11
Bus voltage (pu)	U_1	1.050	1.050	1.050
	U_2	1.066	1.068	1.068
	U_5	1.099	1.100	1.100
	U_8	1.042	1.016	1.016
	U_{11}	1.073	1.081	1.081
	U_{13}	1.046	1.030	1.033
Generated power (MW)	P_{G2}	161.000	161.000	161.000
	P_{G5}	129.000	129.000	129.000
	P_{G8}	67.722	76.550	76.600
	P_{G11}	16.170	16.170	16.170
	P_{G13}	10.500	10.500	10.500
Tap (pu)	$a_{(6-9)}$	0.978	0.942	0.943
	$a_{(6-10)}$	0.969	0.926	0.928
	$a_{(4-12)}$	0.932	0.948	0.942
	$a_{(28-27)}$	0.968	0.900	0.900
UPFC series voltage (pu)	$U_{se(6-2)}$	–	–	0.094
UPFC series angle (deg.)	$\theta_{se(6-2)}$	–	–	99.504
UPFC shunt voltage (pu)	$U_{sh(6-2)}$	–	–	1.017
UPFC shunt angle (deg.)	$\theta_{sh(6-2)}$	–	–	−2.658
Load increase parameter		1.632	1.761	1.766
PDIP	iterations	11	12	10
	time (s)	0.13	0.14	0.12

Case 4: This case is similar to case 2, except that all UPFC control constraints are deactivated. The dispatch cost (802.093 £/h) is even lower than the base case. This result is explained by the increased degrees of freedom furnished by the UPFC device and the absence of all the UPFC constraints.

Case 5: Minimum loss solution without FACTS devices. In this case, all real power generation is assumed to be zero except at bus 1 (slack bus) and at bus 2, where it is fixed at 40 MW. The total real power loss is 16.042 MW.

Case 6: Minimum loss OPF with one UPFC installed in line 12–15 with its shunt branch at bus 12. The UPFC was simulated under normal operation:

$$U_{12} = 1.05 \text{ pu},$$
$$P_{12-15} = -30 \text{ MW},$$
$$Q_{12-15} = -5 \text{ MVAR}.$$

The total real power loss is 21.888 MW. It is higher as compared to case 5 because of the above operational requirements.

Case 7: Constrained power flow solution without FACTS devices. As in case 5, all real power generation is assumed to be zero except at bus 1 (slack bus) and at bus 2 (40 MW). The voltage targets are set at the slack and the voltage-controlled buses as in Freris and Sasson (1968):

$$U_1 = 1.060 \text{ pu}, \quad U_2 = 1.045 \text{ pu}, \quad U_5 = 1.010 \text{ pu}, \quad U_8 = 1.010 \text{ pu},$$
$$U_{11} = 1.082 \text{ pu}, \quad U_{13} = 1.071 \text{ pu}.$$

Moreover, the reactive power generation limits are enforced at the expense of the generation voltages, and the transformer taps are fixed to the values in Table 2 (Freris and Sasson 1968). The power flow solution achieves all target values except at bus 2, where the voltage is set to 1.043 pu. This is because the reactive power generation at this bus reaches its maximum limit of 50 MVAR. The specification of voltage targets and the treatment of transformer taps as fixed variables explain the increased loss (17.562 MW) as compared to case 5.

Case 8: Constrained power flow solution as in case 7 but with an additional UPFC installed in line 12–15 with its shunt branch at bus 12. The UPFC control targets are the same as in case 6. The power flow solution in this case yields a total loss of 23.827 MW, which is greater than that in the minimum loss solution in case 6. Table 2 shows that to satisfy all constraints, the generation voltage targets at buses 1 and 8 have not been exactly achieved.

Case 9: Transfer capability computation without FACTS devices. In this case, the IEEE 30 bus system was divided into two areas as in Zhang (2005), Zhang and Handschin (2002), and Zhang et al. (2006). Area 1 includes buses 1, 2, 3, 4, 5, 6, 7, 8, and 28 whereas the remaining buses are in area 2. The two areas are connected by tie-lines: 4–12, 6–9, 6–10, and 28–27. The transfer capability from area 1 to area 2 is investigated. Table 8 shows that the generators at buses 11 and 13 in area 2 have equal minimum and maximum generation limits, that is, they contribute fixed power injections. With the transformer taps fixed to their values in Table 3 (Freris and Sasson 1968), the load increase parameter is $\lambda = 1.632$. The number of binding branch current magnitude constraints is 3.

Case 10: This case is similar to case 9, except that the transformer taps are allowed to vary. The load increase parameter becomes 1.761 with nine binding branch current magnitude constraints.

Case 11: Transfer capability computation as in case 10 but with an additional UPFC installed in line 6–2 with its shunt branch at bus 6. The load increase parameter slightly increases to 1.766 with a total of ten binding branch current magnitude constraints. In this case, the UPFC does not significantly contribute to increasing the transfer capability limit.

The last two rows in Tables 1–3 show that the different OPF problem types require roughly the same computational effort. It is important to note that the computational efficiency of the proposed formulation has been established in Jabr (2008) by comparing with numerical results obtained from other OPF solvers.

8.1 System Data

Tables 4 and 5 give the details of the branch data and load data, respectively. This data appears originally in Alsac and Stott (1974) and Freris and Sasson (1968). Table 6 includes the generator data (Alsac and Stott 1974) for cases 1–4. For cases 5–8, the MVAR capability constraints are given in Table 7 (Freris and Sasson 1968). Table 8 shows the generator capability limits for cases 9–11, where it is assumed that the generator maximum apparent power rating at buses 1, 2, 5, and 8 occurs at 0.85 power factor lagging.

The branches numbered 11, 12, 15, and 36 in Table 4 are tap-changing transformers with assumed tapping ranges of ±10%. For cases 1–4 and 9–11, the lower bus voltage limits at all buses are 0.95 pu, and the upper limits are 1.1 pu for the generator buses 2, 5, 8, 11, and 13 and 1.05 pu for the remaining load buses, including the slack bus (Alsac and Stott 1974). For cases 5–8, all lower bus voltage limits are set at 0.9 pu and upper limits at 1.1 pu.

The branch MVA ratings in Table 4 are converted to pu current limits, which apply at both the sending and receiving ends of each branch. For each of the UPFC devices, it is assumed that the series and shunt winding impedances of the coupling transformers contain no resistance and a leakage reactance of 0.05 pu. The system base is 100 MVA.

9 Conclusion

The optimal power flow is an important function for the operation and planning of both vertically integrated and restructured power systems. This article details the recent developments that have occurred in the network and power flow controller

Table 4 Branch data

Branch No.	Bus-i	Bus-j	r(pu)	x(pu)	$b_{sh}/2\ (pu)$	Rating (MVA)
1	1	2	0.0192	0.0575	0.0264	130
2	1	3	0.0452	0.1852	0.0204	130
3	2	4	0.0570	0.1737	0.0184	65
4	3	4	0.0132	0.0379	0.0042	130
5	2	5	0.0472	0.1983	0.0209	130
6	2	6	0.0581	0.1763	0.0187	65
7	4	6	0.0119	0.0414	0.0045	90
8	5	7	0.0460	0.1160	0.0102	70
9	6	7	0.0267	0.0820	0.0085	130
10	6	8	0.0120	0.0420	0.0045	32
11	6	9	0.0	0.2080	0.0	65
12	6	10	0.0	0.5560	0.0	32
13	9	11	0.0	0.2080	0.0	65
14	9	10	0.0	0.1100	0.0	65
15	4	12	0.0	0.2560	0.0	65
16	12	13	0.0	0.1400	0.0	65
17	12	14	0.1231	0.2559	0.0	32
18	12	15	0.0662	0.1304	0.0	32
19	12	16	0.0945	0.1987	0.0	32
20	14	15	0.2210	0.1997	0.0	16
21	16	17	0.0824	0.1923	0.0	16
22	15	18	0.1070	0.2185	0.0	16
23	18	19	0.0639	0.1292	0.0	16
24	19	20	0.0340	0.0680	0.0	32
25	10	20	0.0936	0.2090	0.0	32
26	10	17	0.0324	0.0845	0.0	32
27	10	21	0.0348	0.0749	0.0	32
28	10	22	0.0727	0.1499	0.0	32
29	21	22	0.0116	0.0236	0.0	32
30	15	23	0.1000	0.2020	0.0	16
31	22	24	0.1150	0.1790	0.0	16
32	23	24	0.1320	0.2700	0.0	16
33	24	25	0.1885	0.3292	0.0	16
34	25	26	0.2544	0.3800	0.0	16
35	25	27	0.1093	0.2087	0.0	16
36	28	27	0.0	0.3960	0.0	65
37	27	29	0.2198	0.4153	0.0	16
38	27	30	0.3202	0.6027	0.0	16
39	29	30	0.2399	0.4533	0.0	16
40	8	28	0.0636	0.2000	0.0214	32
41	6	28	0.0169	0.0599	0.0065	32
42	10	10	0.0	−5.2600	–	–
43	24	24	0.0	−25.0000	–	–

Table 5 Load data

Bus No.	Load		Bus No.	Load	
	MW	MVAR		MW	MVAR
1	0.0	0.0	16	3.5	1.8
2	21.7	12.7	17	9.0	5.8
3	2.4	1.2	18	3.2	0.9
4	7.6	1.6	19	9.5	3.4
5	94.2	19.0	20	2.2	0.7
6	0.0	0.0	21	17.5	11.2
7	22.8	10.9	22	0.0	0.0
8	30.0	30.0	23	3.2	1.6
9	0.0	0.0	24	8.7	6.7
10	5.8	2.0	25	0.0	0.0
11	0.0	0.0	26	3.5	2.3
12	11.2	7.5	27	0.0	0.0
13	0.0	0.0	28	0.0	0.0
14	6.2	1.6	29	2.4	0.9
15	8.2	2.5	30	10.6	1.9

Table 6 Generator data for cases 1–4

Bus-i	$P_{gi}^{\min}$ (MW)	$P_{gi}^{\max}$ (MW)	$Q_{gi}^{\min}$ (MVAR)	$S_{gi}^{\max}$ (MVA)	c_{0i} (£/h)	c_{1i} (£/MWh)	c_{2i} (£/MW^2h)
1	50	200	−20	250	0.0	2.0	0.00375
2	20	80	−20	100	0.0	1.75	0.0175
5	15	50	−15	80	0.0	1.0	0.0625
8	10	35	−15	60	0.0	3.25	0.00834
11	10	30	−10	50	0.0	3.0	0.025
13	12	40	−15	60	0.0	3.0	0.025

Table 7 Regulated bus MVAR capability for cases 5–8

Bus-i	$Q_{gi}^{\min}$ (MVAR)	$Q_{gi}^{\max}$ (MVAR)
2	−40	50
5	−40	40
8	−10	40
11	−10	24
13	−6	24

Table 8 Generator capability for cases 9–11

Bus-i	$P_{gi}^{\min}$ (MW)	$P_{gi}^{\max}$ (MW)	$Q_{gi}^{\min}$ (MVAR)	$Q_{gi}^{\max}$ (MVAR)
1	0.0	403	−20	250
2	0.0	161	−20	100
5	0.0	129	−15	80
8	0.0	97	−15	60
11	0.0	0.0	16.17	16.17
13	0.0	0.0	10.50	10.50

device modeling in which emphasis is on the extended conic quadratic format. It is shown that the nonlinear constraints in this format are limited to a set of binding rotated quadratic cones and arctangent functions. All other constraints are linear. These include the active/reactive power injection equations together with the power injection models for load flow controllers such as tap-changing transformers, phase-shifting transformers, and unified power flow controllers. The solution of the new OPF formulation via a primal-dual interior-point method is also discussed.

This article can serve as a starting point for future research in this area and as a guide for OPF code production. Numerical results on four types of OPF problems are presented. These results together with the complete specification of the system data are useful as testing benchmarks for future OPF developments.

References

Acha E, Fuerte-Esquivel CR, Ambriz-Pérez H, Angeles-Camacho C (2004) FACTS: modeling and simulation in power networks. Wiley, Chichester

Ajjarapu V, Christy C (1992) The continuation power flow: a tool for steady state voltage stability analysis. IEEE Trans Power Syst 7(1):416–423

Alsac O, Bright J, Prais M, Stott B (1990) Further developments in LP-based optimal power flow. IEEE Trans Power Syst 5(3):697–711

Alsac O, Stott B (1974) Optimal load flow with steady-state security. IEEE Trans Power App Syst 93(3):745–751

Ambriz-Pérez H, Acha E, Fuerte-Esquivel CR, De la Torre A (1998) Incorporation of a UPFC model in an optimal power flow using Newton's method. IEE Proc- Gener Transm Distrib 145(3):336–344

Bakirtzis AG, Biskas PN, Zoumas CE, Petridis V (2002) Optimal power flow by enhanced genetic algorithm. IEEE Trans Power Syst 17(2):229–236

Belegundu AD, Chandrupatla TR (1999) Optimization concepts and applications in engineering. Prentice Hall, New Jersey

British Electricity International (1991) Modern power station practice, vol. L: system operation. Pergamon, Oxford

Burchett RC, Happ HH, Vierath DR (1984) Quadratically convergent optimal power flow. IEEE Trans Power App Syst 103(11):3267–3275

Cañizares CA, Alvarado FL (1993) Point of collapse and continuation methods for large AC/DC systems. IEEE Trans Power Syst 8(1):1–8

Capitanescu F, Glavic M, Ernst D, Wehenkel L (2007) Interior-point based algorithms for the solution of optimal power flow problems. Elec Power Syst Res 77(5–6):508–517

Carpentier J (1962) Contribution a l'étude du dispatching économique. Ser. 8: Bulletin de la Société Française des Electriciens 3:431–447

Chebbo AM, Irving MR (1995) Combined active and reactive dispatch - part 1: problem formulation and solution algorithms. IEE Proc- Gener Transm Distrib 142(4):393–400

Dai Y, McCalley JD, Vittal V (2000) Simplification, expansion and enhancement of direct interior point algorithm for power system maximum loadability. IEEE Trans Power Syst 15(3): 1014–1021

Debs AS (1988) Modern power systems control and operation. Kluwer, Boston

Ejebe GC, Waight JG, Santos-Nieto M, Tinney WF (2000) Fast calculation of linear available transfer capability. IEEE Trans Power Syst 15(3):1112–1116

Freris LL, Sasson AM (1968) Investigation of the load-flow problem. IEE Proc 115(10):1459–1470

Fuerte-Esquivel CR, Acha E (1997) Unified power flow controller: a critical comparison of Newton-Raphson UPFC algorithms in power flow studies. IEE Proc- Gener Transm Distrib 144(5):437–444
Fuerte-Esquivel CR, Acha E, Ambriz-Pérez H (2000) A comprehensive Newton-Raphson UPFC model for the quadratic power flow solution of practical power networks. IEEE Trans Power Syst 15(1):102–109
Gao B, Morison GK, Kundur P (1996) Towards the development of a systematic approach for voltage stability assessment of large-scale power systems. IEEE Trans Power Syst 11(3): 1314–1324
Grainger JJ, Stevenson WD Jr (1994) Power system analysis. McGraw-Hill, New York
Granville S (1994) Optimal reactive dispatch through interior-point methods. IEEE Trans Power Syst 9(1):136–146
Gyugyi L (1992) Unified power-flow control concept for flexible AC transmission systems. IEE Proc- Gener Transm Distrib 139(4):323–331
Hamoud, G (2000) Assessment of available transfer capability of transmission systems. IEEE Trans Power Syst 15(1):27–32
Handschin E, Lehmköster C (1999) Optimal power flow for deregulated systems with FACTS devices. Proc 13th PSCC 2:1270–1276
Irisarri GD, Wang X, Tong J, Mokhtari S (1997) Maximum loadability of power systems using interior point non-linear optimization method. IEEE Trans Power Syst 12(1):162–172
Jabr RA (2003) A primal-dual interior-point method to solve the optimal power flow dispatching problem. Optim Eng 4(4):309–336
Jabr RA (2005) Primal-dual interior-point approach to compute the L_1 solution of the state estimation problem. IEE Proc- Gener Transm Distrib 152(3):313–320
Jabr RA (2007) A conic quadratic format for the load flow equations of meshed networks. IEEE Trans Power Syst 22(4):2285–2286
Jabr RA (2008) Optimal power flow using an extended conic quadratic formulation. IEEE Trans Power Syst 23(3):1000–1008
Jabr RA, Pal BC (2008) AC network state estimation using linear measurement functions. IET Gener Transm Distrib 2(1):1–6
Lai LL, Sinha N (2008) Genetic algorithms for solving optimal power flow problems. In Lee KY, El-Sharkawi MA (ed) Modern heuristic optimization techniques: theory and applications to power systems. Wiley, New Jersey
Lehmköster C (2002) Security constrained optimal power flow for an economical operation of FACTS-devices in liberalized energy markets. IEEE Trans Power Deliv 17(2):603–608
Lin SY, Ho YC, Lin CH (2004) An ordinal optimization theory-based algorithm for solving the optimal power flow problem with discrete control variables. IEEE Trans Power Syst 19(1): 276–286
Min W, Shengsong L (2005) A trust region interior point algorithm for optimal power flow problems. Elec Power Energy Syst 27(4):293–300
North American Electric Reliability Council (1995) Transmission transfer capability. Available ftp://ftp.nerc.com/pub/sys/all_updl/docs/pubs/TransmissionTransferCapability_May1995.pdf
Rider MJ, Paucar VL, Garcia AV (2004) Enhanced higher-order interior-point method to minimise active power losses in electric energy systems. IEE Proc- Gener Transm Distrib 151(4):517–525
Ruíz Muñoz JM, Gómez Expósito A (1992) A line-current measurement based state estimator. IEEE Trans Power Syst 7(2):513–519
Saadat H (1999) Power system analysis. McGraw-Hill, Singapore
Salgado R, Brameller A, Aitchison P (1990) Optimal power flow solutions using the projection method - part 1: theoretical basis. IEE Proc- Gener Transm Distrib 137(6):424–428
Sterling, MJH (1978) Power system control. Peregrinus for the IEE, Stevenage
Sun DI, Ashley B, Brewer B, Hughes A, Tinney WF (1984) Optimal power flow by Newton approach. IEEE Trans Power App Syst 103(10):2864–2880
Torres GL, Quintana VH (1998) An interior-point method for nonlinear optimal power flow using voltage rectangular coordinates. IEEE Trans Power Syst 13(4):1211–1218

Torres GL, Quintana VH (2001) On a nonlinear multiple-centrality-corrections interior-point method for optimal power flow. IEEE Trans Power Syst 16(2):222–228

Tosovic LB (1973) Some experiments on sparse sets of linear equations. SIAM J App Math 25(2):142–148

Wang H, Murillo-Sánchez CE, Zimmerman RD, Thomas RJ (2007) On computational issues of market-based optimal power flow. IEEE Trans Power Syst 22(3):1185–1193

Wei H, Sasaki H, Kubokawa J, Yokoyama R (1998) An interior point nonlinear programming for optimal power flow problems with a novel data structure. IEEE Trans Power Syst 13(3): 870–877

Wood AJ, Wollenberg BF (1996) Power generation, operation, and control. Wiley, New York

Wright SJ (1997) Primal dual interior point methods. SIAM, Philadelphia

Wu YC, Debs AS, Marsten RE (1994) A direct nonlinear predictor-corrector primal-dual interior-point algorithm for optimal power flows. IEEE Trans Power Syst 9(2):876–883

Xiao Y, Song YH, Liu CC, Sun YZ (2003) Available transfer capability enhancement using FACTS devices. IEEE Trans Power Syst 18(1):305–312

Xiao Y, Song YH, Sun YZ (2002) Power flow control approach to power systems with embedded FACTS devices. IEEE Trans Power Syst 17(4):943–950

Zhang XP (2005) Transfer capability computation with security constraints. Proc 15th PSCC, Available http://www.montefiore.ulg.ac.be/services/stochastic/pscc05/papers/fp434.pdf

Zhang XP, Handschin E (2001) Advanced implementation of UPFC in a nonlinear interior-point OPF. IEE Proc- Gener Transm Distrib 148(5):489–496

Zhang XP, Handschin E (2002) Transfer capability computation of power systems with comprehensive modeling of FACTS controllers. Proc 14th PSCC, Available http://www.pscc02.org/papers/s30p02.pdf

Zhang XP, Petoussis SG, Godfrey KR (2005) Nonlinear interior-point optimal power flow method based on a current mismatch formulation. IEE Proc- Gener Transm Distrib 152(6):795–805

Zhang XP, Rehtanz C, Pal B (2006) Flexible AC transmission systems – modelling and control. Springer, Berlin

Algorithms for Finding Optimal Flows in Dynamic Networks

Maria Fonoberova

Abstract This article presents an approach for solving some power systems problems by using optimal dynamic flow problems. The classical optimal flow problems on networks are extended and generalized for the cases of nonlinear cost functions on arcs, multicommodity flows, and time- and flow-dependent transactions on arcs of the network. All parameters of networks are assumed to be dependent on time. The algorithms for solving such kind of problems are developed by using special dynamic programming techniques based on the time-expanded network method together with classical optimization methods.

Keywords Dynamic networks · Minimum cost flow problem · Multicommodity flows · Network flows · Optimal flows

Introduction

This chapter is addressed to the elaboration of methods and algorithms for determining optimal single-commodity and multicommodity dynamic network flows, which are widely used for studying and solving a large class of practical problems, including power systems problems as well as some theoretical problems. Transportation, production and distribution, scheduling, telecommunication, management, and many other problems can be formulated and solved as optimal dynamic flow problems. Dynamic flows can be used to solve problems related to the power transmission or the optimal scheduling of generation resources in power systems as well as problems related to the coordination of operations. Power systems problems are very difficult to solve because of these systems dimensions, complexity, and their dependence on many factors. This article presents an approach to solve some power systems problems by using optimal dynamic flow problems.

M. Fonoberova
Aimdyn Inc, 1919 State St, Santa Barbara, CA 93101, USA
e-mail: mfonoberova@aimdyn.com

S. Rebennack et al. (eds.), *Handbook of Power Systems II*, Energy Systems,
DOI 10.1007/978-3-642-12686-4_2, © Springer-Verlag Berlin Heidelberg 2010

In many practical optimization problems, the factor time is a key ingredient to the problem formulation. In this paper the time-varying flow models that capture the essential properties of flows arising in real-life applications are studied. We consider minimum cost flow problems on networks with demand-supply and capacity functions that depend on time, and cost functions that are nonlinear and depend on both time and flow. We also investigate the dynamic model with transit time functions that depend on the amount of flow and the entering time-moment of flow in the arc. These problems generalize the classical flow problems on static networks and extend some known dynamic optimization models on networks (Ahuja et al. 1993; Aronson 1989; Assad 1978; Cai et al. 2001; Carey and Subrahmanian 2000; Fleischer 2001a; Fleischer 2001b; Fleisher and Skutella 2002; Ford and Fulkerson 1958; Ford and Fulkerson 1962; Glockner and Nemhauser 2002; Hoppe and Tardos 2000; Klinz and Woeginger 1995, 1998; Ma et al. 2004; Pardalos and Guisewite 1991; Powell et al. 1995). To solve the considered dynamic flow problems, we elaborate the corresponding algorithms on the basis of the time-expanded network method.

1 Optimal Dynamic Network Flow Models and Power Industry

The optimal dynamic network flow models can be used for solving various problems in the power industry, including problems related to power generation, transmission, and distribution. To operate a power system, a lot of characteristic and parameters have to be taken into account, among which are equipment capabilities, time factor, optimal cost, and many others (Weber 2005; Wood and Wollenberg 1996). The proposed optimal dynamic flow models can provide an adequate framework for such problems.

A power system is an integrated system consisting of generating plants, transmission lines, distribution facilities, and other facilities that must operate simultaneously in real-time or in a synchronous manner to provide power to consumers (Chambers and Kerr 1996; Denny and Dismukes 2002; McDonald et al. 1997; Pansini 2005). The operation and control of such systems is very complex because of their huge dimensions as well as many interconnected factors that can influence the functionality of the systems (Contreras et al. 2002; Cook et al. 2000; Rajan 2006).

Optimal dynamic flows can be used for the optimal scheduling of generating resources to meet anticipated demand, what includes the scheduling of water, fossil fuels as well as equipment maintenance, and other factors (Batut and Renaud 1992; Feltenmark and Lindberg 1997). With the expansion of power systems and the increase of the number and size of generating units, the higher operability can be attained through properly established interconnections between different components of the system, which can be done by using dynamic network flows.

Optimal dynamic flow models can be a good choice for solving problems related to power transmission from generating stations or power plants to major load centers as well as further distribution of the power to final consumers (Kersting 2006; Kim and Baldick 1997; Short 2003; Weber et al. 2007; Willis 2004; Willis et al. 2000). The power transmission networks have multiple redundant lines between points so that power can be routed through a variety of routes from any power plant to any load center. The transmission problems occupy a very important place in the power industry, because in many cases the capital cost of electric power stations is so high and electric demand is so variable that it is cheaper to import some portion of the variable load than to generate it locally. Wide area transmission grids span across countries and even large portions of continents. The electricity cost can be reduced by allowing multiple generating plants to be interconnected over a large area. However, in this case various techniques have to be applied to provide the functionality of the whole system as well as the efficient and feasible transmission taking into account economic factors, network safety, and redundancy. Optimal dynamic flow problems can be used to find the best route or to minimize the total cost of the transmission. The dynamic flow problems with flow storage at nodes can be used to match the loads with the generation by using the buffering capability.

In general, problems of power generation, transmission, and distribution can be regarded as generalization of optimal multicommodity flow problems. Different producers negotiate with different consumers for the power. It is impossible to model each of such negotiation separately, because of the nature of the problem, and so we have to consider the picture in the whole. Multicommodity dynamic flows can also be used for solving problems in the interconnection between members of power pools. For example, many electric utilities in the contiguous United States and a large part of Canada operate as members of power pools. Each individual utility in such pools operates independently, but has contractual arrangements with other members.

2 Minimum Cost Dynamic Single: Commodity Flow Problems and Algorithms for Their Solving

In this section we formulate and study minimum cost dynamic flow problems on networks with nonlinear cost functions, which depend on flow and time, and demand-supply functions and capacity functions, which depend on time. To solve the considered problems we propose algorithms based on the reduction of dynamic problems to static ones on auxiliary networks. We analyze dynamic problems with transit time functions that depend on flow and time and elaborate algorithms for solving such problems.

2.1 The Minimum Cost Dynamic Flow Problem Formulation

A dynamic network $N = (V, E, \tau, d, u, \varphi)$ is determined by directed graph $G = (V, E)$ with set of vertices V, $|V| = n$, and set of arcs E, $|E| = m$, transit time function τ: $E \to R_+$, demand-supply function d: $V \times \mathrm{T} \to R$, capacity function u: $E \times \mathrm{T} \to R_+$, and cost function φ: $E \times R_+ \times \mathrm{T} \to R_+$. It is considered the discrete time model, in which all times are integral and bounded by horizon T. Time horizon is the time until which the flow can travel in the network, and it defines the set $\mathrm{T} = \{0, 1, \dots, T\}$ of the considered time moments. Time is measured in discrete steps, so that if one unit of flow leaves vertex z at time t on arc $e = (z, v)$, one unit of flow arrives at vertex v at time $t + \tau_e$, where τ_e is the transit time on arc e. The continuous flow model formulations can be found in Fleisher (2000); Fleischer (2001a); Fleisher and Skutella (2002).

In order for the flow to exist, it is required that $\sum_{t\in\mathrm{T}} \sum_{v\in V} d_v(t) = 0$. If for an arbitrary node $v \in V$ at a moment of time $t \in \mathrm{T}$, the condition $d_v(t) > 0$ holds, then this node v at the time-moment t is treated as a source. If at a moment of time $t \in \mathrm{T}$ the condition $d_v(t) < 0$ holds, then the node v at the time-moment t is regarded as a sink. In the case $d_v(t) = 0$ at a moment of time $t \in \mathrm{T}$, the node v at the time-moment t is considered as an intermediate node. In such a way, the same node $v \in V$ at different moments of time can serve as a source, a sink, or an intermediate node.

Without losing generality we consider that the set of vertices V is divided into three disjoint subsets V_+, V_-, V_*, such that we have the following:

V_+ consists of nodes $v \in V$, for which $d_v(t) \geq 0$ for $t \in \mathrm{T}$ and there exists at least one moment of time $t_0 \in \mathrm{T}$ such that $d_v(t_0) > 0$

V_- consists of nodes $v \in V$, for which $d_v(t) \leq 0$ for $t \in \mathrm{T}$ and there exists at least one moment of time $t_0 \in \mathrm{T}$ such that $d_v(t_0) < 0$

V_* consists of nodes $v \in V$, for which $d_v(t) = 0$ for every $t \in \mathrm{T}$

So, V_+ is a set of sources, V_- is a set of sinks, and V_* is a set of intermediate nodes of the network N.

A feasible dynamic flow in network N is a function x: $E \times \mathrm{T} \to R_+$ that satisfies the following conditions:

$$\sum_{e\in E^-(v)} x_e(t) - \sum_{\substack{e\in E^+(v)\\ t-\tau_e\geq 0}} x_e(t-\tau_e) = d_v(t),\ \forall\, t \in \mathrm{T},\ \forall\, v \in V; \qquad (1)$$

$$0 \leq x_e(t) \leq u_e(t),\quad \forall\, t \in \mathrm{T},\ \forall\, e \in E; \qquad (2)$$

$$x_e(t) = 0,\ \forall\, e \in E,\ t = \overline{T - \tau_e + 1, T}, \qquad (3)$$

where $E^-(v) = \{(v, z) \mid (v, z) \in E\}$, $E^+(v) = \{(z, v) \mid (z, v) \in E\}$.

Here the function x defines the value $x_e(t)$ of flow entering arc e at time t. The flow does not enter arc e at time t if it has to leave the arc after time T; this is ensured by condition (3). Restrictions (2) are capacity constraints. Conditions (1) represent flow conservation constraints.

To model transit costs, which may change over time, it is defined as the cost function $\varphi_e(x_e(t), t)$ with the meaning that flow of value $\rho = x_e(t)$ entering arc e at time t will incur a transit cost of $\varphi_e(\rho, t)$. It is assumed that $\varphi_e(0, t) = 0$ for all $e \in E$ and $t \in \mathrm{T}$.

The total cost of the dynamic flow x in the network N is defined as follows:

$$F(x) = \sum_{t \in \mathbb{T}} \sum_{e \in E} \varphi_e(x_e(t), t). \tag{4}$$

The minimum cost dynamic flow problem consists in finding a feasible dynamic flow that minimizes the objective function (4).

2.2 *The Algorithm for Solving the Minimum Cost Dynamic Flow Problem*

In the following, the approach based on the reduction of the dynamic problem to a corresponding static problem is proposed to solve the formulated above problem. It is shown that the minimum cost dynamic flow problem on network $N = (V, E, \tau, d, u, \varphi)$ can be reduced to a minimum cost static flow problem on an auxiliary time-expanded network $N^T = (V^T, E^T, d^T, u^T, \varphi^T)$. The advantage of such an approach is that it turns the problem of determining an optimal flow over time into a classical network flow problem.

The essence of the time-expanded network is that it contains a copy of the vertex set of the dynamic network for each moment of time $t \in \mathbb{T}$, and the transit times and flows are implicit in arcs linking those copies. The network N^T is defined as follows:

1. $V^T: = \{v(t) \,|\, v \in V,\ t \in \mathbb{T}\}$
2. $E^T: = \{e(t) = (v(t), z(t + \tau_e)) \,|\, e \in E,\ 0 \le t \le T - \tau_e\}$
3. $d_{v(t)}{}^T: = d_v(t)$ for $v(t) \in V^T$
4. $u_{e(t)}{}^T: = u_e(t)$ for $e(t) \in E^T$
5. $\varphi_{e(t)}{}^T(x_{e(t)}{}^T): = \varphi_e(x_e(t), t)$ for $e(t) \in E^T$

In the following we construct the time-expanded network N^T for the dynamic network N given in Fig. 1. The set of time moments is $\mathbb{T} = \{0, 1, 2, 3\}$. The transit times on each arc are as follows: $\tau_{e_1} = 1$, $\tau_{e_2} = 1$, and $\tau_{e_3} = 2$. The capacity, demand-supply, and cost functions are considered to be given. The constructed time-expanded network is presented in Fig. 2.

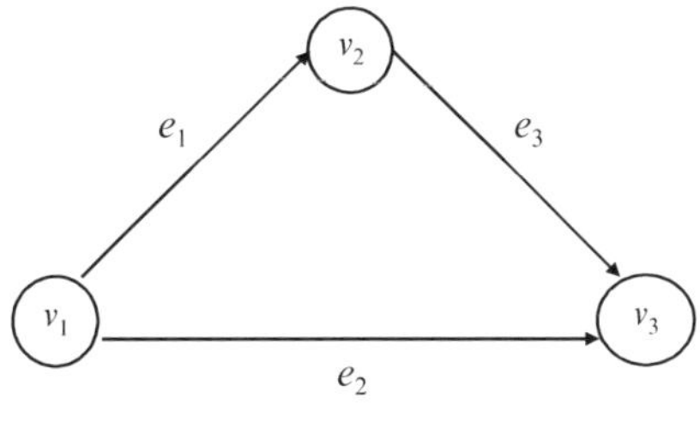

Fig. 1 The dynamic network N

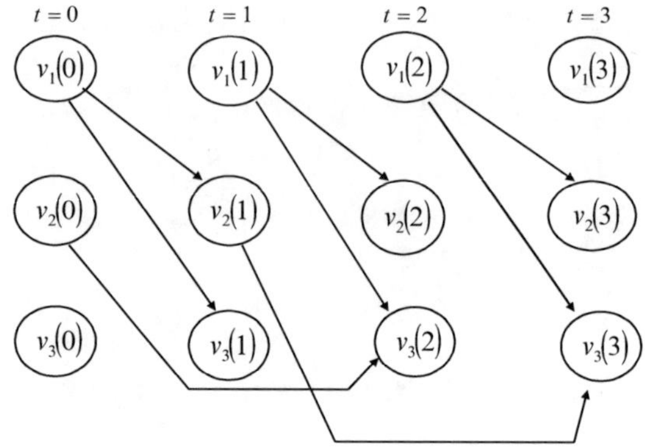

Fig. 2 The time-expanded network N^T

The correspondence between feasible flows in the dynamic network N and feasible flows in the time-expanded network N^T is stated in the following way. Let $x_e(t)$ be a flow in the dynamic network N, then the function x^T is defined as follows:

$$x_{e(t)}{}^T = x_e(t), \ \forall\, e(t) \in E^T, \tag{5}$$

which represents a flow in the time-expanded network N^T.

Lemma 1. *The correspondence (5) is a bijection from the set of feasible flows in the dynamic network N onto the set of feasible flows in the time-expanded network* N^T.

Proof. It is obvious that the correspondence (5) is a bijection from the set of T-horizon functions in the dynamic network N onto the set of functions in the time-expanded network N^T. In the following we have to show that each dynamic flow in the dynamic network N is put into the correspondence with a static flow in the time-expanded network N^T and vice-versa.

Let $x_e(t)$ be a dynamic flow in N, and let $x_{e(t)}{}^T$ be a corresponding function in N^T. Let us prove that $x_{e(t)}{}^T$ satisfies the conservation constraints in the static network N^T. Let $v \in V$ be an arbitrary vertex in N and t, $0 \leq t \leq T - \tau_e$, an arbitrary moment of time:

$$\begin{aligned} d_v(t) \overset{(i)}{=} \sum_{e \in E^-(v)} x_e(t) - \sum_{\substack{e \in E^+(v) \\ t-\tau_e \geq 0}} x_e(t-\tau_e) = \\ = \sum_{e(t) \in E^-(v(t))} x_{e(t)}{}^T - \sum_{e(t-\tau_e) \in E^+(v(t))} x_{e(t-\tau_e)}{}^T \overset{(ii)}{=} d_{v(t)}{}^T. \end{aligned} \tag{6}$$

Note that according to the definition of the time-expanded network, the set of arcs $\{e(t-\tau_e) | e(t-\tau_e) \in E^+(v(t))\}$ consists of all arcs that enter $v(t)$, while the set of arcs $\{e(t) | e(t) \in E^-(v(t))\}$ consists of all arcs that originate from $v(t)$. Therefore, all necessary conditions are satisfied for each vertex $v(t) \in V^T$. Hence, $x_{e(t)}{}^T$ is a flow in the time-expanded network N^T.

Let ${x_{e(t)}}^T$ be a static flow in the time-expanded network N^T and let $x_e(t)$ be a corresponding function in the dynamic network N. Let $v(t) \in V^T$ be an arbitrary vertex in N^T. The conservation constraints for this vertex in the static network are expressed by equality (ii) from (6), which holds for all $v(t) \in V^T$ at all times t, $0 \leq t \leq T - \tau_e$. Therefore, equality (i) holds for all $v \in V$ at all moments of time t, $0 \leq t \leq T - \tau_e$. In such a way $x_e(t)$ is a flow in the dynamic network N.

It is easy to verify that a feasible flow in the dynamic network N is a feasible flow in the time-expanded network N^T and vice-versa. Indeed,

$$0 \leq {x_{e(t)}}^T = x_e(t) \leq u_e(t) = {u_{e(t)}}^T.$$

The lemma is proved. □

Theorem 1. *If x is a flow in the dynamic network N and x^T is a corresponding flow in the time-expanded network N^T, then*

$$F(x) = F^T(x^T),$$

where

$$F^T(x^T) = \sum_{t \in \mathrm{T}} \sum_{e(t) \in E^T} {\varphi_{e(t)}}^T({x_{e(t)}}^T)$$

is the total cost of the static flow x^T in the time-expanded network N^T.

Moreover, for each minimum cost flow x^* in the dynamic network N, there is a corresponding minimum cost flow x^{*T} in the static network N^T such that

$$F(x^*) = F^T(x^{*T})$$

and vice-versa.

Proof. Let $x : E \times T \to R_+$ be an arbitrary dynamic flow in the dynamic network N. Then according to Lemma 1, the unique flow x^T in N^T corresponds to the flow x in N, and therefore we have

$$F(x) = \sum_{t \in \mathrm{T}} \sum_{e \in E} \varphi_e(x_e(t), t) = \sum_{t \in \mathrm{T}} \sum_{e(t) \in E^T} {\varphi_{e(t)}}^T({x_{e(t)}}^T) = F^T(x^T).$$

So, the first part of the theorem is proved.

To prove the second part of the theorem, we again use Lemma 1. Let $x^* : E \times T \to R_+$ be the optimal dynamic flow in N and x^{*T} be the corresponding optimal flow in N^T. Then

$$F(x^*) = \sum_{t \in \mathrm{T}} \sum_{e \in E} \varphi_e(x_e^*(t), t) = \sum_{t \in \mathrm{T}} \sum_{e(t) \in E^T} {\varphi_{e(t)}}^T({x^*_{e(t)}}^T) = F^T(x^{*T}).$$

The converse proposition is proved in an analogous way. □

The following algorithm for solving the minimum cost dynamic flow problem can be proposed:

1. To build the time-expanded network N^T for the dynamic network N
2. To solve the classical minimum cost flow problem on the static network N^T (Ahuja et al. 1993; Bland and Jensen 1985; Ermoliev and Melnic 1968; Goldberg and Tarjan 1987a,b; Hu 1970; Papadimitrou and Steiglitz 1982)
3. To reconstruct the solution of the static problem on the network N^T to the dynamic problem on the network N

Building the time-expanded network and reconstructing the solution of the minimum cost static flow problem to the dynamic one has complexity $O(nT + mT)$. The complexity of step 2 depends on the complexity of the algorithm used for the minimum cost flow problem on static networks. If such an algorithm has complexity $O(f(n', m'))$, where n' is a number of vertices and m' is a number of arcs in the network, then the complexity of solving the minimum cost flow problem on the time-expanded network employing the same algorithm is $O(f(nT, mT))$.

Some specific algorithms are proposed in Lozovanu and Fonoberova (2006) to minimize the size of the auxiliary static network. In the case of uncapacitated dynamic networks with cost functions that are concave with regard to flow value and do not change over time, the problem can be reduced to the minimum cost flow problem on a static network of equal size, not the time-expanded network.

2.3 The Dynamic Model with Flow Storage at Nodes

The previous mathematical model can be extended for the case with flow storage at nodes if we associate a transit time τ_v to each node $v \in V$, which means that the flow passage through this node takes τ_v units of time. If in addition we associate the capacity function $u_v(t)$ and the cost function $\varphi_v(x_v(t), t)$ to each node v, a more general model can be obtained. In this case the problem can be reduced to the previous one by simple transformation of the network where each node v is changed by a couple of vertices v' and v'' connected with directed arc $e_v = (v', v'')$. Here v' preserves all entering arcs and v'' preserves all leaving arcs of the previous network. The transit time $\tau_{e_v} = \tau_v$, the cost function $\varphi_{e_v}(x_{e_v}(t), t) = \varphi_v(x_v(t), t)$, and the capacity function $u_{e_v}(t) = u_v(t)$ are associated to arc e_v.

An important particular case of the minimum cost dynamic flow problem is the one when all amount of flow is dumped into the network from sources $v \in V_+$ at the time-moment $t = 0$ and it arrives at sinks $v \in V_-$ at the time-moment $t = T$. This means that the supply-demand function $d : V \times T \to R$ satisfies the conditions

(a) $d_v(0) > 0, d_v(t) = 0, t = 1, 2, \ldots, T$, for $v \in V_+$

(b) $d_v(T) < 0, d_v(t) = 0, t = 0, 1, 2, \ldots, T - 1$, for $v \in V_-$

So let us consider the minimum cost flow problem on the dynamic network with flow storage at nodes and integral constant demand-supply functions. Let

$N = (V, E, \tau, d, u, \varphi)$ be a given dynamic network, where the demand-supply function $d : V \to R$ does not depend on time. Without losing generality, we assume that no arcs enter sources or exit sinks. In order for a flow to exist, supply must equal demand: $\sum_{v \in V} d_v = 0$.

The mathematical model of the minimum cost flow problem on this dynamic network is the following:

$$\sum_{e \in E^-(v)} \sum_{t=0}^{T} x_e(t) - \sum_{e \in E^+(v)} \sum_{t=\tau_e}^{T} x_e(t - \tau_e) = d_v, \ \forall\, v \in V; \tag{7}$$

$$\sum_{e \in E^-(v)} \sum_{t=0}^{\theta} x_e(t) - \sum_{e \in E^+(v)} \sum_{t=\tau_e}^{\theta} x_e(t - \tau_e) \le 0, \ \forall v \in V_*, \ \forall \theta \in \mathrm{T}; \tag{8}$$

$$0 \le x_e(t) \le u_e(t), \ \ \forall\, t \in \mathrm{T}, \ \forall\, e \in E; \tag{9}$$

$$x_e(t) = 0, \ \forall\, e \in E, \ t = \overline{T - \tau_e + 1, T}. \tag{10}$$

Condition (10) ensures that there is no flow in the network after time horizon T. Conditions (9) are capacity constraints. As flow travels through the network, unlimited flow storage at the nodes is allowed, but any deficit is prohibited by constraint (8). Finally, all demands must be met, flow must not remain in the network after time T, and each source must not exceed its supply. This is ensured by constraint (7).

As earlier we seek for a feasible dynamic flow x that minimizes the total cost:

$$F(x) = \sum_{t \in \mathrm{T}} \sum_{e \in E} \varphi_e(x_e(t), t).$$

We would like to mention that the more general model can be obtained if we define the cost function as also dependent on the flow storage at nodes. In this case the problem can be solved by using the similar approach.

To solve the formulated above minimum cost dynamic flow problem, we use the modified time-expanded network method. The auxiliary static network N^T is constructed as follows:

1. $V^T := \{v(t) | v \in V, t \in \mathrm{T}\}$
2. $V_+^T := \{v(0) | v \in V_+\}$ and $V_-^T := \{v(T) | v \in V_-\}$
3. $E^T := \{(v(t), z(t + \tau_e)) \mid e = (v, z) \in E, \ 0 \le t \le T - \tau_e\} \cup$
   $\{v(t), v(t+1) \mid v \in V, \ 0 \le t < T\}$
4. ${d_{v(t)}}^T := d_v$ for $v(t) \in V_+^T \cup V_-^T$; otherwise ${d_{v(t)}}^T := 0$
5. ${u_{(v(t), z(t+\tau_{(v,z)}))}}^T : = u_{(v,z)}(t)$ for $(v(t), z(t + \tau_{(v,z)})) \in E^T$
   ${u_{(v(t), v(t+1))}}^T : = \infty$ for $(v(t), v(t + 1)) \in E^T$
6. ${\varphi_{(v(t), z(t+\tau_{(v,z)}))}}^T({x_{(v(t), z(t+\tau_{(v,z)}))}}^T) : = \varphi_{(v,z)}(x_{(v,z)}(t), t)$
   for $(v(t), z(t + \tau_{(v,z)})) \in E^T$
   ${\varphi_{(v(t), v(t+1))}}^T({x_{(v(t), v(t+1))}}^T) : = 0$ for $(v(t), v(t + 1)) \in E^T$

If the flow correspondence is the following: ${x_{e(t)}}^T := x_e(t)$, where ${x_{(v(t),v(t+1))}}^T$ in N^T corresponds to the flow in N stored at node v at period of time from t to $t+1$, then the minimum cost flow problem on dynamic networks can be solved by solving the minimum cost static flow problem on the time-expanded network.

2.4 Determining the Minimum Cost Dynamic Flows in Networks with Transit Time Functions that Depend on Flow and Time

In the above dynamic models, the transit time functions are assumed to be constant on each arc of the network. In this setting, the time it takes to traverse an arc does not depend on the current flow situation on the arc and the moment of time. Intuitively, it is clear that in many applications the amount of time needed to traverse an arc of the network increases as the arc becomes more congested, and it also depends on the entering time-moment of flow in the arc. If these assumptions are taken into account, a more realistic model can be obtained. In this model we assume that the transit time function $\tau_e(x_e(t), t)$ is a nonnegative nondecreasing left-continuous step function with respect to the amount of flow $x_e(t)$ for every fixed time-moment $t \in \mathbb{T}$ and an arbitrary given arc $e \in E$. We also consider two-side restrictions on arc capacities $u'_e(t) \leq x_e(t) \leq u''_e(t), \forall\, t \in \mathbb{T}, \forall\, e \in E$, where $u', u'' \colon E \times \mathbb{T} \to R_+$ are lower and upper capacities, respectively.

It is shown (Fonoberova and Lozovanu 2007) that the minimum cost flow problem on dynamic network with transit time functions that depend on the amount of flow and the entering time-moment of flow in the arc can be reduced to a static problem on a special time-expanded network $N^T = (V^T, E^T, d^T, u'^T, u''^T, \varphi^T)$, which is defined as follows:

1. $\overline{V}^T \colon = \{v(t) \mid v \in V,\ t \in \mathbb{T}\}$
2. $\widetilde{V}^T \colon = \{e(v(t)) \mid v(t) \in \overline{V}^T,\ e \in E^-(v),\ t \in \mathbb{T} \setminus \{T\}\}$
3. $V^T \colon = \overline{V}^T \cup \widetilde{V}^T$
4. $\widetilde{E}^T \colon = \{\widetilde{e}(t) = (v(t), e(v(t))) \mid v(t) \in \overline{V}^T$ and corresponding $e(v(t)) \in \widetilde{V}^T,\ t \in \mathbb{T} \setminus \{T\}\}$
5. $\overline{E}^T \colon = \{e^p(t) = (e(v(t)), z(t + \tau_e^p(x_e(t), t))) \mid e(v(t)) \in \widetilde{V}^T,\ z(t + \tau_e^p(x_e(t), t)) \in \overline{V}^T,\ e = (v, z) \in E,\ 0 \leq t \leq T - \tau_e^p(x_e(t), t),\ p \in P_{e,t}$ – set of numbers of steps of the transit time function $\tau_e^p(x_e(t), t)\}$
6. $E^T \colon = \overline{E}^T \cup \widetilde{E}^T$
7. ${d_{v(t)}}^T \colon = d_v(t)$ for $v(t) \in \overline{V}^T$
   ${d_{e(v(t))}}^T \colon = 0$ for $e(v(t)) \in \widetilde{V}^T$
8. ${u'_{\widetilde{e}(t)}}^T := u'_e(t)$ for $\widetilde{e}(t) \in \widetilde{E}^T$
   ${u''_{\widetilde{e}(t)}}^T \colon = u''_e(t)$ for $\widetilde{e}(t) \in \widetilde{E}^T$
   ${u'_{e^p(t)}}^T \colon = \overline{x_e^{p-1}}(t)$ for $e^p(t) \in \overline{E}^T$, where $\overline{x_e^0}(t) = u'_e(t)$
   ${u''_{e^p(t)}}^T \colon = \overline{x_e^p}(t)$ for $e^p(t) \in \overline{E}^T$

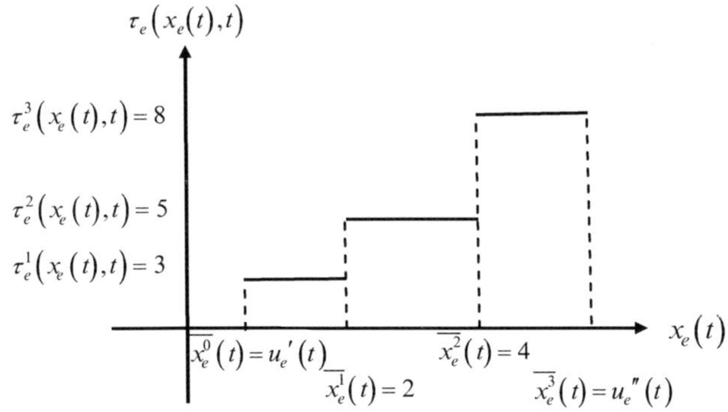

Fig. 3 The transit time function for the fixed moment of time t and the given arc $e = (v, z)$

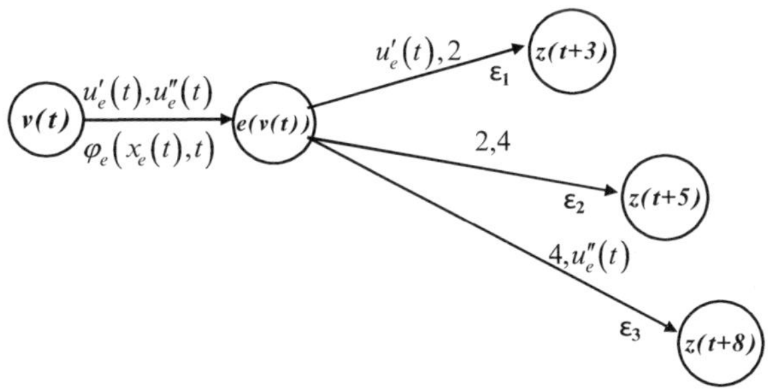

Fig. 4 The part of the constructed time-expanded network N^T for the fixed moment of time t for the arc $e = (v, z)$

9. $\varphi_{\widetilde{e}(t)}{}^T(x_{\widetilde{e}(t)}{}^T)\colon = \varphi_e(x_e(t), t)$ for $\widetilde{e}(t) \in \widetilde{E}^T$

   $\varphi_{e^p(t)}{}^T(x_{e^p(t)}{}^T)\colon = \varepsilon_p$ for $e^p(t) \in \overline{E}^T$, where $\varepsilon_1 < \varepsilon_2 < \cdots < \varepsilon_{|P_{e,t}|}$ are small numbers

Let us consider, for example, the transit time function $\tau_e = \tau_e(x_e(t), t)$, graphic representation of which for the fixed moment of time t and the given arc e is presented in Fig. 3. Here $P_{e,t} = \{1, 2, 3\}$. So, for the fixed moment of time t on the given arc e, the transit time is equal to 3 if the value of flow belongs to interval $[u_e'(t), 2]$; the transit time is equal to 5 if the value of flow belongs to interval $(2, 4]$; the transit time is equal to 8 if the value of flow belongs to interval $(4, u_e''(t)]$.

In Fig. 4, a part of the obtained time-expanded network is presented for the fixed moment of time t for the given arc $e = (v, z)$, with the transit time function in Fig. 3. Lower and upper capacities of arcs are written above each arc and costs are written below each arc.

The solution of the dynamic problem can be found on the basis of the following results.

Lemma 2. *Let $x^T\colon E^T \to R_+$ be a flow in the static network N^T. Then the function $x\colon E \times \mathbb{T} \to R_+$ defined as follows:*

$$x_e(t) = x_{\widetilde{e}(t)}{}^T = x_{e^p(t)}{}^T$$
$$\mathit{for}\ e = (v, z) \in E,\ \widetilde{e}(t) = (v(t), e(v(t))) \in \widetilde{E}^T,$$
$$e^p(t) = (e(v(t)), z(t + \tau_e^p(x_e(t), t))) \in \overline{E}^T,$$
$$p \in P_{e,t}\ \mathit{is\ such\ that}\ x_{\widetilde{e}(t)}{}^T \in (\overline{x_e^{p-1}}(t), \overline{x_e^p}(t)],\ t \in \mathrm{T},$$

which represents a flow in the dynamic network N.

Let x: $E \times \mathrm{T} \to R_+$ be a flow in the dynamic network N. Then the function x^T: $E^T \to R_+$ is defined as follows:

$$x_{\widetilde{e}(t)}{}^T = x_e(t)\ \mathit{for}\ \widetilde{e}(t) = (v(t), e(v(t))) \in \widetilde{E}^T,\ e = (v, z) \in E,\ t \in \mathrm{T};$$
$$x_{e^p(t)}{}^T = x_e(t)\ \mathit{for\ such}\ p \in P_{e,t}\ \mathit{that}\ x_e(t) \in (\overline{x_e^{p-1}}(t), \overline{x_e^p}(t)]$$
$$\mathit{and}\ x_{e^p(t)}{}^T = 0\ \mathit{for\ all\ other}\ p \in P_{e,t}$$
$$\mathit{for}\ e^p(t) = (e(v(t)), z(t + \tau_e^p(x_e(t), t))) \in \overline{E}^T,\ e = (v, z) \in E,\ t \in \mathrm{T},$$

which represents a flow in the static network N^T.

Theorem 2. *If x^{*T} is a static minimum cost flow in the static network N^T, then the corresponding one according to Lemma 2 dynamic flow x^* in the dynamic network N is also a minimum cost flow and vice-versa.*

The proofs of the above lemma and theorem can be obtained by using the arguments similar to the ones in the proofs of Lemma 1 and Theorem 1.

3 Minimum Cost Dynamic Multicommodity Flow Problems and Algorithms for Their Solving

In this section we formulate and investigate minimum cost multicommodity flow problems on dynamic networks. The multicommodity flow problem consists of shipping several different commodities from their respective sources to their sinks through a given network satisfying certain objectives in such a way that the total flow going through arcs does not exceed their capacities. No commodity ever transforms into another commodity, so that each one has its own flow conservation constraints, but they compete for the resources of the common network. In this section we consider the minimum cost multicommodity flow problems on dynamic networks with time-varying capacities of arcs and transit times on arcs that depend on sort of commodity entering them. We assume that cost functions, defined on arcs, are nonlinear and depend on time and flow, and demand-supply functions depend on time. For solving the considered problems, we propose algorithms based on the modification of the time-expanded network method. We also consider dynamic problems with transit time functions that depend on flow and time and propose algorithms for their solving.

3.1 The Minimum Cost Dynamic Multicommodity Flow Problem Formulation

It is considered a dynamic network $N = (V, E, K, \tau, d, u, w, \varphi)$, determined by directed graph $G = (V, E)$, where V is a set of vertices and E is a set of arcs, $K = \{1, 2, \ldots, q\}$ is set of commodities that must be routed through the same network, $\tau: E \times K \to R_+$ is transit time function, $d: V \times K \times \mathrm{T} \to R$ is demand-supply function, $u: E \times \mathrm{T} \to R_+$ is mutual capacity function, $w: E \times K \times \mathrm{T} \to R_+$ is individual capacity function, and $\varphi: E \times R_+ \times \mathrm{T} \to R_+$ is cost function. So, $\tau_e = (\tau_e^1, \tau_e^2, \ldots, \tau_e^q)$ is a vector, each component of which reflects the transit time on arc $e \in E$ for commodity $k \in K$. It is considered the discrete time model, where all times are integral and bounded by horizon T, which defines the set $\mathrm{T} = \{0, 1, \ldots, T\}$ of time moments.

In order for the flow to exist, it is required that $\sum_{t \in \mathrm{T}} \sum_{v \in V} d_v^k(t) = 0, \forall k \in K$. As earlier without losing generality we consider that, for every commodity $k \in K$, the set of vertices V is divided into three disjoint subsets V_+^k, V_-^k, V_*^k, such that we have the following:

V_+^k consists of nodes $v \in V$, for which $d_v^k(t) \geq 0$ for $t \in \mathrm{T}$, and there exists at least one moment of time $t_0 \in \mathrm{T}$ such that $d_v^k(t_0) > 0$

V_-^k consists of nodes $v \in V$, for which $d_v^k(t) \leq 0$ for $t \in \mathrm{T}$, and there exists at least one moment of time $t_0 \in \mathrm{T}$ such that $d_v^k(t_0) < 0$

V_*^k consists of nodes $v \in V$, for which $d_v^k(t) = 0$ for every $t \in \mathrm{T}$

So, V_+^k is a set of sources, V_-^k is a set of sinks, and V_*^k is a set of intermediate nodes for the commodity $k \in K$ in the network N

A feasible dynamic multicommodity flow in the network N is determined by a function $x: E \times K \times \mathrm{T} \to R_+$ that satisfies the following conditions:

$$\sum_{e \in E^-(v)} x_e^k(t) - \sum_{\substack{e \in E^+(v) \\ t - \tau_e^k \geq 0}} x_e^k(t - \tau_e^k) = d_v^k(t), \ \forall t \in \mathrm{T}, \ \forall v \in V, \ \forall k \in K; \tag{11}$$

$$\sum_{k \in K} x_e^k(t) \leq u_e(t), \ \forall t \in \mathrm{T}, \ \forall e \in E; \tag{12}$$

$$0 \leq x_e^k(t) \leq w_e^k(t), \ \ \forall t \in \mathrm{T}, \ \forall e \in E, \ \forall k \in K; \tag{13}$$

$$x_e^k(t) = 0, \ \forall e \in E, \ t = \overline{T - \tau_e^k + 1, T}, \ \forall k \in K. \tag{14}$$

Here the function x defines the value $x_e^k(t)$ of flow of commodity k entering arc e at moment of time t. Condition (14) ensures that the flow of commodity k does not enter arc e at time t if it has to leave the arc after time horizon T. Individual and mutual capacity constraints (13) and (12) are called weak and strong forcing constraints, respectively. Conditions (11) represent flow conservation constraints.

The total cost of the dynamic multicommodity flow x in the network N is defined as follows:

$$F(x) = \sum_{t \in \mathbb{T}} \sum_{e \in E} \varphi_e(x_e^1(t), x_e^2(t), \ldots, x_e^q(t), t). \tag{15}$$

The minimum cost dynamic multicommodity flow problem consists in finding a feasible dynamic multicommodity flow that minimizes the objective function (15).

3.2 The Algorithm for Solving the Minimum Cost Dynamic Multicommodity Flow Problem

To solve the formulated problem, we propose an approach based on the reduction of the dynamic problem to a static problem. It is shown that the minimum cost multicommodity flow problem on network N can be reduced to a static problem on a special auxiliary network N^T. In the case of the minimum cost multicommodity flow problem on dynamic network with different transit times on an arc for different commodities, the auxiliary time-expanded network $N^T = (V^T, E^T, K, d^T, u^T, w^T, \varphi^T)$ is defined in the following way:

1. $\overline{V}^T := \{v(t) \mid v \in V,\ t \in \mathbb{T}\}$
2. $\widetilde{V}^T := \{e(v(t)) \mid v(t) \in \overline{V}^T,\ e \in E^-(v),\ t \in \mathbb{T} \setminus \{T\}\}$
3. $V^T := \overline{V}^T \cup \widetilde{V}^T$
4. $\widetilde{E}^T := \{\widetilde{e}(t) = (v(t), e(v(t))) \mid v(t) \in \overline{V}^T$ and corresponding $e(v(t)) \in \widetilde{V}^T$, $t \in \mathbb{T} \setminus \{T\}\}$
5. $\overline{E}^T := \{e^k(t) = (e(v(t)), z(t + \tau_e^k)) \mid e(v(t)) \in \widetilde{V}^T,\ z(t + \tau_e^k) \in \overline{V}^T$, $e = (v, z) \in E,\ 0 \le t \le T - \tau_e^k,\ k \in K\}$
6. $E^T := \overline{E}^T \cup \widetilde{E}^T$
7. ${d^k_{v(t)}}^T := d_v^k(t)$ for $v(t) \in \overline{V}^T,\ k \in K$

   ${d^k_{e(v(t))}}^T := 0$ for $e(v(t)) \in \widetilde{V}^T,\ k \in K$
8. ${u_{\widetilde{e}(t)}}^T := u_e(t)$ for $\widetilde{e}(t) \in \widetilde{E}^T$

   ${u_{e^k(t)}}^T := \infty$ for $e^k(t) \in \overline{E}^T$
9. ${w^l_{e^k(t)}}^T := \begin{cases} w_e^k(t), & \text{if } l = k \text{ for } e^k(t) \in \overline{E}^T,\ l \in K \\ 0, & \text{if } l \neq k \text{ for } e^k(t) \in \overline{E}^T,\ l \in K \end{cases}$

   ${w^l_{\widetilde{e}(t)}}^T = \infty$ for $\widetilde{e}(t) \in \widetilde{E}^T,\ l \in K$
10. ${\varphi_{\widetilde{e}(t)}}^T({x^1_{\widetilde{e}(t)}}^T, {x^2_{\widetilde{e}(t)}}^T, \ldots, {x^q_{\widetilde{e}(t)}}^T) := \varphi_e(x_e^1(t), x_e^2(t), \ldots, x_e^q(t), t)$ for $\widetilde{e}(t) \in \widetilde{E}^T$

    ${\varphi_{e^k(t)}}^T({x^1_{e^k(t)}}^T, {x^2_{e^k(t)}}^T, \ldots, {x^q_{e^k(t)}}^T) := 0$ for $e^k(t) \in \overline{E}^T$

In the following we construct the time-expanded network N^T for the dynamic network N given in Fig. 1 with set of two commodities $K = \{1, 2\}$, set of time moments $\mathbb{T} = \{0, 1, 2, 3\}$, and transit times $\tau_{e_1}^1 = 2$, $\tau_{e_1}^2 = 1$, $\tau_{e_2}^1 = 1$, $\tau_{e_2}^2 = 3$,

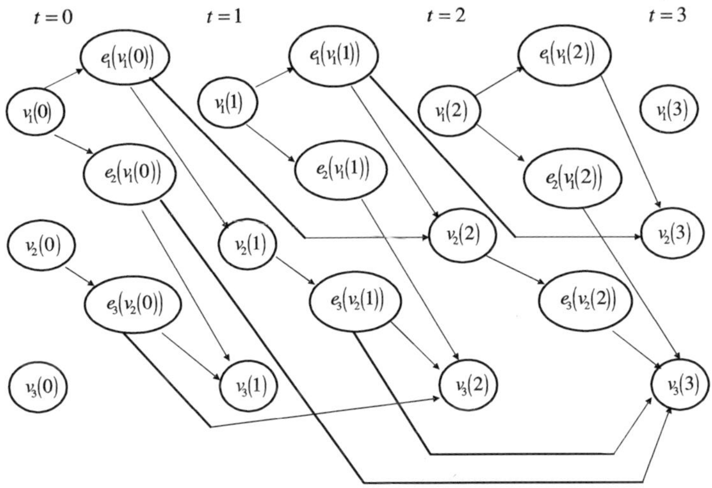

Fig. 5 The time-expanded network (case of different transit times on an arc for different commodities)

$\tau^1_{e_3} = 1$, $\tau^2_{e_3} = 2$. The mutual capacity, individual capacity, demand-supply, and cost functions are considered to be known. The constructed time-expanded network N^T is presented in Fig. 5.

Lemma 3. *Let $x^T: E^T \times K \to R_+$ be a multicommodity flow in the static network N^T. Then the function $x: E \times K \times \mathbb{T} \to R_+$ is defined in the following way:*

$$x^k_e(t) = x^{k\;\;T}_{e^k(t)} = x^{k\;\;T}_{\widetilde{e}(t)}$$
$$\textit{for } e = (v,z) \in E,\ e^k(t) = (e(v(t)), z(t+\tau^k_e)) \in \overline{E}^T,$$
$$\widetilde{e}(t) = (v(t), e(v(t))) \in \widetilde{E}^T,\ k \in K,\ t \in \mathbb{T},$$

which represents a multicommodity flow in the dynamic network N.

Let $x: E \times K \times \mathbb{T} \to R_+$ be a multicommodity flow in the dynamic network N. Then the function $x^T: E^T \times K \to R_+$ is defined in the following way:

$$x^{k\;\;T}_{\widetilde{e}(t)} = x^k_e(t) \textit{ for } \widetilde{e}(t) = (v(t), e(v(t))) \in \widetilde{E}^T,\ e = (v,z) \in E,\ k \in K,\ t \in \mathbb{T};$$
$$x^{k\;\;T}_{e^k(t)} = x^k_e(t);\ x^{l\;\;T}_{e^k(t)} = 0,\ l \neq k$$
$$\textit{for } e^k(t) = (e(v(t)), z(t+\tau^k_e)) \in \overline{E}^T,\ e = (v,z) \in E,\ l,k \in K,\ t \in \mathbb{T},$$

which represents a multicommodity flow in the static network N^T.

Proof. To prove the first part of the lemma, we have to show that conditions (11)–(14) for the defined above x in the dynamic network N are true. These conditions evidently result from the following definition of multicommodity flows in the static network N^T:

$$\sum_{e(t)\in E^-(v(t))} {x_{e(t)}^k}^T - \sum_{e(t-\tau_e^k)\in E^+(v(t))} {x_{e(t-\tau_e^k)}^k}^T = {d_{v(t)}^k}^T, \quad \forall\, v(t) \in V^T,\ \forall k \in K; \tag{16}$$

$$\sum_{k\in K} {x_{e(t)}^k}^T \le {u_{e(t)}}^T,\ \forall e(t) \in E^T; \tag{17}$$

$$0 \le {x_{e(t)}^k}^T \le {w_{e(t)}^k}^T,\ \forall\, e(t) \in E^T,\ \forall k \in K; \tag{18}$$

$${x_{e(t)}^k}^T = 0,\ \forall\, e(t) \in E^T,\ t = \overline{T - \tau_e^k + 1, T},\ \forall k \in K, \tag{19}$$

where by $v(t)$ and $e(t)$ we denote $\overline{v}(t)$ or $\widetilde{v}(t)$ and $\overline{e}(t)$ or $\widetilde{e}(t)$, respectively, against context.

To prove the second part of the lemma, it is sufficient to show that conditions (16)–(19) hold for x^T defined above. Correctness of these conditions results from the procedure of constructing the time-expanded network, the correspondence between flows in static and dynamic networks, and the satisfied conditions (11)–(14).

The lemma is proved. □

Theorem 3. *If x^{*T} is a minimum cost multicommodity flow in the static network N^T, then the corresponding one according to Lemma 3 multicommodity flow x^* in the dynamic network N is also a minimum cost one and vice-versa.*

Proof. Taking into account the correspondence between static and dynamic multicommodity flows on the basis of Lemma 3, we obtain that costs of the static multicommodity flow in the time-expanded network N^T and the corresponding dynamic multicommodity flow in the dynamic network N are equal. To solve the minimum cost multicommodity flow problem on the static time-expanded network N^T, we have to solve the following problem:

$$F^T(x^T) = \sum_{t\in \mathbb{T}} \sum_{e(t)\in E^T} {\varphi_{e(t)}}^T ({x_{e(t)}^1}^T, {x_{e(t)}^2}^T, \ldots, {x_{e(t)}^q}^T) \to \min$$

subject to (16)–(19).

□

In the case of the minimum cost multicommodity flow problem on dynamic network with common transit times on an arc for different commodities, the time-expanded network N^T can be constructed more simply:

1. $V^T := \{v(t) \mid v \in V,\ t \in \mathbb{T}\}$
2. $E^T := \{e(t) = (v(t), z(t + \tau_e)) \mid v(t) \in V^T,\ z(t + \tau_e) \in V^T,\ e = (v, z) \in E,$ $0 \le t \le T - \tau_e\}$

3. ${d^k_{v(t)}}^T := d^k_v(t)$ for $v(t) \in V^T$, $k \in K$
4. ${u_{e(t)}}^T := u_e(t)$ for $e(t) \in E^T$
5. ${w^k_{e(t)}}^T := w^k_e(t)$ for $e(t) \in E^T$, $k \in K$
6. ${\varphi_{e(t)}}^T({x^1_{e(t)}}^T, {x^2_{e(t)}}^T, \ldots, {x^q_{e(t)}}^T) := \varphi_e(x^1_e(t), x^2_e(t), \ldots, x^q_e(t), t)$ for $e(t) \in E^T$

The following lemma and theorem can be considered as particular cases of Lemma 3 and Theorem 3.

Lemma 4. *Let $x^T: E^T \times K \to R_+$ be a multicommodity flow in the static network N^T. Then the function $x: E \times K \times \mathrm{T} \to R_+$ is defined as follows:*

$$x^k_e(t) = {x^k_{e(t)}}^T \; \textit{for } e \in E,\ e(t) \in E^T,\ k \in K,\ t \in \mathrm{T},$$

which represents the multicommodity flow in the dynamic network N.

Let $x: E \times K \times \mathrm{T} \to R_+$ be a multicommodity flow in the dynamic network N. Then the function $x^T: E^T \times K \to R_+$ is defined as follows:

$${x^k_{e(t)}}^T = x^k_e(t) \; \textit{for } e(t) \in E^T,\ e \in E,\ k \in K,\ t \in \mathrm{T},$$

which represents the multicommodity flow in the static network N^T.

Theorem 4. *If ${x^*}^T$ is a minimum cost multicommodity flow in the static network N^T, then the corresponding one according to Lemma 4 multicommodity flow x^* in the dynamic network N is also a minimum cost one and vice-versa.*

In such a way, to solve the minimum cost multicommodity flow problem on dynamic networks we have the following:

1. To build the time-expanded network N^T for the given dynamic network N
2. To solve the classical minimum cost multicommodity flow problem on the static network N^T (Assad 1978; McBride 1998; Castro and Nabona 1996; Castro 2000, 2003; Ermoliev and Melnic 1968; Fleisher 2000).
3. To reconstruct the solution of the static problem on N^T to the dynamic problem on N.

The complexity of this algorithm depends on the complexity of the algorithm used for the minimum cost multicommodity flow problem on the static network. If such an algorithm has complexity $O(f(n', m'))$, where n' is a number of vertices and m' is a number of arcs in the network, then the complexity of solving the minimum cost multicommodity flow problem on the time-expanded network employing the same algorithm is $O(f((n+m)T, m(k+1)T))$, where n is the number of vertices in the dynamic network, m is the number of arcs in the dynamic network, and k is the number of commodities.

3.3 The Reduced Time-Expanded Network for Acyclic Graphs

In this subsection we consider the minimum cost multicommodity flow problem on acyclic dynamic network $N = (V, E, K, \tau, d, u, w, \varphi)$ with time horizon $T = +\infty$ and common transit times on an arc for different commodities. Without losing generality, we assume that no arcs enter sources or exit sinks. Let $T^* = \max\{|L|\} = \max\{\sum_{e \in L} \tau_e\}$, where L is a directed path in the graph $G = (V, E)$. It is not difficult to show that $x_e^k(t) = 0$ for $\forall e \in E, \forall k \in K, \forall t \geq T^*$. This fact allows us to replace the infinite time horizon with the finite one, by substituting T^* for the positive infinity.

In many cases a big number of nodes is not connected with a directed path both to a sink and a source. Removing such nodes from the considered network does not influence the set of flows in this network. These nodes are called irrelevant to the flow problem. Nodes that are not irrelevant are relevant. The static network obtained by eliminating the irrelevant nodes and all arcs adjacent to them from the time-expanded network is called the reduced time-expanded network.

The following algorithm is proposed for constructing the reduced network $N^{rT^*} = (V^{rT^*}, E^{rT^*}, d^{rT^*}, u^{rT^*}, w^{rT^*}, \varphi^{rT^*})$, which is based on the process of elimination of irrelevant nodes from the time-expanded network:

1. To build the time-expanded network N^{T^*} for the given dynamic network N.
2. To perform a breadth-first parse of the nodes for each source from the time expanded-network. The result of this step is the set $V_-(V_-^{T^*})$ of the nodes that can be reached from at least a source in V^{T^*}.
3. To perform a breadth-first parse of the nodes beginning with the sink for each sink and parsing the arcs in the direction opposite to their normal orientation. The result of this step is the set $V_+(V_+^{T^*})$ of nodes from which at least a sink in V^{T^*} can be reached.
4. The reduced network will consist of a subset of nodes V^{T^*} and arcs from E^{T^*} determined in the following way:

$$V^{rT^*} = V^{T^*} \cap V_-(V_-^{T^*}) \cap V_+(V_+^{T^*}),$$

$$E^{rT^*} = E^{T^*} \cap (V^{rT^*} \times V^{rT^*}).$$

5. ${d^{rk}_{v(t)}}^{T^*} := d_v^k(t)$ for $v(t) \in V^{rT^*}$, $k \in K$.
6. ${u^r{}_{e(t)}}^{T^*} := u_e(t)$ for $e(t) \in E^{rT^*}$.
7. ${w^{rk}_{e(t)}}^{T^*} := w_e^k(t)$ for $e(t) \in E^{rT^*}$, $k \in K$.
8. ${\varphi^r{}_{e(t)}}^{T^*}({x^1_{e(t)}}^{T*}, {x^2_{e(t)}}^{T*}, \dots, {x^q_{e(t)}}^{T*}) := \varphi_e(x_e^1(t), x_e^2(t), \dots, x_e^q(t), t)$ for $e(t) \in E^{rT^*}$.

The complexity of this algorithm can be estimated to be the same as the complexity of constructing the time-expanded network. It can be proven by using the similar approach as in Lozovanu and Stratila (2001) that the reduced network can be used in place of the time-expanded network.

We would like to mention that the proposed above approach with some modifications can be used for constructing the reduced time-expanded network for the optimal single-commodity dynamic flow problems and the optimal multicommodity dynamic flow problems with different transit times on an arc for different commodities.

3.4 The Minimum Cost Multicommodity Dynamic Flow Problem with Transit Time Functions that Depend on Flow and Time

In this subsection an approach for solving the minimum cost multicommodity dynamic flow problem with transit time functions that depend on flow and time is proposed. This problem is considered on dynamic networks with time-varying lower and upper capacity functions, time-varying mutual capacity function, and time-varying demand-supply function. It is assumed that cost functions, defined on arcs, are nonlinear and depend on flow and time. The transit time function $\tau_e^k(x_e^k(t), t)$ is considered to be a nonnegative nondecreasing left-continuous step function for each commodity $k \in K$.

The method for solving the minimum cost multicommodity dynamic flow problem with transit time functions that depend on flows and time is based on the reduction of the dynamic problem to a static problem on an auxiliary time-expanded network $N^T = (V^T, E^T, d^T, u^T, w'^T, w''^T, \varphi^T)$, which is defined as follows:

1. $\overline{V}^T := \{v(t) \mid v \in V,\ t \in \mathbb{T}\}$
2. $\widetilde{V}^T := \{e(v(t)) \mid v(t) \in \overline{V}^T,\ e \in E^-(v),\ t \in \mathbb{T} \setminus \{T\}\}$
3. $V^T := \overline{V}^T \cup \widetilde{V}^T$
4. $\widetilde{E}^T := \{\widetilde{e}(t) = (v(t), e(v(t))) \mid v(t) \in \overline{V}^T$ and corresponding $e(v(t)) \in \widetilde{V}^T,\ t \in \mathbb{T} \setminus \{T\}\}$
5. $\overline{E}^T := \{e^{k,p}(t) = (e(v(t)), z(t + \tau_e^{k,p}(x_e^k(t), t))) \mid e(v(t)) \in \widetilde{V}^T,\ z(t + \tau_e^{k,p}(x_e^k(t), t)) \in \overline{V}^T, e = (v, z) \in E,\ 0 \le t \le T - \tau_e^{k,p}(x_e^k(t), t),\ p \in P_{e,t}^k -$ set of numbers of steps of the transit time function $\tau_e^k(x_e^k(t), t), k \in K\}$
6. $E^T := \overline{E}^T \cup \widetilde{E}^T$
7. ${d_{v(t)}^k}^T := d_v^k(t)$ for $v(t) \in \overline{V}^T,\ k \in K$

   ${d_{e(v(t))}^k}^T := 0$ for $e(v(t)) \in \widetilde{V}^T,\ k \in K$
8. ${u_{\widetilde{e}(t)}}^T := u_e(t)$ for $\widetilde{e}(t) \in \widetilde{E}^T$

   ${u_{e^{k,p}(t)}}^T := \infty$ for $e^{k,p}(t) \in \overline{E}^T$

9. $w'^{l}_{e^{k,p}(t)}{}^{T} := \begin{cases} \overline{x_e^{k,p-1}}(t), \text{ if } l=k \text{ for } e^{k,p}(t) \in \overline{E}^T, \ l \in K, \text{ where } \overline{x_e^{k,0}}(t) = w'^{k}_e(t) \\ 0, \quad \text{if } l \neq k \text{ for } e^{k,p}(t) \in \overline{E}^T, \ l \in K \end{cases}$

$w''^{l}_{e^{k,p}(t)}{}^{T} := \begin{cases} \overline{x_e^{k,p}}(t), \text{ if } l = k \text{ for } e^{k,p}(t) \in \overline{E}^T, \ l \in K \\ 0, \quad \text{if } l \neq k \text{ for } e^{k,p}(t) \in \overline{E}^T, \ l \in K \end{cases}$

$w'^{l}_{\widetilde{e}(t)}{}^{T} = -\infty; \ w''^{l}_{\widetilde{e}(t)}{}^{T} = +\infty$ for $\widetilde{e}(t) \in \widetilde{E}^T, \ l \in K$

10. $\varphi_{\widetilde{e}(t)}{}^{T}(x^1_{\widetilde{e}(t)}{}^{T}, x^2_{\widetilde{e}(t)}{}^{T}, \ldots, x^q_{\widetilde{e}(t)}{}^{T}) := \varphi_e(x^1_e(t), x^2_e(t), \ldots, x^q_e(t), t)$ for $\widetilde{e}(t) \in \widetilde{E}^T$

$\varphi_{e^{k,p}(t)}{}^{T}(x^1_{e^{k,p}(t)}{}^{T}, x^2_{e^{k,p}(t)}{}^{T}, \ldots, x^q_{e^{k,p}(t)}{}^{T}) := \varepsilon^{k,p}$ for $e^{k,p}(t) \in \overline{E}^T$, where $\varepsilon^{k,1} < \varepsilon^{k,2} < \cdots < \varepsilon^{k,|P^k_{e,t}|}$ are small numbers

For example, let us consider the transit time functions for an arc $e = (v, z)$ at the moment of time t presented in Figs. 6 and 7, which correspond to commodities 1 and 2, respectively.

The constructed part of the time-expanded network for the fixed moment of time t for the arc $e = (v, z)$ is presented in Fig. 8.

The following lemma and theorem give us the relationship between flows in network N and flows in network N^T.

Lemma 5. *Let $x^T: E^T \times K \to R_+$ be a multicommodity flow in the static network N^T. Then the function $x: E \times K \times \mathbb{T} \to R_+$ is defined in the following way:*

$$x^k_e(t) = x^k_{\widetilde{e}(t)}{}^{T} = x^k_{e^{k,p}(t)}{}^{T}$$
$$\text{for } e = (v, z) \in E, \ \widetilde{e}(t) = (v(t), e(v(t))) \in \widetilde{E}^T,$$
$$e^{k,p}(t) = (e(v(t)), z(t + \tau^{k,p}_e(x^k_e(t), t))) \in \overline{E}^T,$$
$$p \in P^k_{e,t} \text{ is such that } x^k_{\widetilde{e}(t)}{}^{T} \in (\overline{x^{k,p-1}_e}(t), \overline{x^{k,p}_e}(t)], \ t \in \mathbb{T}, \ k \in K,$$

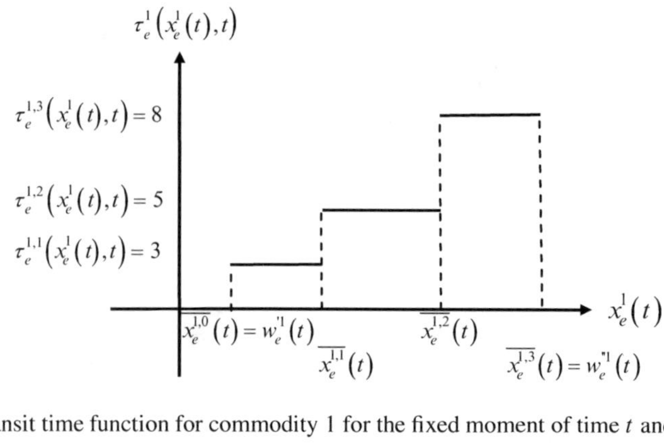

Fig. 6 The transit time function for commodity 1 for the fixed moment of time t and the given arc $e = (v, z)$

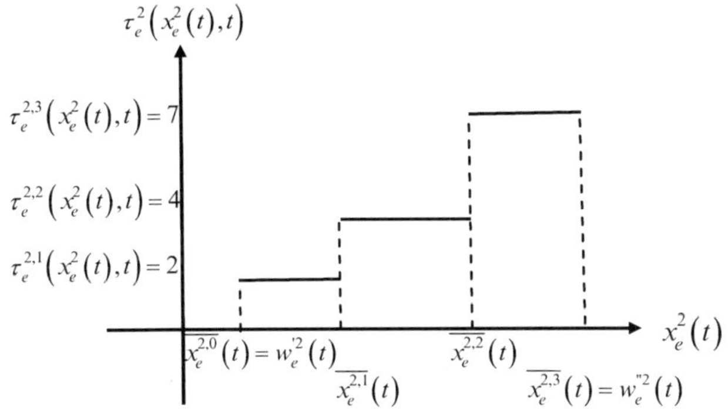

Fig. 7 The transit time function for commodity 2 for the fixed moment of time t and the given arc $e = (v, z)$

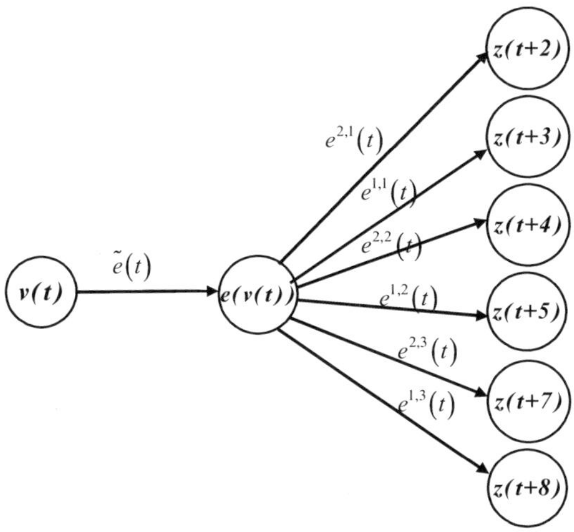

Fig. 8 The part of the constructed time-expanded network N^T for the fixed moment of time t for the arc $e = (v, z)$

which represents a multicommodity flow in the dynamic network N.

Let x: $E \times K \times \mathbb{T} \rightarrow R_+$ be a multicommodity flow in the dynamic network N. Then the function x^T: $E^T \times K \rightarrow R_+$ is defined in the following way:

$$x^{k\;\;T}_{\widetilde{e}(t)} = x_e^k(t)$$

$$for\, \widetilde{e}(t) = (v(t), e(v(t))) \in \widetilde{E}^T,\ e = (v, z) \in E,\ k \in K,\ t \in \mathbb{T};$$

$$x^{l\;\;\;\;T}_{e^{k,p}(t)} = 0,\ l \neq k;$$

$$x^{k}_{e^{k,p}(t)}{}^{T} = x_e^k(t) \, for \, such \, p \in P_{e,t}^k \, that \, x_e^k(t) \in (\overline{x_e^{k,p-1}}(t), \overline{x_e^{k,p}}(t)],$$
$$x^{k}_{e^{k,p}(t)}{}^{T} = 0 \, for \, all \, other \, p \in P_{e,t}^k$$
$$for \, e^{k,p}(t) = (e(v(t)), z(t + \tau_e^{k,p}(x_e^k(t), t))) \in \overline{E}^T, \; e = (v, z) \in E, \; l, k \in K,$$
$$t \in \mathrm{T},$$

which represents a multicommodity flow in the static network N^T.

Theorem 5. *If x^{*T} is a minimum cost multicommodity flow in the static network N^T, then the corresponding one according to Lemma 5 multicommodity flow x^* in the dynamic network N is also a minimum cost one and vice-versa.*

Conclusions

In this chapter we extended and generalized the classical optimal flow problems on networks, which can be used for solving problems related to power generation, transmission, and distribution. We considered optimal single-commodity and multicommodity flow problems on networks when all parameters are time-dependent. To solve the considered problems, we elaborated the corresponding algorithms on the basis of the time-expanded network method.

Acknowledgements I express my gratitude to Dmitrii Lozovanu for close collaboration.

References

Ahuja R, Magnati T, Orlin J (1993) Network flows. Prentice-Hall, Englewood Cliffs

Aronson J (1989) A survey of dynamic network flows. Ann Oper Res 20:1–66

Assad A (1978) Multicommodity network flows: a survey. Networks 8:37–92

Batut J, Renaud A (1992) Daily generation scheduling optimization with transmission constraints: a new class of algorithms. IEEE Trans Power Syst 7(3):982–989

Bland RG, Jensen DL (1985) On the computational behavior of a polynomial-time network flow algorithm. Technical Report 661, School of Operations Research and Industrial Engineering, Cornell University

Cai X, Sha D, Wong CK (2001) Time-varying minimum cost flow problems. Eur J Oper Res 131:352–374

Carey M, Subrahmanian E (2000) An approach to modelling time-varying flows on congested networks. Transport Res B 34:157–183

Castro J (2000) A specialized interior-point algorithm for multicommodity network flows. Siam J Optim 10(3):852–877

Castro J (2003) Solving difficult multicommodity problems with a specialized interior-point algorithm. Ann Oper Res 124:35–48

Castro J, Nabona N (1996) An implementation of linear and nonlinear multicommodity network flows. Eur J Oper Res Theor Meth 92:37–53

Chambers A, Kerr S (1996) Power industry dictionary. PennWell Books, OK

Contreras J, Losi A, Russo M, Wu FF (2002) Simulation and evaluation of optimization problem solutions in distributed energy management systems. IEEE Trans Power Syst 17(1):57–62
Cook D, Hicks G, Faber V, Marathe M, Srinivasan A, Sussmann Y, Thornquist H (2000) Combinatorial problems arising in deregulated electrical power industry: survey and future directions. In: Panos PM (ed) Approximation and complexity in numerical optimization: continuous and discrete problems. Kluwer, Dordrecht, pp. 138–162
Denny F, Dismukes D (2002) Power system operations and electricity markets. CRC Press, FL
Ermoliev I, Melnic I (1968) Extremal problems on graphs. Naukova Dumka, Kiev
Feltenmark S, Lindberg PO (1997) network methods for head-dependent hydro power schedule. In: Panos MP, Donald WH, William WH (eds) Network optimization. Springer, Heidelberg, pp. 249–264
Fleisher L (2000) Approximating multicommodity flow independent of the number of commodities. Siam J Discrete Math 13(4):505–520
Fleischer L (2001a) Universally maximum flow with piecewise-constant capacities. Networks 38(3):115–125
Fleischer L (2001b) Faster algorithms for the quickest transshipment problem. Siam J Optim 12(1):18–35
Fleisher L, Skutella M (2002) The quickest multicommodity flow problem. Integer programming and combinatorial optimization. Springer, Berlin, pp. 36–53
Fonoberova M, Lozovanu D (2007) Minimum cost multicommodity flows in dynamic networks and algorithms for their finding. Bull Acad Sci Moldova Math 1(53):107–119
Ford L, Fulkerson D (1958) Constructing maximal dynamic flows from static flows. Oper Res 6:419–433
Ford L, Fulkerson D (1962) Flows in networks. Princeton University Press, Princeton, NJ
Glockner G, Nemhauser G (2002) A dynamic network flow problem with uncertain arc capacities: formulation and problem structure. Oper Res 48(2):233–242
Goldberg AV, Tarjan RE (1987a) Solving minimum-cost flow problems by successive approximation. Proc. 19th ACM STOC, pp. 7–18
Goldberg AV, Tarjan RE (1987b) Finding minimum-cost circulations by canceling negative cycles. Technical Report CS-TR 107-87, Department of Computer Science, Princeton University
Hoppe B, Tardos E (2000) The quickest transshipment problem. Math Oper Res 25:36–62
Hu T (1970) Integer programming and network flows. Addison-Wesley Publishing Company, Reading, MA
Kersting W (2006) Distribution system modeling and analysis, 2nd edn. CRC, FL
Kim BH, Baldick R (1997) Coarse-grained distributed optimal power flow. IEEE Trans Power Syst 12(2):932–939
Klinz B, Woeginger C (1995) Minimum cost dynamic flows: the series parallel case. Integer programming and combinatorial optimization. Springer, Berlin, pp. 329–343
Klinz B, Woeginger C (1998) One, two, three, many, or: complexity aspects of dynamic network flows with dedicated arcs. Oper Res Lett 22:119–127
Lozovanu D, Fonoberova M (2006) Optimal flows in dynamic networks, Chisinau, CEP USM
Lozovanu D, Stratila D (2001) The minimum-cost flow problem on dynamic networks and algorithm for its solving. Bull Acad Sci Moldova Math 3:38–56
Ma Z, Cui D, Cheng P (2004) Dynamic network flow model for short-term air traffic flow management. IEEE Trans Syst Man Cybern A Syst Hum 34(3):351–358
McBride R (1998) Progress made in solving the multicommodity flow problems. Siam J Optim 8(4):947–955
McDonald JR, McArthur S, Burt G, Zielinski J (eds) (1997) Intelligent knowledge based systems in electrical power, 1st edn. Springer, Heidelberg
Pansini A (2005) Guide to electrical power distribution systems, 6th edn. CRC, FL
Papadimitrou C, Steiglitz K (1982) Combinatorial optimization: algorithms and complexity. Prentice-Hall, Englewood Cliffs, NJ
Pardalos PM, Guisewite G (1991) Global search algorithms for minimum concave cost network flow problem. J Global Optim 1(4):309–330

Powell W, Jaillet P, Odoni A (1995) Stochastic and dynamic networks and routing. In: Ball MO, Magnanti TL, Monma CL, Nemhauser GL (eds) Network routing, vol. 8 of Handbooks in operations research and management science, chapter 3. North Holland, Amsterdam, The Netherlands, pp. 141–295

Rajan GG (2006) Practical energy efficiency optimization. PennWell Books. OK

Short T (2003) Electric power distribution handbook, 1st edn. CRC, FL

Weber C (2005) Uncertainty in the electric power industry: methods and models for decision support. Springer, Heidelberg

Weber C, Marechala F, Favrata D (2007) Design and optimization of district energy systems. In: Plesu V, Agachi PS (eds) 17th European Symposium on Computer Aided Process Engineering – ESCAPE17, Elsevier, Amsterdam

Willis HL (2004) Power distribution planning reference book. CRC Press, FL

Willis HL, Welch G, Schrieber R (2000) Aging power delivery infrastructures. CRC Press, FL

Wood A, Wollenberg B (1996) Power generation, operation and control. Wiley, NY

Signal Processing for Improving Power Quality

Long Zhou and Loi Lei Lai

Abstract The wide spread use of power electronic equipment causes serious current harmonics in electrical power system. Harmonic currents that flow in the electrical power system would cause extra copper loss and immature operation of over current protection devices. Voltage distortion due to harmonic voltage drop in the electrical power distribution system impairs the operation of voltage-sensitive equipment. To improve the electrical power quality and reduce energy wastage in the electrical power distribution system, especially under the deregulated environment, the nature of the harmonics must be identified so that the causes and effects of the harmonics would be studied. Moreover, corrective measures cannot be easily implemented without knowing the characteristics of the harmonics existing in the electrical power system. The chapter presents investigation results obtained from the harmonics analysis and modeling of load forecast and consists of three main sections. The first presents an algorithm based on continuous wavelet transform (CWT) to identify harmonics in a power signal. The new algorithm is able to identify the frequencies, amplitudes, and phase information of all distortion harmonic components, including integer harmonics, sub-harmonics, and inter-harmonics. The second section describes a wavelet transform-based algorithm for reconstructing the harmonic waveforms from the complex CWT coefficients. This is useful for identifying the amplitude variations of the nonstationary harmonics over the estimation period. The third section presents wavelet-genetic algorithm-neural network-based hybrid model for accurate prediction of short-term load forecast. Examples and case study will be used to demonstrate the benefits derived from these approaches.

Keywords Harmonics · Power quality · Power system · Signal processing · Transients · Wavelet

---

L. Zhou (✉)
City University, London, UK
e-mail: long.zhou.1@city.ac.uk

S. Rebennack et al. (eds.), *Handbook of Power Systems II*, Energy Systems,
DOI 10.1007/978-3-642-12686-4_3,

1 Wavelet-based Algorithm for Harmonics Analysis

Current harmonics are predominately produced by nonlinear loads. Traditionally, the distorting frequency components are in majority integer harmonics; the level of sub-harmonics, integer harmonics, inter-harmonics are, however, rising and becoming an important concern (Carbone et al. 1995).

Traditionally, discrete Fourier transform (DFT) implemented by fast Fourier transform (FFT) is used to analyze harmonics contents of a power signal. Short-time Fourier transform and Gabor transform were developed for estimating time-variant harmonic information. These methods have certain limitations as well (Smith 2002; Ramirez 1985; Girgis 1980; Addison 2002).

In this chapter, we present an algorithm based on continuous wavelet transform (CWT) to identify harmonics in a power signal (Pham and Wong 1999). The new algorithm is able to identify the frequencies, amplitudes, and phase information of all distortion harmonic components, including integer harmonics, sub-harmonics, and inter-harmonics.

1.1 Wavelet Transform and Wavelet

CWT is adopted for harmonic analysis because of its ability to identify harmonic frequencies and preserve phase information (Tse 2006; Strang and Nguyen 1996; Teolis 1998).

The wavelet transform of a continuous signal, f(t), is defined as (Teolis 1998)

$$Wf(u,a) = \langle f, \varphi_{u,a} \rangle = \int_{-\infty}^{+\infty} f(t) \frac{1}{\sqrt{a}} \varphi^* \left(\frac{t-u}{a} \right) dt \tag{1}$$

where $\varphi^*(t)$ is the complex conjugate of the wavelet function φ (t), a is the dilation parameter (scale) of the wavelet, and u is the translation parameter of the wavelet.

Compared to the traditional definition of the complex Morlet wavelet (CMW), the CMW adopted in this chapter is modified slightly by changing the scaling factor $1/\sqrt{a} - 1/a$. The simplified CMW (Huang et al. 1999); (Misiti et al. 1996) is adopted in the algorithm for harmonic analysis, defined as

$$\varphi\left(\frac{t}{a}\right) = \frac{1}{\sqrt{a}} \frac{1}{\sqrt{\pi f_b}} e^{-\left(\frac{t}{a}\right)^2 / f_b} e^{j 2\pi f_c \left(\frac{t}{a}\right)} \tag{2}$$

where f_b is the bandwidth parameter and f_c is the center frequency of the wavelet.

Because of the analytic nature, CMW is able to separate amplitude and phase information.

From the classical uncertainty principle, it is well known that localization in one domain necessarily comes at the cost of localization in the other. The time-frequency localization is measured in the mean squares sense and is represented as a Heisenberg box. The area of the Heisenberg box is limited by

$$\Delta t \; x \; \Delta\omega \geq \frac{1}{2}, \tag{3}$$

where $\Delta\omega$ is the frequency resolution, and

Δt is the time resolution.

For a dilated CMW,

$$\Delta\omega = \frac{1}{a\sqrt{f_b}}, \; \& \; \Delta t = \frac{a\sqrt{f_b}}{2}. \tag{4}$$

The CMW achieves a desirable compromise between time resolution and frequency resolution, with the area of the Heisenberg box equal to 0.5.

The filter banks produced by the CMW have very good frequency resolution at low frequencies. The frequency resolution becomes poorer and poorer as the wavelet center frequency increases. The bandpass filters at large dilations would have large lobe heights in frequency domain, by the factor of $\sqrt{a}$. This is not a desirable feature for harmonic analysis of power signals as current harmonics have smaller magnitudes at high frequencies. If the bandpass filters have lobe heights inversely proportional to frequencies, the small magnitudes of higher order harmonics would further be scaled by the magnitudes of the bandpass filters, resulting in erroneous estimation of harmonic frequencies, amplitudes and phases. Figure 1 shows the filter banks produced for the CMW.

The definition of the CMW shown in (2) is modified slightly, with the scaling factor $1/\sqrt{a}$ changed to 1/a, as in (5):

$$\phi\left(\frac{t}{a}\right) = \frac{1}{a}\frac{1}{\sqrt{\pi f_b}} e^{-\left(\frac{t}{a}\right)^2/f_b} e^{j2\pi f_c\left(\frac{t}{a}\right)}. \tag{5}$$

The Fourier transform of (5) is given as

$$\Phi(af) = e^{-\pi^2 f_b (af - f_c)^2}. \tag{6}$$

As seen in Fig. 2, filter banks generated by (6) exhibit the same lobe height for all harmonic frequencies in frequency domain. The time spread and frequency spread of the modified CMW are the same as the original CMW. The modified CMW is more suitable for estimating harmonics with very small amplitudes compared to adjacent harmonics, commonly found in sub-harmonics, inter-harmonics and higher order harmonics. The modified CMW described by (5) will be used for the subsequent harmonic analysis applications.

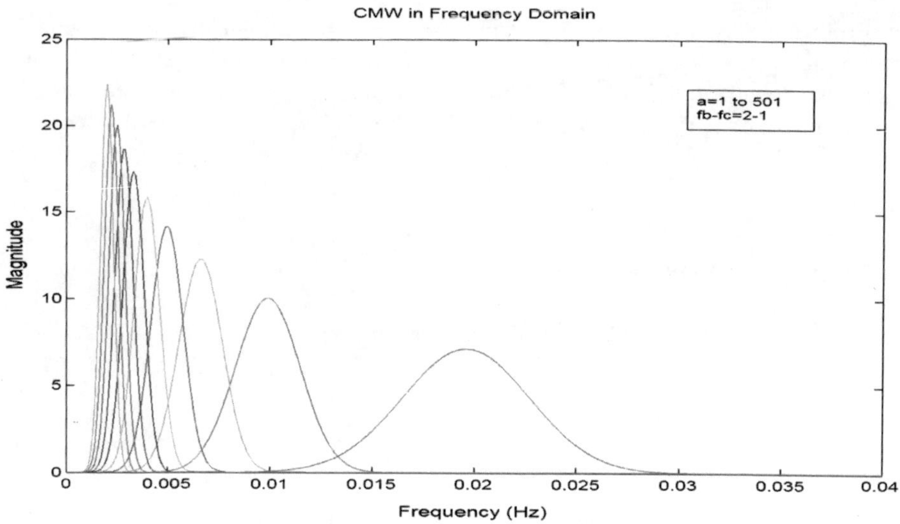

Fig. 1 Filter banks by CMW

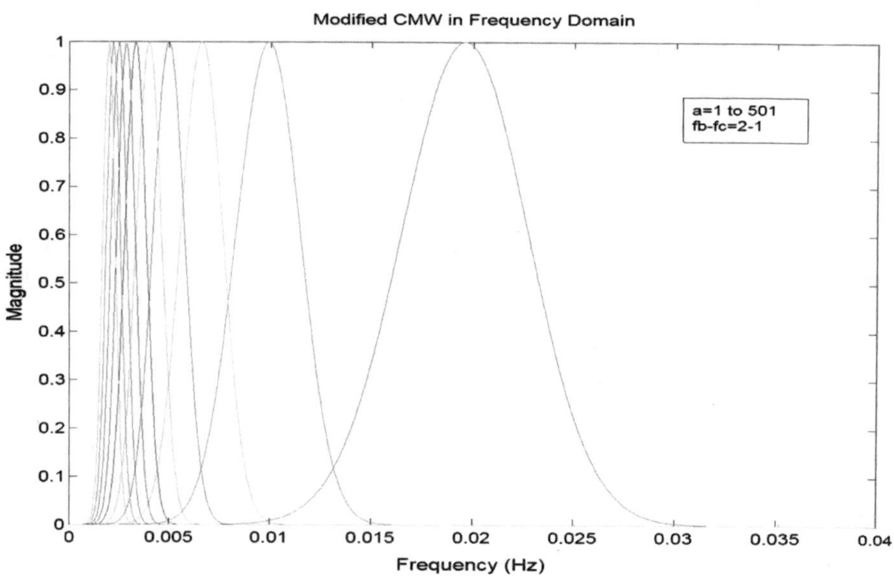

Fig. 2 Filter banks by the modified CMW

1.2 Harmonics Detection

As discussed in (Tse and Lai 2007), the harmonic frequencies contained in a power signal can be determined by the wavelet ridges. With the modified CMW suggested in (5), the corresponding harmonics amplitudes would be determined directly by

$$A(u) = 2\left|Wf(u,a)\right|, \tag{7}$$

where $|Wf(u,a)|$ is the wavelet coefficient at location u and scale a.

The CMW, given its analytic nature, can preserve signal phase information. The phase information contained in the wavelet coefficients is termed 'instantaneous phase'. Each sampled data point in the time signal has a wavelet coefficient related to it.

Consider the two sinusoids defined as

$$\begin{aligned} v_1 &= A_1 \cos(\omega_1 t + \theta_1) \\ v_2 &= A_2 \cos(\omega_2 t + \theta_2); \end{aligned} \tag{8}$$

their phase difference at any time t is

$$\angle v_2 - \angle v_1 = \omega_2 t + \theta_2 - (\omega_1 t + \theta_1) = (\omega_2 - \omega_1)\, t + \theta_2 - \theta_1. \tag{9}$$

The phase difference has two terms in (9). The first term depends on the frequency difference between the two sinusoids and is a function of time. The second term is the phase difference at time t = 0 and is time-invariant, termed as 'initial phase difference'.

Given a harmonic signal of frequency ω and an initial phase angle θ, let θ_w be the instantaneous phase obtained from the wavelet coefficient at time t. θ_w at time t is related to θ by

$$\theta = \theta_w - 2\pi f t, \tag{10}$$

$$t = \frac{n-1}{f_s}. \tag{11}$$

Therefore, from (10) and (11),

$$\theta = \theta_w - 2\pi f \left(\frac{n-1}{f_s}\right). \tag{12}$$

The initial phase angle θ, that is, the phase angle of the harmonic signal at time t = 0 (n = 1), is

$$\begin{aligned} \theta &= \theta_w - 2\pi f(0) \\ \theta &= \theta_w. \end{aligned} \tag{13}$$

The wavelet coefficients at data ends of the signal data would be distorted by edge effect (Addison 2002). Therefore, wavelet coefficients that are sufficiently away from the data ends of the sampled signal data should be used for initial phase angle estimation.

From (9), the initial phase difference between any two harmonics is given by

$$\theta_2 - \theta_1 = (\theta_{w2} - \theta_{w1}) - 2\pi\,(n-1)\,\frac{(f_2 - f_1)}{f_s}. \tag{14}$$

The wavelet coefficient generated for the data at the center position of the signal is used for phase estimation.

1.3 Discrimination of Adjacent Frequencies

Figure 3 shows the plot of Fourier transform of two modified CMWs with center frequencies of f_c/a_1 and f_c/a_2, respectively, in the frequency domain, obtained from (6). The two peaks in Fig. 3 must be separated apart sufficiently to discriminate adjacent frequencies.

From (Tse and Lai 2007), the $f_c\sqrt{f_b}$ of the modified CMW should satisfy the following condition to achieve the discrimination:

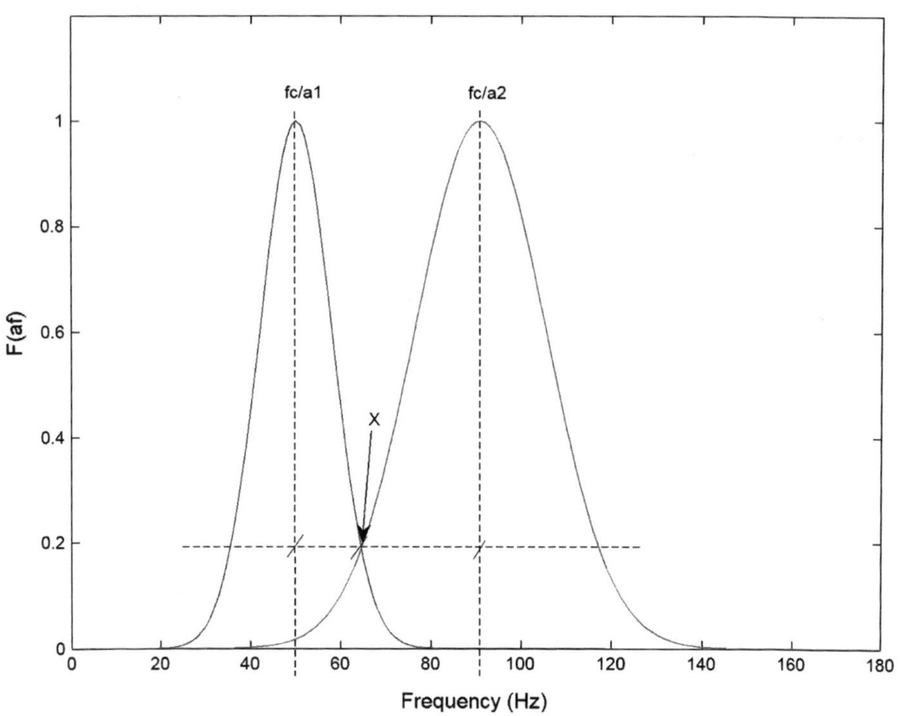

Fig. 3 Fourier Transform of modified CMWs with $f_c = f_1 \& f_2$

$$f_c\sqrt{f_b} \geq \frac{\sqrt{|In(x)|}}{\pi}\left|\frac{f_1+f_2}{f_1-f_2}\right|. \tag{15}$$

The x in (15) should ideally be very small for accurate harmonic frequency and amplitude estimation and would depend on the following:

(a) The relative amplitudes of the harmonics
(b) The accuracy demanded in the amplitude estimation

The selection of the x value would need to be compromised with the signal length to be used for analysis. It is found that x = 0.1 is sufficient for accurate frequency and amplitude detection if the amplitudes of adjacent frequencies do not differ very much from each other, which gives

$$\frac{\sqrt{|In(0.1)|}}{\pi} = 0.483 \approx 0.5. \tag{16}$$

1.4 Sampling Frequency and Signal Length

It is well known that the sampling frequency must be at least twice the highest frequency component in the signal to avoid aliasing. Nyquist frequency is defined as equal to half of the sampling frequency.

From Fig. 3, the filters produced by the CMW have finite bandwidth, which is liable to aliasing if the sampling frequency is set close to the center frequency of the dilated CMW. From (15), the highest frequency f_H that the filter band would cover is

$$f_H = \frac{f_c}{a} + \frac{1}{a}\frac{\sqrt{|\ln(x)|}}{\pi}\frac{1}{\sqrt{f_b}}. \tag{17}$$

By setting f_H as the Nyquist frequency, that is, $f_s/2$, and by (17),

$$f_s \geq 2f\left[1 + \frac{2\sqrt{|\ln(x)|}}{\pi\, f_c\sqrt{f_b}}\right]. \tag{18}$$

The minimum sampling frequency is dependent on the $f_c\sqrt{f_b}$ of the modified CMW.

From (16), and with $f_c\sqrt{fb} \geq 1.14$, (18) gives

$$f_s \geq 2f\left[1 + \frac{0.483}{1.13}\right] = 2.855f. \tag{19}$$

Therefore, with the modified CWM, the sampling frequency lies between 2 and 3 times of the highest harmonic frequency in the harmonic signal. The larger the factor $f_c\sqrt{f_b}$, the lower would be the minimum sampling frequency.

The signal length is controlled by the time width of the wavelet required to achieve a mean zero value. It is estimated that the signal length is given as

$$T = b\left(\frac{f_c\sqrt{f_b}}{2}\right)\frac{1}{f}, \tag{20}$$

where b is a constant.

When very accurate estimation in both frequency and amplitude is required, a larger value of b is to be used. In any case, the longer the time window, the better would be the estimation of the frequency and amplitude of the harmonics. On the other hand, for time event localization, the shorter the time window, the better would be the localization of the time event. Hence, a balance should be struck between the two.

The value of b is chosen as 10. Therefore, the time width, T, of the CMW is determined as

$$T \geq \frac{5f_c\sqrt{f_b}}{f}. \tag{21}$$

1.5 The Proposed Harmonics Detection Algorithm

The proposed WT-based harmonic detection algorithm is implemented by FFT in Matlab software. Figure 4 shows the flow chart of the computational algorithm for computing wavelet coefficients by FFT (Addison 2002).

As discussed earlier, the setting of the modified CMW is determined by the separation between adjacent frequencies. Assume that the frequency separation is represented as f_{sep}. Let $f_1 = f_H$ and $f_2 = f_H - f_{sep}$, in (15), and putting $x = 0.1,\ y = 0.5$, and $f_b = 10000$,

$$f_c \geq \frac{1}{100}\left(\frac{f_H}{f_{sep}} - \frac{1}{2}\right), f_H \geq f_{sep}. \tag{22}$$

The minimum sampling frequency to avoid aliasing is estimated from (18) and (22) as

$$f_s \geq \frac{4f_H^2}{2f_H - f_{sep}}. \tag{23}$$

At high harmonic frequencies, $2f_H >> f_{sep}$, (23) becomes

$$f_s \geq 2f_H,$$

which agrees with the classical sampling theorem.

From (21) and (22),

$$T \geq 5\left(\frac{1}{f_{sep}} + \frac{1}{2f_L}\right), \tag{24}$$

where f_L is the lowest harmonic frequency in the signal.

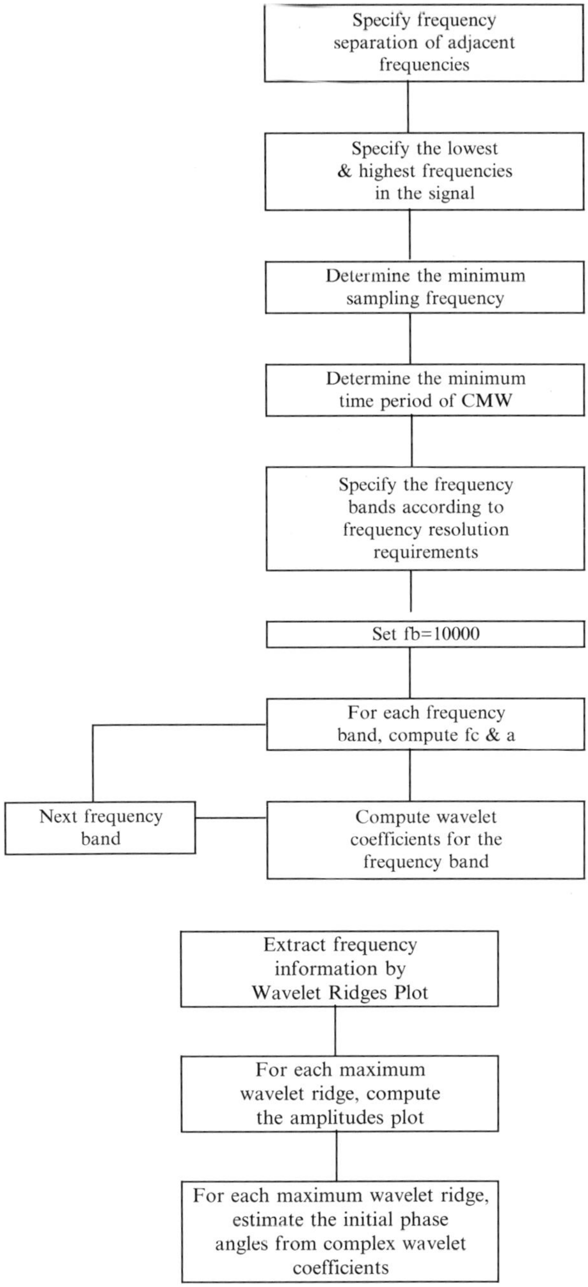

Fig. 4 Flow chart for harmonics detection

The minimum signal time length is dependent on the lowest harmonic frequency in the harmonic signal and the frequency separation between adjacent frequencies. Equation (24) confirms the common understanding (Chan 1995) that the closer spaced the frequencies of two harmonic frequencies, the longer the signal length required to distinguish their frequencies.

The minimum dilation scale is determined by the minimum sampling frequency without aliasing. The maximum dilation scale is determined by the minimum signal length for accurate harmonic analysis. For finer harmonic frequency estimation, the required frequency resolution is determined first; f_c is calculated and the corresponding scales are determined accordingly for decomposition.

1.6 Synthesized Waveform Analysis

Two tests for synthesized waveform analysis are conducted. In the first test, the WT-based algorithm is used to estimate a synthesized harmonic signal containing sub-harmonics. In the second test, the WT-based algorithm is used to estimate a synthesized harmonic signal containing harmonics and inter-harmonics.

1.6.1 Sub-harmonics Test

The harmonic signal contains the sub-harmonic components as shown in Table 1. The synthesized harmonic signal waveform is shown in Fig. 5.

The highest harmonic frequency is 50 Hz, the minimum separation frequency f_{sep} is set as 3.8 Hz. The minimum sampling frequency and the minimum signal length are determined by (23) and (24):

$$f_s \geq \frac{4f_H^2}{2f_H - f_{sep}} = \frac{4x49.8^2}{2x49.8 - 3.6} = 103.34Hz \tag{25}$$

$$T \geq 5\left(\frac{1}{f_{sep}} + \frac{1}{2f_L}\right) = 5\left(\frac{1}{3.6} + \frac{1}{2x13}\right) = 1.581s \tag{26}$$

The sampling frequency is chosen as 105 Hz and the minimum signal time length is chosen as 1.6 s.

Table 1 Sub-harmonics in synthesized signal

Frequency (Hz)	Amplitude	Phase Angle (degree)
13	0.01	8
20.5	0.32	12
25.6	0.87	−15
36.6	0.75	−20
46.2	0.33	−10
49.8	1	0

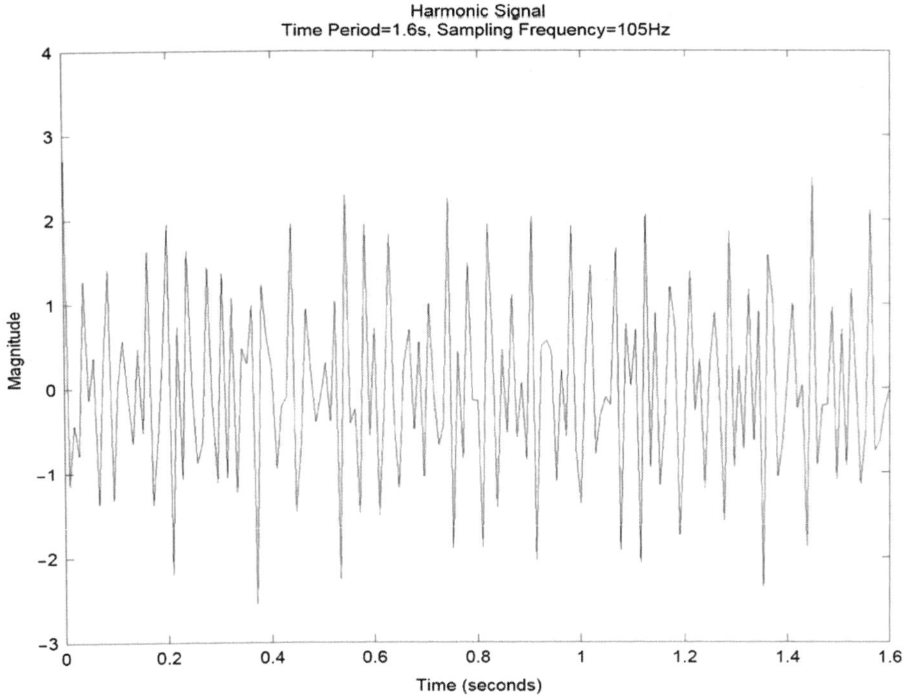

Fig. 5 Waveform of synthesized sub-harmonics signal

FFT is used to estimate the harmonics of the same harmonic signal. The FFT number used is 1,050 for a frequency resolution of 0.1 Hz. A hamming window is applied to reduce the effect of discontinuities at the window edges and frequency leakages.

The sub-harmonics in the synthesized signal estimated by both the WT-based algorithm and the FFT are tabulated in Table 2. On frequency estimation, both FFT and WT-based algorithm are very accurate. FFT would have small frequency detection error when the amplitude of the harmonic is comparatively small. On amplitudes and initial phase detection, WT-based algorithm is better than FFT.

1.6.2 Inter-Harmonics and Harmonics Test

The harmonic signal contains the harmonic and inter-harmonic components as shown in Table 3. The waveform of the synthesized harmonic signal is shown in Fig. 6.

Table 2 Estimation results by WT-based algorithm and FFT

WT-based harmonic analysis estimation			FFT		
Freq. (Hz)	Amp.	Init. phase (deg.)	Freq. (Hz)	Amp.	Init. phase (deg.)
13	0.01	8	12.9	0.014	30.86
20.5	0.32	12	20.5	0.323	12.17
25.6	0.87	−15	25.6	0.867	−14.75
36.6	0.75	−20	36.6	0.747	−19.94
46.2	0.3301	−10.01	46.2	0.328	−9.1
49.8	1	0	49.8	0.993	0.253

Table 3 Inter-harmonics in synthesized signal

Freq. (Hz)	Amp.	Initial phase (deg.)	Freq. (Hz)	Amp.	Initial phase (deg.)
49.5	311	0	1,336.5	20	8
102	288	10	1,435.5	18	−8
148.5	280	−15	1,534.5	15	20
247.5	225	−12	1,633.5	13	−30
346.5	180	−20	1,732.5	11	12
445.5	155	−14	1,831.5	9	−8
544.5	130	30	1,930.5	7	9
643.5	102	36	2,029.5	5.5	9
742.5	80	42	2,128.5	3	35
811	76	11	2,160	2.2	−4
841.5	62	−28	2,227.5	3	−32
940.5	53	5	2,326.5	2	7
1,039.5	32	7	2,425.5	1	2
1,138.5	30	15	2,524.5	0.5	1
1,237.5	26	−15			

The power supply authority specifies that the supply frequency of 50 Hz has a maximum variation of ±2%, that is, ±1 Hz. The lowest harmonic frequency that would be found in the harmonic signal is therefore 49 Hz. The WT-based algorithm is intended to detect up to the 51st harmonics. The 51st harmonic may have the highest frequency of 2,601 Hz when the fundamental frequency is 51 Hz. The minimum frequency range being specified for the WT-based algorithm is therefore from 49 to 2,601 Hz.

The minimum separation frequency f_{sep} is set as 30 Hz. The minimum sampling frequency and the minimum signal length are determined by (23) and (24) as follows:

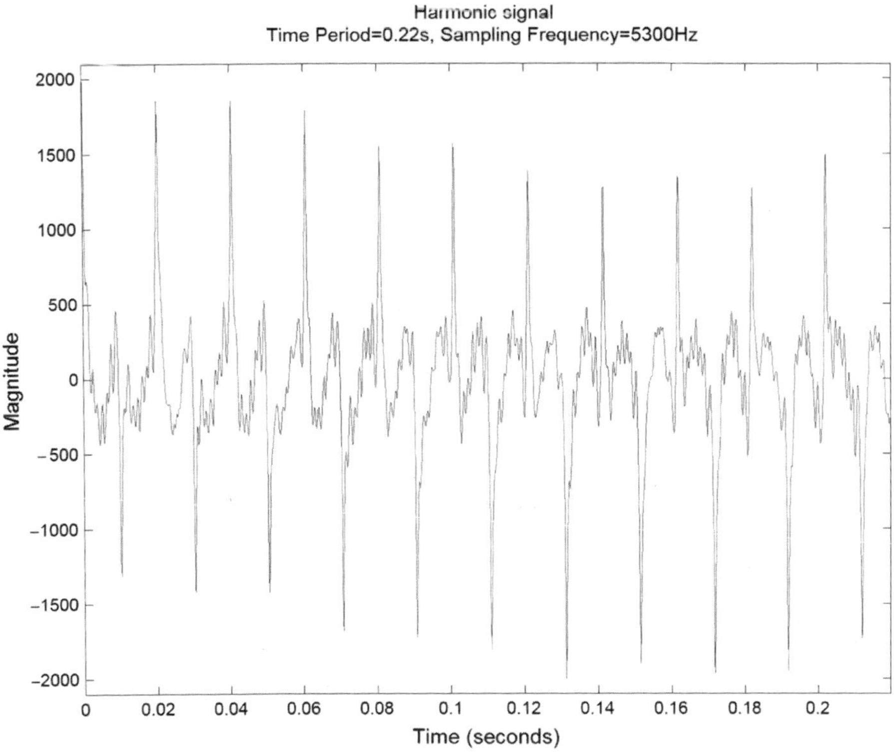

Fig. 6 Waveform of synthesized inter-harmonics signal

$$f_s \geq \frac{4f_H^2}{2f_H - f_{sep}} = \frac{4x2601^2}{2x2601 - 30} = 5232.2Hz \approx 5233Hz.$$

$$T \geq 5\left(\frac{1}{f_{sep}} + \frac{1}{2f_L}\right) = 5\left(\frac{1}{30} + \frac{1}{2x49}\right) = 0.218s \approx 0.22s.$$

The sampling frequency is chosen as 5,300 Hz and the minimum signal time length is chosen as 0.22 s.

FFT is used to estimate the harmonics of the synthesized signal. The FFT number used is 10,600 for frequency resolution of 0.5 Hz. The Hamming window is being used to reduce edge effects and frequency leakages.

The harmonics in the synthesized signal estimated by both the WT-based algorithm and the FFT are tabulated in Table 4.

The harmonic frequencies estimated by WT-based algorithm are practically 100% correct. The highest amplitude detection error found is only 0.2% and the highest initial phase detection error is 2.5%.

Table 4 Estimation results by WT-based algorithm and FFT

WT-based harmonic analysis estimation			FFT		
Freq. (Hz)	Amp.	Init. phase (deg.)	Freq. (Hz)	Amp.	Init. phase (deg.)
49.5	311	0	49.5	310.3	0.12
102	288	10	102	288.5	9.75
148.5	280	−15	148.5	280.6	−15.12
247.5	225	−12	247.5	224.7	−12.1
346.5	180	−20	346.5	179.6	−20.2
445.5	155	−14	445.5	154.8	−14.3
544.5	130	30	544.5	129.5	29.76
643.5	102	36	643.5	101.6	35.69
742.5	80	42	742.5	79.8	41.64
811	76	11	811	76.16	10.93
841.5	62	−28	841.5	62.38	−28.66
940.5	53	5	940.5	53.15	4.4
1039.5	32	7	1,039.5	32.18	6.22
1,138.5	30	15	1,138.5	30.12	14.33
1,237.5	26	−15	1,237.5	26.25	−15.55
1,336.5	20	7.99	1,336.5	20.15	7.05
1,435.5	18	−8.01	1,435.5	18.28	−8.78
1,534.5	15	19.99	1,534.5	15.17	18.89
1,633.5	13	−30.01	1,633.5	13.29	−30.17
1,732.5	11	11.98	1,732.5	11.17	10.92
1,831.5	9.003	−8.01	1,831.5	9.249	−8.93
1,930.5	7.003	8.98	1,930.5	7.233	7.42
2,029.5	5.503	8.98	2,029.5	5.75	7.46
2,128.5	3.002	34.94	2,129	3.207	12.71
2,160	2.204	−4.04	2,160	2.447	−5.46
2,227.5	3.004	−31.98	2,227.5	3.243	−31.1
2,326.5	2.004	6.95	2,326.5	2.243	4.07
2,425.5	1.001	1.95	2,425.5	1.278	−0.95
2,524.5	0.5041	0.96	2,524.5	0.7768	−0.17

The harmonic frequencies estimated by FFT algorithm are practically 100% correct. However, the amplitudes and initial phase angles estimated by FFT are erroneous. For amplitude detection, the highest error is 27.8%, while the highest initial phase detection error is as high as 148%. The errors in amplitudes and initial phases at harmonics with small amplitudes are very significant. It can be observed from the results that FFT is quite good at frequency estimation. This is expected as FFT is a frequency domain analysis tool. As WT is regarded as a time-frequency domain analysis tool, it can be used to estimate frequency, amplitude and phase with high accuracy.

1.7 *Experimental Results*

Two tests are conducted. The first test dealt with the analysis of field waveforms obtained from the red-phase input current to a three-phase variable speed drive supplying a submersible water pump. The waveform of the input current to the VSD with its output frequency at 30 Hz is being analyzed. The second test is conducted to the current of a single-phase final circuit supplying power to a number of single loads containing an electronic dimmer-controlled lamp bulb, fluorescent luminaries complete with electronic ballasts, a hairdryer and an air compressor motor.

1.7.1 VSD Input Current

Given the nature of a six-pulse three-phase VSD, the input current would contain 5th, 7th, 11th, 13th, 17th, and 19th ... harmonics. The input current waveform is sampled at 10 kHz. A low pass filter with a cutoff frequency of 4 kHz is applied. By checking zero crossing, the average fundamental frequency of the harmonic signal for a time period of 0.6s, that is, 30 cycles of the fundamental frequency, is 49.983 Hz. Two cycles of the input current waveform is shown in Fig. 7. The estimation results by both WT-based algorithm and FFT are shown in Table 5.

Besides the characteristic harmonics of a six-pulse three-phase VSD, the harmonic signal contains triplen harmonics and even harmonics. The triplen harmonics may be produced by the single-phase control circuit. The even harmonics are of very small amplitudes, which may be due to waveform asymmetry produced by load changes and the thyristor switching, or errors introduced during measurements.

Comparing the odd harmonics, both the FFT and the WT-based algorithm give nearly identical estimated harmonic frequencies. The WT-based algorithm is only slightly better than FFT. For the estimation of even harmonics, which have only very small amplitudes, the WT-based algorithm is more accurate.

1.7.2 Single-Phase Circuit Input Current

Figure 8 shows the waveform of the current.

By checking zero crossing, the average fundamental frequency of the harmonic signal for a time period of 0.6 s, that is, 30 cycles of the fundamental frequency is 50.026 Hz. The estimation results by both WT-based algorithm and FFT are tabulated in Table 6.

It can be seen that the harmonic signal contains odd harmonics and even harmonics. The even harmonics are of small amplitudes, which may be due to waveform asymmetry produced by load changes and switched mode power supply switching, or errors introduced during measurements.

Both WT-based algorithm and FFT are able to estimate the harmonic frequencies accurately. In a strict sense, the WT-based algorithm is slightly better but the difference is very small. The amplitudes of the harmonics estimated by both WT-based

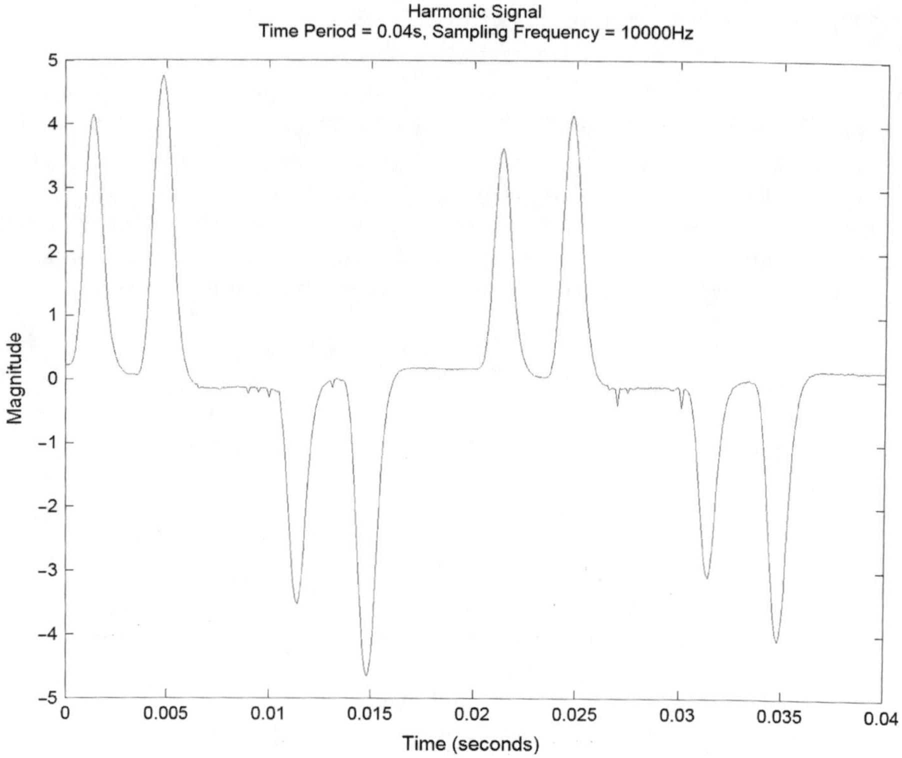

Fig. 7 Waveform of two cycles of VSD input current

algorithm and FFT are comparable. However, there are larger deviations in initial phase estimations.

2 Wavelet-based Algorithm for Nonstationary Power System Waveform Analysis

Traditionally, short-time Fourier transform (STFT) and Gabor transform (GT) were used for harmonic studies of nonstationary power system waveforms, which are basically Fourier transform (FT)-based methods. These methods have certain limitations (Robertson and Camps 1996). To overcome the limitations in these existing methods, wavelet transform (WT)-based algorithm has been developed to estimate the frequency and time information of a harmonic signal (Pham and Wong 1999; Tse 2006; Strang and Nguyen 1996; Teolis 1998; Huang et al. 1999; Misiti et al. 1996; Mallet 1998; Carmona et al. 1999; Carmona et al. 1997; Tse and Lai 2007; Chan 1995; Robertson and Camps 1996; Pham and Wong 2001). Because of the

Table 5 Estimation by WT-based algorithm and FFT

Harmonic No.	Expected frequency (Hz)[a]	WT-based algorithm				FFT			
		Freq. (Hz)	% Error	Amplitude	Initial phase (degree)	Frequency (Hz)	% Error	Amplitude	Initial phase (degree)
1	49.983	49.98	0.006	3.7372	−77.61	49.98	0.006	3.7348	−77.64
2	99.966	100.02	−0.054	0.1094	−68.73	100.14	−0.174	0.1111	−81.82
3	149.949	149.96	−0.007	0.5336	26.11	149.96	−0.007	0.536	26.26
4	199.932	199.9	0.016	0.0901	130.08	199.76	0.086	0.09	147.33
5	249.915	249.92	−0.002	2.7426	153.95	249.92	−0.002	2.742	154.01
7	349.881	349.88	0	1.8322	−8.39	349.88	0	1.8323	−8.33
9	449.847	449.9	−0.012	0.1506	56.01	449.9	−0.012	0.1508	56.07
11	549.813	549.82	−0.001	0.6762	−133.37	549.82	−0.001	0.6761	−133.27
13	649.779	649.78	0	0.2324	74.99	649.78	0	0.2322	75.1
15	749.745	749.8	−0.007	0.0507	50.02	749.78	−0.005	0.0514	52.29
17	849.711	849.76	−0.006	0.046	163.05	849.74	−0.003	0.0459	165.34
19	949.677	949.68	0	0.065	−23.73	949.68	0	0.0648	−23.91

[a] Note: Based on a fundamental frequency of 49.983 Hz

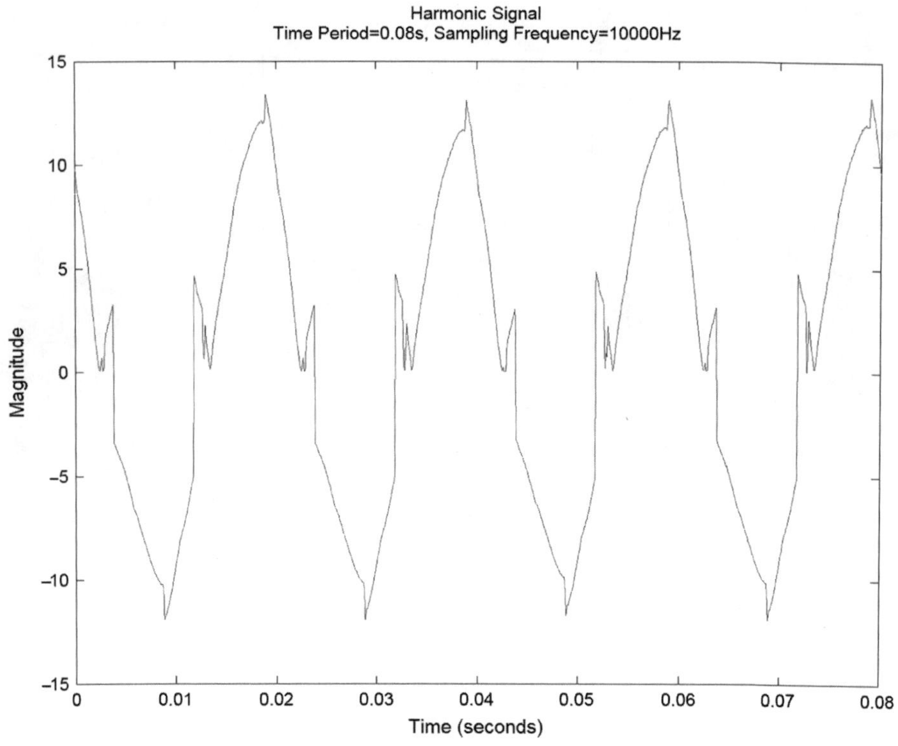

Fig. 8 Waveform of four cycles of single-phase input current

shifting and scaling operations of the WT, it is most suitable to problems involving nonstationary power system harmonic waveforms.

It presents a WT-based algorithm for reconstructing the harmonic waveforms from the complex CWT coefficients. This is useful for identifying the amplitude variations of the nonstationary harmonics over the estimation period.

Discrete wavelet transform (DWT) is receiving increasing attention due to its fast and efficient computation algorithm and the ability to extract signal features. Applications are found in power system transient analysis and fault detections where time information of the signal is to be analyzed (Chan et al. 2000; Lai et al. 1999; Butler-Purry and Bagriyanik 2003). DWT is able to divide harmonic components of a signal in frequency bands for ready inspection, but is liable to aliasing problems and amplitude distortions without careful selection of frequency bands.

Complex continuous wavelet transform (CCWT) is adopted for harmonic analysis, which can identify harmonic frequencies and preserve time and phase information (Pham and Wong 1999, 2001). The simplified CMW (Addison 2002) is chosen for harmonics analysis due to its smooth and harmonic-like waveform. The CMW has the Heisenberg box area of 0.5 and is able to achieve the best compromise in both time and frequency resolution.

Table 6 Estimation by WT-based algorithm and FFT

Harmonic No.	Expected frequency (Hz)[a]	Proposed algorithm				FFT			
		Frequency (Hz)	% Error	Amplitude	Initial phase (deg.)	Frequency (Hz)	% Error	Amplitude	Initial phase (deg.)
1	50.026	50.02	0.012	10.3274	34.52	50.02	0.012	10.3269	34.49
3	150.078	150.08	−0.001	1.4587	56.71	150.1	−0.015	1.4595	54.54
4	200.104	200.08	0.012	0.7041	173.66	200.08	0.012	0.702	173.45
5	250.130	250.14	−0.004	0.9474	6.68	250.16	−0.012	0.9497	4.49
6	300.156	300.14	0.005	0.5211	72.4	300.18	−0.008	0.5207	67.62
7	350.182	350.18	0.001	0.6644	−105.87	350.18	0.001	0.6658	−105.69
8	400.208	400.16	0.012	0.2018	−6.56	400.3	−0.023	0.2026	−21.96
9	450.234	450.24	−0.001	0.8836	161.28	450.24	−0.001	0.8824	161.2
11	550.286	550.28	0.001	0.6169	85.6	550.3	−0.003	0.6154	83.16
13	650.338	650.30	0.006	0.1681	−25.5	650.4	−0.010	0.1687	−36.81
14	700.364	700.36	0.001	0.373	−168.56	700.38	−0.002	0.3718	−170.73
15	750.390	750.28	0.015	0.0408	−85.81	750.48	−0.012	0.0408	−105.58
16	800.416	800.4	0.002	0.3209	105.95	800.44	−0.003	0.3184	101.25
17	850.442	850.44	0	0.2543	−69.93	850.48	−0.004	0.2562	−74.19
18	900.468	900.44	0.003	0.1654	21.48	900.52	−0.006	0.1645	12.24

[a] Note: Based on a fundamental frequency of 50.026 Hz

Scalogram and wavelet ridges are used to extract frequency and amplitude information from wavelet coefficients computed from CCWT (Carmona et al. 1999). The WT-based algorithm by using CMW is able to estimate harmonic frequencies accurately with a shorter harmonic signal as compared to DFT (Pham and Wong 2001).

Compared to the traditional definition of the CMW, the CMW adopted in this chapter is modified slightly, by changing the scaling factor $1/\sqrt{a} - 1/a$, as in (27) below:

$$\phi\left(\frac{t}{a}\right) = \frac{1}{a}\frac{1}{\sqrt{\pi f_b}}e^{-\left(\frac{t}{a}\right)^2/f_b}e^{j2\pi\omega_c\left(\frac{t}{a}\right)}, \tag{27}$$

where f_b is the bandwidth parameter, ω_c is the wavelet center frequency and a is the dilation factor.

The Fourier Transform of (27) is

$$\Phi(af) = e^{-\pi^2 f_b (af - f_c)^2}, \tag{28}$$

where $f_c = \omega_c/(2\pi)$.

The filter banks generated by (28) exhibit the same lobe height for all harmonic frequencies in the frequency domain. The modified CMW is more suitable for estimating harmonics with very small amplitudes compared to adjacent harmonics, commonly found in sub-harmonics, inter-harmonics and higher order harmonics.

2.1 Waveform Reconstruction Algorithm

The CCWT by using the modified CWM in (27) is given as

$$Wf(u,a) = \int_{-\infty}^{+\infty} f(t)\frac{1}{a}\frac{1}{\sqrt{\pi f_b}}e^{-\left(\frac{t-u}{a}\right)^2/f_b}e^{-j\omega_c\left(\frac{t-u}{a}\right)}dt. \tag{29}$$

The real part and imaginary part of the complex wavelet coefficients generated by (29) are shown in (30) and (31), respectively.

$$Re\ [Wf(u,a)] = \int_{-\infty}^{+\infty} f(t)\frac{1}{a}\frac{1}{\sqrt{\pi\ f_b}}e^{-\left(\frac{t-u}{a}\right)^2/f_b}cos\left[\frac{\omega_c}{a}(t-u)\right]dt \tag{30}$$

$$\mathrm{Im}\ [Wf(u,a)] = -\int_{-\infty}^{+\infty} f(t)\frac{1}{a}\frac{1}{\sqrt{\pi\ f_b}}e^{-\left(\frac{t-u}{a}\right)^2/f_b}\sin\left[\frac{\omega_c}{a}(t-u)\right]dt. \tag{31}$$

Given a harmonic signal represented as

$$f(t) = A\,cos\,\omega t, \tag{32}$$

the real part of the complex CWT coefficient for (32) from (30) is

$$Re\ [Wf(u,a)] = \frac{A}{a\sqrt{\pi\ f_b}}\int_{-\infty}^{+\infty} e^{-\left(\frac{t-u}{a}\right)^2/f_b} \cos\omega t \cos\left[\frac{\omega_c}{a}(t-u)\right]dt. \tag{33}$$

By change of variable, (31) becomes

$$Re\ [Wf(u,a)] = \frac{A}{a\sqrt{\pi\ f_b}}\int_{-\infty}^{+\infty} e^{-\left(\frac{t}{a}\right)^2/f_b} \cos[\omega(t+u)] \cos\frac{\omega_c}{a}t\,dt. \tag{34}$$

The complex wavelet coefficient will be the largest when the frequency of the harmonic component is equal to the wavelet center frequency at the dilation a. Substitute $\omega = \omega_{c/a}$ into (32),

$$Re\ [Wf(u,a)] = \frac{A}{a\sqrt{\pi f_b}}\int_{-\infty}^{+\infty} e^{-\left(\frac{t}{a}\right)^2/f_b} (\cos\omega u \cos^2\omega t - \sin\omega u \sin\omega t \cos\omega t)dt. \tag{35}$$

The imaginary part of the complex CWT coefficient for (32) from (31) is calculated similarly as

$$Im\ [Wf(u,a)] = -\frac{A}{a\sqrt{\pi f_b}}\int_{-\infty}^{+\infty} e^{-\left(\frac{t}{a}\right)^2/f_b} \cos[\omega(t+u)] \sin\omega\,t\,dt. \tag{36}$$

Shifting the imaginary part of the complex CWT coefficient by 90° backward in time, (36) becomes

$$\begin{aligned}&\mathrm{Im}\left[Wf\left(u+\frac{\pi}{2\omega},a\right)\right]\\ &= \frac{A}{a\sqrt{\pi f_b}}\int_{-\infty}^{+\infty} e^{-\left(\frac{t}{a}\right)^2/f_b} (\cos\omega u \sin^2\omega t + \sin\omega u \sin\omega t \cos\omega t)dt.\end{aligned} \tag{37}$$

Adding (35) and (37) produces

$$\mathrm{Re}[Wf(u,a)] + \mathrm{Im}\left[Wf\left(u+\frac{\pi}{2\omega},a\right)\right] = A\cos\omega u. \tag{38}$$

Equation (38) verifies mathematically that for a sinusoidal waveform of sufficient length, the waveform can be fully reconstructed by adding the real part of the corresponding complex CWT coefficients to the imaginary part of the corresponding complex CWT coefficients being shifted backward in time by 90°. Since the reconstruction is time invariant, the instantaneous phase of the harmonic component is preserved in the reconstructed waveform.

The flowchart of the proposed power system waveform reconstruction algorithm is shown in Fig. 9.

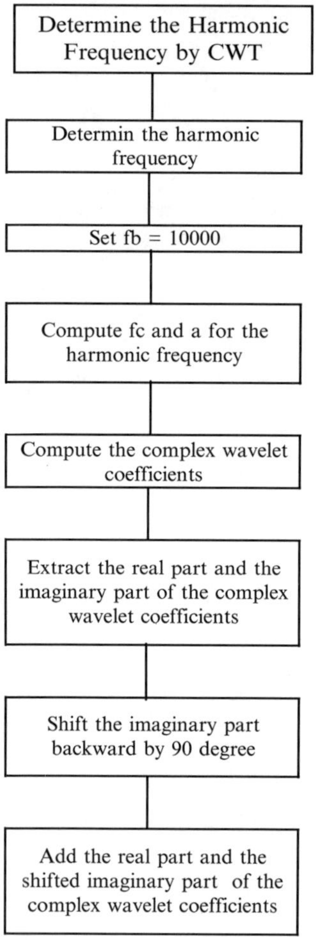

Fig. 9 Flow chart of the WT-based waveform reconstruction algorithm

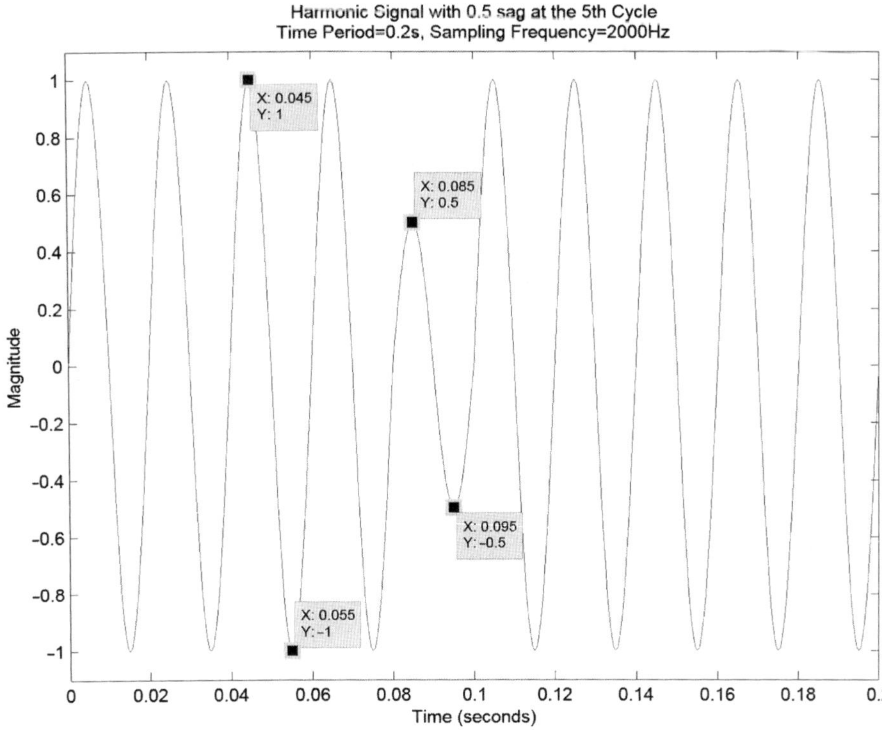

Fig. 10 Synthesized waveform of 50 Hz containing one cycle of sag (Test A)

2.2 *Synthesized Waveforms Reconstruction*

Two tests are conducted to verify the effectiveness of the proposed WT-based waveform reconstruction algorithm.

Test A: A waveform described by (39) with unit amplitude contains a sag of 0.5 at the fifth cycle. The synthesized waveform is shown in Fig. 10:

$$f(t) = \cos(2\pi 50t - 90^{o}). \qquad (39)$$

Test B: A waveform at a fundamental frequency of 50 Hz with slowly varying amplitudes as shown in Fig. 11 is mixed with harmonics described by (40). The synthesized waveform is shown in Fig. 12:

$$0.5\cos(2\pi 150t) + 0.3\cos(2\pi 250t + 45^{o}) + 0.2\cos(2\pi 350t) \qquad (40)$$

All synthesized waveforms are of 0.2 s duration and are sampled at 2,000 Hz. The frequency information of the synthesized waveforms is estimated by the algorithm proposed in Butler-Purry and Bagriyanik (2003).

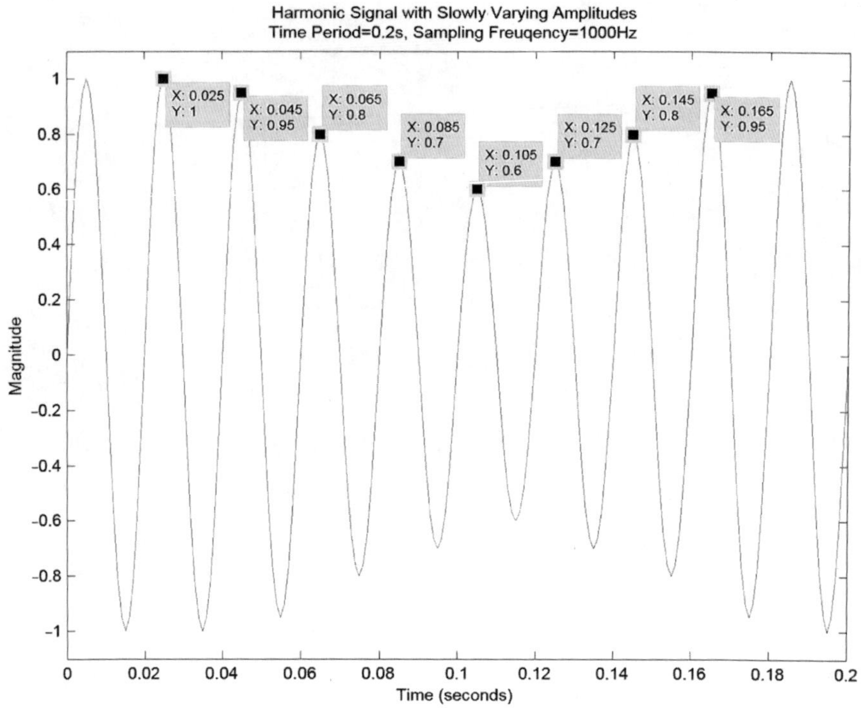

Fig. 11 Synthesized waveform of 50 Hz with slowly varying amplitudes

2.2.1 Test A

Table 7 and Figure 13 show a comparison of the synthesized waveform and the reconstructed waveform. The reconstructed waveform coincides with the synthesized waveform. Because of the finite support of the modified CMW and the abrupt change in the amplitude of the waveform at the sag, the reconstructed waveform has only a small error of approximately 2% at the negative peak of the sag.

2.2.2 Test B

Table 7 and Figure 14 show the comparison of the synthesized fundamental waveform with slowly varying amplitudes and the reconstructed waveform at the fundamental frequency. The reconstruction of the synthesized waveform at the fundamental frequency is nearly exact for most parts of the waveform. The reconstruction is not affected by the presence of harmonic frequencies.

2.3 DWT-Based Waveform Reconstruction

Waveform reconstruction by DWT is also conducted to the two synthesized waveforms in Sect. 1.4 for comparison. The Daubechies Db-8 type wavelet was chosen

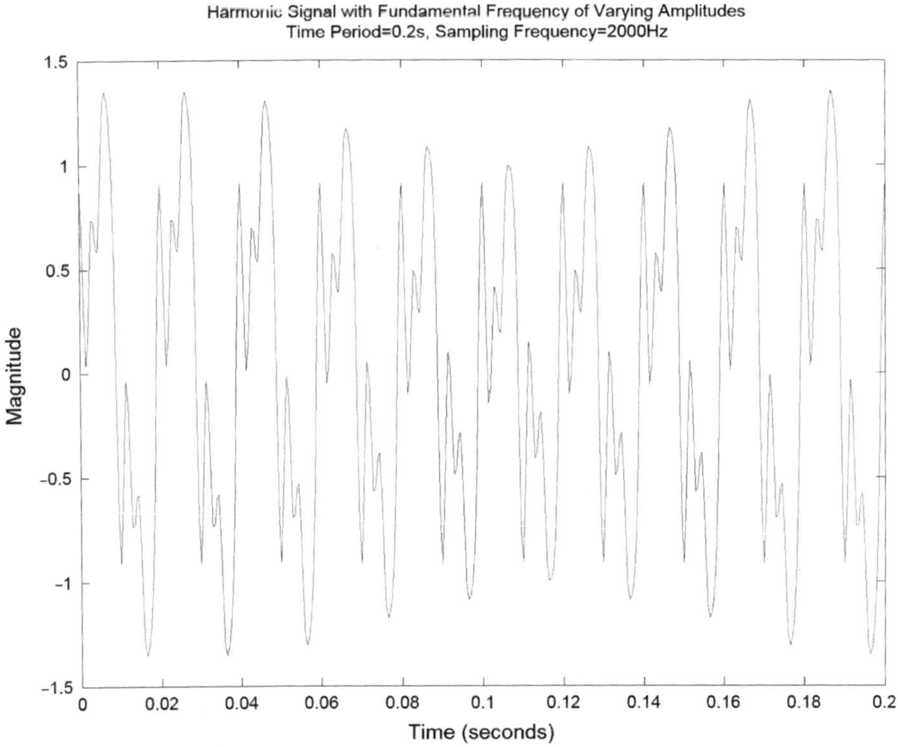

Fig. 12 Synthesized waveform in Fig. 3 mixed with harmonics (Test B)

as the mother wavelet. The synthesized signals were decomposed up to the third scale. With a sampling frequency of 2,000 Hz, the frequency bands at each scale are shown in Table 8. The synthesized signal (s), three detail coefficients (d_1, d_2, d_3) and the third approximation coefficients (a_3) are displayed graphically in each test.

2.3.1 Test A

Figure 15 shows the DWT analysis results of the synthesized waveform. The reconstructed waveform is shown at the third approximation.

2.3.2 Test B

Figure 16 shows the DWT analysis results of the synthesized waveform. The reconstructed waveform of the fundamental component is shown at the third approximation.

It can be seen from Table 7 that when the synthesized waveform contains a single frequency component only, DWT is able to reconstruct the variation in

Table 7 Comparison of waveform reconstruction by DWT and the proposed algorithm

	Synthesized signal		DWT-based reconstruction				CWT-based reconstruction			
	Set amplitude		Estimated amplitude		% Error		Estimated amplitude		% Error	
	+ peak	− peak	+ peak	− peak	+ peak	− peak	+ peak	− peak	+ peak	− peak
(A) Synthesized waveform at 50 Hz with a sag at the fifth cycle										
5th cycle	0.5	−0.5	0.5241	−0.4909	4.82%	1.82%	0.5	−0.5117	0	2.34%
(B) Synthesized waveform containing 50 Hz fundamental component with slowly varying amplitudes and harmonics										
3rd cycle	0.95	−0.95	1.022	−0.9696	7.58%	2.06%	0.944	−0.9217	0.63%	2.98%
4th cycle	0.8	−0.8	0.8764	−0.8215	9.55%	2.69%	0.794	−0.7791	0.75%	2.61%
5th cycle	0.7	−0.7	0.788	−0.7228	12.57%	3.26%	0.694	−0.6791	0.86%	2.99%
6th cycle	0.6	−0.6	0.7003	−0.6177	16.72%	2.95%	0.594	−0.6087	1%	1.45%
7th cycle	0.7	−0.7	0.7892	−0.714	12.74%	2%	0.6939	−0.7087	0.8%7	1.24%
8th cyle	0.8	−0.8	0.8779	−0.8105	9.74%	1.31%	0.7939	−0.8161	0.76%	2.01%
9th cycle	0.95	−0.95	1.02	−0.9607	7.37%	1.13%	0.9438	−0.9513	0.65%	0.14%

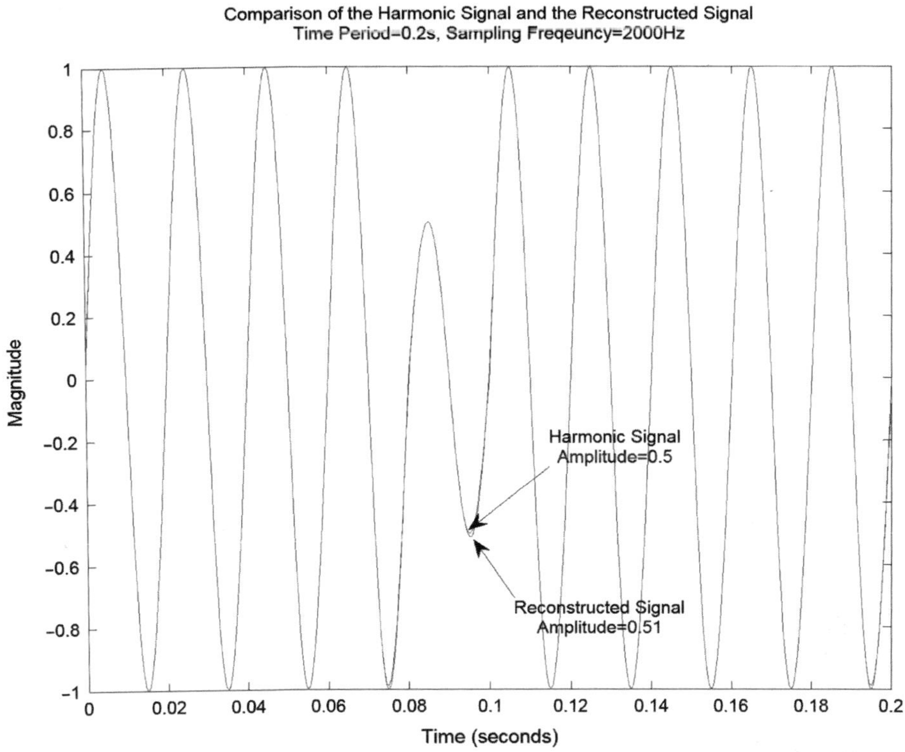

Fig. 13 Synthesized waveform vs. reconstructed waveform (Test A)

waveform amplitudes quite accurately. The proposed WT-based waveform reconstruction algorithm performs better over the DWT on estimation of positive peaks, but is comparable with DWT on negative peaks reconstruction.

When the synthesized waveform contains fundamental component of slowly varying-amplitudes mixed with harmonics, the waveform of the fundamental frequency component reconstructed by DWT is erroneous, while the proposed WT-based waveform reconstruction algorithm is able to reconstruct the waveform accurately. Furthermore, the proposed WT-based waveform reconstruction algorithm is able to reconstruct any harmonic components easily and efficiently, as will be shown in Sect. 1.6. DWT, on the other hand, would introduce aliasing and waveform distortions with inappropriate frequency band assignment.

2.4 *Experimental Results*

Experiments are conducted to the following two field waveforms to validate the robustness of the proposed WT-based waveform reconstruction algorithm. The frequency information of the field waveforms are estimated by the algorithm proposed in Butler-Purry and Bagriyanik (2003).

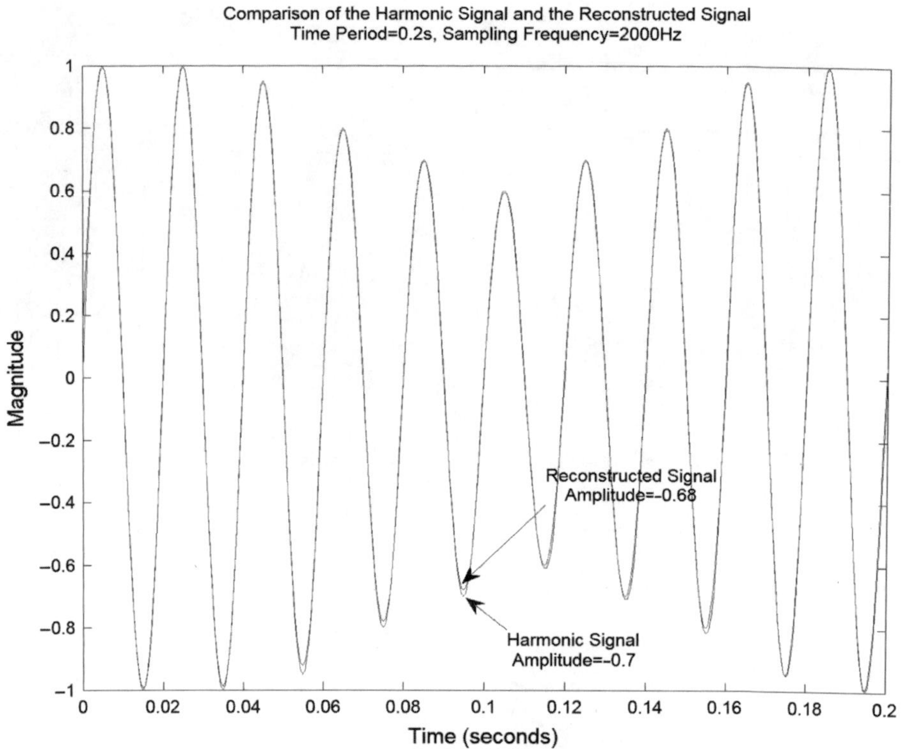

Fig. 14 Synthesized waveform vs. reconstructed fundamental waveform (Test B)

Test A: A harmonic waveform captured from the red-phase input current to a three-phase VSD supplying a submersible water pump. The output voltage of the VSD is set at 20 Hz.

Test B: A harmonic waveform captured from the input current of a single-phase LV final circuit supplying power to an electronic dimmer-controlled lamp bulb, fluorescent luminaries complete with electronic ballasts, a hairdryer and an air compressor motor.

2.4.1 Test A

Figure 17 shows two cycles of the field harmonic signal. Figure 18 shows the reconstructed fundamental component waveform and the fifth harmonic waveform together with the field harmonic signal.

The reconstructed current waveform at fundamental frequency coincides exactly with the field harmonic waveform in terms of time location and frequency. The proposed reconstruction algorithm is also able to represent the variations in amplitude of the current waveform at fundamental frequency.

The reconstructed fifth harmonic waveform coincides with the peaks of the field harmonic waveform. The reconstruction algorithmis also able to represent

Table 8 Frequency bands at each scale

Scale	Frequency band (Hz)
d1	500–1,000
d2	250–500
d3	125–250
a3	0–125

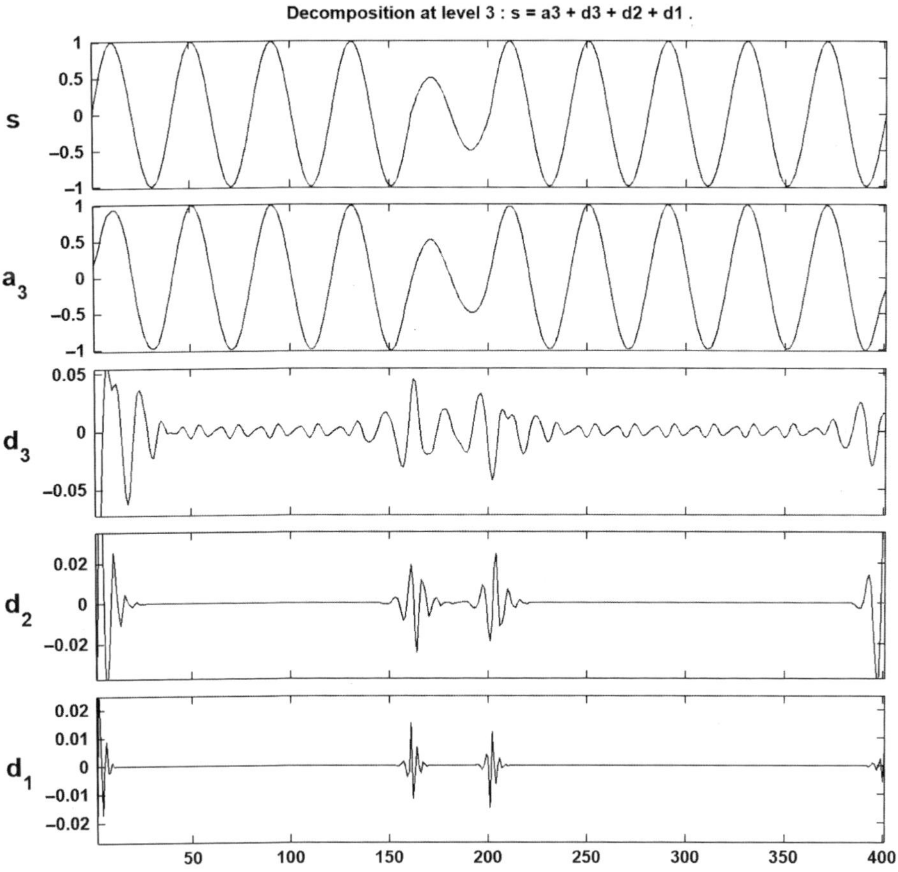

Fig. 15 Reconstructed waveform at the approximation (sag at fifth cycle)

the variations in amplitude of the fifth harmonic waveform. The proposed reconstruction algorithm can also be used to represent the amplitude variations of other harmonic frequencies.

2.4.2 Test B

Figure 19 shows four cycles of the field harmonic signal. Figure 20 shows the reconstructed current waveform at fundamental frequency and the third harmonic current waveform together with the field harmonic waveform.

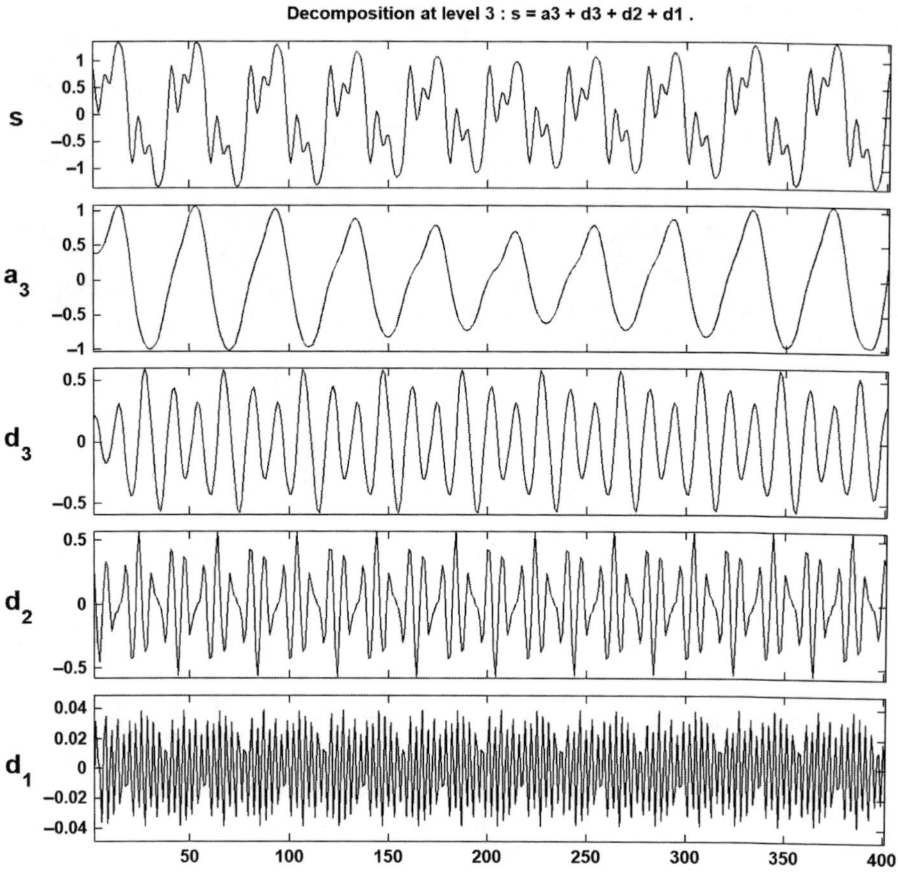

Fig. 16 Reconstructed fundamental waveform at the approximation (slowly varying amplitudes)

The reconstructed current waveform at fundamental frequency coincides exactly with the field harmonic signal in terms of time location and frequency. The reconstruction algorithm is also able to represent the variations in amplitudes of both the fundamental frequency current waveform and the third harmonic current waveform.

3 Wavelet-GA-ANN Based Hybrid Model for Accurate Prediction of Short-term Load Forecast

Deregulation of power utility industry being a reality today, which has resulted into the competition in every aspects in power systems; be it in power generation, or in transmission or in energy consumption, professional management of electric energy is of utmost importance. Many power systems not only are being pushed to their limits to meet their customers' demands, but also spend a lot of resources in their

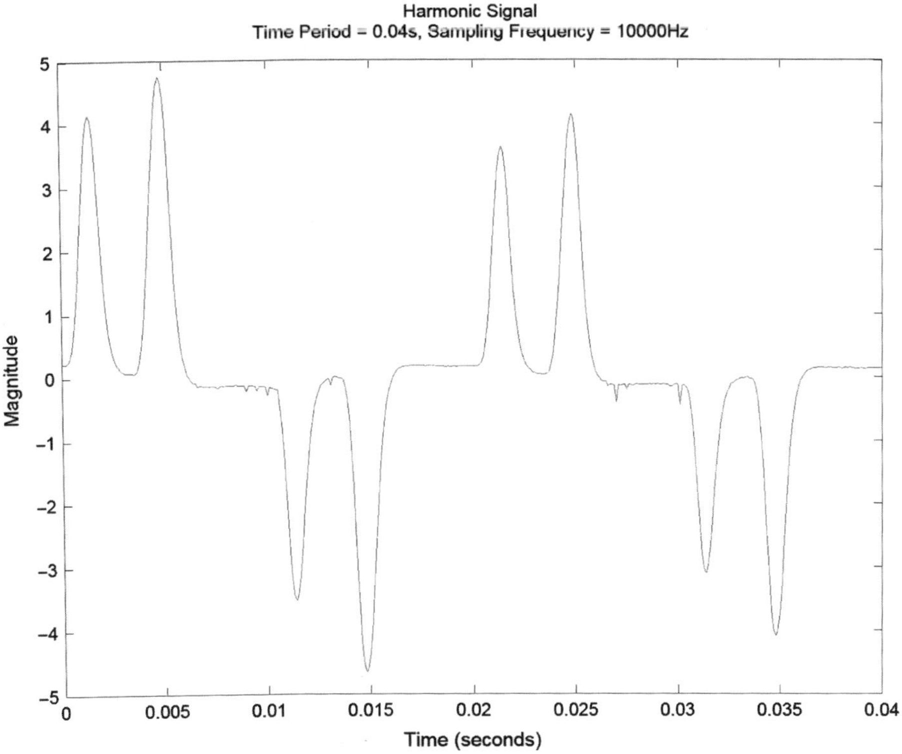

Fig. 17 Waveform of two cycles of VSD input current

operation scheduling. Furthermore, power systems need to operate even at higher efficiency in a deregulated electricity market, whereby the generating companies (Gencos) and distribution companies (Discos) have to compete to maximize their profits. Accurate prediction of load consumption pattern is becoming very important function to a utility company, as it is needed to support for wiser management decisions. A forecast that exceeds the actual load may lead to extra power being generated and therefore may result in excessive investment in a power plant that is not fully utilized. On the other hand, a forecast that is too low may lead to some revenue loss due to loss of opportunity of selling power to neighboring utilities. Hence, accurate electricity load forecasting (LF), including very short-term, short-term, mid-term, and long-term, plays a vital role in ensuring adequate electricity generation to meet the customer's demands in the future. LF also helps to build up cost effective risk management plans for the participating companies in the electricity market. Consequently, good operational, planning and intelligent management decision making, such as economic scheduling of generation capacity, scheduling of fuel purchase, ability to avoid unnecessary start-ups of generating units, planning the scheduling of peaking power, buying or selling electricity at best price, and scheduling of ancillary services, all of them can be carried out based on accurate

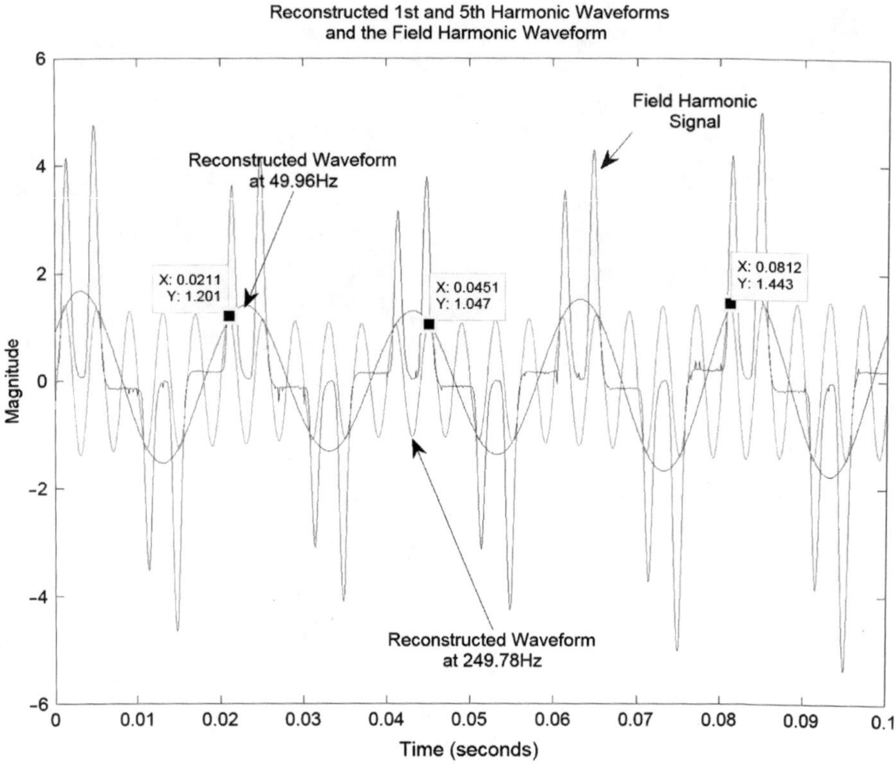

Fig. 18 Field harmonic signal vs. reconstructed fundamental and fifth harmonic waveform

LF, which forecasts the load of a few minutes, hours, days, weeks, months ahead. The aim of LF is to predict future electricity demand based on historical load data and currently available data.

To facilitate accurate load-forecasting analysis, a robust noise filtering and trend analysis algorithm must be used to enable effective eventual automation of the analysis of large volumes of data generated by the monitoring and recording of load consumption readings in any particular system. Currently, several forecasting schemes utilize artificial intelligence (AI) methods like ANN and GA to perform load-forecasting tasks. The common problem with such a method is that an AI scheme is only as intelligent as the program that trains it. This in turns depends heavily on the reliability of the training data collected. If such training data were in the first place corrupted by noise, it would mean that pre-processing of such data would be necessary. All these add to the implementation cost and set-up time. A good trend analysis scheme should be able to de-noise the electrical noise inherent in the data, and disregard portions of data where monitoring devices might have failed, giving lower resolution readings as a result of, and be able to take a macro view of the trend while preserving temporal information. The analysis of nonstationary signals like load consumption data often involves a compromise between

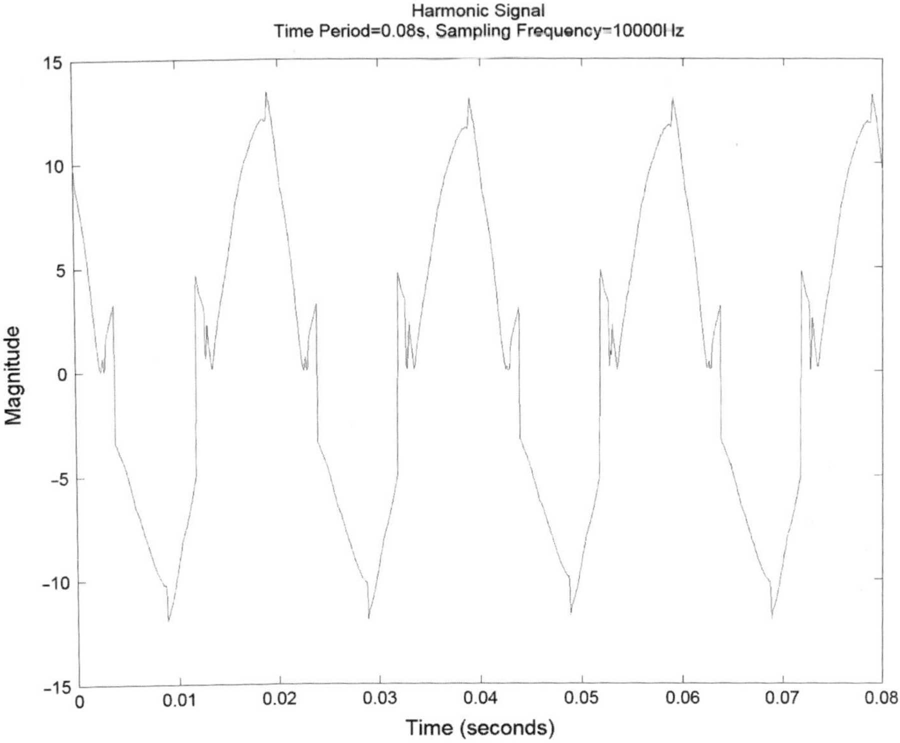

Fig. 19 Waveform of four cycles of single-phase input current

how well important transients can be located and how finely evolutionary behaviors can be detected. Extremely noisy data poses a problem to the operator as how to ascertain the amount of noise in the retained high frequency transient data (Santos et al. 2006).

The interest in applying neural networks to electric load forecasting began more than a decade ago. Artificial neural networks based methods for forecasting have shown ability to give better ability in dealing with the nonlinearity and other difficulties in the modeling of the time series data. ANNs have been applied recently in the area of time-series forecasting due to their flexibilities in data modeling (Ranaweera et al. 1996; Bastian et al. 1999). Most of the approaches reported since are based on the use of an MLP network as an approximator of an unknown nonlinear relation. There have been some pioneering works on applying wavelet techniques together with ANN to time series forecasting (Geva 1998; Oonsivilai and El-Hawary 1999; Renaud et al. 2002; Lin 2006; Kim et al. 2002). Among ANN-based forecasting methods, radial basis function (RBF) networks have been widely used primarily because of their simple construction and easier training is as compared to multi-layer perceptrons (MLPs) in addition to their capability in inferring the hidden relationship between input and desired target patterns. This capability is attributed

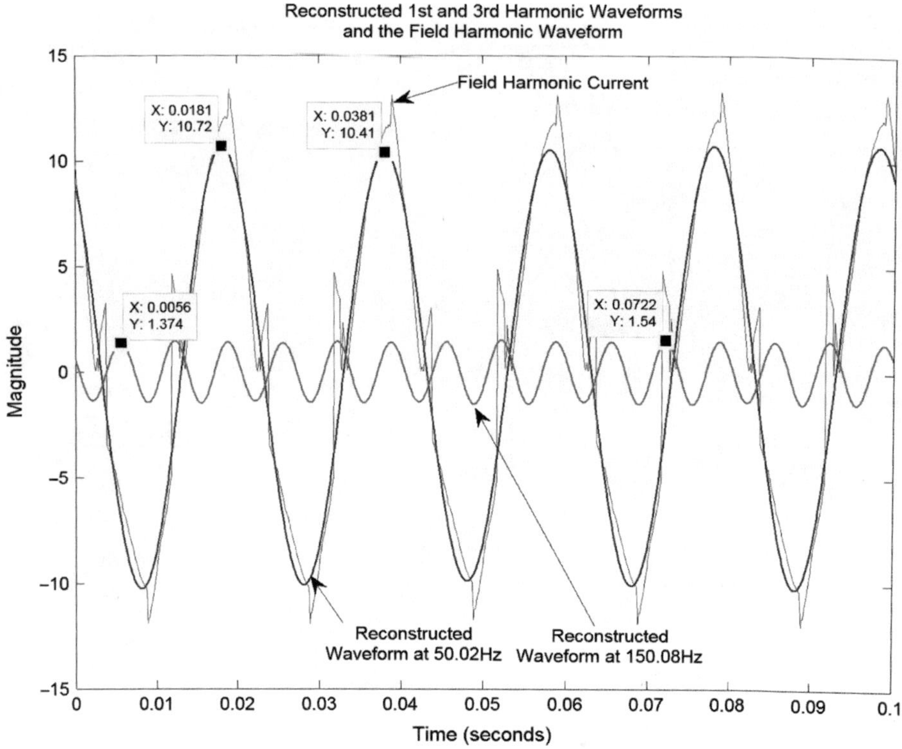

Fig. 20 Field harmonic signal vs. reconstructed fundamental and third harmonic waveform

to its property that it can approximate any continuous function to any degree of accuracy by constructing localized radial basis functions. From the standpoint of preserving characteristics of different classes, this local approximation approach has the advantage over the global approximation approach of multilayer perception networks.

As large amounts of historical load patterns are needed in a typical load-forecasting algorithm, even low sampling rates of one sample per minute generates a huge amount of data. Hence, the effective compression of large data and faithful reconstruction of original signal from compressed data is a major challenge for time series data. Also, when an ANN, especially RBF network, is trained with huge data (with noise), it may result into not only a big network model and very time consuming training but also that the network may fail to capture the true features in the data. With the development of wavelet transforms, the difficulty of effective data compression and faithful retrieval of original data can be well tackled. This tempted researchers to try RBF networks model combined with wavelet transformed data for capturing useful information on various time scales. These strategies approximate a time-series at different levels of resolution using multi-resolution decomposition. Recent works (Satio and Beylkin 1993) stress on the use of shift invariant wavelet transforms, which is an auto correlation shell representation technique for making the analysis of time series data easier. This technique is employed to reconstruct singles after wavelet decomposition. With the help of this technique, a time series can be expressed as an additive combination of the wavelet coefficients at different resolution levels. These data are then applied to build neural-wavelet-based forecasting models to predict electricity demand as from the data obtained from a real electricity market.

Autocorrelation shell representation based wavelet transform is used to approximate short-term load forecast (STLF) at different levels of resolution using multiresolution decomposition. This decomposed data is used for training the RBF network for predicting the wavelet coefficients of future loads. RBF networks optimized with the help of floating point genetic algorithm (FPGA). This technique is then applied to build neural-wavelet-based forecasting models to predict electricity demand as from the data obtained from a real electricity market.

3.1 *Wavelet Transforms in Load Forecast*

Wavelet transforms (Daubechies 1992; Soman and Ramachandran 2004), though known previously, has gained much attention only recently. It has been exploited in many fields such as seismic studies, image compression, signal processing processes, and mechanical vibrations. The flexible time-scale representations of wavelet transform has found its place in many applications that traditionally used modified forms of Fourier transforms (FT) such as STFT and the Gabor transforms. Its impressive temporal content and frequency isolation features have tempted researchers to use them in the area of power systems analysis.

Wavelet transforms provide a useful decomposition of a signal, or time series, so that faint temporal structures can be revealed and handled by nonparametric models. They have been used effectively for image compression, noise removal, object detection, and large-scale structure analysis, among other applications.

3.1.1 Time Series and Wavelet Decomposition in Load Forecasting

The CWT of a continuous function produces a continuum of scales as output. On the other hand, input data is usually discretely sampled, and furthermore, a dyadic or twofold relationship between resolution scales is both practical and adequate. The latter two issues lead to the discrete transform. Figure 21 shows the wavelet decomposition.

Wavelet decomposition provides a way of analyzing a signal in both time and frequency domains. For a suitably chosen mother wavelet function ψ, a function f can be expanded as

$$f(t) = \sum_{j-{-\infty}}^{\infty} \sum_{k-{-\infty}}^{\infty} w_{jk} 2^{i/2} \varphi(2^j t - k), \tag{41}$$

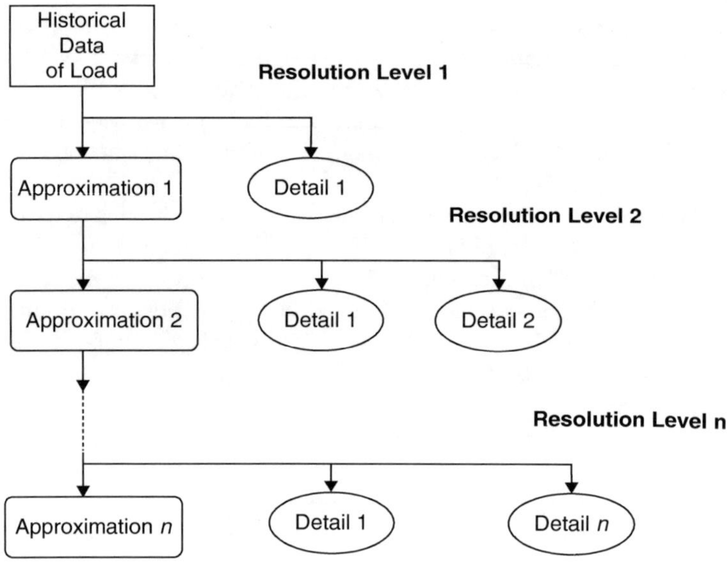

Fig. 21 Wavelet decomposition process

where the functions $\psi(2^j t - k)$ are all orthogonal to each other. The coefficients w_{jk} give information about the behavior of the function f concentrating on the effects of scale around 2–j near time t × 2 – j. This wavelet decomposition of a function is closely related to a similar decomposition (the discrete wavelet transform, DWT) of a signal observed in discrete time.

It is well known that DWT has many advantages in compressing a wide range of signals observed in the real world. However, in time series analysis, DWT often suffers from a lack of translation invariance. This means that DWT-based statistical estimators are sensitive to the choice of origin. The output of a discrete wavelet transform can take various forms (Benaouda et al. 2006). Traditionally, a triangle (or pyramid in the case of two-dimensional images) is often used to represent all that is worth considering in the sequence of resolution scales. Such a triangle comes about as a result of decimation or the retaining of one sample out of every two. The major advantage of decimation is that just enough information is retained to allow exact reconstruction of the input data. Therefore, decimation is ideal for effective compression. However, it can be easily shown that the storage required for the wavelet-transformed data is exactly the same as is required by the input data. The computation time for many wavelet transform methods is also linear in the size of the input data, that is, $O(n)$ or n-length input time series. Also, with the decimated form of output, it is less easy to visually or graphically relate information at a given time point at different scales. More problematic is their lack of shift invariance. This means that if the last few values of the input time series are deleted, then the wavelet-transformed, decimated output data will be quite different from heretofore.

One way to solve this problem at the expense of greater storage requirements is by means of a redundant or nondecimated wavelet transform.

A nondecimated wavelet transform based on an n-length input time series, then, has an n-length resolution scale for each of the resolution levels of interest. Therefore, information at each resolution scale is directly related to each time point. This results in shift invariance. Finally, the extra storage requirement is by no means excessive. An à *trous* algorithm is used to realize the shift-invariant wavelet transforms. Such transforms are based on the so-called auto-correlation shell representation (Satio and Beylkin 1993) by dilations and translations of the auto-correlation functions of compactly supported wavelets.

By definition, the auto-correlation functions of a compactly supported scaling function ϕ(x) and the corresponding wavelet Ψ(x) are as follows:

$$\begin{aligned} \varphi(x) &= \int_{-\infty}^{\infty} \varphi(y)\varphi(y-x)dy \\ \psi(x) &= \int_{-\infty}^{\infty} \psi(y)\psi(y-x)dy, \end{aligned} \tag{42}$$

The set of functions $\{\overline{\psi}_{j,k}(x)\}_{1\leq j\leq n_0, 0\leq k\leq N-1}$ and $\{\overline{\phi}_{n_0,k}(x)\}_{0\leq k\leq N-1}$ is called an auto correlation shell, where

$$\begin{aligned} \overline{\psi}_{j,k}(x) &= 2^{-j/2}\,\psi(2^{-j}(x-k)) \\ \overline{\varphi}_{n_0,k}(x) &= 2^{-n_0/2}\,\varphi(2^{-n_0/2}(x-k)). \end{aligned} \tag{43}$$

A set of filters $P = \{p_k\}_{-L+1\leq k\leq L-1}$ and $Q = \{q_k\}_{-L+1\leq k\leq L-1}$ can be defined as

$$\begin{aligned} \frac{1}{\sqrt{2}}\varphi\left(\frac{x}{2}\right) &= \sum_{k=-L+1}^{L-1} p_k\varphi(x-k) \\ \frac{1}{\sqrt{2}}\psi\left(\frac{x}{2}\right) &= \sum_{k=-L+1}^{L-1} q_k\varphi(x-k). \end{aligned} \tag{44}$$

Using the filters P and Q, the pyramid algorithm for expanding into the auto-correlation shell can be obtained as

$$c_j(k) = \sum_{l=-L+1}^{L-1} p_l c_{j-1}(k+2^{j-1}l) \tag{45}$$

$$w_j(k) = \sum_{l=-L+1}^{l=L-1} q_l c_{j-1}(k+2^{j-1}l). \tag{46}$$

These shell coefficients obtained from (45) and (46) can then be used to directly reconstruct the signals. Given smoothed signal at two consecutive resolution levels, the detailed signal can be derived as:

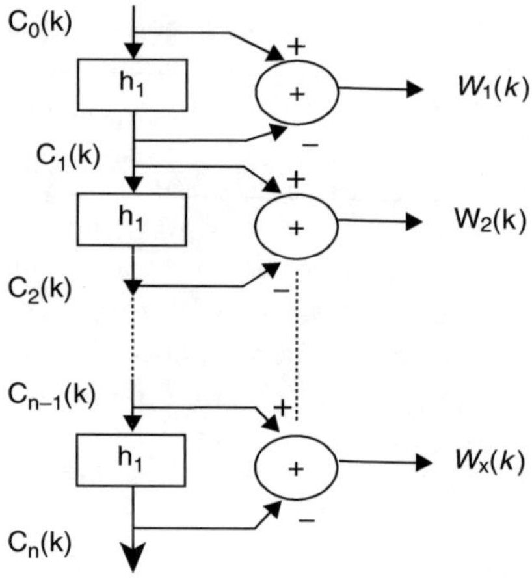

Fig. 22 Wavelet transform of a time-series signal

$$w_j(k) = \sqrt{2}c_{j-1}(k) - c_j(k). \tag{47}$$

The process of generating wavelet coefficient series is further illustrated with the block diagram as shown in Fig. 22.

Then the original signal $c_0(k)$ can be reconstructed from the coefficients $\{w_j(k)\}_{1\leq j\leq n_0, 0\leq k\leq N-1}$ and residual $\{c_{n_0}(k)\}_{0\leq k\leq N-1}$:

$$c_0(k) = 2^{-n_0/2}c_{n_0}(k) + \sum_{j=1}^{n_0} 2^{-j/2}w_j(k) \tag{48}$$

for k = 0, . . . , N − 1, where $c_{n_0}(k)$ is the final smoothed signal.

To make more precise predictions the most recent data shall be used. In case of adaptive learning, the previous data is penalized with forgetting factors. The time-based à tours filters similar to that of are used to deal with the boundary condition. Figure 23 shows the wavelet recombination process.

3.2 *Radial Basis Networks*

An RBF is a function that has in-built distance criterion with respect to a center (Park and Sandberg1991). A typical RBF neural network consists of three layers (input, hidden, output). The activation of a hidden neuron is determined in two steps: The first is to compute the distance (usually the Euclidean norm) between the input vector and a center c_i that represents its hidden neuron; second, a function, that is usually bell shaped, is applied, using the obtained distance to get the final activation of the hidden neuron. In the present case, the well known Gaussian function G(x) is used:

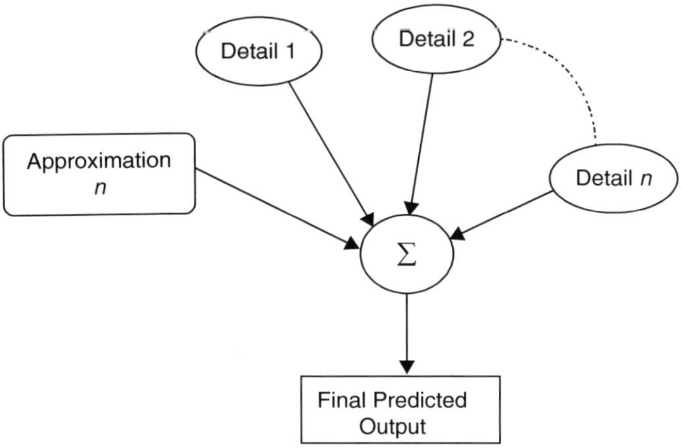

Fig. 23 Wavelet recombination process

$$G(x) = \exp\left(-\frac{\|x - c_i\|^2}{2\sigma^2}\right). \tag{49}$$

The parameter σ is called unit width (spread factor) and is determined using the GA. All the widths in the network are fixed to the same value and these results in a simpler training strategy. The activation of a neuron in the output layer is determined by a linear combination of the fixed nonlinear basis functions, that is,

$$F(x) = w_0 + \sum_{i=1}^{M} w_i \varphi_i(x), \tag{50}$$

where $\phi_i(x) = G(\|x - c_i\|)$ and w_i are the adjustable weights that link the output nodes with the appropriate hidden neurons and w_0 is the bias weight. These weights in the output layer can then be learnt using the least-squares method.

The present work adopts a systematic approach to the problem of center selection. Because a fixed center corresponds to a given regressed in a linear regression model, the selection of RBF centers can be regarded as a problem of subset selection. The orthogonal least squares (OLS) method (Chen et al. 1991) can be employed as a forward selection procedure that constructs RBF networks in a rational way. The algorithm chooses appropriate RBF centers one by one from training data points until a satisfactory network is obtained. Each selected center minimizes the increment to the explained variance of the desired output, and so ill-conditioned problems occurring frequently in random selection of centers can automatically be avoided. In contrast to most learning algorithms, which can work only if a fixed network structure has first been specified, the OLS algorithm is a structural identification technique, where the centers and estimates of the corresponding weights can be simultaneously determined in a very efficient manner during learning. OLS learning procedure generally produces an RBF network smaller than a randomly selected RBF network. Because of its linear computational procedure at the output layer, the RBF is shorter in training time compared to its back propagation counter part.

A major drawback of this method is associated with the input space dimensionality. For large numbers of input units, the number of radial basis functions required can become excessive. If too many centers are used, the large number of parameters available in the regression procedure will cause the network to be over sensitive to the details of the particular training set and result in poor generalization performance (overfit).

The present work uses a floating point GA-based algorithm for optimizing the centers and spread factors.

3.2.1 A Hybrid Neural-Wavelet Model for Short-term Load Prediction

The proposed hybrid neural-wavelet model for short-term load prediction is shown in Fig. 24. Given the time series f(n), n = 1, ..., N, the aim is to predict the l-th sample ahead, f(N + l), of the series. As a special case, l = 1 stands for single step prediction. For each value of l, separate prediction architecture is trained accordingly. The hybrid scheme basically involves three stages (Geva 1998). At the first stage, the time series is decomposed into different scales by autocorrelation shell decomposition; at the second stage, each scale is predicted by a separate

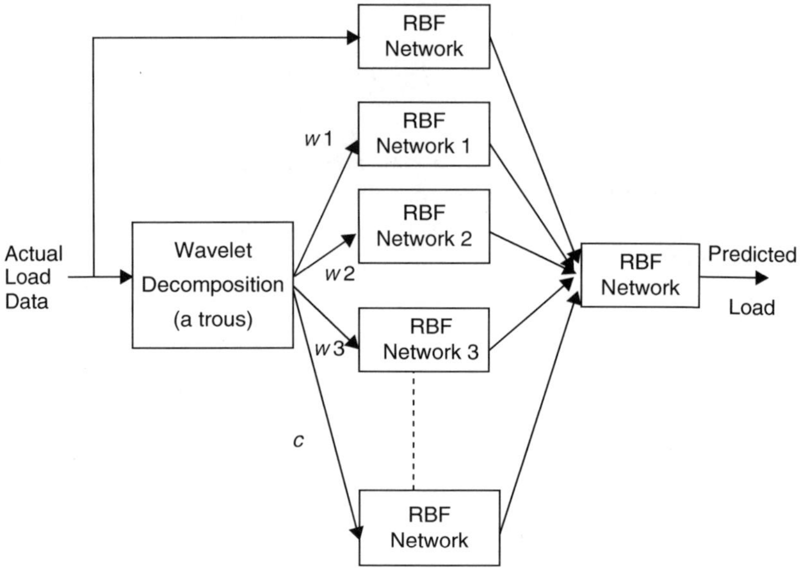

Fig. 24 Overview of the neural-wavelet multiresolution forecasting system. $w_1, \ldots, w_k$ are wavelet coefficients, c is the residual coefficient series

RBF network; and at the third stage, the next sample of the original time series is predicted by another RBF network using the different scale's prediction.

For time series prediction, correctly handling the temporal aspect of data is one of the primary concerns. The time-based à trous transform as described earlier provides a simple but robust approach. Here we introduce an à trous wavelet transform based on the autocorrelation shell representation for the prediction model usage. This approach is realized by applying (47) and (48) to successive values of t. As an example, given an electricity demand series of 1,008 values, we hope to extrapolate into the future with 1 or more than 1 subsequent value. By the time-based à trous transform, we simply carry out a wavelet transform on values $x_1 - x_{1008}$. The last values of the wavelet coefficients at time-point t = 1,008 are kept because they are the most critical values for forecasting system. $w_1, \ldots, w_k$ are wavelet coefficients, c is the residual coefficient series.

Repeat the same procedure at time point t = 1,009, 1,010... repeatedly. It empirically determines the number of resolution levels J, mainly depending on the inspection of smoothness of the residual series for a given J. Much of the high-resolution coefficients are noisy. Prior to forecasting, we get an over complete, transformed dataset.

In Fig. 25, it shows the behavior of the four-wavelet coefficients over 1,008 points for a load series. Note that the data have been normalized for wavelet analysis. Normalization of data is an important stage for training the neural network. The normalization of data not only facilitates the training process but also helps in shaping the activation function. It should be done such that the higher values should not suppress the influence of lower values and the symmetry of the activation function is retained. The input load data is normalized between the minimum value, −1, and the maximum value, +1, by using the formula

$$\left(\frac{Actualvalue - Minimum}{Maximum - Minimum}\right) \times (Maximum - Minimum) + Minimum. \tag{51}$$

The load data should be normalized to the same range of values. The original time series and residual are plotted at the top and bottom in the same figure, respectively. As the wavelet level increases, the corresponding coefficients become smoother. As we will discuss in the next section, the ability of the network to capture dynamical behavior varies with the resolution level.

At the second stage, a predictor is allocated for each resolution level and the following wavelet's coefficients $w_i^j(t)$; $j = 0, \ldots, J$; $I = 1, \ldots, N$ are used to train the predictor. All networks used to predict the wavelets' coefficients of each scale are of similar feed forward RBF perceptrons with D input units, one hidden layer with radial basis function as an activation function and one linear output neuron. Each unit in the networks has an adjustable bias. The D inputs to the jth network are the previous samples of the wavelets' coefficients of the jth scale. In the proposed model implementation, each network is trained by the orthogonal least squares (OLS) method, which can be employed as a forward selection procedure that constructs RBF networks in a rational way. The procedure for designing neural network

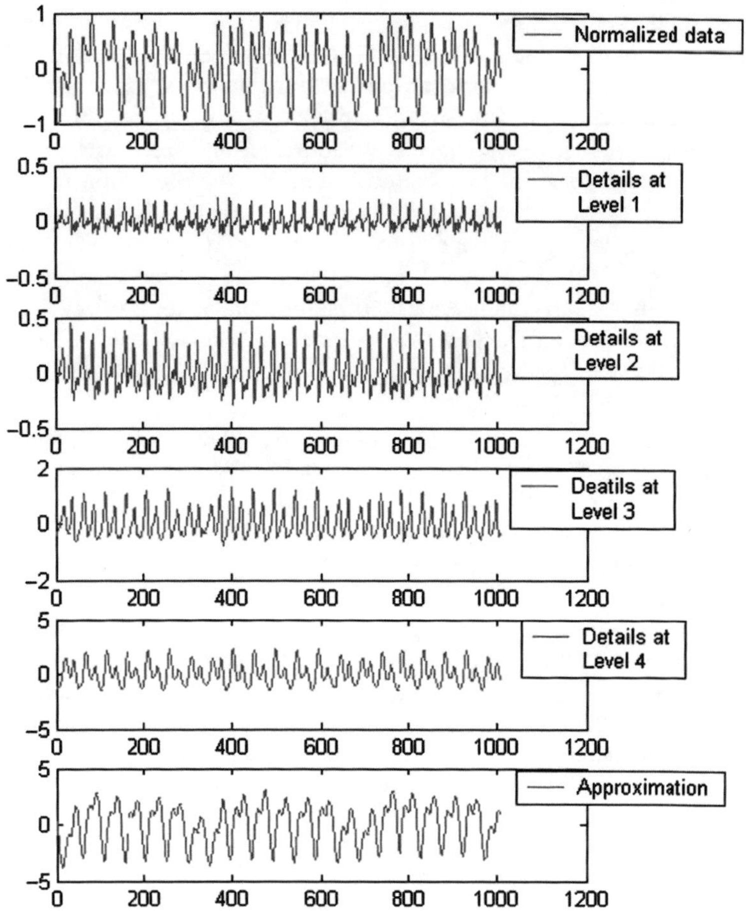

Fig. 25 Illustrations of the à trous wavelet decomposition of a series of electricity demand

structure essentially involves selecting the input, hidden, and output layers. At the third stage, the predicted results of all the different scales $\hat{w}^j_{N+i}(t)$, $j = 0, \ldots, J$ are appropriately combined. Here we discussed and compared three methods of combination. In the first method, we simply applied the linear additive reconstruction property of the à trous, see (46). The fact that the reconstruction is additive allows the predictions to be combined in an additive manner. For comparison purpose, a plain RBF was also trained and tested for original time series, denoted as RBF, without any wavelet preprocessing involved.

The target selection is an important issue in applying neural networks to time series forecasting. A neural network, whose output neurons are reduced from two to one, will have half the number of network weights required. It also carries with important consequences for the generalization capability of the network. A single output neuron is the ideal case, because the network is focused on one task and there is no danger of conflicting outputs causing credit assignment problems in the output layer. Accordingly, it is preferred to have a forecasting strategy, which proceeds separately for each horizon in the second stage.

The target selection is an important issue in applying neural networks to time series forecasting. A neural network, whose output neurons are reduced from two to one, will have half the number of network weights required. It also carries with important consequences for the generalization capability of the network. A single output neuron is the ideal case, because the network is focused on one task and there is no danger of conflicting outputs causing credit assignment problems in the output layer. Accordingly, it is preferred to have a forecasting strategy, which proceeds separately for each horizon in the second stage.

3.3 Experimental Results

The proposed model is tested with two sets of historical data containing the electricity load for the month of July 2005 and month of July 2006, on a half-hourly basis; both sets of electricity load data of Queensland. The sets of electricity load data are downloaded from the NEMMCO website (NEMMCO 2005).

The simulation results are obtained through the use of four different programs. These programs were written in MATLAB command line in association with MATLAB toolboxes on wavelet and neural network. Programs are run in a PC of Pentium IV, 256 MB RAM, 3.2 GHz.

Before the wavelet decomposition technique (à trous) is applied, the sets of historical load data are first normalized.

The model is evaluated based on its prediction errors. A successful model would yield an accurate time-series forecast. The performance of the model is hence measured using the absolute percentage error (APE), which is defined as

$$APE = \left(\frac{|x_i - y_i|}{x_i}\right) \times 100, \qquad (52)$$

where x_i is the actual values and y_i is the predicted values at time instance i. This error measure is more meaningfully represented as an average and standard deviation (SD) over the forecasting range of interests. Additional measure of the error is defined from the cumulative distribution function as the 90th percentile of the absolute percentage error, which provides an indication of the behavior of the tail of the distribution of errors and indicates that only 10% of the errors exceed this value.

The forecasting results from the different forecasting schemes are presented in Table 9. The RBF network is optimized using FPGA in terms of the number of inputs, centers, and spread factor. The number of neurons in the hidden layer is auto-configured by the OLS algorithm. Table 9 shows that the à trous wavelet transform system with adaptive combination coefficients for summing up the wavelet coefficients forecasting is the best in seven step ahead forecasting for the testing

Table 9 Load forecast performance on testing data on APE measure for FPGA optimized spread factor and input

Scheme number	1	2	3	4	5	6	7
μ_R	5.00	5.11	5.66	6.14	6.73	7.20	7.85
$\sigma^2{}_R$	0.7	1.1	1.6	2.3	2.9	3.6	3.4
η_R	0.086	0.093	0.107	0.123	0.138	0.150	0.156
μ_w	1.43	1.13	1.19	1.53	2.22	1.55	1.14
$\sigma^2{}_w$	0.057	.0662	0.052	0.065	0.083	0.183	0.175
η_w	0.024	0.0216	0.022	0.025	0.033	0.036	0.034

$\sigma^2{}_R \times 10^{-3}$, $\sigma^2{}_w \times 10^{-3}$ are the spread factors and subscript R refers to the results with only RBF networks while subscript w refers to the results with hybrid wavelet-RBF model

data, with regards to the mean, variance, and percentile over the absolute percentage error (APE).

3.3.1 Parameters for FPGA Algorithm

Population Size = 40
Maximum Iterations = 30

Operators for FPGA:

1. Heuristic crossover
2. Uniform mutation
3. Normalized geometric select function

4 Conclusions

The chapter has demonstrated the use of wavelet techniques and its integration with neural networks for power quality analysis and assessment and load forecasting. A computational algorithm for identifying integer harmonics and noninteger harmonics by using wavelet-based transform has been reported. Better results have also been achieved as compared with that obtained from conventional techniques. Under the electricity deregulated environment, it is important to have innovation in engineering so that power system will operate in a much more cost effective, reliable, and secure way. This chapter has demonstrated some of the techniques that have a high potential to be used in the near future for real-life applications.

Acknowledgements The authors thank ISAP in granting the copyright release for using materials in the paper entitled Wavelet-GA-ANN based Hybrid Model for Accurate Predication of Short-term Load Forecast.

References

Carbone R, Testa A, Menniti D, Morrison RE, Delancy E (1995) Harmonic and inter-harmonic distortion in current source type inverter drives. IEEE Trans Power Deliv 10(3):1576–1583
Smith SW (2002) Digital signal processing: a practical guide for engineers and scientists. Newnes
Ramirez RW (1985) The FFT: fundamentals and concepts. Prentice-Hall, Englewood Cliffs, NJ
Girgis AA (1980) A quantitative study of pitfalls in the FFT. IEEE Trans Aerosp Electron Syst AES-16(4):434–439
Addison PS (2002) The Illustrated wavelet transform handbook. Institute of Physics Publishing Ltd., Bristol and Philadelphia
Pham VL, Wong KP (1999) Wavelet-transform-based algorithm for harmonic analysis of power system waveforms. IEE Proc Gener Transm Distrib 146(3):249–254
Tse NCF (2006) Practical application of wavelet to power quality analysis. Proceedings of the 2006 IEEE Power Engineering Society General Meeting, IEEE Catalogue Number 06CH37818C, June CD ROM
Strang G, Nguyen T (1996) Wavelets and filter banks. Wellesley-Cambridge, Wellesley, MA
Teolis A (1998) Computational signal processing with wavelets. Birkhauser
Huang S-J, Hsieh C-T, Huang C-L (1999) Application of morlet wavelets to supervise power system disturbances. IEEE Trans Power Delivery 14(1):235–243
Misiti M, Misiti Y, Oppenheim G (1996) Wavelet toolbox for use with matlab. The Mathworks, pp.6–43
Mallet S (1998) A wavelet tour of signal processing. Academic
Carmona RA, Hwang WL, Torresani B (1999) Multiridge detection and time-frequency reconstruction. IEEE Trans signal process 47(2):480–492
Carmona RA, Hwang WL, Torresani B (1997). Characterization of signals by the ridges of their wavelet transforms. IEEE Trans signal process 45(10):2586–2590
Tse NCF, Lai LL (2007) Wavelet-based algorithm for signal analysis. EURASIP J Adv Signal Process 2007 Article ID 38916
Chan YT (1995) Wavelet basics. Kluwer, Royal Military College of Canada
Robertson DC, Camps OI (1996) Wavelets and electromagnetic power system transients. IEEE Trans PWRD-11 11(2):1050–1058
Pham VL, Wong KP (2001) Antidistortion method for wavelet transform filter banks and nonstationary power system waveform harmonic analysis. IEE Proc Gener Transm Distrib 148(2)
Chan WL, So ATP, Lai LL (2000) Harmonics load signature recognition by wavelets transforms. Proceedings of the International Conference on Electric Utility Deregulation, Restructuring and Power Technologies DRPT2000, IEEE Catalog Number 00EX382, pp. 666–671
Lai LL, Chan WL, Tse CT, So ATP (1999) Real-time frequency and harmonic evaluation using artificial neural networks. IEEE Trans Power Delivery 14(1):52–59
Butler-Purry KL, Bagriyanik M (2003) Characterization of transients in transformers using discrete wavelet transforms. IEEE Trans Power Syst 18(2):648–656
Addison PS (2002) The Illustrated wavelet transform handbook. Institute of Physics Publishing Ltd., Bristol and Philadelphia
Carmona RA, Hwang WL, Torresani B (1999) Multiridge detection and time-frequency reconstruction. IEEE Trans Signal Process 47(2):480–492
Ranaweera DK, Karady GG, Farmer RG (1996) Effect of probabilistic inputs on neural network-based electric load forecasting. IEEE Trans Neural Works 7(6):1528–1532
Bastian JN, Zhu J, Banunarayana V, Mukerji R (1999) Forecasting energy prices in a competitive market. IEEE Comput Appl Power 13(3):40–45
Geva AB (1998) Scalenet - Multiscale neural-network architecture for time series prediction. IEEE Trans Neural Netw 9(5):1471–1482
Oonsivilai A, El-Hawary ME (1999) Wavelet neural network based short term load forecasting of electric power system commercial load. IEEE Proc 3:1223–1228

Renaud O, Starck J-L, Murtagh F (2002) Wavelet-based forecasting of short and long memory time series

Lin C-J (2006) Wavelet neural networks with a hybrid learning approach. J Inf Sci Eng 22: 1367–1387

Kim C-I, In-keun Y, Song YH (2002) Kohonen neural network and wavelet transform based approach to short-term load forecasting. Electric Power Syst Res 63:169–176

Chen S, Cowan CFN, Grant PM (1991) Orthogonal least squares learning algorithm for radial basis function networks. IEEE Trans Neural Netw 2:302–309

Satio N, Beylkin G (1993) Multiresolution representations using the auto-correlation functions of compactly supported wavelets. IEEE Trans Signal Process 41(12):3584–3590

Santos PJ, Martins AG, Pries AJ, Martins JF, Mendes RV (2006) Short-term load forecast using trend information and process reconstruction. Int J Energy Res 811–822

Benaouda D, Muratagh F, Starck JL, Renaud O (2006) Wavelet-based nonlinear multiscale decomposition model for Electricity load forecasting. Neurocomputing 70:139–154

Daubechies I (1992) Ten Lectures on Wavelets. CBMS-NSF, SIAM

Soman KP, Ramachandran KI (2004) Insight into wavelets from theory to practice. Prentice-Hall of India

Park J, Sandberg IW (1991) Universal approximation using radial basis function networks. Neural Comput 3(2):246–257

Australian National Electricity Market Management Company Limited (NEMMCO) (2005) website, http://www.nemmco.com.au/

Transmission Valuation Analysis based on Real Options with Price Spikes

Michael Rosenberg, Joseph D. Bryngelson, Michael Baron, and Alex D. Papalexopoulos

Abstract The presence of optionality in the generation and transmission of power means that valuing physical and financial assets requires using option theory, which in turn requires studying stochastic processes appropriate for the description of power prices. Power prices are much more volatile than other commodity prices and exhibit interesting behavior such as regime switching between normal and spiked states. The probability distributions underlying such stochastic process provide an input for price forecasts, which are based on price history. They also provide an input into valuation of transmission and transmission options in cases where the implied market-based measures of volatilities and correlations are lacking. Combining information from this analysis of stochastic processes for power prices with the Black–Scholes framework for option valuation, specifically using that framework to calculate the value of spread options, yields methods for calculating the value of transmission as well as for calculating the value of financial transmission options, which also depend on spread of power prices. The three main techniques for obtaining the option value include analytical approaches, binomial-type trees (finite difference methods), and Monte Carlo simulations. Each of these techniques presented in the paper has its own advantages and disadvantages and is complementary to the other two, providing independent validation and quality control for transmission valuation algorithms.

Keywords Options · Power · FTRs · Price · Spikes · Transmission · Valuation · Congestion.

1 Introduction

Energy companies often need to move energy from one location to another. These transactions involve the transmission of certain quantity of power from a generator location to a load location. Another example of a transaction is the transportation of

A.D. Papalexopoulos (✉)
ECCO International, Inc., 268 Bush Street, Suite 3633, San Francisco, CA 94104, USA
e-mail: alexp@eccointl.com

S. Rebennack et al. (eds.), *Handbook of Power Systems II*, Energy Systems,
DOI 10.1007/978-3-642-12686-4_4,

certain volume of gas via pipeline from a gas storage location to a power plant location. Crude oil often has to be transported to refineries by tankers and then through pipelines to the power plants. Also, energy companies convert one commodity into another. An example of such conversion include burning of fuels, such as natural gas, coal, or oil, to produce electricity at power plants, or the refinery operations refining crude oil into its products.

Common to all aforementioned transactions is the move of a commodity from a place of relative abundance and lower price to a place of scarcity and higher price or conversion of a cheaper commodity into a more expensive one. This makes the operation of transporting or converting the commodity profitable. Moreover, the operation itself is not a zero-sum game and often is worth a lot of money for its potential to continue the delivery or conversion and turning a profit. The value of this operation depends on the costs of delivery or conversion and the difference between the prices at two locations or the prices of two commodities. Once the payoff function is established, one can price the transaction using methods described in this paper, and hence establish a value of the transportation and conversion, whether represented by physical assets or financial instruments.

The financial and energy trading community considers these transactions as spread obligations or spread options. A variety of other names has also been used; these include exchange, transportation, tolling, spark-spread obligations, or options, depending on the commodities exchanged and the nature of transaction. The so-called structured deals can combine a number of simpler transactions into one complex deal.

This paper will focus on the physical transmission of power and associated financial instruments. The examples of such transactions include the following:

- Physical rights to transmit power on transmission lines
- Financial Transmission Rights (FTRs)/Congestion Revenue Rights (CRRs)/ Transmission Congestion Contracts (TCCs)
- Tolling agreements (options)
- FTR options

The paper will focus on the valuation of such transmission-based instruments. We will consider obligations, contrast them with options, and highlight the impact of the optionality on the value of the asset or the financial instrument. We will also examine the impact of the price process model on the valuation of obligations and options.

Post electric deregulation, the transmission assets are owned by Independent Transmission Companies or separated by a "firewall" from the commercial operations to ensure equal access. In the areas where transmission is scheduled by the Independent System Operators (ISOs) or Regional Transmission Organizations (RTOs), market participants, such as generators or Load Serving Entities (LSEs), can enter only into financial transactions on transmission of power within the respective control areas. In regions where the vertically integrated structure still prevails, the market participants may combine financial instruments on transmission with physical reservations of capacity on the transmission lines.

Financial obligations or options on transmission may be available from a centralized market or available only from other market participants. The latter are obtained over-the-counter (OTC). One example of the centralized auction of transmission obligations and options is the FTR/CRR/TCC auction run by an ISO or a RTO with the purpose to auction off the transmission capacity of the network they control to various market participants. The participants use these instruments to hedge their exposure to locational price differential at various locations in the transmission grid when they move power for generation locations or trading hubs, or load aggregation points (LAPs) to load locations. Many OTC transactions in the market place that are outside the domain of an ISO or a RTO are tailored to individual needs of generators and load. Examples of such OTC transactions include tolling agreements (options) to transport power.

The rest of the paper is organized as follows. The next section reviews models of commodity prices in general and power prices in particular. The most appropriate and widely used model is the single factor model. One possible extension is the multifactor family of models, which we briefly describe in the next section as well. The most important step for the purpose of transmission valuation is the introduction of spikes into the commodity prices. This issue is mostly acute in markets where the probability of surges in power prices occasionally changes the dynamics of the marketplace. After the model for the price process is established we proceed to describe the valuation methods, starting with simpler models, which include simplifying approximations and proceeding towards the presentation of models that provide an exact solution. We also discuss the incorporation of spikes into the solution.

Finally, we close with the discussion of the numerical solutions for the methods we have introduced. The simplest approaches have closed form solutions and thus can be evaluated in terms of standard functions. The exact solution requires numerical integration and should be done with proper controls to ensure convergence. One of the most intuitive numerical schemes is a binomial tree approach that builds on a similar approach for valuations of derivatives contingent on one underlying asset or commodity. Finally, we present the Monte Carlo method, which provides a simulation-based approach to the valuation of spread options.

2 Behavior of Commodity Prices

2.1 Models of Spot and Forward Commodity Prices

The single factor model for energy prices is mathematically the simplest approach for modeling energy prices. As the name for this approach suggests, there is only one source of uncertainty in the prices that can be modeled as geometric mean-reverting processes of log type (Schwartz 1997) with time varying mean

$$dS = \eta(m(t) - \ln S)Sdt + \sigma_0 Sdz \tag{1}$$

where $S = S(t)$ is the stochastic process of energy spot prices, $m(t)$ is the general trend, η is reversion rate, σ_0 is the standard deviation of the error term (volatility), and dz is the associated white noise process. This single source of uncertainty is the volatility of spot prices. More information about the properties and reasons for the choice of this process can be found in Rosenberg et al. (2002b). It can be shown (Clewlow and Strickland 1999) that this assumption about the spot price, combined with the fact that in a risk-neutral world the futures price of a commodity is equal to its expected spot price, leads to the following equation for the futures commodity price at time T:

$$dF(T) = F(T)\sigma(T)dz \tag{2}$$

where F is the futures price and σ is the corresponding volatility. Equation (2) is the right candidate for the description of future prices. The geometric Brownian motion with zero drift means that there is no preference as to the direction of the change of future prices in the risk-neutral world. In other words, all currently available information was absorbed by the market into the current price of the futures. Again, we must distinguish between the risk-neutral distribution under which valuation takes place and the true/objective/statistical distribution. That true/objective/statistical distribution may have a nonzero drift owing to the presence of a market price of risk (MPR) (detailed general discussions of the MPR can be found in Hull 2002 and Bjork 2004, and a discussion of this in energy markets can be found in Ronn 2002),

$$dF(T) = \mu_1(F,T)dt + F(T)\sigma(T)dz \tag{3}$$

where μ_1 is the nonrandom drift. The variable that determines a particular form of future $F(T)$ in (2) and (3) is the volatility term structure $\sigma(T)$. One of the simplest assumptions consistent with empirical observations for certain period of time is the exponentially decaying volatility term structure:

$$\sigma(T) = \sigma_0 e^{-\eta T} \tag{4}$$

Moreover, the assumption of geometric mean-reverting process (1) requires that the rate of decay is equal to the mean-reverting parameter and, in the limit $T \to 0$, the volatility of future converges to the volatility of spot (Clewlow and Strickland 1999). In practice, the observed nonzero asymptotic value of volatility term structure is the evidence of more complex forward curve dynamics than is allowed by one factor model (Smith and Schwartz 2003). We shall not concern ourselves with these details as they are not important for the purpose of this paper.

2.2 Stochastic Modeling of Price Spikes

The time plot on Fig. 1 reveals a rather irregular behavior of daily electricity prices. One can clearly notice sharp and relatively short-term surges of prices – spikes. They are caused by an abrupt increase in the demand of energy, which may be a result of a combination of extreme conditions. Behavior of electricity prices during

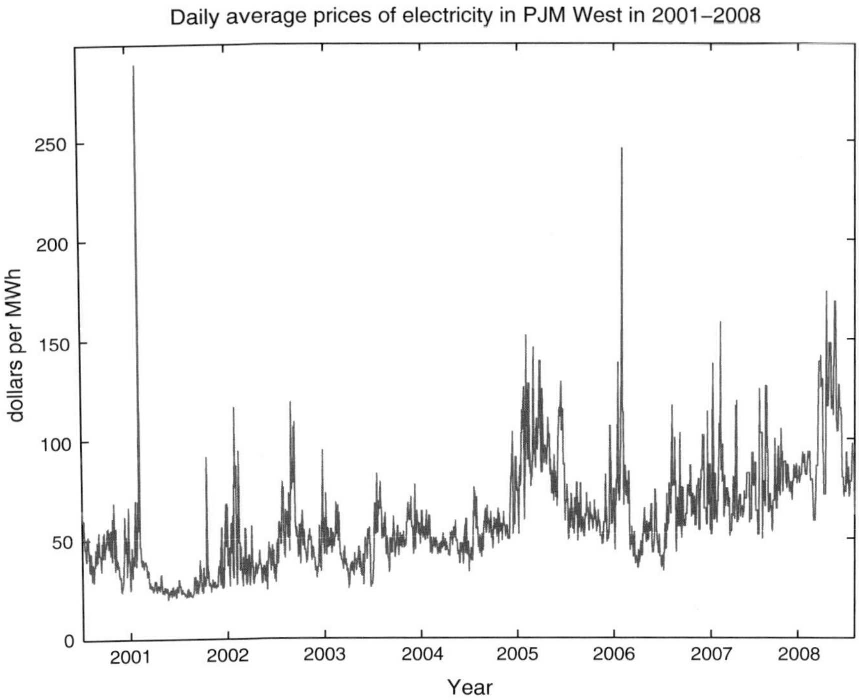

Fig. 1 Daily electricity prices in Pennsylvania–New Jersey–Maryland West region (Data Source: Intercontinental Exchange – theice.com)

spikes is dramatically different from the periods between spikes. Spikes last from several hours to several days, but they can lead to a tremendous price increase (Baron et al. 2001, 2002).

We conclude that, for the process of electricity prices with spikes, (1–3) require different sets of parameters for its two phases, the spike periods and interspike periods. This leads to a *multistate* model (Baron 2004), which represents a nonstationary stochastic process consisting of homogeneous segments (spikes and interspike periods) separated by *change points*. Each change point marks the moment when one segment ends and the next one begins. To model such a process, one has to determine the distribution of prices during each type of segments and the stochastic mechanism of switching between the segments.

2.2.1 Interspike Segments (Regular Mode)

At any time t, (1) yields a *lognormal* distribution of prices with the density

$$f_0(x) = \frac{1}{x\tau_0\sqrt{2\pi}} \exp\{-(In\, x - m(t))^2/2\tau_0^2 \qquad (5)$$

with $m(t)$ and τ_0 representing the time dependent mean and volatility. This is a homogeneous (stationary) segment of a stochastic process that is located between two successive change points (Spokoiny 2009). Moreover, since only a finite number of observations can be available, any data satisfying model (1) forms *an autoregressive process of order* 1 or AR(1) (Dixit and Pindyck 1994, Section 3.B). That is, electricity prices $S(t)$ and $S(t-1)$ are connected by the stochastic *difference* equation

$$\ln S(t) - m(t) = \phi\{\ln S(t-1) - m(t-1)\} + \sigma_0 Z(t) \tag{6}$$

for a Gaussian white noise sequence $Z(t)$ and $t = 1, 2, \ldots$. Parameters of this equation are related to (1) and (4) through the formulas $\phi = e^{-\eta}$ and $\sigma_0 = \tau_0\sqrt{1-\phi^2}$. Here ϕ, σ_0, and τ_0, are parameters of the named distributions: ϕ is the first autoregressive coefficient that is related to the reversion rate, τ_0 is the standard deviation of log-price at any time, and σ_0 is the error standard deviation of the AR(1) process of log-prices.

2.2.2 Spikes with Varying Magnitudes (Spike Mode)

According to Fig. 1, electricity prices during one spike can differ very significantly from those during another spike. Different spikes are characterized by different location parameters μ that show the magnitude of a spike. Variable values of μ can be viewed as realizations of a random variable M. This yields a *Bayesian model* (where a parameter has its own distribution, see, e.g., Carlin and Louis 2008) with a lognormal *conditional* density of prices:

$$f_1(x|\mu) = \frac{1}{x\tau_1\sqrt{2\pi}} \exp\left\{-(\ln x - \mu)^2/2{\tau_1}^2\right\}, \tag{7}$$

conditioned on parameter μ, and a normal *prior* distribution of μ,

$$\pi(\mu) = \frac{1}{\omega\sqrt{2\pi}} \exp\{-(\mu - \theta)^2/2\omega^2\}$$

Parameters of this prior distribution have the following meaning. The prior mean, θ, is the average magnitude of spikes, while the prior standard deviation, ω, quantifies the variability of spike magnitudes and how much the spikes can differ from each other.

This Bayesian model reflects the fact that electricity prices are dependent on the same spike (because they are based on the same parameter μ), but independent between different spikes. This reflects the real situation. Indeed, when the price of electricity reaches a rather high mark during a spike, it stays around that level almost until the end of the spike, not affecting, however, the next spike. This Bayesian model can be developed further by letting the prior mean θ of the distribution of

μ to depend on time, $\theta = \mathbb{E}(\mu) = \theta(t)$. This time trend is high during the peak season, gradually reducing during the shoulder months.

2.2.3 Change of Modes

Transitions between the segments are governed by a *Markov chain* with the *transition probability matrix:*

$$P = \begin{pmatrix} 1 - p(t) & p(t) \\ q & 1 - q \end{pmatrix}, \tag{8}$$

where transitions from a regular state to a spike occur with probability $p(t)$, and transitions from a spike to a regular state occur with probability q. Consequently, $\{1 - p(t)\}$ is the probability that a day with no spike is followed by another day with no spike, and $\{1 - q\}$ is the probability for a spike to extend for one more day. Noticeably, $p(t)$ is a time-dependent probability because the chance of a new spike has a seasonal and weekly component. For example, it increases during a peak season and practically vanishes on weekends. On the other hand, the spike-ending probability q is constant, confirming geometric distribution of spike durations.

Estimation of parameters in such a multistate process requires statistical *change-point detection* techniques (Basseville and Nikiforov 1993; Zacks 2009, Chap. 9) to separate the segments as well as standard tools for estimating parameters of segments. Parameters of each state are estimated separately by standard available tools. Monthly and weekly components as well as the general trend are estimated by the method of weighted least squares to reduce the influence of possible spikes on estimates. Then, an autoregressive time series model is fit to detrended log-prices, which are residuals obtained after the estimated trend is subtracted from the observed log-prices. This results in estimates of the autoregressive coefficient and residual variance (volatility). Substituting the estimates of ϕ, η, σ_0, and τ_0 into (4) and (5), we obtain the interspike distribution of electricity prices. This is all we need for the initial estimation of change points between the segments by the sequential cumulative sum algorithm (CUSUM, as in Basseville and Nikiforov 1993; Khan 2009; Zacks 2009, Sect. 9.3). Estimated change points partition the observed time series into spike and interspike segments. Given this, the interspike parameters can be re-estimated based on interspike segments only. Also, we can estimate the spike parameters. Under the assumed Bayesian model, the unconditional distribution of detrended log-prices during spikes is normal with parameters θ and $\omega^2 + \tau_1^2$. Consequently, θ is estimated by the sample mean of all the residuals during spikes and ω^2 and τ_1^2 as the between and within mean square in the one-way analysis of variance during spikes. With the refined parameter estimates, the CUSUM algorithm can be repeated for more accurate estimation of change points. This iterative scheme will be concluded when a new round of CUSUM makes no changes in the obtained change-point estimates. For more details, see Baron (2004).

3 Valuation of Obligations and Options

This section reviews the methodology for valuing obligations and options on transmission. These two classes of transactions require drastically different valuation methods. We begin by looking into valuation of obligations.

3.1 Valuation of Transmission Obligations

Transmission obligation is either a financial gain or loss derived from holding a financial transaction structured to equal the difference between the prices at two locations and may be either positive or negative at any moment in time, depending on which locational price is higher. Examples of such transactions include (a) Financial Transmission Rights (FTRs)/Congestion Revenue Rights (CRRs)/Transmission Congestion Contracts (TCCs) and (b) Basis spread transactions (going long futures or forwards at one location and shorting the same amount of futures or forwards on the same commodity at another location).

If the forwards for each location in the obligation are liquidly traded in the market, the expected value of the difference is currently observed by a holder of the obligation and is equal to the difference of the observed prices. If the forwards/futures are not liquidly traded, such as FTRs and similar nodal contracts, the best estimate of the expected value of the spread can be estimated from forecasting the future spot prices and inferring for the MPR (see the MPR discussion in Ronn 2002). Once both parameters are estimated, the expected value of spread on futures or forwards may be computed. Such value would represent the expected market-transacted price for the purchase or sale of the spread, if such transaction did take place.

The value of such obligations to each market participant depends on his risk aversion, which, unlike MPR, is idiosyncratic to each participant and partly depends on the purpose for which he enters into such transactions. Therefore, one should not confuse the equilibrium or equilibrium-estimated price with the worth of the instrument to individual participant. This is true for obligations, as well as options to which we now turn to.

3.2 Valuation of Spread Options

In this section we discuss methods for valuing options for energy producing and delivery systems. Option valuation is much more complex than the valuation of obligations described in the previous subsection. Given the variety of options and models for energy commodities, such an undertaking could easily become unwieldy. We shall solve this problem by focusing on a specific kind of option, *viz.*, an option on the spread between two futures, and a specific kind of model for the futures prices, *viz.*, single factor geometric Brownian motion. In this subsection we try to illustrate a large number of mathematical techniques that are useful for valuing *any* option on an energy commodity. Given the practical emphasis of this paper, we make no claims of mathematical rigor or completeness, but instead we concentrate on sketching the most basic information and reference sources with more complete information.

We use a simple model of futures prices. Later in the paper we will move beyond the simple model analyzed here to discuss the valuation of spread options when there are spikes in the prices of the underlying asset. A paper that analyzes a spread option with a very simple generalization of single factor geometric Brownian motion in the underlying asset is Alexander and Scourse (2004). The correlation coefficient, which plays a major role in the following analysis, is less well suited for describing non-normal distributions. An influential and lucid discussion of these problems is in Embrechts et al. (1999). A popular alternative to the use of correlation coefficients to describe multivariate probability distributions is the use of copula functions. Cherubini and Luciano (2002) use copulas to analyze a general spread option. Gregoire et al. (2008) describe the use of copulas to model price dependence in energy markets. We hope that the choice of a simple, but very important, option and a simple, but very flexible and useful, model for futures prices will assist the reader clearly understand these useful valuation techniques. Further, we hope that they will provide sufficient foundation to understand and analyze more complex and realistic models, such as the models we describe later in the paper and the models described in the above references.

A versatile and conceptually straightforward method of valuing options consists of deriving and solving a differential equation for the option value. Here we shall display and discuss the differential equation for the value of an option on two futures contracts. As mentioned earlier, we are modeling the behavior of the futures prices by a simple, single factor model. We denote the two futures prices by F_1 and F_2, and we assume that the futures prices have volatilities of σ_1 and σ_2, respectively. With these assumptions, the stochastic differential equations,

$$dF_1(t,T) = \mu_1(F_1, F_2, t)dt + \sigma_1(t,T)F_1(t,T)dz_1(t) \tag{9}$$
$$dF_2(t,T) = \mu_2(F_1, F_2, t)dt + \sigma_2(t,T)F_2(t,T)dz_2(t) \tag{10}$$

where z_1 and z_2 represent standard Wiener processes, describe the behavior of the futures prices. The relation between the two Weiner processes z_1 and z_2 accounts for the correlation between the returns of the futures prices. To describe this correlation more precisely, we define the operator $\mathbb{E}[x]$ to represent the expectation (average) of x; then the correlation coefficient, ρ, of the returns of the two futures is defined so that

$$\mathbb{E}[dz_1(t)dz_2(t)] = \rho(t)dt \tag{11}$$

This correlation coefficient is the usual correlation coefficient between price returns that trading organizations compute. We denote by $V(F_1, F_2, t)$ the value of a financial instrument that depends on the values of the futures prices F_1 and F_2 and the time t. With this model and the assumption of no-arbitrage, the value of the financial instrument must satisfy the differential equation

$$\frac{\partial V}{\partial t} + \frac{1}{2}\sigma_1^2 F_1^2 \frac{\partial^2 V}{\partial F_1^2} + \rho\sigma_1\sigma_2 F_1 F_2 \frac{\partial^2 V}{\partial F_1 \partial F_2} + \frac{1}{2}\sigma_2^2 F_2^2 \frac{\partial^2 V}{\partial F_2^2} - rV = 0 \qquad (12)$$

It is important to note that, although here we will focus on spread options, this differential equation (12) is applicable to *any* financial instrument with a value that depends only on the value of two futures prices and time. A derivation of the differential equation (12) based on the methods that would be required to perfectly hedge the options can be found in Rosenberg et al. (2002a), and a more mathematical derivation, based on the Feynman–Kacs formula, can be found in Carmona and Durrleman (2003). (Pedagogical discussions for the Feyman–Kacs formula, which can be used to relate risk-neutral expectations to solutions of partial differential equations, can be found in Bjorck (2004) and Shreve (2004)). A more general version of (12) for n futures or assets is derived in Bjorck (2004), and can also be found in the standard text of Hull (2002).

What differentiates between the many possible financial instruments that have this property and therefore have a value that is described by (12)? To answer this question we note that we have not completely specified the mathematical problem by showing that the financial instrument must satisfy (12); to complete the specification of the mathematical problem we must also specify what happens at the boundaries of the regions where $V(F_1, F_2, t)$ is defined. These boundary conditions differentiate between different financial instruments that depend on the values of two futures prices and time. In the case of a spread option, the relevant boundary is the expiration time, T, of the option and the boundary value is the payoff, $P(F_1, F_2)$. To make these ideas concrete, consider an option to exchange a_1 futures contracts of type 1 for a_2 futures contracts of type 2 plus a fixed sum K, which we will call the strike price, all at an agreed upon future time, the expiration time of the option, which we denote by T. This is a spread option and it satisfies (12) for the value of a general financial instrument, and in addition the value of the option at expiration is known to be

$$V(F_1, F_2, T) = P(F_1, F_2) = \max[a_1 F_1(T) - a_2 F_2(T) - K, 0] \qquad (13)$$

Note that K can be positive or negative. For future reference, also notice that the differential equation (12) is time homogeneous, by which we mean that it is not changed by the transformation: $t \to t + \delta t$. The only time that matters in the value of the option is the time until expiration, $\tau \equiv T - t$. In this paper we are concerned with European-type exercise options. For these options the boundary conditions are given by the payoff of the options at expiration. Most energy options are European-type exercise.

3.2.1 Closed-form Solutions

Valuing a spread option becomes much easier if the strike equals to zero, that is, $K = 0$. In that case, as Rosenberg et al. (2002a) prove, a simple dimensional analysis suggests that the option value has the form

$$V(F_1, F_2, t) = F_2 h(z, t) \tag{14}$$

where we have defined as the ratio of the values of the two futures prices

$$z = \frac{F_1}{F_2} \tag{15}$$

In terms of the new variable z, the payoff of the exchange option, given in (13), becomes

$$P(F_1, F_2) = a_1 F_2 \max\left[z - \frac{a_2}{a_1}, 0\right] = F_2 \varphi(z) \tag{16}$$

where we have defined the function $\varphi(z)$ as

$$\varphi(z) = a_1 \max\left[z - \frac{a_2}{a_1}, 0\right] \tag{17}$$

where a_1 and a_2 are the number of futures contracts that are being exchanged. (Good introductions to dimensional analysis include Barenblatt (1996), Barenblatt (2003), Bridgeman (1931), and Wilson (1990) Using the chain rule of calculus, the differential equation (12) for the value of the option becomes a differential equation for $h(z)$,

$$\frac{\partial h}{\partial t} + \frac{1}{2}\tilde{\sigma}^2 z^2 \frac{\partial^2 h}{\partial z^2} - rh = 0 \tag{18}$$

where we have defined the effective volatility

$$\tilde{\sigma} = \sqrt{\sigma_1^2 - 2\rho\sigma_1\sigma_2 + \sigma_2^2} \tag{19}$$

where σ_1 and σ_2 are the volatilities of the two futures prices and ρ is the correlation coefficient of the returns of the two futures prices. The boundary condition for $h(z)$ at the expiration time T is given by

$$h(z, T) = \varphi(z) = a_1 \max\left[z - \frac{a_2}{a_1}, 0\right] \tag{20}$$

Now notice that the differential equation (18) for $h(z, t)$ together with the boundary condition (20) describe a_1 vanilla call options on a futures price z, having a strike price of a_2/a_1 and a volatility of $\tilde{\sigma}$ (Hull 2002), and so the solution of the differential equation can be obtained directly from the Black-76 formula for an option on a futures contract (derived in Black 1976):

$$h(z,t) = a_1 e^{-r\tau}\left[zN(d_1) - \left(\frac{a_2}{a_1}\right)N(d_2)\right] \tag{21}$$

where we have defined

$$d_1 = \frac{1}{\tilde{\sigma}\sqrt{\tau}}\log\left(\frac{a_1 z}{a_2}\right) + \frac{\tilde{\sigma}\sqrt{\tau}}{2} \tag{22}$$

$$d_2 = \frac{1}{\tilde{\sigma}\sqrt{\tau}}\log\left(\frac{a_1 z}{a_2}\right) - \frac{\tilde{\sigma}\sqrt{\tau}}{2} \tag{23}$$

and $N(x)$ represents the cumulative normal distribution. We remind the reader again that $\tau = T - t$ represents the time until the option expires. Substituting this expression for $h(z,t)$ into (14) for $V(F_1, F_2, t)$ and using the definition (15) of z yields the desired value of an option to exchange two futures contracts,

$$V(F_1, F_2, t) = e^{-r\tau}\left[a_1 F_1 N\left(d_1'\right) - a_2 F_2 N\left(d_2'\right)\right] \tag{24}$$

where

$$d_1' = \frac{1}{\tilde{\sigma}\sqrt{\tau}}\log\left(\frac{a_1 F_1}{a_2 F_2}\right) + \frac{\tilde{\sigma}\sqrt{\tau}}{2} \tag{25}$$

$$d_2' = \frac{1}{\tilde{\sigma}\sqrt{\tau}}\log\left(\frac{a_1 F_1}{a_2 F_2}\right) - \frac{\tilde{\sigma}\sqrt{\tau}}{2} \tag{26}$$

Equations (24)–(26) represent the value of the exchange option and are simply the Margrabe formula (Margrabe 1978) modified so that they apply to futures contracts. The Black-76 formula and the Margrabe formula are both standard results. Clear derivations of both can be found in Hull (2002).

3.2.2 Solutions for Nonzero Strike

The next level of complexity for exchange options is the option to exchange with a fixed strike. The simplification we used to produce the Margrabe formula (24)–(26) does not work for this case. In terms of the dimensional analysis of Rosenberg et al., this simplification occurred because all three of the variables or parameters in the problem that had dimensions of money were explicitly present in the differential equation. These three variables could combine to produce only two dimensionless variables, and so rewriting the differential equation in terms of the two dimensionless variables reduced the number of variables in the equation. In an option with a fixed strike, there exists another parameter with the dimensions of money, *viz.*, the fixed strike. This parameter occurs in the payout, and hence the boundary condition for the differential equation, but not in the differential equation itself, so that dimensional analysis cannot be used to reduce the number of variables in the differential equation. The value of a spread option with a nonzero strike can be solved exactly; however, the expressions are more complex than the Margrabe formula that gives the solution with a zero strike. Specifically, the value of the option can be expressed as a double integral. This solution is derived explicitly in Rosenberg et al. (2002b) and in Carmona and Durrleman (2003). (Mathematically oriented readers can just note that (12) can be turned into a linear second order differential equation with constant coefficients with the substitution $x_i = \log(F_i/K)$. Such equation is amenable to a solution with the method of Fourier transforms.) We just give the results here. The value of the spread option with the payoff of (13) at time T is given by the formula

$$V(F_1, F_2, \tau) = \frac{e^{-r\tau}}{2\pi\sigma_1\sigma_2\tau\sqrt{1-\rho^2}} \int_0^\infty \int_0^\infty \max\left[a_1 F_1' - a_2 F_2' - K, 0\right]$$

$$\times \exp\left[-\frac{y_1^2}{2\sigma_1^2\tau\left(1-\rho^2\right)} - \frac{\rho y_1 y_2}{\sigma_1\sigma_2\tau\left(1-\rho^2\right)} - \frac{y_2^2}{2\sigma_2^2\tau\left(1-\rho^2\right)}\right] \frac{dF_1'}{F_1'} \frac{dF_2'}{F_2'} \qquad (27)$$

where we have retained the notation $\tau = T - t$ and used the definitions

$$y_i = \log\left(\frac{F_i}{F_i'}\right) - \frac{\sigma_i^2\tau}{2} \qquad (28)$$

of y_1 and y_2 to keep the expression (27) from becoming unwieldy. Equation (27) is the desired *exact* expression for the value of the option to exchange with a fixed strike. It is in the form of a double integral that must be evaluated numerically; however, this numerical evaluation is straightforward, as we discuss below. On a practical note, the double integral in (27) for the option value can be reduced to a set of single integrals of a normal probability density times the cumulative normal distribution of a complicated function. This reduction makes the numerical evaluation of the value of the option much faster, easier, more accurate, and more reliable. The reduction involves completing the square in the exponential function several times.

The ability to express the value of an option as an integral, such as (27), represents a significant step towards valuing and hedging the option. Although integral representations do not have the ease of use of closed form solutions, they are significantly easier to use and work with than binomial trees, finite difference solutions, or Monte Carlo solutions. When closed form solutions cannot be obtained, integral representations, when they can be obtained, have three main advantages over the other methods. First, because an integral representation is still, in essence, an analytical expression for the value of the option, it retains many of the advantages of closed form solutions. For example, the standard procedures of differential calculus can be used to calculate integral representations of the Greeks. Second, there is a large literature on computing integrals numerically and many easy to implement, powerful, and well-understood algorithms are available. Almost every elementary numerical analysis text has a section on numerical integration, such as, to take a very arbitrary and limited selection, the books of Dahlquist and Bjork (1974), Acton (1970), and

Stoer and Bulirsch (1980). The book of Judd (1998) is oriented toward applications to economics, and the truly amazing Press et al. (1992) includes a working computer code. One way to think about numerical integration is that it is the simplest differential equation, and so it is reasonable that the algorithms for solving it should be simpler and more powerful than the algorithms for more complex differential equations. Binomial trees and finite difference methods are essentially methods for solving these more general kinds of differential equations. Third, it is relatively easy to write an algorithm that evaluates an integral to a given, predetermined accuracy. The basic idea is that one evaluates the integral numerically using a certain number of points, checks for numerical convergence, and if the convergence is not reached, one just evaluates the integral with more points. A good, practical discussion of this procedure is given by Press et al. (1992), who give both a clear mathematical explanation and a computer code.

3.2.3 Monte Carlo

One application of Monte Carlo simulation is its use as a mathematical technique for numerically solving stochastic differential equations. In the financial context Monte Carlo simulation is used for pricing options on multiple assets, such as spread options. Below we outline the steps for the use of Monte Carlo simulation to price such options:

1. Generation of large number of scenarios at the expiration date for the prices of two underlying assets. This has to be done in a manner that is consistent with the option volatilities, current futures/forwards prices, and a correlation between the prices (or returns). The joint distribution of asset prices at any time in the future can be fully described by their initial price levels, correlation, and volatilities, under the assumption that they follow geometric Brownian motion processes.
2. Evaluation of option payoff at expiration for every simulated scenario. Each scenario consists of a possible randomly drawn pair of assets' prices.
3. Calculation of the fair value or expected, discounted option payoff by first taking the average of all future payoffs, calculated at the previous step, and then discounting it to the valuation date.

Monte Carlo is a powerful and flexible method for solving problems in option valuation, hedging, and risk management. It is frequently discussed in general books on option modeling, such as Clewlow and Strickland (1998), Hull (2002), and Tavella (2002). Two books devoted solely to Monte Carlo methods in finance are Glasserman (2004), which has thorough and clear discussions of basic issues in using the Monte Carlo method in finance, and Jackel (2002), which provides good insights of more specialized (but useful) topics such as low-discrepancy sequences. Books on numerical solution of stochastic differential equations, such as Kloeden and Platen (1994), also provide information that is useful for financial applications of Monte Carlo simulations.

The spread option is an option on the difference between two commodity prices F_1, F_2, which follow geometric Brownian motion, described by the system of

(9) and (10) with instantaneous correlation ρ in (11). To price this option via Monte Carlo simulation, we need to derive a discrete time version of the system of (9)–(11) under risk-neutral measure, that is, $\mu_1(F_1, F_2, t) = \mu_2(F_1, F_2, t) = 0$:[1]

$$\begin{cases} dF_1(t,T) = \sigma_1(t,T)F_1(t,T)dz_1(t) \\ dF_2(t,T) = \sigma_2(t,T)F_2(t,T)dz_2(t) \end{cases} \tag{29}$$

The following steps detail a discretization scheme for (29).

Application of Ito's Lemma to the logarithms of prices in (29) gives

$$\begin{cases} d\ln F_1(t,T) = -\dfrac{\sigma_1^2(t,T)dt}{2} + \sigma_1(t,T)dz_1(t) \\ d\ln F_2(t,T) = -\dfrac{\sigma_2^2(t,T)dt}{2} + \sigma_2(t,T)dz_2(t) \end{cases} \tag{30}$$

where $F_1(T)$ and $F_2(T)$ are the assets' futures prices for delivery time T. Random variables z_1 and z_2 come from the standard bivariate normal distribution with correlation ρ.

Simulation of (30) requires the following discretizations:

$$\begin{cases} \ln F(t+\Delta t) - \ln F(t) = -\dfrac{1}{2}\sigma_1^2\Delta t + \sigma_1\Delta z_1 \\ \ln F(t+\Delta t) - \ln F(t) = -\dfrac{1}{2}\sigma_2^2\Delta t + \sigma_2\Delta z_2 \end{cases} \tag{31}$$

Integrating (31) from 0 to T and using the definition of implied volatility,

$$\sigma_{imp}^2(0,T) = \frac{1}{T}\int_0^T \sigma^2(\tau,T)d\tau, \tag{32}$$

We finally obtain

$$\begin{cases} F_{1,T} = F_1 e^{-.5\sigma_{imp\,1}^2 T + \sigma_{imp\,1} z_1 \sqrt{T}} \\ F_{2,T} = F_2 e^{-.5\sigma_{imp\,2}^2 T + \sigma_{imp\,2} z_2 \sqrt{T}} \end{cases} \tag{33}$$

Correlated standard normal random variables z_1 and z_2 are derived by simulating and combining independent standard normal variables ε_1 and ε_2 as follows:

$$\begin{aligned} z_1 &= \varepsilon_1 \\ z_2 &= \rho\varepsilon_1 + \sqrt{1-\rho^2}\varepsilon_2 \end{aligned}$$

(This transformation is a special case of the Cholesky decomposition method for obtaining correlated standard normal random variables. This method is discussed

[1] Discussion of various discretizing choices can be found in Kloeden and Platen (1994).

in most of the Monte Carlo references given at the beginning of this section). Note that the system of (33) allows us to simulate the prices at expiry without simulating the entire price path as long as we observe or can compute implied volatilities from now until the exercise. Finally, for each pair of simulated prices we evaluate the option's payoff and discount it to the present, that is, evaluate the following expression: $e^{-rT}\max(F_{2,T} - F_{1,T} - X, 0)$. The average of these numbers will be the fair value of the option V. Specifically,

$$V = e^{-rT}\frac{1}{N}\sum_{i=1}^{N}\max(F_{2,T} - F_{1,T} - X, 0)$$

Note that this solution yields an approximate price. By increasing the number of scenarios N, the accuracy of the result could be improved. The main disadvantage of simple Monte Carlo simulation is that its accuracy increases only as the square root of the number of simulations.

There exist many techniques (like variance reduction and quasi-random techniques) that make it possible to improve accuracy and decrease computation time at the same time. Press et al. (1992) contains a good overview of both methods.

3.2.4 Binomial Spread Options with a Strike

Another method for pricing spread options is the multidimensional binomial method. Kamrad and Ritchken (1991) were the first to suggest it. The presentation in this section applies the key elements of this method and considerations of Clewlow and Strickland (1998) for stocks to futures prices. The binomial model assumes that the asset price follows a binomial process, that is, at any step it can either go up or down with a given probability. Thus it assumes that the asset price has a binomial distribution. On the other hand, many financial models assume that the distribution of asset's returns is normal. The binomial option pricing models make use of the fact that as the number of observations/trials in the binomial distribution increases, the assets' returns approach normal distribution.

There are many ways one can build a price evolution tree that preserves the distribution properties of the assets. Computationally, an addition operation is more efficient than multiplication. In what follows we present a widely used additive binomial tree method. Since the price process represented as an additive process in logarithms is

$$\log\, F_i(t + \Delta t) = \log\, F_i(t) + \xi_t,$$

the tree is built for the logarithms of the assets' prices ($x_i = \log(F_i)$).

Since the risk neutral process for x_i is $dx_i = -\frac{1}{2}\sigma_i^2 dt + \sigma_i dz_i$ (cf. 30), the discrete version of the previous equation will be $\Delta x_i = -\frac{1}{2}\sigma_i^2 \Delta t + \sigma_i\sqrt{\Delta t}$ (cf. 31). The discrete time binomial model for x_1 and x_2 is illustrated in Table 1 and Fig. 2. Assuming equal up and down jumps, each of the variables can either go up by Δx_i or down by $-\Delta x_i$. At any time step, the probability that logarithms of both prices go up is p_{uu} and down is p_{dd}; first asset goes up, but second down is p_{ud}, second

Table 1 The joint normal random variable (**X1 (t), X2 (t)**) is approximated by a pair of multinomial discrete normal variables having the following distribution

X1 (t)	X2 (t)	Probability
Δx_1	Δx_2	p_{uu}
Δx_1	$-\Delta x_2$	p_{ud}
$-\Delta x_1$	Δx_2	p_{du}
$-\Delta x_1$	$-\Delta x_2$	p_{dd}

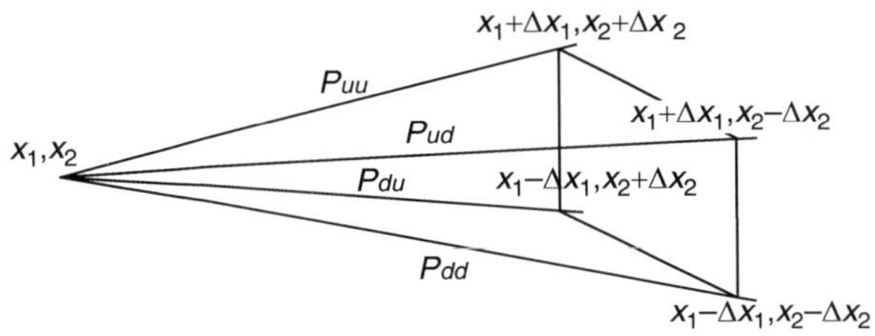

Fig. 2 One branch of the additive two-variable binomial tree for logarithms of assets' prices

goes up, but first goes down is p_{du}. The sum of these probabilities is one, as they contain all four possible outcomes of price movements.

The next step is to establish jump sizes (Δx_1 and Δx_2) and four probabilities so that they match mean, variance, and correlation of the bivariate normal distribution of logarithms of assets' prices. For that we need to solve the system of equations that matches first and second moments of binomial and analytical representations. It is an algebraic system of six equations with six unknowns:

$$\begin{cases} \mathrm{E}(\Delta x_1) = \Delta x_1(p_{uu} - p_{du} + p_{ud} - p_{dd}) = -\frac{1}{2}\sigma_1^2 \Delta t \\ \mathrm{E}(\Delta x_1) = \Delta x_1^2(p_{uu} + p_{du} + p_{ud} + p_{dd}) = \sigma_1^2 \Delta t \\ \mathrm{E}(\Delta x_2) = \Delta x_2(p_{uu} + p_{du} - p_{ud} - p_{dd}) = -\frac{1}{2}\sigma_2^2 \Delta t \\ \mathrm{E}(\Delta x_2) = \Delta x_2^2(p_{uu} + p_{du} + p_{ud} + p_{dd}) = \sigma_2^2 \Delta t \\ \mathrm{E}(\Delta x_1, \Delta x_2) = \Delta x_1 \Delta x_2(p_{uu} - p_{du} - p_{ud} + p_{dd}) = \rho\sigma_1\sigma_2\Delta t \\ p_{uu} + p_{du} + p_{ud} + p_{dd} = 1 \end{cases} \tag{34}$$

The solution to this system of equations for a given time step and given changes in logarithms of both assets' futures prices is

$$\begin{cases} \Delta x_1 = \sigma_1\sqrt{\Delta t} \\ \Delta x_2 = \sigma_2\sqrt{\Delta t} \\ p_{uu} = \dfrac{\Delta x_1 \Delta x_2 - 0.5\Delta x_2 \sigma_1^2 \Delta t - 0.5\Delta x_1 \sigma_2^2 \Delta t + \rho\sigma_1\sigma_2\Delta t\sigma_2^2}{4\Delta x_1 \Delta x_2} \\ p_{du} = \dfrac{\Delta x_1 \Delta x_2 + 0.5\Delta x_2 \sigma_1^2 \Delta t - 0.5\Delta x_1 \sigma_2^2 \Delta t - \rho\sigma_1\sigma_2\Delta t\sigma_2^2}{4\Delta x_1 \Delta x_2} \\ p_{ud} = \dfrac{\Delta x_1 \Delta x_2 - 0.5\Delta x_2 \sigma_1^2 \Delta t + 0.5\Delta x_1 \sigma_2^2 \Delta t - \rho\sigma_1\sigma_2\Delta t\sigma_2^2}{4\Delta x_1 \Delta x_2} \\ p_{dd} = \dfrac{\Delta x_1 \Delta x_2 + 0.5\Delta x_2 \sigma_1^2 \Delta t + 0.5\Delta x_1 \sigma_2^2 \Delta t + \rho\sigma_1\sigma_2\Delta t\sigma_2^2}{4\Delta x_1 \Delta x_2} \end{cases} \tag{35}$$

Now we have all the information necessary to build a two-dimensional tree and calculate the value of the option. The algorithm proceeds as follows:

A user sets up the granularity of the tree N, that is, the number of steps between valuation and expiration dates of the option. The larger this number is, the more accurate the solution will be. Thus, there will be N branching processes, at every time step $\Delta t = T/N$, where T is the time to expiry for a given option. Triplex (i, j, k) refers to the nodes on the tree, where i is a time step, j is the first asset's price level, k is the second assets price level. Since the tree is built for logarithms of assets' prices (x_1, x_2), we convert them to assets' prices at each node of the tree via these simple transformations:

$$\begin{cases} F_{1,i,j,k} = F_1 e^{(2j-i)\Delta x_1} \\ F_{2,i,j,k} = F_2 e^{(2k-i)\Delta x_2} \end{cases}$$

Separately, each asset in the binomial tree follows a one-dimensional process that can be approximated by the one-dimension tree (see Fig. 3). At the same time, their joint distribution characteristic – instantaneous correlation – reveals itself in their joint probabilities of (35).

We initialize asset prices at expiration time (nodes (N, j, k) of the two-variable binomial tree or (N, j) and (N, k) of the individual assets' one-dimensional trees). At this step we estimate option values at expiration, utilizing asset prices computed at the previous step. For all j and k of the final nodes (N, j, k) of the two-variable binomial tree, the option price $C(N, j, k)$ at expiration is $\max(F_2[k]-F_1[j]-X, 0)$. Next we start to iterate option values $C(i, j, k)$ by stepping back through the tree from the time node (N-1) to 0. Option value at each time step i and price levels (j, k) is a discounted, weighted (by probabilities) average of the four option values at the nodes that are connected to a current node.

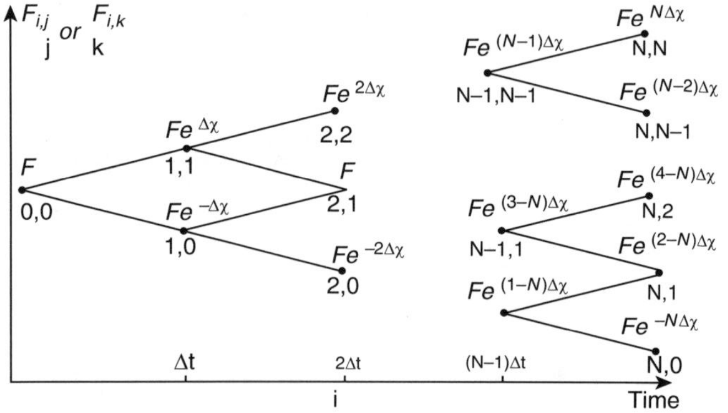

Fig. 3 Individual asset's prices at time node i and price level j for the first asset or k for the second

Specifically,

$$C(i,j,k) = e^{-r\Delta t}\begin{pmatrix} p_{uu}C(i+1,j+1,k+1) + p_{du}C(i+1,j-1,k+1) + \\ + p_{ud}C(i+1,j+1,k-1) + p_{dd}C(i+1,j-1,k-1) \end{pmatrix}$$

Fair value of the spread option at valuation date is given by the option value at the very first node of the tree, that is, $C(0,0,0)$.

The above-described algorithm can be easily modified to accommodate the American exercise. American options can be exercised at any time during their life. Thus at every time step, a holder of the option has a choice to hold it to expiry (European feature) or to exercise it and receive the option payoff, which is $\max(F_2[k] - F_1[j] - X, 0)$ at every time node i (American feature). The last feature can easily be incorporated into our algorithm at the iteration stage 4. Specifically, the American option price at time node i and price levels j and k becomes

$$C(i,j,k) = \max\left\{ e^{-r\Delta t}\begin{pmatrix} p_{uu}C(i+1,j+1,k+1) + p_{du}C(i+1,j-1,k+1) + \\ + p_{ud}C(i+1,j+1,k-1) + p_{dd}C(i+1,j-1,k-1) \end{pmatrix}, \right.$$
$$\left. F_2[k] - F_1[j] - X \right\}$$

All other steps remain unaffected. One can also improve the efficiency of the program by premultiplying probabilities by the one step discount factor $e^{-r\Delta t}$.

4 Valuation in the Presence of Spikes

The presence of spikes creates a new complexity in the valuation of obligations and options. In theory, the Black–Scholes analysis assumes continuous dynamic replication of a portfolio. In practice, however, any such rebalancing incurs transaction costs. These costs force hedging to become discrete and hence imperfect. Therefore, in practice all portfolios bear some risk. This situation is exacerbated during the power price spikes when even approximate dynamic replication becomes impossible. The implications of this difficulty are discussed, in the context of jump rather than spike processes, in the original paper of Merton (1976) on valuing options, where the asset returns have jumps, and in the excellent textbook of Shreve (2004). This difficulty further limits the applications of risk-neutrality assumption for option valuation. In this section, we reduce the use of risk-neutral valuation techniques to the net present value (NPV) analysis. The reason for this is twofold: to illustrate the effects of spikes in the simplest possible way and to provide a base from which readers can develop more sophisticated and realistic valuation techniques.

Now, let us account for the possibility of *spikes*. Analysis of commodity prices S_1, S_2 and futures prices F_1, F_2 at both locations relies on a *multistate* model introduced earlier in the paper. In particular, transitions between the states (spikes and interspike segments) are governed by the transition probability matrix $p = p(t)$

specified in (8). The process appears in one state or the other with probabilities computed from this matrix. For a day that is T days away from the current day t, the matrix of transition probabilities is computed as

$$P^{(t,T)} = P(t+T|t) = \prod_{k=0}^{T-1} P(t+k) = P(t)P(t+1)\ldots P(t+T-1).$$

Element $P_{i.i}^{(t,T)}$ of this 2-by-2 matrix equals the conditional probability of state j, T days ahead, given the current state i (i,j = 0,1). For example, if there is a spike on day t, then the model predicts a spike on day (t + T) with probability $P_{1.1}^{(t,T)}$ and no spike on that day with probability $P_{1.0}^{(t,T)}$. Consequently, the density of prices on that day is a *mixture*

$$f_{t,T}(x) = P_{i,0}^{(t,T)} f_0(x) + P_{i,1}^{(t,T)} f_1(x), \tag{36}$$

where densities $f_0(x)$ and $f_1(x)$ are given by (5) and (7). Expectation of any function h with respect to this mixture density is the weighted average

$$E\{h(X)\} = \int h(x) f_{t,T}(x)dx = P_{i,0}^{(t,T)} \int h(x) f_0(x)dx + P_{i,1}^{(t,T)} \int h(x) f_1(x)dx. \tag{37}$$

In our context, $X = (F_1, F_2)$, and the role of $E_T\{h(X)\}$ is played by the value $V(F_1, F_2, T) = E_T\{h(F_1, F_2)\}$ of a financial instrument, which is based on both F_1 and F_2 (spread option, transportation option) at the expiry date T. Therefore,

$$V_T(F_1, F_2, t) = P_{i,0}^{(t,T)} V_T^{(0)}(F_1, F_2, t) + P_{i,1}^{(t,T)} V_T^{(1)}(F_1, F_2, t),$$

where $V_T^{(0)}(F_1, F_2, t)$ and $V_T^{(1)}(F_1, F_2, t)$ are values of the financial instruments computed for each mode (regular mode and spikes), respectively.

The joint distribution of commodity prices during spikes at two locations, to be used in (36), is *bivariate lognormal*, with correlation coefficient ρ between the two locations. To evaluate the strength of the spike effect, we fitted the introduced multistate model to a 5-year long sequence of day-ahead electricity prices in ERCOT North and a 3-year long sequence in ERCOT South. ERCOT is the Independent System Operator (ISO) for the State of Texas.

Estimation of the log-price mean value $m(t)$ includes the general linear trend, weekday effects, and seasonal components, showing the long-term growth of prices (at the average rate of 13% per year, according to the ERCOT data), variation during a week, and variation during a year (Fig. 4).

Following the model developed for the interspike distribution in (5), detrended log-prices $X_{j,t} = S_j(t) - m_j(t)$ for $j = 1, 2$ at two locations are correlated AR(1) processes

$$X_{1t} = \phi_1 X_{1,t-1} + Z_{1t} \; and \; X_{2t} = \phi_2 X_{2,t-1} + Z_{2t},$$

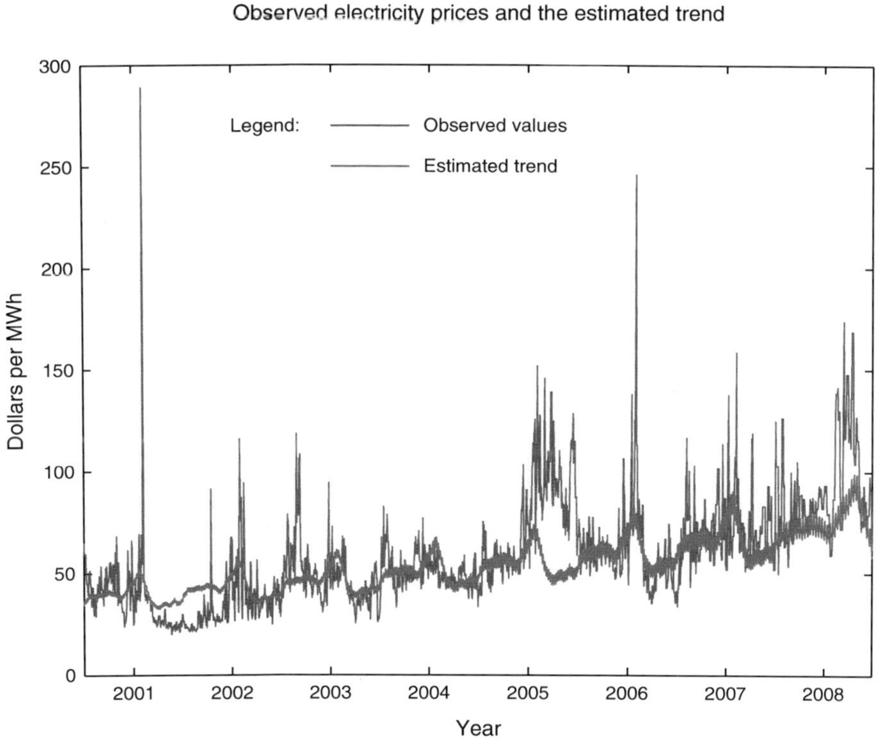

Fig. 4 Electricity prices in ERCOT North and their estimated trend Source: *ICE Trade the World* site (https://www.theice.com)

where (Z_{1t}, Z_{2t}) is a bivariate white noise, $Var(Z_{1t})\sigma_1^2$, $Var(Z_{2t})\sigma_2^2$, and $Cov(Z_{1t}, Z_{2t}) = \rho\sigma_1\sigma_2$. Parameters σ_i, ϕ_i of each process X_{it}, $j = 1, 2$ are estimated by standard methods for time series analysis (see Brockwell and Davis (1991, 2009). To estimate the correlation coefficient ρ, we notice that

$$Cov(X_{1t}, X_{2t}) = Cov\left(\sum_{n=0}^{\infty} \phi_1^n Z_{1t-n}, \sum_{n=0}^{\infty} \phi_2^n Z_{2t-n}\right) = \frac{\rho\sigma_1\sigma_2}{1-\phi_1\phi_2},$$

from where the estimator of ρ based on n days of data is

$$\hat{\rho} = \frac{(1-\hat{\phi}_1)(1-\hat{\phi}_2)\sum_{t=1}^{n}(X_{1t}-\bar{X}_1)(X_{2t}-\bar{X}_2)/n}{\hat{\sigma}_1\hat{\sigma}_2}.$$

For ERCOT North and ERCOT South, the time series parameters are estimated as $\hat{\phi}_1 = 0.9307$, $\hat{\phi}_2 = 0.9592$, $\hat{\sigma}_1 = 0.0742$, $\hat{\sigma}_2 = 0.0729$, and the correlation coefficient is $\hat{\rho} = 0.8211$.

During each spike, the mean vector (μ_1, μ_2) is generated from a bivariate normal distribution with (hyper-) parameters $(\theta_1, \theta_2, \omega_1, \omega_2, \rho_\mu)$, where ρ_μ is the correlation coefficient between the spike means μ_1 and μ_2. Conditioned on (μ_1, μ_2), the vector of detrended log-prices for each spike for the two regions has a bivariate normal distribution with parameters $(\mu_1, \mu_2, \tau_1, \tau_2, \rho^{sp})$. Unconditionally, we have

$$\begin{aligned} Cov(X_{1t}, X_{2t}) &= E\{Cov(X_{1t}, X_{2t}|\mu_1, \mu_2)\} + Cov\{E(X_{1t}|\mu_1), E(X_{2t}|\mu_2)\} \\ &= \rho^{sp}\omega_1\omega_2 + \rho_\mu \tau_1 \tau_2. \end{aligned}$$

Then, the spike-mode distribution of detrended log-prices at two locations is bivariate normal with the unconditional correlation coefficient $\frac{\rho^{sp}\omega_1\omega_2+\rho_\mu\tau_1\tau_2}{\sqrt{(\tau_1^2+\omega_1^2)(\tau_2^2+\omega_2^2)}}$.

With all the parameters estimated, we obtain the joint density (36) of electricity prices for the two regions for any given day. This allows us to make forecasts in the form of a predictive joint density as well as the marginal densities at each location. A 3-year forecast of densities of prices for each day of the year 2011 is given on Fig. 5.

For each day of the year, the cross-section of this graph gives a probability density of electricity prices. During the peak season, the probability of low prices reduces, and a heavy tail of high prices appears. It accounts for the possibility of spikes.

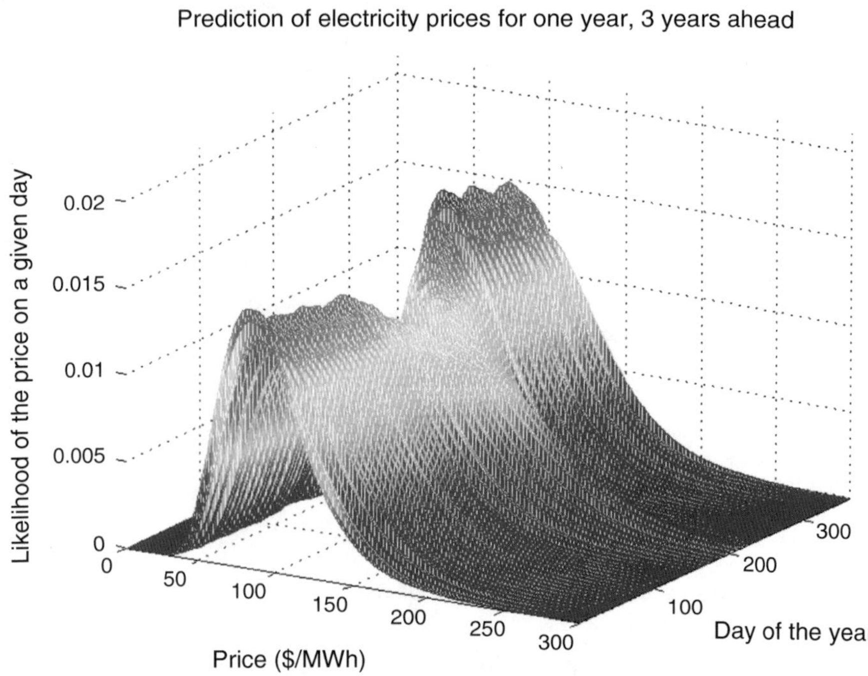

Fig. 5 Predictive densities of electricity prices for each day of the year, 3 years ahead

In Rosenberg et al. (2002a), the authors estimated the effect of accounting for spikes in the estimation of expected value of the spread option payoff. On the example of PJM West–PJM East spread, the use of a multistate model with spikes yielded additional 7% value as compared with the model that did not use spikes and estimated a single lognormal distribution for price returns. This result demonstrated the importance of accounting for spikes in the probability density when computing the values of spread-type instruments.

5 Conclusions

This paper described physical commodity transportation and transmission instruments that have one thing in common: they depend on the spread of locational or cross-commodity prices and thus can be modeled by a family of spread options. As this dependency suggests, the value of the instrument in any future moment depends not only on the level of prices but also on their future distribution function. The inclusion of spikes changes this distribution and hence produces a different value for these assets. The correct calibration and inclusion of these sudden and big deviations from the normal level of prices is very important for correct valuations and mark-to-market procedures. There are a number of methods that may be utilized to solve for the value of assets once the price process is established. They range from approximate solutions under some simplifying assumptions that are quite easy to compute to the exact solution that is reducible to single integrals. Numerical methods include Monte-Carlo and Binomial method and have a different degree of robustness. Monte-Carlo always converges and requires a simple algorithm, but may be slow. Binomial approach requires more analytics and programming, but yields payoffs in the speed of the computation.

References

Acton FS (1970) Numerical methods that work. Harper and Row, New York

Alexander C, Scourse A (2004) Bivariate normal mixture spread option valuation. Quant Finance 4:637–648

Barenblatt GI (1996) Scaling, self-similarity, and intermediate asymptotics. Cambridge University Press, Cambridge

Baronblatt GI (2003) Scaling. Cambridge University Press, Cambridge

Baron MI (2002) Bayes and asymptotically pointwise optimal stopping rules for the detection of influenza epidemics. In: Gatsonis C, Kass RE, Carriquiry A, Gelman A, Higdon D, Pauler DK and Verdinelli I (eds) Case studies of bayesian statistics, vol. 6. Springer, New York

Baron M (2004) Sequential methods for multistate processes. In: Mukhopadhyay N, Datta S and Stattopadhyay S (eds). Application of sequential methodologies. Marcel Dekker, New York

Baron M, Rosenberg M, Sidorenko N (2001) Modelling and prediction with automatic spike detection. Energy Power Risk Manage 36–9

Baron M, Rosenberg M, Sidorenko N (2002) Divide and conquer: forecasting power via automatic price regime separation. Energy Power Risk Manage 70–3

Basseville M, Nikiforov IV (1993) Detection of abrupt changes: theory and application. PTR Prentice-Hall, Englewood Cliffs

Bjork T (2004) Arbitrage theory in continuous time, 2nd edn. Oxford University Press, Oxford, UK

Black F (1976) The pricing of commodity contracts. J Financ Econ 3:167–179

Bridges CB, Winquist AG, Fukuda K, Cox NJ, Singleton JA, Strikas RA (2000) Prevention and Control of Influenza, Morbidity and Mortality Weekly Reports, Centers for Disease Control and Prevention, 49:1–38

Bridgman PW (1931) Dimensional analysis. Yale University Press, New Haven

Brockwell PJ, Davis RA (2nd printing edition, 2009), (1991) Time series: theory and methods. Springer, New York

Buckingham E (1914) On physically similar systems; illustration of the use of dimension equations. Phys Rev 4:345–376

Calvin WH (2002) A brain for all seasons: human evolution and abrupt climate change. University of Chicago Press, Chicago, IL

Carlin BP, Louis TA (2008) Bayesian methods for data analysis, 3rd edn. Chapman & Hall/CRC, Boca Raton, FL

Carmona R, Durrleman V (2003) Pricing and hedging spread options. SIAM Rev 45(4):627–685

Cherubini U, Luciano E (2002) Bivariate option pricing using copulas. Appl Math Finance 9:69–86

Clewlow L, Strickland C (1998) Implementing derivatives models. Wiley, New York

Clewlow L, Strickland C (1999) Energy derivatives: pricing and risk management. Lacima

Cloth P, Backofen R (2000) Computational molecular biology. Wiley, Chichester, England

Dahlquist G, Bjork A (1974) Numerical Methods (trans: Anderson N). Prentice-Hall, Englewood Cliffs, NJ

Dixit AK, Pindyck RS (1994) Investment under uncertainty. Princeton University Press, Princeton, NJ

Durbin S, Eddy S, Krogh A, Mitchison G (1998) Biological sequence analysis. Probabilistic models of proteins and nucleic acids. Cambridge University Press, Cambridge MA

Embrechts P, McNeil A, Straumann D (1999) Correlation: pitfalls and alternatives. Risk 12:69–71

Glasserman P (2004) Monte Carlo methods in financial engineering. Springer, New York

Granott N, Parziale J (2002) Microdevelopment: A process-oriented perspective for studying development and learning. In: Granott N and Parziale J (eds) Microdevelopment: Transition processes in development and learning. Cambridge University Press, Cambridge, pp. 1–28

Gregoire V, Genst C, Gendron M (2008) Using copulas to model price dependence in energy markets. Energy Risk 5:58–64

Hull JC (2002) Options, futures, and other derivatives, 5th edn. Prentice-Hall, Upper Saddle River, NJ

Jackel P (2002) Monte Carlo methods in finance. Wiley, New York

Judd KL (1998) Numerical methods in economics. MIT, Cambridge, MA

Kamrad B, Ritchken P (1991) Multinomial approximating models for options with k state variables. Manage Sci 37(12):1640–1652

Khan R (2009) Distributional properties of CUSUM stopping times and stopped processes. Sequential Anal 28(2):175–187

Kloeden PE, Platen E (1994) Numerical solution of stochastic differential equations through computer experiments. Springer, New York

Lai TL (1995) Sequential changepoint detection in quality control and dynamical systems. J R Stat Soc B 57:613–658

Lai TL (2001) Sequential analysis: some classical problems and new challenges. Stat Sin 11: 303–408

Margrabe W (1978) The value of an option to exchange one asset for another. J Finance 33: 177–186

Merton RC (1976) Option pricing when underlying stock returns are discontinuous. J Financ Econ 3:125–144, reprinted in Merton RC, Continuous Time Finance, Blackwell, Oxford, 1990

Montgomery DC (1997) Introduction to statistical quality control, 3rd edn. Wiley, New York
Piaget J (1970) Piaget's theory, In: Mussen PH (ed) Carmichael's manual of child psychology. Wiley, New York, pp. 703–732
Press WH, Teukolsky SA, Vetterling WT, Flannery BP (1992) Numerical recipes in C, 2nd edn. Cambridge University Press, Cambridge
Papalexopoulos A, Treinen R (2005) Important practical considerations in designing an FTR market. Presented at the IEEE PES Meeting, St. Petersburg, Russia
Rabiner LR (1989) A tutorial on Hidden Markov Models and selected applications in speech recognition. IEEE Proc 77:257–285
Ronn EI (2002) Introduction. In: Ronn EI (ed) Real options and energy management: using options methodology to enhance capital budgeting decisions. Risk Books, London
Rosenberg M, Bryngelson JD, Sidorenko N, Baron M (2002a) Price spikes and real options: transmission valuation, In: Ronn EI (ed) Real options and energy management: using options methodology to enhance capital budgeting decisions. Risk Books, London
Rosenberg M, Bryngelson JD, Baron M (2002b) Probability and stochastic calculus: review of probability concepts, In: Ronn EI (ed) Real options and energy management: using options methodology to enhance capital budgeting decisions. Risk Books, London
Spokoiny V (2009) Multiscale local change point detection with applications to value-at-risk. Ann Stat 37(3):1405–1436
Shreve S (2004) Stochastic calculus for finance II: continuous time models. Springer, New York
Stoer J, Bulirsch R (1980) Introduction to numerical analysis. Springer, Berlin
Tavella D (2002) Quantitative methods in derivative pricing. Wiley, New York
Wilson EB (1990) An introduction to scientific research. Dover, New York
Zacks S (2009) Stage-wise adaptive designs. Wiley, Hoboken, NJ

Part II
Forecasting in Energy

Short-term Forecasting in Power Systems: A Guided Tour

Antonio Muñoz, Eugenio F. Sánchez-Úbeda, Alberto Cruz, and Juan Marín

Abstract In this paper, the three main forecasting topics that are currently getting the most attention in electric power systems are addressed: load, wind power and electricity prices. Each of these time series exhibits its own stylized features and is therefore forecasted in a very different manner. The complete set of forecasting models and techniques included in this revision constitute a guided tour in power systems forecasting.

Keywords Electricity markets · Electricity price forecasting · Short term load forecasting · Time series models · Wind power forecasting

1 Introduction

Energy has always been one of the most active areas in forecasting due to its major role for the effective and economic operation of power systems. However, since the liberalization of the electricity industry in the 1990s and the expansion of renewable sources of energy at the beginning of the twenty-first century, the interest in forecasting has spread from System Operators to Electric Utilities, energy traders, independent power producers and consumers, etc.

Electricity demand forecasters are at the core of many operational processes, including power system planning, scheduling and control. In the new competitive framework, load forecasts are additionally used by market agents as a reference variable for strategic bidding and are a fundamental driver of electricity prices. Furthermore, the liberalization of the electricity sector has stimulated the research on zonal load forecasting, where not only the aggregated load of the complete power system but also the loads of predefined transmission and distribution zones needs to be predicted to properly operate the network. In addition, energy retailers and large

A. Muñoz (✉)
Instituto de Investigación Tecnológica, Escuela Técnica Superior de Ingeniería – ICAI, Universidad Pontificia Comillas, C/Alberto Aguilera 23, Madrid 28015, Spain
e-mail: Antonio.Munoz@iit.upcomillas.es

S. Rebennack et al. (eds.), *Handbook of Power Systems II*, Energy Systems,
DOI 10.1007/978-3-642-12686-4_5,

consumers need local forecasts to optimize the trading of this energy in the wholesale market. Recent advances in short-term load forecasting have been focused on modelling multiple seasonalities, the treatment of weather variables and calendar effects.

Wind power generation has become in many countries the main source of uncertainty for the operation of power systems. The impacts of increased levels of wind penetration range from power systems security assessment issues to fluctuations in electricity prices. System operators and market agents have been forced to cope with this new source of volatility in a very short time, and they have found wind power forecasting a very useful and indispensable tool for that purpose. The latest works in this area include spatial correlation, ensemble forecasting and density forecasting methods.

Since market clearing prices are obtained by crossing stepwise supply and demand curves constructed from aggregated bids, electricity prices in day-ahead markets are generally erratic and ill-behaved. In this context, electricity market agents assume a more intense risk exposure than in the traditional framework. To hedge against this risk, participants typically carry out part of their transactions through other financial markets, such as future markets or bilateral contracts. In this environment, participants need to develop efficient tools to obtain accurate energy spot price forecasts. In recent years, electricity price forecasting models are moving from univariate to multivariate, from single to multiple regime switching models and from point forecasts to interval and density forecasts.

These three forecasting topics, load, wind power and electricity prices, are covered in the next sections. The main features of actual time series are stressed, the key forecasting trends are presented and a gentle bibliography is provided for each of them.

2 Electricity Load Forecasting

Short-term load forecasting has gradually become the central piece in the electricity industry. In the traditional centralized framework, power systems were planned, designed and operated as a whole, using short-term load forecasting (STLF) mainly to ensure the reliability of supply. As electricity cannot be stored, the instantaneous generation must match the demand being taken from the system. To ensure this balance between demand and supply, as well as the security and quality of electricity supply, STLF is required. These forecasts provide the basis for generation and maintenance scheduling, and they can be used to estimate load flows to prevent the system from suffering major disturbances.

After competition has been introduced in the electricity sector, market participants need accurate load forecasts to minimize the volumetric risk associated with the energy trading process. In fact, a reduction of the forecast errors by a fraction of a percent can lead to substantial increases in trading profits. For example, according to Bunn and Farmer (1985) and Soares and Medeiros (2008), an increase of only 1%

in the forecast error (in 1984) caused an increase of 10 million pounds in operating costs per year for one electric utility in the UK.

Although STLF is synonym of hourly load data and forecasting horizons up to 24 h, there also exists literature where the sample period reduces to half-hourly data (e.g. Bunn and Farmer (1985)) or the forecasting horizon increases up to 7 days (e.g. Amaral et al. (2008)). More recently, there are a few published papers on the so-called very short-term load forecasting. For example, in Taylor (2008) methods for forecasting up to 30 min ahead are evaluated using minute-by-minute British data.

2.1 Features of Electricity Load Time Series

There exists a set of empirical findings in the electricity load series that have been systematically reported in the literature. These stylized facts of the load series are essential features of broad generalization. In fact, many of the presented characteristics are shared by most national system load in the world.

Basically, the load series displays trend with different levels of seasonality (annual, weekly, daily), short-term dynamics, calendar events dependence and non-linear effects of meteorological variables. The load trend is usually associated to economic and demographic factors, whereas the other features are related to climate variations and human behaviour.

Figure 1 provides different views of the hourly electricity load in Spain. The first panel depicts 7 years of data, illustrating the usual positive trend of an annual seasonality. Panel (b) focuses on 1 year of hourly data, where the weekly seasonal cycle is clearly visible. The calendar effects on electricity consumption become apparent, especially in summer vacations (mainly during August in Spain) and Christmas time, as well as the effect of temperature. The consumption is usually higher in winter (electric heating) and summer (air conditioning) than in autumn and spring. Finally, graphs (c) and (d) zoom in a time window of three winter and summer weeks, respectively. The systematic decrease in the load on Saturdays and Sundays, due to the different levels of activity not only in industrial and commercial sectors, but also in the behaviour of households on holidays, is also a crucial characteristic of demand series. These two plots reveal significant differences between winter and summer intra-day load patterns, especially during peak hours.

These features are further illustrated in Fig. 2, where eight daily patterns have been automatically obtained by the k-means clustering algorithm (Kaufman and Rousseeuw 1990). These mean load profiles have been classified in terms of seasons and days of the week, the main drivers of the routinary human activity. During the winter the hourly peak loads in Spain are usually in the evening, at 9PM, when lighting loads are at maximum. However, during the summer, the peak loads in working days are at 12 AM.

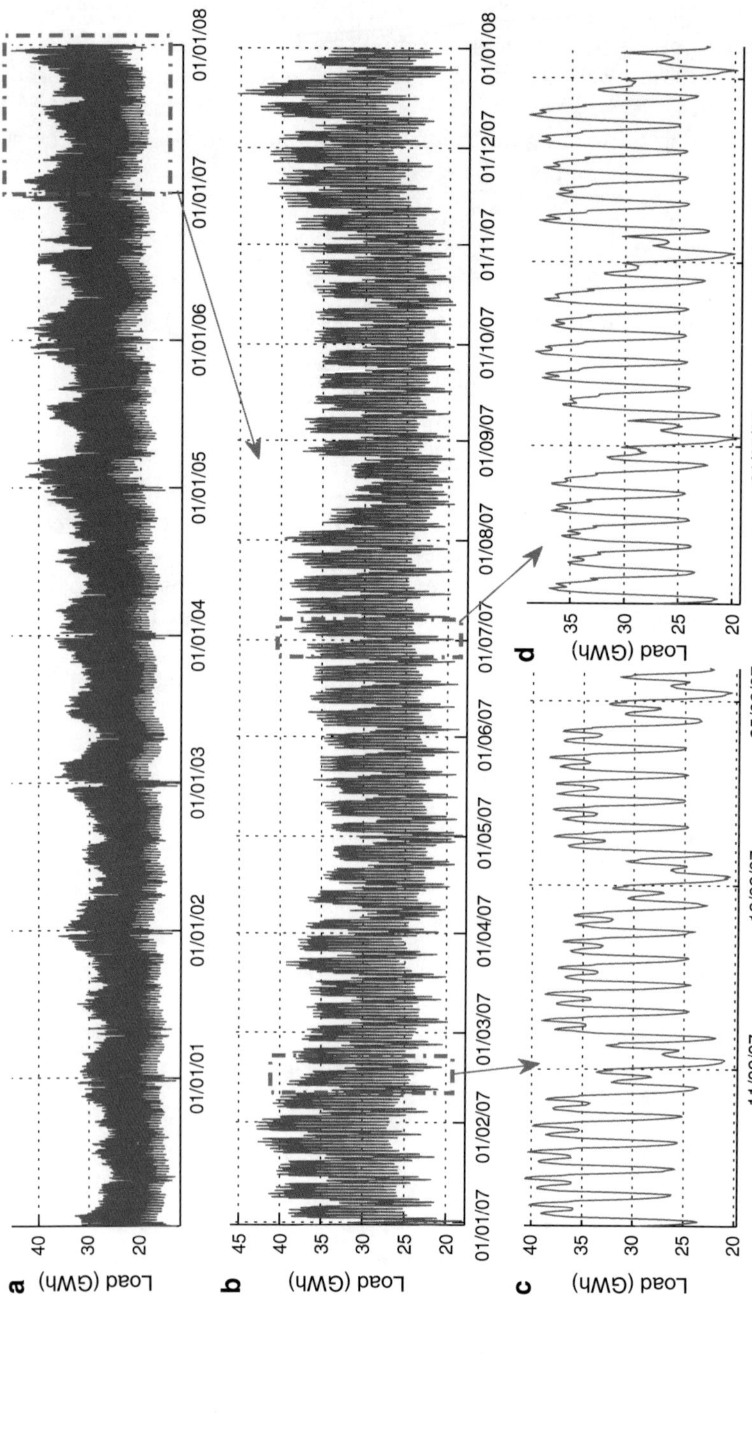

Fig. 1 Hourly electricity demand in Spain: (**a**) From 1 January 2000 to 31 December 2007; (**b**) From 1 January 2007 to 31 December 2007; (**c**) Three winter weeks (2007); (**d**) Three summer weeks (2007)

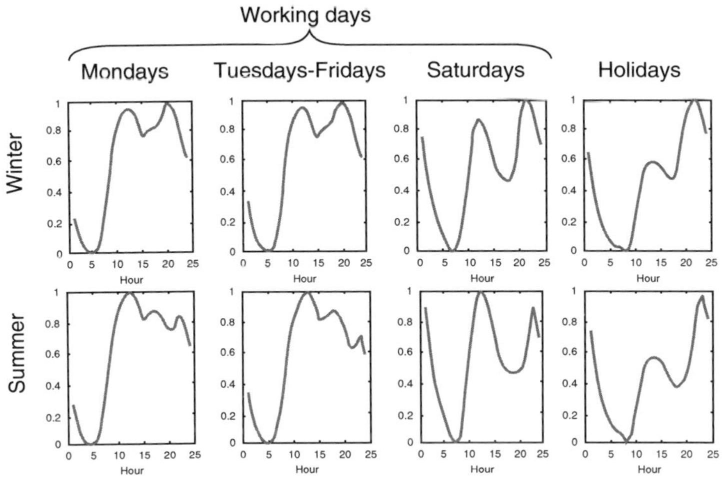

Fig. 2 Normalized intra-day load profiles for the Spanish electricity load

2.2 *Modelling Electricity Load Features*

Before reviewing the forecasting approaches proposed in the literature, we would like to mention several important issues related to the main features of electricity load series. Note that many of these characteristics are strongly linked to physical factors, motivating STLF approaches based on well-known relationships between load and exogenous variables.

2.2.1 Dealing with Trend and Annual Seasonality

Several approaches have been proposed in the literature to take the load trend into account, although most papers in the STLF literature consider taking first-order differences of the load series to handle the load trend. On the other hand, most of the annual fluctuations exhibited by the electricity load are mainly governed by climate conditions, such as the outdoor temperature or the number of daylight hours. Although these systematic fluctuations exist in the load series, in STLF the forecast lead times is substantially shorter than the length of the annual cycle; therefore, it could be surmised that methods that make no attempt to model the annual seasonality may be adequate. For that reason, most of the proposed models for STLF ignore the existing annual cycle, focusing on the intra-week and intra-day cycles (e.g. Weron (2006) and Taylor (2008)).

Alternatively, some authors have proposed recently the explicit modelling of the trend and the annual seasonality. For example, in Soares and Medeiros (2008) the trend in the load is modelled as a deterministic function of the gross domestic product, whereas in Dordonnat et al. (2008) local linear trends are estimated for each hour of the day. Concerning the annual cycle, in Dordonnat et al. (2008) and Soares and Medeiros (2008), it is modelled as a combination of sines and cosines, as in a Fourier decomposition.

2.2.2 Dealing with Intra-week and Intra-day Patterns

When modelling an hourly (or half-hourly) load series, the systematic shape of the load curve for each day together with the existing within-week seasonal cycle are two relevant characteristics that need to be modelled appropriately. Note that this intra-day shape smoothly changes between the seasons and between weekdays and weekends (Fig. 2).

In STLF there exist two main approaches for dealing with intra-day profiles: using a single-equation model for all the hours or using multiple-equation models with different equations for the different hours of the day. The first approach allows applying models that are able to model the dynamics of both intra-week and intra-day patterns, such as the double seasonal ARIMA model or the exponential smoothing method for double seasonality (see, e.g. Taylor (2003)). Another extended approach to capture the intra-day pattern consists in treating each hour as a separate time series. In its most simple version, this approach uses 24 independent models specified over a daily time scale. More sophisticated versions include vector models where the equations for the different hours are linked. This strategy has been adopted by articles, including Ramanathan et al. (1997), Cottet and Smith (2003), Dordonnat et al. (2008) or Soares and Medeiros (2008). Note that according to Cancelo et al. (2008), although there is some controversy about the best approach, most authors prefer to model each hour as a different series.

2.2.3 Dealing with Weather Variables

It is well-known that meteorological conditions have a significant influence on electricity loads. Derived factors from weather variables such as temperature, solar radiation, humidity, wind speed, cloudiness, or rainfall have been used as exogenous variables in the literature to improve load forecasting. However, according to Weron (2006), the load prediction survey (Hippert et al. 2001a) indicated that most of the considered research publications made use of temperature (19 out of 22), but only six of them made use of additional weather parameters. A review of more recent papers confirms that the majority of authors assume that temperature is the main weather variable, ignoring the additional effects on the load of the other weather variables. Note that sometimes the reason for this simplification is the unavailability of reliable recordings. For example, in Soares and Medeiros (2008), no weather data is included as exogenous variables due to the existence of deficiencies in the temperature observations (outliers and missing values).

Focusing on the relationship between electricity load and outdoor air temperature, various authors have reported similar correlation patterns (Engle et al. 1986;

Sailor and Muñoz 1997; Valor et al. 2001; Pardo et al. 2002; Moral-Carcedo and Vicéns-Otero 2005; Cancelo et al. 2008). Although this relationship depends on the climate characteristics of the geographical area to which the load data refer, principal studies indicate that it is highly complex due to several reasons. Basically, the load increases both for decreasing and increasing temperatures, the response being asymmetric and clearly non-linear due to the use of electric heating appliances in winter and air conditioners in summer. There exist also differences between working days and holidays, which change with the time of the year (Cancelo et al. 2008). Furthermore, there is a dynamic effect due to the physical inertia of buildings, as well as saturation effects because of the limited capacity of the installed heating and cooling appliances.

Figure 3 shows the relationship between the filtered electricity load[1] of a distribution area in the north of Spain and an average daily mean temperature of this

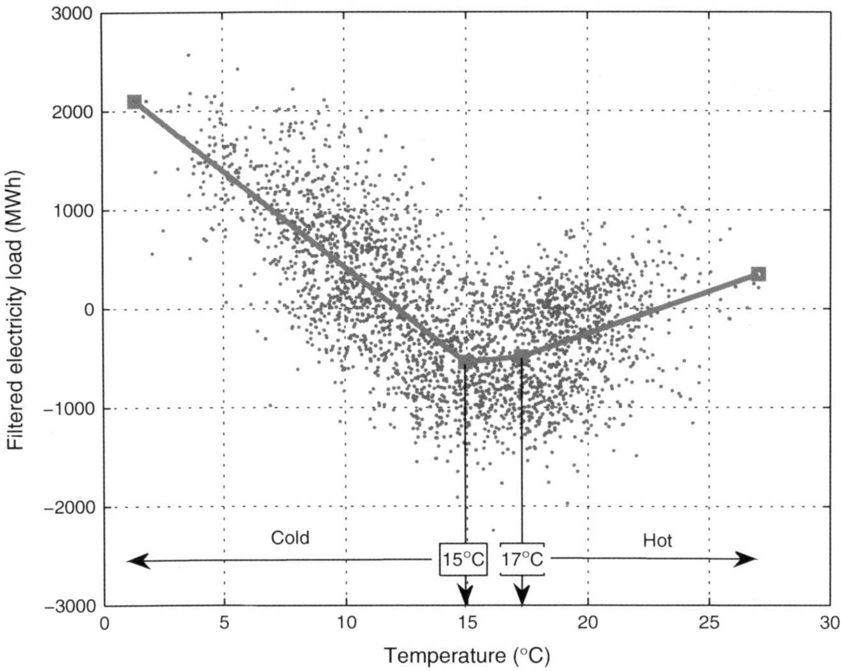

Fig. 3 Non-linear relationship between the (filtered) electricity load of a distribution area of the north of Spain and the average daily mean temperature of this region

[1] To show the response of load to temperature variations, the electricity load has been previously filtered out by removing both the trend and the calendar effects, see Moral-Carcedo and Vicéns-Otero (2005) for further details.

region.[2] To highlight this relationship, a Linear Hinges Model has been adjusted (Sánchez-Úbeda and Wehenkel 1998). This piecewise linear model summarizes the non-linear response of electricity load to temperature by automatically segmenting the temperature axis into three regions: cold (for temperatures lower than 15°C), neutral (between 15 and 17°C) and hot (for temperatures higher than 17°C).

The non-linear relationship between load and temperature is modelled in the literature by computing several degree-days transformations of the observed temperatures, usually daily high and low outdoor air temperatures.[3] The main idea of this approach is to segment the variation of temperatures into several derived variables. The heating degree-days (HDD) and the cooling degree-days (CDD) are defined as the difference between the actual temperature and a reference temperature:

$$HDD_t = \max(T_R - T_t, 0); CDD_t = \max(T_t - T_R, 0), \quad (1)$$

where T_t is the mean outdoor air temperature at time t and T_R is a reference temperature (e.g. 65°F in the USA, 15°C in Spain). If T_t is above T_R, there are no HDD, whereas if the T_t is less than T_R, there are no CDD. Thus, these degree days try to measure the intensity and duration of cold and heat in winter and summer days, respectively. Note that the reference temperature should be adequately selected to separate correctly the cold and heat branches of the load–temperature relationship.

The usual approach to STLF uses the forecasted weather scenario as an input. If multiple scenarios for the future value of a weather variable (e.g. outdoor air temperature) are available, it is possible to use the so-called ensemble approach (Taylor and Buizza 2003), where multiple load forecasts are computed from different temperature, wind speed and cloud cover scenarios, and combined to produce not only a single point load prediction, but also load prediction intervals.

2.2.4 Dealing with Calendar Events

The industrial, commercial and residential activity patterns differs from holidays to working days, leading to systematic variations in the electricity load.

Although in practice the unusual consumption on special days, including public holidays such as 1 January, is usually obtained by means of judgemental methods, there exist more refined approaches. The most common approach consists in building different models for normal and special days of the week (e.g. one model for Monday, Tuesday–Thursday, Friday, Sunday and Saturday) (Ramanathan et al. 1997). The main drawback of this approach is the treatment of special holidays, as long weekends or the first of January.

---

[2] Usually a regional average temperature is built as a weighted average of temperatures recorded at different observatories that represent the different climatic regions covering the whole territory of interest. The weightings used should reflect the consumption volume of each region, and they are usually obtained from population figures.

[3] The mean outdoor temperature is also very frequent in the literature, and it is calculated from the maximum and minimum daily values.

Within the single model approach, the treatment of special days such as Easter, Christmas, or public holidays is usually carried out by means of dummy variables.[4] All in all, the number of dummies representing the different types of days can be large. For example, in Moral-Carcedo and Vicéns-Otero (2005) hundreds of dummy variables are used to deal with calendar effects. In this reference the calendar effect is finally collected by a unique variable (the so-called "working day effect"), which represents the effect of calendar in the load of a particular day as a percentage of the electricity load on a representative day of the week (e.g. Wednesday).

2.3 *Electricity Load Forecasting Models*

A large variety of methods and models have been applied to STLF during several decades. In spite of the numerous trials for finding a superior model, we have to agree with some authors (Taylor et al. 2006; Piras and Buchenel 1999) that no sophisticated method is clearly better than others. In this section we briefly review the most interesting and promising models.

The literature on STLF contains a variety of models that can be roughly classified into two main categories: (1) statistical time series approaches and (2) Artificial intelligence-based techniques. Within the statistical time series approaches, univariate and multivariate models can be distinguished. Expert systems, artificial neural networks (ANN), fuzzy logic and support vector machines (SVM) are the main artificial intelligence paradigms applied to STLF. A non-exhaustive overview of the most relevant models of these two groups is presented in the following sections.

2.3.1 Statistical Time Series Analysis-based Models

Statistical time series models, pioneered by Box–Jenkins (1976) and Holt–Winters (1960) methods, have been present since the dawn of electric load forecasting. Statistical methods can be classified into univariate and multivariate. Univariate methods, as ARIMA and exponential smoothing models, are usually applied in the literature as reference models for very short horizons (see Taylor et al. (2006) for an empirical comparison of univariate methods in STLF). Multivariate methods using explanatory variables, often expressed as ARIMAX and linear transfer function models with non-linear transformations of the input variables, are necessary to anticipate the effects of climatology and calendar on consumption habits (see Bunn (1982) and Weron (2006) for extended reviews). In the next paragraphs, the most common models in STLF are introduced.

[4] The working days after and before a holiday could require additional dummies.

Seasonal ARIMA Models (SARIMA)

The standard form of a multiplicative double seasonal ARIMA $(p, d, q) \times (P_1, D_1, Q_1)_{24} \times (P_2, D_2, Q_2)_{168}$ model for hourly time series is given by

$$\begin{aligned}\phi_p(B)\Phi_{P_1}(B^{24})\Phi_{P_2}(B^{168})\nabla^d\nabla_{24}^{D_1}\nabla_{168}^{D_2}L_t \\ = C + \theta_q(B)\Theta_{Q_1}(B^{24})\Theta_{Q_2}(B^{168})\varepsilon_t,\end{aligned} \tag{2}$$

with

$$\phi_p(B) = 1 - \phi_1 B - \ldots - \phi_p B^p \tag{3}$$
$$\Phi_P(B^S) = 1 - \Phi_1 B^S - \ldots - \Phi_P B^{PS} \tag{4}$$
$$\theta_q(B) = 1 - \theta_1 B - \ldots - \theta_q B^q \tag{5}$$
$$\Theta_Q(B^S) = 1 - \Theta_1 B^S - \ldots - \Theta_Q B^{QS}, \tag{6}$$

where L_t is the output variable, that is, the hourly electricity load, C is a constant term and ε_t is an independent and identically distributed random noise. B is a backshift operator ($BL_t = L_{t-1}$), and $\phi_p(B), \Phi_{P_1}(B^{24}), \Phi_{P_2}(B^{168})$ and $\theta_q(B), \Theta_{Q_1}(B^{24}), \Theta_{Q_2}(B^{168})$ are backshift operator polynomials, of orders p, P_1, P_2, q, Q_1, Q_2 respectively, modelling the regular, daily and weekly autoregressive and mean average effects, respectively. $\nabla^d, \nabla_{24}^{D_1}, \nabla_{168}^{D_2}$ are difference operators ($\nabla^d L_t = (1-B)^d L_t$ and $\nabla_S^D L_t = (1-B^S)^D L_t$) describing the differences applied to induce stationary in integrated processes. The Box–Jenkins methodology (Box and Jenkins 1976) offers a systematic identification procedure that has been widely tested in many areas.

In present days, univariate ARIMA models are basically being used as benchmarks for comparative purposes (see, e.g. Moghram and Rahman (1989), Taylor and McSharry (2008), Taylor (2008), Soares and Medeiros (2008)).

Exponential Smoothing for Double Seasonality

The application of exponential smoothing models to hourly data requires an extension of the standard Holt–Winters formulation to accommodate the two seasonalities (daily and weekly). This extension was proposed by James W. Taylor in Taylor (2003) and further investigated in Taylor et al. (2006) and Taylor (2008):

$$\mathit{Level} \quad S_t = \alpha(L_t/(D_{t-24}W_{t-168})) + (1-\alpha)(S_{t-1} + T_{t-1}) \tag{7}$$
$$\mathit{Trend} \quad T_t = \gamma(S_t - S_{t-1}) + (1-\gamma)T_{t-1} \tag{8}$$
$$\mathit{Seasonality}\ 1 \quad D_t = \delta(L_t/(S_t W_{t-168})) + (1-\delta)D_{t-24} \tag{9}$$
$$\mathit{Seasonality}\ 2 \quad W_t = \omega(L_t/(S_t D_{t-24})) + (1-\omega)W_{t-168} \tag{10}$$
$$\widehat{L}_t(k) = (S_t + kT_t)D_{t-24+k}W_{t-168+k}, \tag{11}$$

where $\widehat{L}_t(k)$ is the k-step ahead load prediction made at time t, and α, γ, δ and ω are model parameters.

Linear Transfer Functions models (LTF)

The LTF model has the following general expression for hourly data (daily and weekly seasonalities):

$$\begin{aligned} L_t &= C + v_1(B)X_{1t} + \cdots + v_n(B)X_{nt} \\ &\quad + \frac{\theta_q(B)\Theta_{Q_1}(B^{24})\Theta_{Q_2}(B^{168})}{\phi_p(B)\Phi_{P_1}(B^{24})\Phi_{P_2}(B^{168})\nabla^d\nabla_{24}^{D_1}\nabla_{168}^{D_2}}\varepsilon_t, \end{aligned} \tag{12}$$

where the definitions made for the ARIMA model hold, and $v_1, \cdots, v_n$ represent a family of linear transfer functions able to capture a wide number of impulse responses with few parameters (see Pankratz (1991) for more details). They can be expressed as

$$v(B) = \frac{(w_0 + w_1 B + w_2 B^2 + \cdots + w_s B^s)B^b}{1 + \delta_0 + \delta_1 B + \delta_2 B^2 + \cdots + \delta_r B^r}. \tag{13}$$

Predictions are therefore obtained as linear combinations of past and present values of actual and predicted loads and, if available, of other exogenous variables as temperatures and calendar effects (see Engle et al. (1986)). The success of these models lays in a well-established identification and diagnostic checking methodology (see Box and Jenkins (1976) and Pankratz (1991)), a reduced number of parameters that can be easily interpreted, and their representation capabilities (ARIMA, ARMAX and Dynamic Regression models fit within the general formulation of LTF models). In fact, many load forecasting models reported in the literature can be expressed with transfer function models. Their main differences are related to the treatment of weather and calendar effects and the introduction of switching regimes through periodic and smooth transition models (see Pardo et al. (2002), Moral-Carcedo and Vicéns-Otero (2005), Cancelo et al. (2008), Dordonnat et al. (2008), Amaral et al. (2008), Taylor and McSharry (2008)).

2.3.2 Artificial Intelligence Based Models

The history of AI has determined the evolution of non-statistical techniques applied to STLF (Bansal and Pandey 2005). Expert systems were applied for the first time in 1988 (Rahman and Bhatmagar 1988) and improved during the 1990s by modelling calendar effects and the non-linear relationship of load and temperature (see e.g. Ho et al. (1990) and Rahman and Hazim (1993)). These knowledge-based models are rule-based techniques, where the rules are derived from human experts and not directly from data.

In the 1990s a great amount of research was devoted to the empirical application of ANN to STLF (see Hippert et al. (2001b) for a comprehensive review of ANNSTLF models published between 1991 and 1999). Most of the papers published in that period used feed-forward multilayer perceptrons (MLP) as non-linear multivariate regression models (e.g. Park et al. (1991), Chen et al. (1992), Chow and Leung (1996), Czernichow et al. (1996), Khotanzad et al. (1997)). The main characteristic of these black-box models is that their internal structure is not defined a priori, being obtained automatically from the in-sample data during the training process. They benefit from the universal approximation capability, but at the same time they suffer the risk of overfitting (this controversial issue is addressed in Hippert et al. (2005)). Recurrent networks, a class of neural network that not only operate on an input space but also on an internal state space, have been used to model the recurrent dynamics of electricity load time series (e.g. Choueiki et al. (1997), Vermaak and Botha (1998)).

The application of neural networks to STLF mixed up in the early twenty-first century with other soft computing techniques as fuzzy logic and evolutionary computation (Kodogiannis and Anagnostakis (2002); Ling et al. (2003); Senjyu et al. (2005)). These methods were introduced as an attempt to improve the learning abilities and interpretability of neural networks. SVM, derived from the statistical learning theory developed by Vapnik (1995), have also yielded promising results (Mohandes 2002; Pai and Hong 2005).

2.3.3 Error Measures for Electricity Load Forecasting

A vast array of accuracy measures can be used to evaluate the performance of forecasting methods (see e.g. DeGooijer and Hyndman (2005) for the most commonly used measures). However, in the context of STLF literature, a reduced number of measures have been used to present load forecasting errors. The most popular one is the mean absolute percentage error (MAPE), as well as the root mean squared error (RMSE). For period $[1, N]$ they are given by

$$MAPE = \frac{1}{N}\sum_{t=1}^{N}\frac{|L_t - \widehat{L_t}|}{L_t}100 \tag{14}$$

$$RMSE = \sqrt{\frac{1}{N}\sum_{t=1}^{N}(L_t - \widehat{L_t})^2}, \tag{15}$$

where L_t and $\widehat{L_t}$ are the actual and predicted loads for time t.

The MAPE captures the proportionality between the forecast absolute error and the actual load, enabling a simple interpretation by those in the industry, where it has become a standard. Nevertheless, the MAPE has problems when the series has values close to (or equal to) zero, as noted by different authors (e.g. Makridakis et al. (1998), p. 45).

It is well known that without an explicit comparative test using the same data set, it is not possible to deduce general conclusions of how effective a forecasting procedure is. In spite of this, when dealing with hourly load series, the typical out-of-sample MAPE for 24 steps ahead ranges from 1 to 4%, whereas the reported results for the one step ahead range from 1 to 2%.

3 Wind Power Forecasting

In the recent years, because of the huge raise in the wind power capacity installed in an increasing number of countries all over the world, the necessity of models and tools to predict wind energy has become more and more important. The main reasons for this growth are oriented to reduce the dependence and the environmental impacts caused by the use of fossil fuels. This is the case of the principal producer of wind energy, the European Union, with a wind power capacity of 66 GW,[5] with 55% of the world total wind power capacity in 2008 and with the objective of covering around 20% of the electricity consumption in 2010 (Giebel et al. 2003; IDAE 2003). As a consequence of this growth, utilities and regulators have been forced to cope with the unpredictability and volatility of wind to integrate this "adolescent" technology in our "mature" power systems. Without wind power forecasting, wind energy would have never attained the penetration levels observed in present days.

Wind power forecasting applications can be classified into four main groups according to the prediction horizon:

1. Ultra short-term (seconds range): For wind turbine control applications
2. Very short term (from minutes to 1 h ahead): For power system security assessment applications
3. Short term (from 1 to 48 h ahead): This is the prediction horizon required for the operation of wind energy in day-ahead electricity markets, unit commitment, economic dispatch and short term maintenance planning
4. Medium and long term (up to several years): Maintenance planning, generation planning and energy policy models require wind energy scenarios as input. Simulation techniques are usually applied for that purpose

This chapter will be devoted to the third group of applications, taking into account that wind energy is being traded in many day-ahead electricity markets as any other source of energy. On the one hand, wind energy producers have to predict their hourly resources 1 day in advance to sell their predicted energies in the daily market. On the other hand, the system operator has to predict the aggregated wind generation in the whole system to estimate the effective electricity demand that has to be covered by others resources.

---

[5] Source GWEC

3.1 Features of Wind Power Time Series

Basically, wind power is highly correlated with wind speed, although a complex physical relationship exists between wind speed and generated power. This energy conversion process leads to non-linear and high-volatile wind power time series (see Fig. 4). On the one hand, the main source of non-linearity is related to the shape of wind turbine power curves (see Fig. 5). On the other hand, high volatility is inherent to the wind dynamics. These features make wind power forecasting a very difficult task, requiring very complex physical and statistical models. Short-term accuracy is significantly improved when forecasting the aggregated production of a portfolio of wind farms, in the same manner as utilities group them to send an aggregated bid to the electricity market. This fact is illustrated in Fig. 4, where the normalised hourly wind energy production of a portfolio of wind farms (integrating more than 1 GW, plot (a)) is compared to the individual time series of four wind farms of the

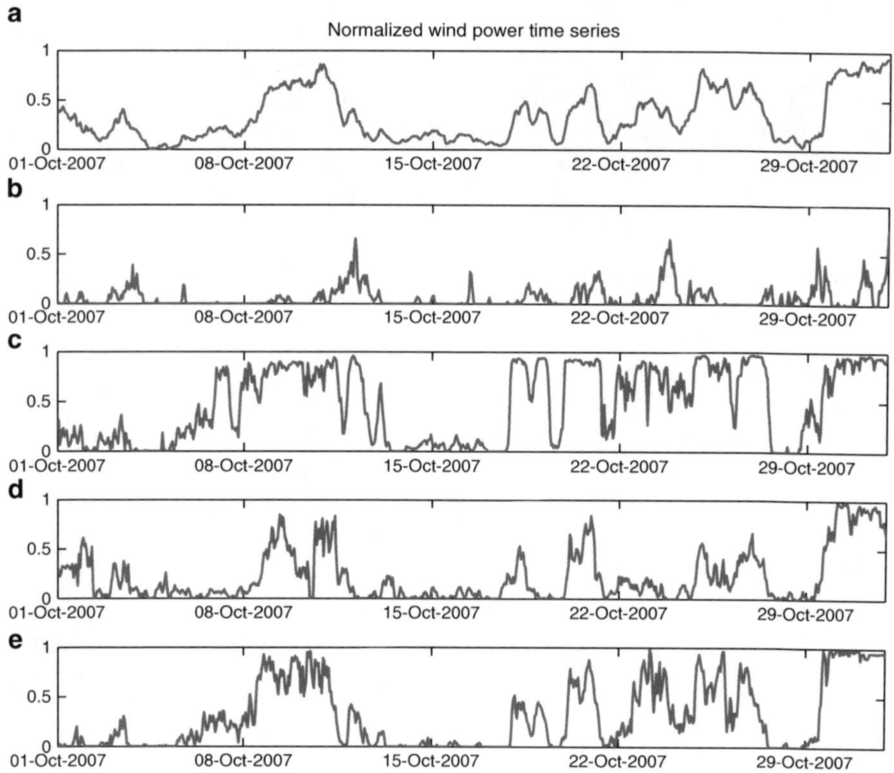

Fig. 4 Wind power time series (all hourly wind energy production time series have been normalized by the corresponding rated power: (**a**) Portfolio of wind farms (**b**) 20 MW wind farm (**c**) 25 MW wind farm (**d**) 50 MW wind farm (**e**) 20 MW wind farm)

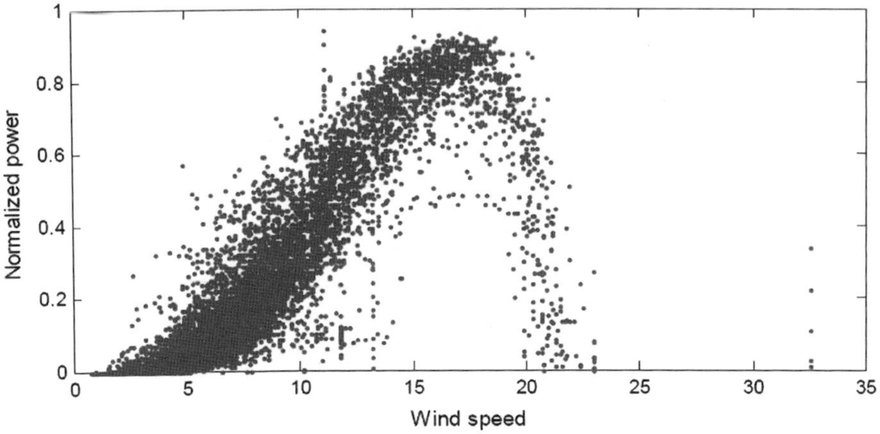

Fig. 5 Empirical power curve of a 30 MW wind farm

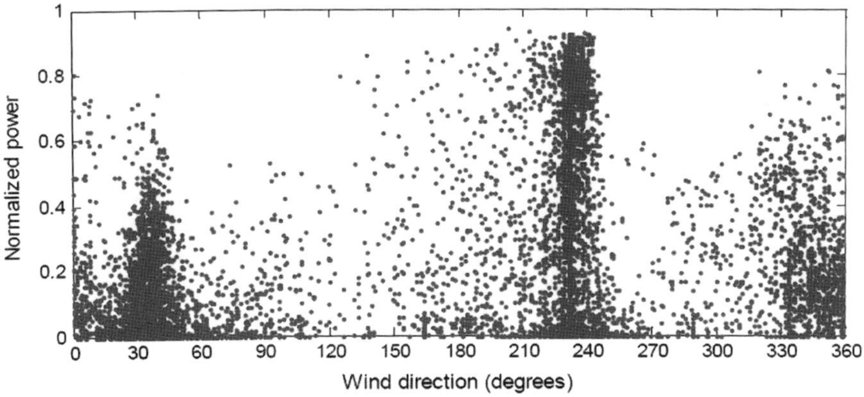

Fig. 6 Wind-speed direction vs. energy production in a wind farm

same portfolio (less than 50 MW each, plots (b–e)). The portfolio time series is significantly smoother than the individual plots.

Figure 5 shows the measured power curve of a 30 MW wind farm. It can be observed that it significantly differs from the sum of theoretic power curves supplied by wind turbine manufacturers. The non-linearity of this curve, combined with the high volatility of the wind dynamics, determines the main characteristics of wind power time series.

An empirical 30 MW wind farm output power vs. wind direction plot is shown in Fig. 6. A few predominant wind directions can easily be identified on this plot, as well as the fact that wind direction does not explain wind power by its own. When used as input variable, wind direction is always combined with wind speed.

3.2 Wind Power Forecasting Models

Most approaches to short-term wind power forecasting (see Giebel et al. (2003) and Landberg et al. (2003) for a detailed review) can be roughly classified into two groups according to the type of model used in the process: physical or statistical. In both cases a second distinction may be done according to the predicted output variable: wind speed (m/s) or power output (MW).

In this context, physical or numerical weather prediction (NWP) models use physical considerations that govern meteorological phenomena to reach to the best possible estimate of the local wind speed before using model output statistics (MOS) to reduce the remaining error. The statistical approach tries to learn (from data) functional relationships between a set of explanatory variables, including NWP estimates and online power measurements, and the desired output. The statistical approach has the advantage of being computationally much less expensive than the continuous re-evaluation of NWP models. In practice, models combining NWP estimates and online power measurements have shown to be the most accurate solution for the short term.

With respect to physical models, Landberg (1999) has shown that a simple NWP and physical downscaling approach is effectively linear, thereby being very easily amenable to improvements through MOS. Whether it is generally better to follow the complete model chain (first predicting wind at ground level, then production with the wind park specific power curve) depends – given the availability of data – on the forecast horizon: using auto-regressive models, Jensen (1994) showed that the use of wind speed predictions as explanatory variable is important for up to 8–12 h ahead. For longer prediction horizons, use of separate wind speed forecasts offered no advantage over direct wind power prediction.

Within the statistical time series approach, ARIMA models have been applied to wind power forecasting by Guoyang et al. (2005) and Sfetsos (2000). Torres et al. reported in Torres et al. (2005) that an ARMA model improved the persistence model by 12–20% for a forecasting horizon of 10 h, but produced worse forecasts for the next hour. Schwartz (Milligan et al. 2003), who applied a class of ARMA models to both wind speed and wind power output from wind farms in the US, found that the improvements he achieved over persistence were strongly dependent on the considered month. Costa et al. (2007) proposed the use of the Kalman filters for 5-min sampled wind power time series.

Many researchers have applied ANN for short-term wind power forecasting with promising results. Regarding the network structure, simple networks seem to be most suitable for precise forecasts (only six neurons in the hidden layer were used in Beyer et al. (1994)). Besides production data of the predicted wind farm, data from upwind parks can be used in an ANN approach to further improve the predictions, as was shown in Alexiadis et al. (1999). This strategy, which is known as spatial correlation, was also applied in Yang et al. (2005) and Damousis et al. (2004). In Kariniotakis et al. (1996), the authors apply an algorithm to automatically optimize the architecture of the network. Li (2001) uses an ANN for diagnostic purposes, as lower-than-expected wind power can be an early indicator for maintenance.

Barbounis (2006) compares the performance of feed-forward and recurrent neural networks, finding a significant advantage of the second ones in their case study. Sfetsos (2001) argues that hourly averaged wind measures are not able to represent the structure of the wind time series. He uses an ANN for 10-min prediction, obtaining better results than the ones obtained with hourly data.

Apart from ANN, fuzzy logic and SVM have also been applied in wind power forecasting. In Damousis et al. (2004), fuzzy variables are used to model membership to different wind scenarios. In Gou-Rui (2007), SVM are proposed to solve some of the problems of neural networks, such as over-fitting and local minimums.

A quite different approach is proposed by Ismael Sánchez in Sanchez (2006), which describes the procedure that is currently implemented in SIPREÓLICO, a wind energy prediction tool that is part of the online management of the Spanish Peninsular system. In this paper, two different sets of models are used: dynamic linear models and non-parametric models. The estimations of these models are then combined using an adaptive combination of forecasts with time varying weights.

Most efforts are now being devoted to the design of hybrid models, the combination of forecasts and ensemble forecasting. In Sideratos and Hatziargyriou (2007), neural networks and fuzzy models are integrated in a new hybrid model. Several NWP models are used in Giebel et al. (2005) to estimate the uncertainty of wind forecasts. The basic assumption is that if the different model members are differing widely, then the forecast is very uncertain, while close model tracks mean that this particular weather situation can be forecasted with good accuracy. Furthermore, it has been empirically confirmed since a long time (Newbold and Granger 1974) that a simple combination of forecasts works better in many cases than any of the models themselves.

3.2.1 Error Measures for Wind Power Forecasting

Comparing the different models is quite difficult as the performance strongly depends on the geographic situation of the wind farm (constant or volatile winds, onshore or offshore) as well as the quality and quantity of available input data. Furthermore, different reference models (like persistence[6] (Nielsen et al. 1998) or meteorological mean[7]) and error measures (mean absolute error (MAE) or mean square error (MSE), relative to wind farm installed capacity or relative to average production) are used in practice.

The most difficult case, where the highest errors are achieved independently of the method, corresponds to stand-alone onshore wind farms. In that case, the reported MAE, for prediction horizons from 3 to 10 h, ranges from 30 to 50% of the

[6] The persistence model acts on the assumption that the produced energy remains constant during the prediction horizon and is hard to beat for short prediction horizons of up to 3 h.

[7] The estimated value of the meteorological mean model is the historic average production.

energy production, depending on the complexity of the terrain. By grouping wind farms in portfolios, errors can be reduced to the range of 15–20% of energy production. The best results are obtained by integrating the results of NWP with online measures through black-box or grey-box models such as ANN (some knowledge of the wind power properties is used to tune the grey-box models to the specific domain).

4 Forecasting Electricity Prices

The new competitive framework, imposed by the worldwide liberalization of the electricity industry, has forced agents operating in wholesale electricity markets to follow the long-term, medium-term and short-term spot price movements in order to trade this new commodity on regulated markets or through bilateral contracts. In many countries, the key component of the wholesale market is a day-ahead auction (Green 2008), where sellers and buyers submit their bids in terms of prices and quantities for the 24 h (or 48 half-hour in some markets) of the next day. The hourly (or half-hourly) marginal prices are then obtained at the intersection of aggregated supply and demand curves. Many short-term decision-making processes in this framework, as the design of bidding strategies, require accurate predictions of the spot prices.

4.1 Factors Affecting Electricity Prices

As stated in Karakatsani and Bunn (2004),

> due to the idiosyncrasies of wholesale electricity markets, spot price dynamics are only partially understood.

These idiosyncrasies include the instantaneous nature of the commodity, the steeply nature of the supply function due to the presence of different generation technologies, the oligopolistic nature of these markets, complex market designs, frequent regulatory interventions and market structure changes.

However, electricity load is usually the most important factor affecting the behaviour of electricity prices time series around the world. This is an obvious fact, as market clearing prices are obtained by crossing supply and "quasi-inelastic" demand curves constructed from aggregated bids. Therefore, with stable remainder factors, an increase of demand will entail an increase of the spot price, and factors affecting the price, as economic activity, daily and hourly cycles or temperature fluctuations, are captured in a certain degree through this variable. System operators usually provide demand predictions to market agents as an input to their bidding planning processes. Notice how PJM prices are mimicking the demand in the particular case shown in Fig. 7. Nevertheless, as stated in Bunn and Karakatsani (2003), spot electricity prices display in general a rich structure much more

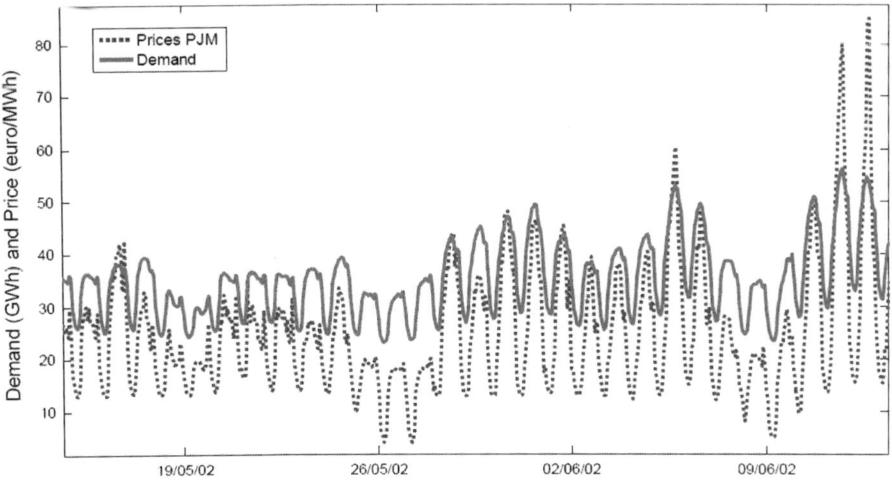

Fig. 7 Price (euro/MWh) and demand (GWh) in PJM

complicated than a simple functional rescaling of demand to reflect the marginal costs of generation.

On the supply side, the generation mix of each specific market is a determinant factor to explain the spot price. In Spain, for example, wind generation has become a very meaningful variable, due to its high volatility, high penetration level[8] and a special regulatory treatment which leads wind generators to act as price takers. In the Nord Pool, a very outstanding variable is the amount of water resources available for power generation. In general, the way in which suppliers construct their bids is strongly determined by the production costs of the generation technologies covering the demand. Therefore fuel costs, emissions costs, reserves, power exports and imports and maintenance outages, among others, determine the resulting electricity price. However, it is important to note that when hourly (or half-hourly) short-term predictions are involved, some of these variables are not useful to explain the high frequency fluctuations of the prices. Others variables, for example hydropower generation, are difficult to predict or are not available in the short time for every market participant.

A more detailed and exhaustive description of the explanatory variables affecting spot market prices can be found in Karakatsani and Bunn (2008), Bunn (2004) and Li et al. (2005).

[8] Eleven percent of the Spanish demand in 2008 was satisfied by wind generators. The maximum hourly demand in 2008 was attained on December 15th (42,961MW) and 18% of this demand was fulfilled with wind energy.

4.2 Features of Electricity Price Time Series

Electricity price time series in deregulated electricity markets are generally considered to be erratic and ill-behaved (see Knittel and Roberts (2005), Weron (2006) and Escribano et al. (2002)). The main features that characterize this behaviour are mean reversion, seasonal effects (daily, weekly and annual), different intra-day and intra-week patterns, calendar effects, time-varying volatility, fat-tailed and skew distributions, and extreme values. Note some of these features, as the daily and weekly patterns, the time-varying volatility, or the extreme values, in Fig. 2.

The most distinctive of these features is the presence of extreme values or spikes. These spikes are unanticipated extreme changes in the time series within a very short period of time. During this period, the price dramatically increases and then drops back to the previous level. They are normally observed when demand is high (peak hours) and are present in most of the electricity markets, though not with the same intensity. The lower picture in Fig. 8 illustrates some spikes that occurred in the Victoria market. The price value reached the value of 8,000 cE/MWh, two orders of magnitude above its normal level.

These spikes are usually due to a combination of different factors. On the one hand, the non-storability of electricity causes most expensive technologies to establish the price during periods of high demand. Moreover, generation outages or transmission failures can make these situations worse. In this context, the participant strategies play a very important role. On the other hand, as the electricity is an essential commodity for many buyers, they are willing to pay almost any price to secure the supply of power.

In the modelling context, it is an advisable practice to previously eliminate the spikes from the in-sample data. Otherwise, these extreme values could dramatically affect the value of the final parameters and make the model capture properly neither the usual behaviour nor the spikes. Some parametric methods have been developed to automatically detect and correct outliers (see Maravall (2005) and Chang et al. (1988)). Other ways to prevent the negative effects of spikes in the fitting process are pre-filtering using the wavelet transform (Conejo et al. 2005) or damping the prices above a certain threshold (Weron 2006). However, spikes are not incorrect values of the price time series, and well specified models should be able to model their occurrence. Jump diffusion and regime-switching models (Hamilton 2005) including at least one spike generation process have been developed for that purpose. These models will be described in more detail in Sect. 4.3.2.

Another important feature, related with spikes, is the special underlying distribution of the electricity price stochastic process. The empirical distributions obtained through the traditional Gaussian fitting framework rarely accomplish with the normality (or log-normality) assumption. There is a lot of literature pointing out the fact that electricity price distributions are heavy-tailed and skewed (see, e.g. Knittel and Roberts (2005) and Weron (2006)). Furthermore, the time-varying volatility of day-ahead market prices prevents the fulfilment of the assumption of homokedasticity. GARCH models (Generalized Auto Regressive Conditional Heterokedastic) are an attempt to capture this dynamic volatility as a function of variances of previous time

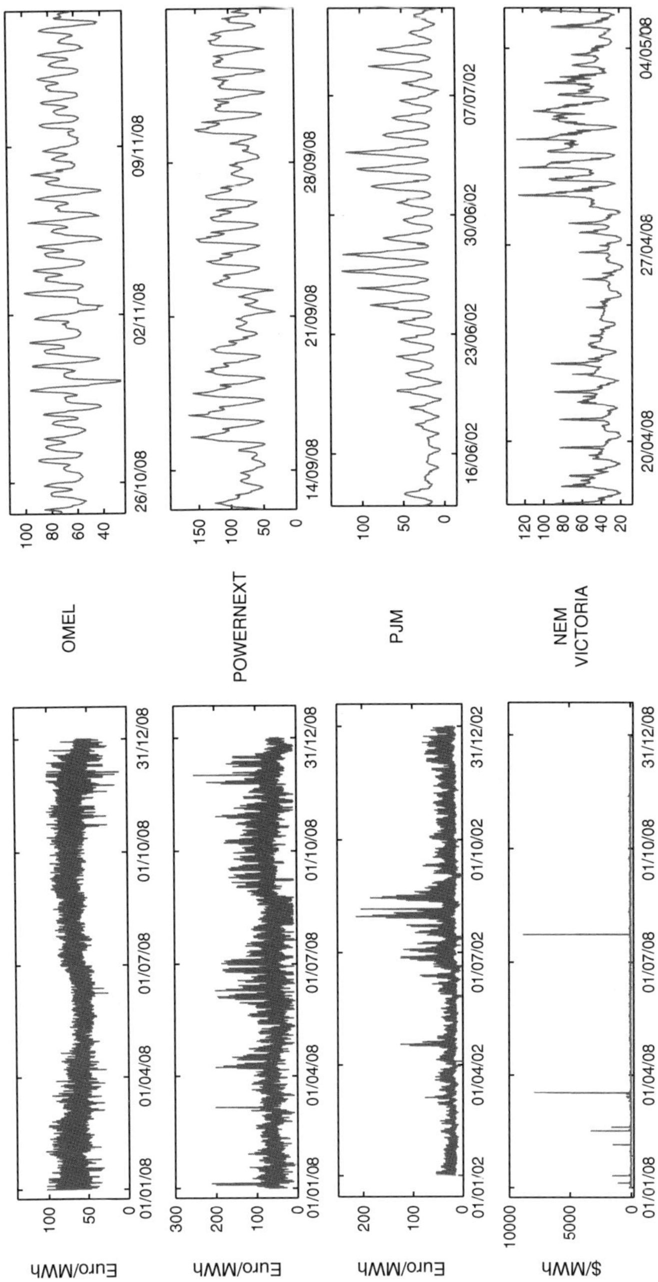

Fig. 8 Prices in OMEL (Spain), POWERNEXT (France), PJM (Pennsylvania, New Jersey, Maryland) and NEM (Victoria)

periods. However, models with GARCH noise have not shown evidences of better forecasting ability in price prediction than their homokedastic equivalent (Misiorek et al. 2006; Knittel and Roberts 2005). Recent advances trying to find a satisfactory distribution include the work of Weron (2008), where a set of non-Gaussian distributions are tested on the price time series of different electricity markets. In Weron and Misiorek (2008), a set of semi-parametric models whose density functions are estimated through kernel estimators is compared with their Gaussian parametric counterparts, concluding that semiparametric models lead to better point and interval forecasts. In Panagiotelis and Smith (2008), a VARX (Vector autoregressive) model specified through a sparse coefficients matrix and skew t distributions is proposed.

4.3 Electricity Price Forecasting Models

Many different modelling approaches have been proposed in the literature to deal with short-term electricity price forecasting. Most of these techniques may be roughly divided into three categories: (1) quantitative models, (2) models coming from the statistical time series analysis domain and (3) artificial intelligence approaches. In the following sections, an overview of the most relevant models of these three groups is presented. The overview does not pretend to be exhaustive, and so readers interested in a detailed taxonomy in short-term price prediction are referred to the published works in Weron (2006), Bunn and Karakatsani (2003), Li et al. (2005) and Mateo et al. (2005).

4.3.1 Quantitative Models

Quantitative models are generally used with the primary aim of derivatives valuation and risk management (see Pilipovic (1998)). It is important to note that they are not pretended to be accurate forecasting tools, but to capture the main characteristics of electricity prices. They include mean reversion and jump-diffusion models (see Knittel and Roberts (2005), Deng et al. (2001), Lucia and Schwartz (2002), de Jong and Huisman (2003), and Geman and Roncoroni (2006)), conditional heteroskedastic models (Escribano et al. 2002; Wilkinson and Winsen 2002) and hybrid models, which introduce fundamental explanatory variables, as demand, fuel costs or generation capacity, in the above models (Barlow 2002; Burger et al. 2004; Anderson and Davison 2008).

The mathematical formulation of these models can be illustrated with the general expression of jump diffusion models, which are described by stochastic differential equations of the form (Weron 2006)

$$\mathrm{d}P_t = \mu(P_t, t)\mathrm{d}t + \sigma(P_t, t)\mathrm{d}W_t + \mathrm{d}q(P_t, t), \qquad (16)$$

where P_t stands for the price at time t, $\mu(P_t, t)$ is a drift term that induces mean reversion to a stochastic or deterministic long term mean, $\sigma(P_t, t)$ is a volatility term, W_t is a Brownian motion process and $q(P_t, t)$ is a Poisson pure jump process that produces infrequent but large jumps.

4.3.2 Statistical Time Series Analysis-based Models

Within the statistical time series analysis framework, a lot of research activity has been developed for modelling and forecasting the day-ahead spot prices of world-wide liberalized electricity markets. The models more frequently used are detailed below.

Linear Transfer Functions Models (LTF)

Some initial contributions are the papers by Nogales et al. (2002) and Carnero et al. (2003), which pointed out that dynamic regression and transfer function models are very accurate methods for the Spanish and Californian market. These conclusions are corroborated in Conejo et al. (2005b) and extended to others electricity markets in Zareipour et al. (2006a), Bunn and Karakatsani (2003), Misiorek et al. (2006).

The general expression of LTF models was introduced in (12). ARFIMA (Autoregressive fractionally integrated moving average) models are also obtained from LTF by permitting fractional integration orders (parameters $d, D_1 and D_2$). They are used in Koopman et al. (2007) to model the long memory dynamics of different European electricity market prices.

The LTF formulation can be easily extended to the multiple-input multiple-output (MIMO) case. These models provide an alternative way to forecast the 24 hourly prices of the next day. Each hourly component of the daily price vector can be differently explained through daily vectors of delayed prices and explanatory variables. Moreover, the daily price vector is predicted one-step ahead, that is, in the same manner the day-ahead market clearing process occurs. It is important to note the difference when considering the hourly approach, in which the 24 h of the next day are predicted in a sequential manner from hour 1 to hour 24, and therefore the prediction uncertainty of each hour affects the following ones. In Panagiotelis and Smith (2008), a VARX model is used to forecast the Australian electricity spot prices.

Regime-Switching Models

ARIMA and LTF models have been extended to incorporate non-linearities. This is the case of threshold auto-regressive (TAR) models, in which a different set of parameters is applied at each time according to the value of an observable variable.

These models are tested for short term electricity price forecasting in Misiorek et al. (2006), Weron and Misiorek (2008) and Bunn and Karakatsani (2003), where competitive results are reported for both point and interval forecasting. Periodic time series models are similar to threshold models, but in this case a different set of parameters is determined for each pre-defined season. Koopman et al. (2007) provides a comparison of alternative univariate time series models that are advocated for the analysis of seasonal data. Electricity prices of several European markets are investigated and modelled, taking into account the day of the week through periodic models. In Guthrie and Videbeck (2002), periodic models are used to capture the intra-day dynamics of New Zealand electricity prices.

The general expression of a LTF regime-switching process with m seasonalities is given by

$$P_t = C_{R_t} + v_{1,R_t}(B)X_{1t}, +\ldots + v_{n,R_t}(B)X_{nt} + \frac{\theta_{q,R_t}(B)\Theta_{Q_1,R_t}(B^{S_1})\cdots\Theta_{Q_m,R_t}(B^{S_m})}{\phi_{p,R_t}(B)\Phi_{P_1,R_t}(B^{S_1})\cdots\Phi_{P_m,R_t}(B^{S_m})\nabla^d\nabla_{S_1}^{D_1}\cdots\nabla_{S_m}^{D_m}}\varepsilon_t, \tag{17}$$

where $R_t = \{1, \cdots, K\}$ stands for the season index (case of periodic models) or the regime determined by an observable variable (case of threshold models) at time t, K being the number of seasons or regimes and having been the remainder parameters previously defined.

Markov regime-switching models are an alternative to threshold and periodic models, where the different underlying processes are not directly determined by an observable variable, but by a set of hidden exclusive states and a probability law that governs the transition from one regime to another. In the context of short-term price prediction, they have been used for modelling spikes and other anomalous behaviors. Related works can be found in Weron (2008), Bunn and Karakatsani (2003) and Huisman (2008).

4.3.3 Artificial Intelligence-based Models

The techniques coming from this field of knowledge differ from the previous ones, in that the model structure is not specified a priori, but is instead determined from data. Because of this fact, they are also known as non-parametric techniques. A large number of heterogeneous techniques lay within this category but ANN have received most attention in short-term electricity price forecasting. Other important techniques within this group which will not be covered in more detail are fuzzy Logic (Niimura and Nakashima 2001), weighted nearest neighbours (Lora et al. 2007), similar-day based methods (Mandal et al. 2007), multi-variable adaptative regression splines (Zareipour et al. 2006b), SVM (Gao et al. 2007) and dimensionality reduction (Chen et al. 2007).

ANNs are used in this context as multivariate non-linear regression models with universal function approximation capabilities (Cybenko 1989). Their main drawbacks are the risk of overfitting and their lack of interpretability.

One of the first attempts to tackle the problem of price prediction through feed-forward neural networks is presented in Szkuta and Sanabria (1999). In this work, actual and delayed system loads and reserves, and delayed prices are selected as network input variables to predict the spot prices in the Victoria electricity market. The multilayer perceptron is also used in Gao et al. (2000) to predict the prices and quantities of the Californian day-ahead energy market. In this case, historical prices, system loads, fuel costs, power imports and exports, temperatures, and hour and weekday seasonal indexes are considered as input variables. Feed-forward networks are also applied in Pino et al. (2008) after a previous selection of the training samples through an adaptative resonance theory (ART) neural network to predict Spanish electricity prices. In Pindoriya et al. (2008) and Andalib et al. (2008), feed-forward wavelet neural networks, in which wavelet functions are used as activation functions of the hidden-layer neurons (usually referred as "wavelons"), are considered for the Spanish, PJM and Ontario markets. In Rodríguez and Anders (2004) and Amjady (2006), fuzzy logic techniques and neural networks are combined in fuzzy neural networks to predict the prices in Ontario and Spain, respectively. In Mandal et al. (2007), a feed-forward neural network is used to correct the price curve obtained from a similar day approach.

More complex architectures have also been tested to cope with this task. For instance, Elman recurrent neural networks are used in Andalib et al. (2008) and Hong and Hsiao (2002). In both cases the ability to model fast non-linear variations and good generalization performance of this architecture are pointed out. Switching-regime models based on neural networks are applied in Mateo et al. (2005). In this work, an input–output hidden Markov model is proposed to model the discrete changes in competitors' strategies. The model provides not only the point prediction but also the probability density function, which is conditioned to input variables and the previous state of the system.

4.3.4 Error Measures for Electricity Price Forecasting

Because of the erratic and changing behaviour of electricity prices, error measures are often considered for each week of the validation period. The weekly mean absolute percentage error (WMAPE) is commonly used in the literature to test the accuracy of price prediction models (see, e.g. Amjady (2006), Pindoriya et al. (2008), Nogales et al. (2002)). However, to avoid the adverse effects of prices close to zero, weekly-weighted mean absolute error (WMAE) and weekly root mean square error (WRMSE) are also frequently considered (see, e.g. Conejo et al. (2005), Weron and Misiorek (2008), Mandal et al. (2007)). These error measures are given by (P_h and $\widehat{P_h}$ are the actual and predicted prices for hour h)

$$WMAPE = \frac{1}{168}\sum_{h=1}^{168}\left|\frac{P_h - \widehat{P_h}}{P_h}\right| \tag{18}$$

$$WMAE = \frac{\sum_{h=1}^{168}|P_h - \widehat{P_h}|}{\sum_{h=1}^{168} P_h} \tag{19}$$

$$WRMSE = \sqrt{\frac{1}{168}\sum_{h=1}^{168}(P_h - \widehat{P_h})^2}. \tag{20}$$

The results with these measures along the literature vary a lot depending on the considered market and period of time. The Spanish electricity day-ahead market is one in which less volatility and number of spikes are observed. In this market, the reported results for the WMAPE range from 6 to 25%, depending on the considered week.

5 Conclusions

This paper has presented a survey in short-term forecasting of the three random variables that are currently getting the most attention in electric power systems: electricity loads, wind power and spot prices. Short-term forecasting in this context refers to prediction horizons that range from 1 h to 1 week, as hourly or half-hourly data is considered in the three cases. The survey has covered three main topics: (1) the special features that characterize demand, wind power and electricity price time series, (2) a classification of the modelling approaches and (3) the error measures used to quantify the forecasting accuracy.

Short-term load forecasting is a well established discipline and is in the core of many operational processes, including power system planning, scheduling and control. The key aspects to be considered when forecasting electricity loads are the treatment of multiple seasonalities, the non-linear effect of temperature on electricity consumption and the complexity of calendar effects.

The increasing integration of wind power in electricity systems has forced system operators and market agents to cope with this new source of volatility by investing in wind power forecasting tools and integrating them into their operational systems. Short-term forecasts are needed for power system security assessment applications, the operation of wind energy in day-ahead electricity markets, unit commitment, economic dispatch and maintenance planning. The high volatility of wind dynamics and the non-linear physical relationship between wind speed and generated power lead to very complex wind power time series. Recent contributions in this area include spatial correlation, ensemble forecasting and density forecasting methods.

The worldwide liberalization of the electricity industry has intensified the risk exposure of electricity generators and retailers. Many decision-making processes in this framework, as the design of bidding strategies or the pricing of derivatives, require accurate spot price predictions. Electricity price time series are generally erratic and ill-behaved, contaminated by spikes and non-Gaussian distributions. Electricity price forecasting models are moving from univariate to multivariate, from single to multiple regime switching models and from point forecasts to interval and density forecasts.

The complete set of models and techniques included in this paper, accompanied by a selected bibliography, constitute a guided tour in power systems forecasting.

References

Electricity Load Forecasting

Amaral LF, Souza RC, Stevenson M (2008) A smooth transition periodic autoregressive (STPAR) model for short-term load forecasting. Int J Forecast 24:603–615

Bansal RC, Pandey JC (2005) Load forecasting using artificial intelligence techniques: a literature survey. Int J Comput Appl Technol 22(2–3):109–119

Box G, Jenkins G (1976) Time series analysis: forecasting and control, revised edition. Holden-Day, Oakland, California

Bunn DW (1982) Short-term forecasting: a review of procedures in the electricity supply industry. J Oper Res Soc 33:533–545

Bunn DW, Farmer ED (1985) Comparative models for electrical load forecasting. Int J Forecast 2(4):501–505

Cancelo JR, Espasa A, Grafe R (2008) Forecasting the electricity load from one day to one week ahead for the Spanish system operator. Int J Forecast 24:588–602

Chen ST, Yu DC, Moghaddamjo AR (1992) Weather sensitive short-term load forecasting using nonfully connected artificial neural network. IEEE Trans. Power Syst 7(3):1098–1105

Choueiki MH, Mount-Campbell CA, Ahalt SC (1997) Implementing a weighted least squares procedure in training a neural network to solve the short-term load forecasting problem. IEEE Trans. Power Syst 12(4):1689–1694

Chow TWS, Leung CT (1996) Neural network based short-term load forecasting using weather compensation. IEEE Trans Power Syst 11(4):1736–1742

Cottet R, Smith M (2003) Bayesian modeling and forecasting of intraday electricity load. J Am Stat Assoc 98:839–849

Czernichow T, Piras A, Imhof K, Caire P, Jaccard Y, Dorizzi B, Germond A (1996) Short term electrical load forecasting with artificial neural networks. Eng Intell Syst 2:85–99

De Gooijer JG, Hyndman RJ (2005) 25 years of IIF time series forecasting: a selective review. Monash Econometrics and Business Statistics Working Papers 12/05, Monash University, Department of Econometrics and Business Statistics

Dillon TS, Sestito S, Leung S (1991) An adaptive neural network approach in load forecasting in power systems. ANNPS'91 6(2):442–449

Dordonnat V, Koopman SJ, OOms M, Dessertaine A, Collet J (2008) An hourly periodic state space model for modelling French national electricity load. Int J Forecast 24:566–587

Engle RF, Granger CWJ, Rice J, Weiss A (1986) Semiparametric estimates of the relation between weather and electricity sales. J Am Stat Assoc 81(394):310–320

Hippert HS, Pedreira CE, Souza RC (2001a) Neural networks for short-term load forecasting: a review and evaluation. IEEE Trans Power Syst 16(1):44–55

Hippert HS, Pedreira CE, Souza RC (2001b) Neural networks for short-term load forecasting: a review and evaluation. IEEE Trans Power Syst 16(1):44–55

Hippert HS, Bunn DW, Souza RC (2005) Large neural networks for electricity load forecasting: are they overfitted? Int J Forecast 21(3):425–434

Ho KL, Hsu YY, Chen CF, Lee TE, Liang CC, Lai TS, Chen KK (1990) Short term load forecasting of Taiwan power system using a knowledge-based expert system. IEEE Trans Power Syst 5(4):1214–1221

Kaufman L, Rousseeuw PJ (1990) Finding groups in data: an introduction to cluster analysis. Wiley, New York

Khotanzad A, Afkhami-Rohani R, Tsun-Liang L, Abaye A, Davis M, Maratukulam DJ (1997) ANNSTLF: a neural-network-based electric load forecasting system. IEEE Trans Neural Networks 8(4):835–846

Kodogiannis VS, Anagnostakis EM (2002) Soft computing based techniques for short-term load forecasting. Fuzzy Set Syst 128(3):413–426

Ling SH, Leung FHF, Lam HK, Yim-Shu L, Tam PKS (2003) A novel genetic-algorithm-based neural network for short-term load forecasting. IEEE Trans Ind Electron 50(4):793–799

Makridakis S, Wheelwright SC, Hyndman RJ (1998) Forecasting: methods and applications, 3rd edn. Wiley, New York. ISBN 0-471-53233-9

Moghram I, Rahman S (1989) Analysis and evaluation of five short-term load forecasting techniques. IEEE Trans Power Syst 4(4):1484–1491

Mohandes M (2002) Support vector machines for short-term electrical load forecasting. Int J Energ Res 26:335–345

Moral-Carcedo J, Vicéns-Otero J (2005) Modelling the non-linear response of Spanish electricity demand to temperature variations. Energ Econ 27:477–494

Pai P-F, Hong W-C (2005) Support vector machines with simulated annealing algorithms in electricity load forecasting. Energ Convers Manag 46(17):2669–2688

Pardo A, Meneu V, Valor E (2002) Temperature and seasonality influences on Spanish electricity load. Energ Econ 24:55–70

Park D, El-Sharkawi M, Marks R (1991) Electric load forecasting using an artificial neural network. IEEE Trans Power Syst 6(2):442–449

Piras A, Buchenel B (1999) A tutorial on short term load forecasting. Eng Intell Syst 1:41–47

Rahmam S, Bhamagar R (1988) An expert-system based algorithm for short-term load forecasting. IEEE Trans PWRS 3:392–399

Rahman S, Hazim O (1993) A generalized knowledge-based short-term load-forecasting technique. IEEE Trans Power Syst 8(42):508–514

Ramanathan R, Engle R, Granger CWJ, Vahid-Araghi F, Brace C (1997) Short-run forecasts of electricity loads and peaks. Int J Forecast 13(2):161–174

Sailor DJ, Muñoz JR (1997) Sensitivity of electricity and natural gas consumption to climate in the USA – Methodology and results for eight states. Energy 22(10):987–998

Sánchez-Úbeda EF, Wehenkel L (1998) The Hinges model: a one-dimensional continuous piecewise polynomial model. In: Information Processing and Management of Uncertainty in Knowledge-Based Systems (IPMU'98), vol. 2. Paris, France, pp. 878–885

Senjyu T, Mandal P, Uezato K, Funabashi T (2005) Next day load curve forecasting using hybrid correction method. IEEE Trans Power Syst 20(1):102–109

Soares LJ, Medeiros MC (2008) Modeling and forecasting short-term electricity load: a comparison of methods with a application to Brazilian data. Int J Forecast 24:630–644

Taylor JW (2003) Short-term electricity demand forecasting using double seasonal exponential smoothing. J Oper Res Soc 54:799–805

Taylor JW (2008) An evaluation of methods for very short-term load forecasting using minute-by-minute British data. Int J Forecast 24:645–658

Taylor JW, Buizza R (2003) Using weather ensemble predictions in electricity demand forecasting. Int J Forecast 19:57–70

Taylor JW, McSharry PE (2008) Short-term load forecasting methods: an evaluation based on european data. IEEE Trans Power Syst 22:2213–2216

Taylor JW, de Menezes LM, McSharry PE (2006) A comparison of univariate methods for forecasting electricity demand up to a day ahead. Int J Forecast 22:1–16
Valor E, Meneu V, Caselles V (2001) Daily air temperature and electricity load in Spain. J Appl Meteorol 40:1413–1421
Vapnik VN (1995) The nature of statistical learning theory. Springer, Heidelberg
Vermaak J, Botha EC (1998) Recurrent neural networks for short-term load forecasting. IEEE Trans Power Syst 13(1):126–132
Winters PR (1960) Forecasting sales by exponentially weighted moving averages. Manag Sci 6(3):324–342

Wind Power Forecasting

Alexiadis MC, Dokopoulos PS, Sahsamanoglou HS (1999) Wind speed and power forecasting based on spatial correlation models. IEEE Trans Energ Convers 14:836–842
Barbounis TG, Theocharis JB, Alexiadis MC, Dokopoulos PS (2006) Long-term wind speed and poweer forecasting using local recurrent neural network models. IEEE Trans Energ Convers 21(1):273–284
Beyer HG, Degner T, Haussmann J, Hoffman M, Rujan P (1994) Short term prediction of wind speed and power output of a wind turbine with neural networks. In: 2nd Eur. Congress Intelligent Techniques Soft Computing. Thessaloniki, Greece, pp. 349–352
Costa A, Crespo A, Navarro J, Lizcano G, Madsen H, Feitosa E (2006) A review of the young history of the wind power short-term prediction. Renew Sustain Energ Rev 12(6):1725–1744
Damousis IG, Alexiadis MC, Theocharis JB, Dokopoulos PS (2004) A fuzzy model for wind speed prediction and power generation in wind parkas using spatial correlation. Energ convers 19(2):352–361
Giebel G, Kariniotakis G (2007) Best practice in short-term forecasting. A users Guide. EWEC. Milan
Giebel G, Brownsword R, Kariniotakis G (2003) The state-of-the-art in short-term prediction of wind power. a literature overview. ANEMOS Project http://anemos.cma.fr
Giebel G, Barger J, Landberg L, Nielsen HA, Nielsen TS, Madsen H, Sattler K, Feddeersen H, Vedel H, Tofting J, Kruse L, Voulund L (2005) Wind power prediction using ensembles Riso National Laboratory. Riso-R-1527
Gou-Rui J (2007) Wind speed forecasting based on support vector machine with forecasting error estimation. Proceedings of the 6th International Conference on Machine learning and Cybernetics.
Gouyang W, Yang X, Shasha W (2005) Discussion about short-term forecast of wind speed on wind farm. Jilin Electric Power 181:23–35
Hindman RJ, Koehler AB (2006) Another look at measures of forecast accuracy. Int J Forecast 22:679–688
IDAE (2003) Estudio sobre Predicción Eólica en la Unión Europea. http://www.idae.es
Jensen US, Pelgrum E, Madsen H (1994) The development of a forecasting model for the prediction of wind power production to be used in central dispatch Centres. 5th European Wind Energy Conference, vol. 1. Thessaloniki, Greece, pp. 353–356
Kariniotakis GN, Stavrakakis GS, Nogaret EF (1996) Wind power forecasting using advanced neural network models. IEEE Trans Energ Convers 11(4):762–767
Landberg L (1999) A mathematical look at a physical power prediction model. Wind Energ 1:23–28
Landberg L, Giebel G, Nielsen HA, Nielsen TS, Madsen H (2003) Short-term prediction – an overview. Wind Energ 6(3):273–280
Lei M, Shiyan L, Chuanwen J, Hongling L, Zhang Y (2009) A review on the forecasting of wind speed and generated power. Renew Sustain Energ Rev 13:915–920

Li S, Wunsch DC, O'Hair EA, Giesselmann MG (2001) Using neural networks to estimate wind turbine power generation. IEEE Trans Energ Convers 16(3):276–282

Luenberger DG (1984) Linear and nonlinear programming, 2nd edn. Addison-Wesley, MA

Milligan M, Schwartz MN, Wan Y (2003) Short-term prediction – an overview. National Renewable Energy Laboratory, Golden, USA

Newbold P, Granger CWJ (1974) Experience with forecasting univariate time series and the combination of forecasts (with Discussion). J Roy Stat Soc A 137:131–146

Nielsen TS, Joensen A, Madsen H (1998) A new reference for Wind Power forecasting. Wind Energ 1:29–34

Sánchez I (2006) Short term prediction of wind energy production. Int J Forecast 22:43–56

Sfetsos A (2000) A comparison of various techniques applied to mean hourly wind speed time series. Renew Energ 21(1):23–25

Sfetsos A (2001) A novel approach for the forecasting of mean hourly wind speed time series. Renew Energ 27:163–174

Sideratos G, Hatziargyriou ND (2007) An advanced statistical method for wind power forecasting. IEEE Trans Power Syst 22:258–265

Torres JL, García A, de Blas M, de Francisco A (2005) Forecast of hourly average wind speed with ARMA models in Navarre. Sol Energ 79:65–77

Yang X, Xiao Y, Chen S-Y (2005) Wind speed and generated power forecasting. Proceedings of CSEE, vol. 25, pp. 1–5

Electricity Price Forecasting

Amjady N (2006) Day-ahead price forecasting of electricity markets by a new fuzzy neural network IEEE Trans Power Syst 21(2):887–896

Andalib A, Safavieh E, Atry F (2008) Multi-step ahead forecasts for electricity prices using NARX: a new approach, a critical analysis of one-step ahead forecasts. Energ Convers Manag 50(3):739–747

Anderson CL, Davison M (2008) A hybrid system-econometric model for electricity spot prices: considering spike sensitivity to forced outage distributions. IEEE Trans Power Syst 23(3): 927–937

Barlow MT (2002) A diffusion model for electricity prices. Math Finance 12(4):287–298

Box GP, Jenkins GM (1976) Time series analysis. Forecasting and control. Holden Day, California

Bunn DW (2004) Modelling prices in competitive electricity markets. Wiley, Chichester

Bunn DW, Karakatsani N (2003) Forecasting electricity prices. Report, London Business School

Burger M, Klar B, Mller A, Schindlmayr G (2004) A spot market model for the pricing of derivatives in electricity markets. Quant Finance 4:109–122

Carnero MA, Koopman SJ, Ooms M (2003) Periodic heteroskedastic RegARFIMA models for daily electricity spot prices. Tinbergen Institute Discussion Paper, TI 2003-071/4

Chang I, Tiao GC, Chen C (1988) Estimation of time series parameters in the presence of outliers. Technometrics 30:193–204

Chen J, Deng S-J, Huo X (2007) Electricity price curve modeling by manifold learning, Tech. Rep., Georgia Institute of Technology. http://www2.isye.gatech.edu/statistics/papers/07-02.pdf

Conejo A, Plazas M, Espinola R, Molina A (2005) Day-ahead electricity price forecasting using the wavelet transform and ARIMA models. IEEE Trans Power Syst 20(2):1035–1042

Conejo AJ, Contreras J, Espinola R, Plazas MA (2005) Forecasting electricity prices for a day-ahead pool-based electric energy market. Int J Forecast 21(3):435–462

Cybenko G (1989) Approximation by superpositions of a sigmoidal function. Mathematics of Control, Signals and Systems 2:303–314

Deng S, Johnson B, Sogomonian A (2001) Exotic electricity options and the valuation of electricity generation and transmission assets. Decis Support Syst 30:383–392

Escribano A, Pea JI, Villaplana P (2002) Modeling electricity prices: international evidence. W.P. 02–27, Universidad Carlos III de Madrid
Gao F, Guan X, Cao X-R, Papalexopoulos A (2000) Forecasting power market clearing price and quantity using a neural network method. In: Power Engineering Summer Meet, Seattle, WA, pp. 2183–2188
Gao C, Bompard E, Napoli R, Cheng H (2007) Price forecast in the competitive electricity market by support vector machine. Physica A 382:98–113
Geman H, Roncoroni A (2006) Understanding the fine structure of electricity prices. J Bus 79:1225–1261
Green R (2008) Electricity wholesale markets: designs now and in a low-carbon future. The Energy Journal, Special Issue. The Future of Electricity: Papers in Honor of David Newbery, pp. 95–124
Guthrie G, Videbeck S (2002) High frequency electricity spot price dynamics: an intra-day markets approach. New Zealand Institute for the Study of Competition and Regulation, Tech. Rep.
Hamilton JD (2005) Regime-switching models. Prepared for: Palgrave Dictionary of Economics
Hong YY, Hsiao CY (2002) Locational marginal price forecasting in deregulated electricity markets using artificial intelligence. Proc Inst Elect Eng Gen Transm Distrib 149(5):621–626
Huisman R (2008) The influence of temperature on spike probability in day-ahead power prices. Energ Econ 30:2697–2704
de Jong CM, Huisman R (2003) Option formulas for mean-reverting power prices with spikes. Energ Power Risk Manag 7:12–16
Karakatsani N, Bunn D (2004) Modelling stochastic volatility in high-frequency spot electricity prices. London Business School Working Paper
Karakatsani N, Bunn D (2008) Forecasting electricity prices: the impact of fundamentals and time-varying coefficients. Int J Forecast 24:764–785
Knittel CR, Roberts MR(2005) An empirical examination of restructured electricity prices. Energ Econ 27:791–817
Koopman SJ, Ooms M, Carnero MA (2007) Periodic seasonal Reg-ARFIMA-GARCH models for daily electricity spot prices. J Am Stat Assoc 102:16–27
Li G, Liu C-C, Lawarree J, Gallanti M, Venturini A (2005) State-of-the-art of electricity price forecasting. In CIGRE/IEEE PES Symposium on Congestion Management in a Market Environment. Place of Publication, New Orleans, LA, USA
Lora AT, Santos JMR, Exposito AG, Ramos JLM, Santos JCR (2007) Electricity market price forecasting based on weighted nearest neighbors techniques. IEEE Trans Power Syst 22(3):1294–1301
Lucia JJ, Schwartz ES (2002) Electricity prices and power derivatives: evidence from the nordic power exchange. Rev Derivatives Res 5:5–50
Mandal P, Senjyu T, Urasaki N, Funabashi T, Srivastava AK (2007) A novel approach to forecast electricity price for pjm using neural network and similar days method. IEEE Trans Power Syst 22(4):2058–2065
Maravall A (2005) Brief description of the programs, Banco de Espaa (2002)
Mateo A, Muñoz A, García-González J (2005) Modeling and forecasting elctricity prices with input/output hidden Markov models. IEEE Trans Power Syst 20(1):13–24
Misiorek A, Trueck S, Weron R (2006) Point and interval forecasting of spot electricity prices: linear vs. non-linear time series models. Stud Nonlinear Dynam Econometrics 10:1
Niimura T, Nakashima T (2001) Deregulated electricity market data representation by fuzzy regression models. IEEE Trans Syst Man Cybern C Appl Rev 31(3):320–326
Nogales FJ, Contreras J, Conejo AJ, Espínola R (2002) Forecasting next-day electricity prices by time series models. IEEE Trans Power Syst 17(2):342–348
Panagiotelis A, Smith A (2008) Bayesian density forecasting of intraday electricity pirices using multivariate skew t distributions. Int J Forecast 24(4):710–727
Pankratz A (1991) Forecasting with dynamic regresion models. Wiley Series in Probability and Mathematical Statistics, New York
Pilipovic D (1998) Energy risk: valuing and managing energy derivatives. McGraw Hill, New York

Pindoriya NM, Singh SN, Singh SK (2008) An adaptive wavelet neural network-based energy price forecasting in electricity markets. IEEE Trans Power Syst 23(3):1423–1432

Pino R, Parreno J, Gómez A, Priore P (2008) Forecasting next-day price of electricity in the Spanish energy market using artificial neural networks. Eng Appl Artif Intell 21(1):53–62

Rodríguez CP, Anders GJ (2004) Energy price forecasting in the Ontario competitive power system market. IEEE Trans Power Syst 19(1):366–374

Szkuta BR, Sanabria LA (1999) Electricity price short-term forecasting using artificial neural networks. IEEE Trans Power Syst 14(3):851–857

Weron R (2006) Modeling and forecasting electricity loads and prices: a statistical approach. Wiley, Chichester

Weron R (2008) Heavy-tails and regime-switching in electricity prices. Working paper

Weron R, Misiorek A (2008) Forecasting spot electricity prices: a comparison of parametric and semiparametric time series models. Int J Forecast 24:744–762

Wilkinson L, Winsen J (2002) What we can learn from a statistical analysis of electricity prices in New South Wales. Electricity J 15(3):60–69

Zareipour H, Canizares CA, Bhattacharya K (2006a) Application of public-domain market information to forecast Ontario's wholesale electricity prices. IEEE Trans Power Syst 21(4):1707–1717

Zareipour H, Bhattacharya K, Cañizares CA (2006b) Forecasting the hourly Ontario energy price by multivariate adaptive regression splines. In: IEEE PES General Meeting

State-of-the-Art of Electricity Price Forecasting in a Grid Environment

Guang Li, Jacques Lawarree, and Chen-Ching Liu

Abstract The purpose of electricity price forecasting is to estimate future electricity prices, particularly locational marginal prices (LMP), with consideration to both security and capacity constraints in a grid environment. Electricity price forecasting is vital to both market participants and market operators in wholesale electricity markets. Electricity price forecasts are used to assist the decision making of market participants on bidding submissions, asset allocations, bilateral trades, transmission and distribution planning, and generation construction locations. Electricity price forecasts are also used by market operators to uncover possible market power.

The inaccuracy of electricity price forecasting is due to problems associated with volatility of prices, interpretability of explanatory variables, and underlying impacts of power grid security. This study classifies forecasting techniques common in the literature based on their objective, concept, time horizon, input–output specification, and level of accuracy. Thus the state-of-the-art of electricity price forecasting is described in this study. This survey facilitates the validation, comparison, and improvements of specific or combined methods of price forecasting in competitive electricity markets.

Moreover, this study demonstrates a hybrid forecasting system, which combines fuzzy inference system and least-squares estimation. The proposed mechanism is applied to the day-ahead electricity price forecasting of an actual security-constrained, wholesale electricity market. This hybrid forecasting system provides both accuracy and transparency to electricity price forecasts. The forecasting information is also interpretable with respect to the fuzzy representations of selected inputs.

Keywords Day-ahead energy market · Electricity price forecasting · Fuzzy inference system (FIS) · Grid environment · Intelligent systems · Least-squares estimation (LSE) · Locational marginal prices (LMPs) · Price forecasting · Price volatility · Time series

---

G. Li (✉)
Market Operations Support, Electric Reliability Council of Texas, 2705 West Lake Drive, Taylor, TX 76574
e-mail: gli@ercot.com

S. Rebennack et al. (eds.), *Handbook of Power Systems II*, Energy Systems,
DOI 10.1007/978-3-642-12686-4_6,

1 Introduction

In the context of competitive electricity markets, electricity price forecasting is performed with respect to various time scopes: short term, medium term, and long term.

Short-term price forecasting analyzes the day-ahead or real-time electric power markets. Market participants implement their forecasts to optimize either profits or utilities. On the supply side, supply entities use short-term forecasts to decide on market bidding positions for profit maximization. On the demand side, electricity retailers and large consumers can determine a load schedule to maximize the utilities of their electricity energy purchases, or deploy self-generation capability to hedge against market price volatility (Nogales et al. 2002). Therefore, accurate short-term price forecasting benefits market participants in terms of profits or costs (Szkuta et al. 1999).

Medium-term price forecasting is useful for decision making of bilateral trades of energy, capacity, and ancillary services (Szkuta et al. 1999). Price forecasts are used to evaluate contracts of monthly operation schedules during the medium-term time-frame, which is approximately 6 months to 1 year. Generation entities use medium-term forecasts to optimize the portfolio between bilateral contracts and electricity markets for maximum profits. On the demand side, load service entities maximize utilities by allocating their portfolios or contracting for price difference (Nogales et al. 2002). Long-term electricity price forecasting is used mainly for transmission and distribution planning, power plant investment, and inter-regional electricity exchanges (Szkuta et al. 1999).

Electricity price forecasting techniques fall into categories of statistical techniques, simulation techniques, and equilibrium techniques. Statistical techniques perform regressions based on historical data; simulation techniques compute power flows, which observe transmission constraints, resource physical limits, resource operating costs, and load profiles of electric power grids (Bastian et al. 1999); and equilibrium techniques utilize game theory models to analyze the interactivities of market participants (Bunn 2000). For the purpose of hedging risks of price volatility, risk assessment is associated with electricity price forecasting, for instance, confidence interval prediction and volatility analysis (Dahlgren et al. 2001; Deb et al. 2000).

This study demonstrates a mixture of two statistical techniques: fuzzy inference system (FIS) and least-squares estimation (LSE) (Li et al. 2007). It is applied to the day-ahead LMP forecasting. This hybrid system provides both accuracy and transparency to electricity price forecasts. It is also interpretable in the context of fuzzy logic (Li et al. 2007).

The remaining of this paper is organized as follows. Section 2 reviews existing techniques of electricity price forecasting; Sect. 3 describes the classification of the literature based on the objective, time horizon, input–output specification, and level of accuracy; Sect. 4 provides case studies, which compare state-of-the-art techniques of electricity price forecasting; Sect. 5 illustrates a mixture of FIS–LSE, and compares it with prevailing forecasting techniques described in Sects. 3 and 4; Sect. 6 presents the conclusion.

2 State-of-the-Art Techniques of Electricity Price Forecasting

The state-of-the-art techniques of electricity price forecasting are categorized in Fig. 1 (Li et al. 2005). Statistical techniques, simulation techniques, and equilibrium techniques have been implemented for electricity price forecasting over time scopes of short, medium, and long terms.

2.1 Statistical Techniques

Statistical techniques forecast electricity prices by modeling the correlation behaviors between prices and explanatory factors. Therefore, statistical techniques do not require intensive data about grid constraints and resource characteristics. Given appropriate explanatory factors, statistical techniques can be applied to all time horizons of short-, medium-, and long-term electricity price forecasting. Some statistical models can provide volatility information of price forecasts. Statistical techniques are clustered into five types: time series modeling, econometric techniques, intelligent systems applications, statistical learning, and volatility analysis. The literature of each type is reviewed in this section.

2.1.1 Time Series Modeling

Time series techniques formulate the time-domain sequence of electricity prices based on their historical behaviors and underlying contexts. The typical instances of time series in the literature are as follows. A combination of time series and stochastic process is presented in Kian and Keyhani (2001). It is formulated with various explanatory variables such as load profiles, outage schedules, reliability reserve plan, inelastic property of demand, and predictable bidding behaviors. A threshold autoregressive switching (TARSW) and linear auto-regression (AR) model is studied in Stevenson et al. (2001) with a de-noising component of wavelet filter. In Ni and Luh (2001), an AR model combined with Bayesian-based classification is proposed to analyze the probability density functions (PDF) of market clearing prices. Nogales et al. implemented dynamic regression and transfer function AR models to forecast day-ahead forward-market electricity prices (Nogales et al. 2002). Auto regressive integrated moving average (ARIMA) was presented for electricity price forecasting and its performance could be improved with a pre-process of wavelet transform filtering (Conejo et al. 2005; Contreras et al. 2003). With the assumption of varying price variance, general autoregressive conditional heteroscedasticity

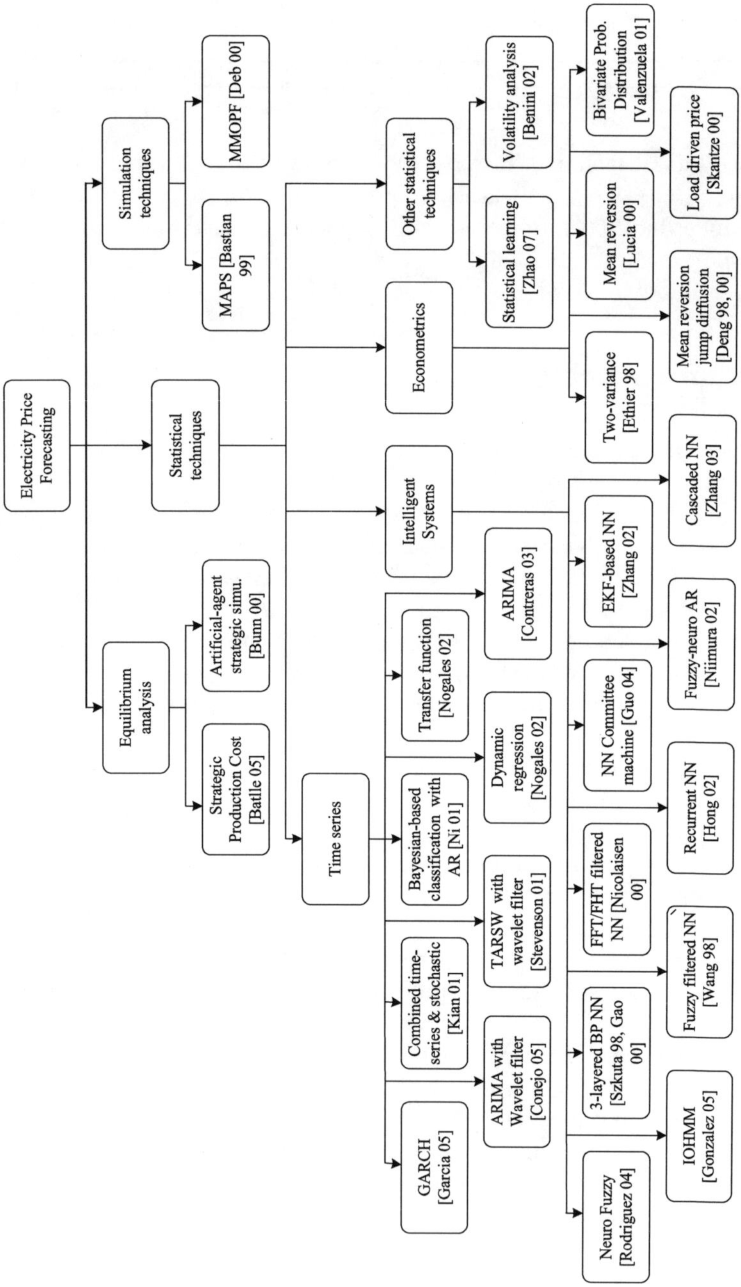

Fig. 1 Categorization of state-of-the-art techniques of electricity price forecasting

(GARCH) models output price forecasts based on historical and future price variances (Garcia et al. 2005). A time series model considers neighboring electricity market data in Zareipour et al. (2006). Seasonal ARIMA (SARIMA), a variant of ARIMA, is presented in Olsson and Soder (2008). This SARIMA has been combined with discrete Markov processes for real time balancing market price forecasting (Olsson and Soder 2008).

2.1.2 Econometric Techniques

Econometric techniques are based on transition dynamics such as mean reversion processes. The typical instances of econometrics in the literature are as follows. A two-variance technique models discontinuous price jumps by combining mean reversion and regime switching in Ethier and Mount (1998). Then it was extended to a three mean-reversion jump-diffusion model of multiple jumps, regime-switching, and stochastic volatility (Deng 2000, 1998). A two-factor model formulates the stochastic context of prices as a mixture of short-term mean reversion and long-term equilibrium (Lucia and Schwartz 2002). It is noticed that this model considers both short- and long-term time scopes. Not only temporal factors, the demand and supply factors are also introduced in econometric techniques. For example, a load-driven price method analyzes the closed-loop price dynamics based on the supply-demand interplay in a multi-market environment in Skantze et al. (2000). Another stochastic model is designed to evaluate the bivariate probability distribution of electricity prices in Valenzuela and Mazumdar (2001). The correlation between price spikes and system conditions, forced outages, and load patterns are used in econometric price forecasting in Anderson and Davison (2008), Valenzuela and Mazumdar (2001).

2.1.3 Intelligent System Applications

Intelligent system applications are implementation of artificial neural networks (ANN), fuzzy logic, or their mixtures such as fuzzy neural network and neuro-fuzzy method (Amjady 2006). Rather than modeling elaboration, intelligent system techniques apply sample learning of universal input–output mapping. Two typical price forecasting instances of ANN are multilayer perceptron ANN (Szkuta et al. 1999) and recurrent neural networks (RNN) (Hong and Hsiao 2002). The prediction accuracy and off-line training performance of ANN can be improved by preprocessing, over-fitting prevention, period partition, and outlying price truncation (Gao et al. 2000; Hong and Hsiao 2002; Nicolaisen et al. 2000; Wang and Ramsay 1998). However, accurate extrapolation beyond the training scope is still an open issue for ANN (Bunn 2000). A group of ANNs can be in the form of streamline cascading or committee machine (Guo and Luh 2004; Zhang et al. 2003). Some variants of ANN exist in the literature. For example, Input/Output Hidden Markov Model (IOHMM) utilizes two ANNs with respect to underlying market states (Gonzalez et al. 2005).

Another variant of feed-forward neural network (FFNN) adopted wavelet transform as its activation function (Pindoriya et al. 2008). Besides ANN, fuzzy logic applications, such as FIS (Li et al. 2007) and adaptive neuro-fuzzy inference system (ANFIS) (Rodriguez and Anders 2004), achieve electricity price forecasting as well.

2.1.4 Statistical Learning Methods

Similar to intelligent systems, statistical learning techniques are also nonparametric approaches. Kernel learning and manifold learning are the typical instances of statistical learning in the literature. A data mining-based approach of support vector machine (SVM) and probability classifier has been used to forecast occurrence of price spikes in Zhao et al. (2007). A heteroscedastic variance formulation has been designed in Zhao et al. (2008). It evaluates the confidence intervals for the price forecast from SVM. Nonlinear dimensionality reduction, such as manifold learning techniques for locally linear embedding, has also been adopted for electricity price forecasting (Chen et al. 2008).

2.1.5 Volatility Analysis

Volatility analysis has been combined in such statistical techniques as time series models and intelligent systems applications. To emphasize its distinct applications to risk assessment, volatility analysis is described as a separate cluster of statistical techniques in the literature. To assess price risks, volatility analysis is associated with electricity price forecasting in several ANN variants (Alvarado and Rajaraman 2000; Bastian et al. 1999; Benini et al. 2002; Botto 1999; Dahlgren et al. 2001). Price uncertainty, such as confidence intervals or possible ranges, has been modeled by Extended Kalman Filter (EKF) or fuzzy AR model in ANN price forecasting (Niimura et al. 2002; Zhang and Luh 2005). To evaluate the confidence interval, PDF is computed in Conejo et al. (2005). Dahlgren et al. estimated the volatility as the average standard deviation of a random walking formulation of electricity prices in Dahlgren et al. (2001). An Exponential GARCH (EGARCH) specification has been used to model the conditional mean and conditional volatility of electricity prices (Chan and Gray 2006). This EGRCH applied extreme value theory to model the tails of the return distribution (Chan and Gray 2006). The market price of risk, that is, the ratio of excess return to standard deviation has been related to electricity futures prices in models of the Term Structure Of Volatility (TSOV) (Kolos and Ronn 2008).

2.2 *Simulation Techniques*

From the view of power engineering, simulation techniques perform production computation, such as unit commitment and economic dispatch. Simulation techniques respect various grid constraints such as transmission capability, resource constraints, production costs, and load forecasts. To accurately capture chronological grid snapshots, simulation techniques require intensive data about system constraints and resource characteristics. For example, market assessment and portfolio strategies (MAPS) model evaluates locational marginal prices (LMP) (Bastian et al. 1999). Another multi-commodity multi-area optimal power flow (MMOPF) model runs Monte Carlo simulation for prediction of prices and their volatility (Deb et al. 2000).

2.3 Equilibrium Techniques

In the context of specific market structure and clearing mechanisms, equilibrium techniques interpret the strategic interactions among market participants. It is useful for evaluation of long-term price trends. The strategic simulation using artificial agents is an example of equilibrium techniques (Bunn 2000). Based on the participants cost structure and expected competitive behavior, profit maximizing strategic bidding curves can be derived and solved for the prediction of market equilibrium prices. Various oligopoly models – such as Bertrand, Cournot, and supply function equilibrium (SFE) – are used to forecast electricity prices and their variance (Ruibal and Mazumdar 2008).

3 Input–Output Specifications of Electricity Price Forecasting Techniques

Electricity price forecasting techniques fall into categories of statistical techniques, simulation techniques, and equilibrium techniques. Each category has its own specific input data and output results.

3.1 Statistical Techniques

In a statistical formulation, the outputs focus on forecasts of market clearing prices (MCP) and relevant statistical property such as price confidence interval and volatility. The data specification of a statistical model is illustrated below. Table 1 summarizes a typical input specification of statistical technique (Li et al. 2005). The output specification includes MCPs, market clearing quantities (MCQs), confidence interval, and volatility.

Table 1 Input specification of a statistical technique

Item	Explanation
Historical hourly MCPs and MCQs	Future hourly MCPs and MCQs the outputs are correlated with historical hourly MCPs and MCQs in temporal patterns.
Temporal indices	Hourly MCPs and MCQs vary in temporal cycles. Hour, weekday/weekend, and month indices can be used to indicate the daily, weekly, and seasonal patterns.
Historical loads and day-ahead load forecast	Hourly MCQs are highly correlated with system load profiles.
Price elasticity of demand	The price elasticity of demand indicates the sensitivity of the demand with respect to changes in price.
Temperature	Temperature is the critical measure for the weather impacts on loads.
Fuel cost	The generation cost depends on fuel prices. The generation cost influences the supply offers in electricity markets.
Power import and export	In a particular area, the power import and export through bilateral contracts affect the equilibrium in electricity markets.
Transmission congestion volume	The existence of transmission congestion results in the need to deliver power above system marginal cost, and possible market powers, which will affect the variation of MCPs.
Historical hydro production and hydro energy prediction	The available hydro capacity influences the variation of MCP.
System power reserves	In the forms of opportunity costs, system power reserves affect the available resources of generation and hence the MCP.

A variant: Zonal/locational statistical techniques

A zonal or locational statistical technique is based on the treatment to zonal or locational data (Kian and Keyhani 2001). Its outputs focus on prediction of zonal prices and relative statistical property such as zonal confidence interval. The input–output specification of a zonal/locational statistical technique is summarized in Table 2 (Li et al. 2005).

3.2 Simulation Techniques

A simulation technique recreates the operation scheduling of a power system in terms of production cost optimization. Such a simulation requires information about transmission network security, generation resource constraints, and production costs (startup costs, minimum generation costs, and fuel cost curves). The input specification of a typical simulation technique is summarized in Table 3 (Bastian et al. 1999; Li et al. 2005). The output specification of a simulation technique is summarized in Table 4 (Bastian et al. 1999).

Table 2 Input–output specification of a zonal/locational statistical technique

Category	Item
Input	• Zonal day-ahead hourly prices of energy • Zonal spot price (real-time price) of energy • Zonal hourly load • Zonal hourly generation • Zonal spinning reserve requirement • Group transmission congestion prices • Generator outage maintenance schedule on capacity and period of outage • Transmission outage maintenance schedule on capacity and period of outage • Calculated zonal demand-elasticity values
Output	• Zonal prices • Zonal confidence interval

Table 3 Input specification of a typical simulation technique

Item	Explanation
Load profiles	• Load forecasts, load curves, and nonconforming loads
Resource attributes	• Zonal prices • Unit profile: capacity, heat rates, ownership, decision-making entity • Maintenance: maintenance schedule and forced outage rates • Operation: quick start capability and minimum down/up time • Fixed and variable operation and maintenance costs • Fuel data: fuel cost and fuel types • Environmental data: emission data and environmental factors
Transmission elements	• Network topology: solved AC power flow, phase angle regulator, and HVDC • Limits: thermal ratings, voltage limits, stability limits, and inter-tie limits • Outage schedule: maintenance schedules of transmission lines and transformers
Hydro attributes	• Unit profile: capacity, minimum rating, ownership, decision-making entity • Dispatch: dispatch strategy and available energy • Storage: storage tank limits and maximum pumping rating • Synchronized condensing: generation mode or spinning mode
Ancillary services	• System requirements on regulation up, regulation down, responsive reserves, and nonspinning
Transmission transactions	• Transferring amount, routing, and wheeling charges
Other	• Inflation pattern, and emergency cost

3.3 Equilibrium Techniques

Equilibrium techniques apply game theory and economic models to determine strategic activity and possible market power. Market power is the ability of market participants to receive awards above marginal costs (Bunn 2000). The input specification of an equilibrium analysis model focuses on market factors such as market shares, market clearing mechanisms, and contracting patterns. Production costs and

Table 4 Output specification of a simulation technique

Item	Explanation
LMPs	LMPs at generators and loads
Unit commitment and dispatch	Hourly dispatch, hourly emission, capacity factors, number of startups, and online durations
Security Monitoring	Hourly line flow, congestion cost, and identification of binding/violated constraints

Table 5 Input specification of an equilibrium technique

Item	Explanation
Market factors	• Market shares • Market clearing mechanisms • Contracting patterns • Market participant bidding strategies
Physical factors	• Production costs • Load Profiles • Generation reliability

other physical factors have also to be taken into account for an equilibrium model (Ruibal and Mazumdar 2008). The output specification includes long-term price trends and predicted market behaviors. The input specification of an equilibrium technique is summarized in Table 5 (Li et al. 2005; Ruibal and Mazumdar 2008). The output specification of an equilibrium technique include long-term price trends and predicted market behaviors.

4 Comparing Existing Statistical Techniques for Electricity Price Forecasting

The popular statistical forecasting techniques fall into two categories: time series techniques and intelligent system applications. Stochastic models, such as autoregressive moving-average (ARMA) (Bastian et al. 1999; Conejo et al. 2005; Contreras et al. 2003; Li et al. 2005; Ni and Luh 2001; Nogales et al. 2002; Stevenson et al. 2001) and GARCH models (Garcia et al. 2005), are common time series techniques being used for electricity price forecasting. Stochastic models require elaborate specifications to estimate the parameters of disturbance correlation. On the other hand, intelligent system applications perform universal input–output mapping without parametric formulation. Intelligent system applications such as neural networks (Gao et al. 2000; Gonzalez et al. 2005; Guo and Luh 2004; Hong and Hsiao 2002; Nicolaisen et al. 2000; Rodriguez and Anders 2004; Szkuta et al. 1999; Wang and Ramsay 1998; Zhang et al. 2003) and fuzzy logic (Rodriguez and Anders 2004) have been applied to electricity price forecasting. However, their accuracy is still an open question for the scenario in which actual prices are beyond the training scope.

This section investigates several prevailing time series techniques and intelligent system applications: ANFIS, ARMA, ANN, and GARCH. Test scenarios are configured for these statistical models with respect to day-ahead electricity price forecasting, and then their performances are compared. Case 1 sets up ANFIS forecasting with input preprocessing of linear transformation for LMP and logarithmic transformation for load. In the formulated ANFIS, the output is a linear combination of membership values of the inputs (Rodriguez and Anders 2004). Case 2 and Case 3 set up ARMA forecasting with the general form of dynamic regression and transfer function models (Nogales et al. 2002) and ARIMA model (Contreras et al. 2003). Case 4 applied cascaded-forward neural networks (CFNN), which is an instance of three-layered ANN with back-propagation (Szkuta et al. 1999). Case 5 sets up the forecasting with Elman-type RNN (Hong and Hsiao 2002). Case 6 is the application of feed-forward neural networks (FFNN) (Szkuta et al. 1999). GARCH methodology is elaborated in Case 7, Case 8, Case 9, and Case 10 (Garcia et al. 2005). The availability of historical prices determines the GARCH (1, 3) configuration in Case 7 and Case 8 but the GARCH (0, 3) configuration in Case 9 and Case 10.

The study cases are based on publicly available information of load-average-weighted day-ahead LMPs in the PECO zone in 2004 PJM market.[1] Table 6 demonstrates the configurations and computation times of 10 scenarios being investigated (Li et al. 2005). Notice that the load input is the historical PECO zonal load due to the unavailability of load forecast information. This substitution is feasible, given the reality of approximately 2% load forecast error.

Each scenario is tested with corresponding toolboxes in MATLAB for four typical weeks in 2004. These four weeks correspond to four seasons. The winter week is from February 23rd to 29th of 2004; the spring week is from May 17th to 23rd of 2004; the summer week is from August 23rd to 29th of 2004; the fall week is from November 22nd to 28th of 2004. The training data for each scenario is the 48-day period previous to the test period (Conejo et al. 2005). Based on same training data and test data, the specifications such as NN structure, ARMA orders, training methods, and input transformation have been manipulated to approach the best accuracy for each tested model. For ARMA or GARCH models in Case 2, Case 3, Case 7, Case 8, Case 9, and Case 10, both the autoregressive model order and the moving-average model order are set as 24, and the maximum number of iterations is 400. For NN models in Case 4, Case 5, and Case 6, the network training method is gradient descent with momentum and adaptive learning rate back-propagation. Case 1 uses a Sugeno-type FIS for ANFIS training with a grid partition. The number of membership functions per input is 3. The type of membership function for each input is Gaussian curve. In Case 2, the maximum number of objective function evaluations is 31,000. In Case 3, the maximum number of objective function evaluations is 55,000. Case 4 created a two-layer cascade-forward network. The first layer has 20 neurons with hyperbolic tangent sigmoid transfer function, and the second layer has

[1] PJM website. Available: http://www.pjm.com/.

Table 6 Configurations and computation times of ten test scenarios

Scenario	Technique	Input pre-processing	Timestamp of historical price input for study hour h	Timestamp of load input for study hour h	Processing time (min) (Typical Week in Nov. 2004)
Case 1	ANFIS	Logarithm (load only)	h-24, h-48, h-168	h	9.97
Case 2	ARMA	None	h-1~h-24, h-48, h-120, h-144, h-168, h-169, h-192, h-193	h, h-1, h-24, h-168	0.86
Case 3	ARMA	None	h-24~h-49, h-120, h-144, h-168, h-169, h-192, h-193	h, h-1, h-24, h-168	8.02
Case 4	CFNN	Logarithm	h-24, h-168	h	0.94
Case 5	ELMAN	Logarithm	h-24, h-168	h	4.92
Case 6	FFNN	Logarithm	h-24, h-48, h-120, h-144, h-168, h-169, h-192, h-193	h, h-1, h-24, h-168	0.62
Case 7	GARCH (1,3)	None	h-1~h-24, h-48, h-120, h-144, h-168, h-169, h-192, h-193	h, h-1, h-24, h-168	0.7
Case 8	GARCH (1,3)	Logarithm	h-1~h-24, h-48, h-120, h-144, h-168, h-169, h-192, h-193	h, h-1, h-24, h-168	0.68
Case 9	GARCH (0,3)	None	h-24~h-49, h-120, h-144, h-168, h-169, h-192, h-193	h, h-1, h-24, h-168	6.98
Case 10	GARCH (0,3)	Logarithm	h-24~h-49, h-120, h-144, h-168, h-169, h-192, h-193	h, h-1, h-24, h-168	7.45

one neuron with linear transfer function. The maximum number of epochs to train is 1,000, and the performance goal is 0.01. Case 5 created an Elman back-propagation network. It has 20 hidden neurons with hyperbolic tangent sigmoid transfer function, and one output neuron with linear transfer function. The maximum number of epochs to train is 500, and the performance goal is 0.01. Case 6 created a two-layer feed-forward back-propagation network. The first layer has 20 neurons with hyperbolic tangent sigmoid transfer function, and the second layer has one neuron with linear transfer function. The maximum number of epochs to train is 10,000, and the performance goal is 0.001. In Case 7 and Case 8, the GARCH model order is set as 1 and the ARCH model order is set as 3. The maximum number of objective function evaluations is 31,000. In Case 9 and Case 10, the GARCH model order is set as 0 and the ARCH model order is set as 3. The maximum number of objective function evaluations is 55,000.[2]

To compare the computing times, the CPU processing time of each scenario for the November week in 2004 is summarized in Table 6. All computations are implemented in MATLAB version 7.0 on a PC with a CPU speed of 3.0 GHz. The observation is that the execution of ANFIS requires more time than any other scenarios. In general, neural networks applications require least processing times. Time series techniques such as ARMA and GARCH stay in between.

Two commonly used criteria are applied to measure the accuracy of price forecasting: root mean square error (RMSE) (Contreras et al. 2003; Garcia et al. 2005; Nogales et al. 2002) and mean absolute percentage error (MAPE) (Contreras et al. 2003; Nogales et al. 2002). RMSE and MAPE are computed as follows:

$$RMSE = \sqrt{\frac{1}{N}\sum_{i=1}^{N}(\hat{p}_i - p_i)^2} \tag{1}$$

$$MAPE = \frac{100}{N}\sum_{i=1}^{N}\frac{|\hat{p}_i - p_i|}{p_i} \tag{2}$$

N is the count of price samples,

$$p_i, i = 1, \ldots, N$$

are actual prices, and

$$\hat{p}_i, i = \ldots, N$$

are price forecasts.

Table 7 demonstrates the RMSE and MAPE of ten test scenarios (Li et al. 2005).

Case 3 in the August week is infeasible. The cause is nonstationary AR polynomial. Nonstationary time series are generally differenced to be modeled as an ARMA process. Otherwise, ARIMA models should probably be used instead of ARMA models. However, to be compared with GARCH models in Case 7 and Case 9, ARMA models are studied in Case 2 and Case 3, and not ARIMA models. Case 2 and Case 3 have exactly the same transformations, ARMA orders, and explanatory variables as those in Case 7 and Case 9, respectively. For further studies, the details of ARIMA modeling applied to electricity price forecasting can be found in Conejo et al. (2005), Contreras et al. (2003), and a detailed comparison between ARIMA and GARCH applications to electricity prices exists in Garcia et al. (2005). Notice that Case 2, Case 7, and Case 8 use historical data of the hour interval previous to the study hour. Load forecasting models are more realistic when combined with previous forecasts. However, electricity prices are more volatile than loads.

[2] MATLAB, the Math Works. Available: http://www.mathworks.com/.

Table 7 RMSE and MAPE in the forecasting of PECO zonal day-ahead LMPs in PJM 2004

Scenario	February RMSE(MAPE)	May RMSE(MAPE)	August RMSE(MAPE)	November RMSE(MAPE)	Average RMSE(MAPE)
Case 1	7.33 $/MWh (11.7%)	8.49 $/MWh (11.7%)	7.11 $/MWh (15.2%)	5.06 $/MWh (10.9%)	7.11 $/MWh (12.4%)
Case 2	6.54 $/MWh (15.3%)	5.61 $/MWh (9.6%)	2.48 $/MWh (6.6%)	6.30 $/MWh (13.3%)	5.48 $/MWh (11.2%)
Case 3	5.95 $/MWh (13.5%)	6.57 $/MWh (9.7%)	N/A	8.19 $/MWh (19.8%)	6.97 $/MWh (14.3%)
Case 4	4.92 $/MWh (8.8%)	7.15 $/MWh (8.9%)	3.09 $/MWh (7.6%)	8.17 $/MWh (17.3%)	6.16 $/MWh (10.7%)
Case 5	6.85 $/MWh (11.4%)	7.76 $/MWh (10.4%)	3.55 $/MWh (7.9%)	6.30 $/MWh (14.7%)	6.31 $/MWh (11.1%)
Case 6	5.23 $/MWh (9.5%)	6.27 $/MWh (8.8%)	3.97 $/MWh (9.4%)	5.06 $/MWh (10.7%)	5.20 $/MWh (9.6%)
Case 7	4.35 $/MWh (9.0%)	5.37 $/MWh (8.7%)	2.45 $/MWh (6.3%)	6.21 $/MWh (13.0%)	4.80 $/MWh (9.3%)
Case 8	6.09 $/MWh (12.1%)	5.56 $/MWh (8.3%)	2.66 $/MWh (7.2%)	6.53 $/MWh (14.1%)	5.43 $/MWh (10.4%)
Case 9	5.77 $/MWh (13.2%)	8.26 $/MWh (10.9%)	3.20 $/MWh (8.5%)	5.70 $/MWh (11.4%)	6.00 $/MWh (11.0%)
Case 10	6.19 $/MWh (11.9%)	6.82 $/MWh (10.0%)	3.15 $/MWh (8.4%)	6.83 $/MWh (15.4%)	5.94 $/MWh (11.4%)

Thus this configuration is less realistic for price forecasting of day-ahead electricity market because only previous forecasts could be fed into the forecasting model. A day-ahead electricity market is cleared for all bids and offers of 24 hourly intervals in the operating day. The forecasting error would be accumulated for later intervals. Because of error accumulation, the forecasts of later intervals such as hour 23 and hour 24 would be less accurate than early intervals such as hour 1 and hour 2. The result shows that the availability of historical prices within 24 hours in Case 2, Case 7, and Case 8 does not improve the forecasting accuracy significantly if these cases are compared with Case 3, Case 9, and Case 10. This observation is consistent with the operation process of the day-ahead market, in which market participants submit their bidding decisions before all the market clearing results within 24 hour intervals are available.

Focusing on realistic and complete scenarios of Case 1, Case 4, Case 5, Case 6, Case 9, and Case 10, (Case 2, Case 7, and Case 8 are less realistic for day-ahead price forecasts, and Case 3 is incomplete), the average RMSE ranges from 5.2 $/MWh to 7.1 $/MWh, and average MAPE ranges from 9.6% to 12.4%. This fact indicates that, given basic inputs of loads and historical prices, the studied techniques forecast at a similar level of accuracy.

In all the 10 test scenarios being studied, Case 1, Case 5, and Case 9 are selected for illustration. Case 1 is a FIS model, Case 5 is a NN model, and Case 9 is an ARMA/GARCH model (Li et al. 2005). They are realistic for day-ahead price forecasting. Figure 2 demonstrates the price forecasts of Case 1, Case 5, and Case 9

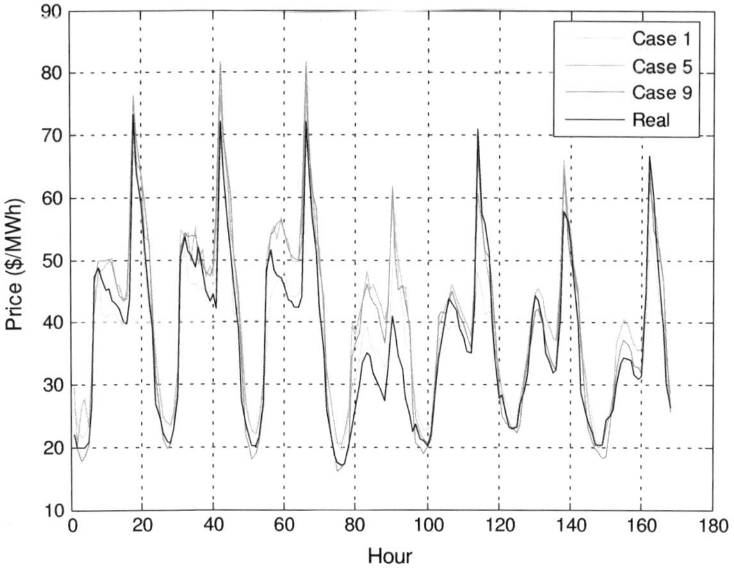

Fig. 2 Illustration of price forecasts of Case 1, Case 5, and Case 9 against actual PECO zonal LMPs in the week of 22nd–28th November 2004

against actual PECO zonal LMPs in the week of 22nd–28th November 2004 (Li et al. 2005).

5 Implementations of Electricity Price Forecasting in a Grid Environment

In this section, FIS and time series methods are combined for day-ahead LMP forecasting with respect to the congestive conditions in a grid environment (Li et al. 2007). FIS is implemented due to its advantage of transparency and interpretability. Least-squares estimation (LSE) is adopted as a simplified parameterization of time series methodology such as AR models. Zonal/area-wide loads and day-ahead transmission constraints are inputs that provide information on demands and congestions, respectively (Li et al. 2007).

5.1 Fuzzy Inference System

Intelligent system applications are implemented for electricity price forecasting due to their adaptability and machine learning capability. There is a tradeoff between interpretability and accuracy for traditional intelligent systems such as ANN and

Decision Tree (DT) (Nauck 2000). The forecasting accuracy of ANN can be superior to that of DT, but DT is more interpretable than an ANN.

A FIS is an instance of fuzzy logic for universal input–output mapping. FIS utilizes fuzzy membership functions to evaluate the intermediate states between discrete states, hence is able to handle the concept of partial truth instead of absolute truth (Wang and Mendel 1992). Figure 3 demonstrates an example of fuzzy membership functions of day-ahead prices.

Adaptive fuzzy systems, such as ANFIS and neuro-fuzzy methods, intend to combine the advantages of ANN and fuzzy logic. They iterate the learning process as ANN does. ANFIS has straightforward linear output functions similar to that of a general FIS, while Neuro-Fuzzy systems are basically an ANN embedded with fuzzy processing (Nauck 2000).

Wang et al. described the mathematical formulation of FIS in detail (Wang and Mendel 1992). The input–output transformation carried out by FIS has been specified as well in Wang and Mendel (1992). FIS is able to formulate heuristic market factors of electricity prices. The study in this section shows that FIS provides interpretable heuristic knowledge as well as adequate accuracy. In this section, Wang–Mendel learning algorithm is utilized to develop a FIS for day-ahead electricity price forecasting (Canazza et al. 2005; Li et al. 2005; Wang and Mendel 1992). This approach is heuristic in interpretation and simple in model architecture. The fuzzy rule base of FIS is a common framework, which combines both numerical and verbal information. The numerical information is acquired from measurements, and the source of verbal information is either interpretation of numerical information or the experience of subject matter experts. Wang–Mendel learning algorithm is competitive to the accuracy of ANN. Therefore, this implementation of FIS is able to provide both interpretability and accuracy due to its linguistically interpretable rule base and competitive accuracy (Wang and Mendel 1992). The interpretability of the FIS model has been exploited in Guillaume (2001), Guillaume and Charnomordic (2004), Wang et al. (2005), and Yen et al. (1998). In addition, FIS constructs the rule base with a one-time process of training samples. It is more efficient than any iterative training techniques of time series, such as ARMA and GARCH, and intelligent systems, such as ANN and neuro-fuzzy systems.

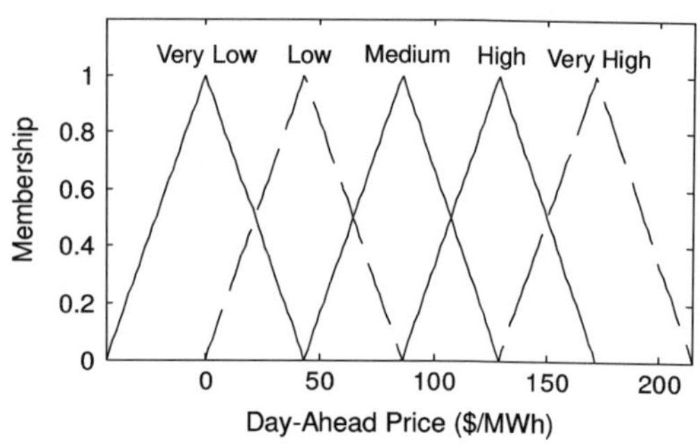

Fig. 3 A fuzzy membership function of day-ahead prices

Table 8 A verbal rule base of simple FIS price forecasting

Rule	Antecedent 1 week partition	Antecedent 2 previous-day price	Antecedent 3 area-wide load	Consequent price forecast	Degree of certainty
1	Early	Low	Low	Very Low	0.82
2	Early	Low	Medium	Low	0.56
3	Early	Medium	Low	Low	0.55
4	Early	Medium	Medium	Medium	0.88
5	Early	Medium	High	Medium	0.35
6	Middle	Low	Low	Very Low	0.72
7	Middle	Low	Medium	Low	0.53
8	Middle	Medium	Low	Low	0.39
9	Middle	Medium	Medium	Medium	0.89
10	Middle	Medium	High	Medium	0.50
11	Middle	High	Medium	Medium	0.67
12	Late	Low	Low	Very Low	0.63
13	Late	Low	Medium	Low	0.46
14	Late	Medium	Low	Low	0.42
15	Late	Medium	Medium	Low	0.74
16	Late	Medium	High	Medium	0.35
17	Late	High	Medium	High	0.24
18	Late	High	High	Medium	0.17

FIS provides a transparent verbal rule base. This rule base is able to contain both numerical information such as historical prices and load demands, as well as verbal indicators such as weekday/weekend and peak-hour/off-peak-hour. Furthermore, the rules can be adjusted manually to reflect human expert knowledge. As a result, the rule base of FIS supports the features of both transparency and interpretability. Additionally, FIS supports flexibility with respect to predefining membership functions.

The inputs and outputs of FIS are referred to as antecedents and consequents, respectively. The verbal rules of FIS are clauses of logic relationships between antecedents and consequents. The rule base collects the relationship between antecedents and consequents from sampling process. Degree of certainty (DOC) is the measure by which an antecedent or consequent fits one or more membership functions for the studied sample. The sample DOC is the DOC multiplication of all antecedents and consequents. For samples with duplicated set of antecedents, the fuzzy rule base keeps only the sample with maximum DOC (Wang and Mendel 1992). This maximum sample DOC is associated with the rule; hence, it is called the rule DOC. After all, the rule base is created as a list of rules, which is composed of membership functions of each variable. The confidence level to this rule is indicated by the corresponding rule DOC. Table 8 demonstrates a verbal rule base of simple FIS price forecasting. The visible contour of nonlinear relationship among week partition, previous-day prices, and Day-Ahead (DA) price forecasts is illustrated in Fig. 4 (Li et al. 2007).

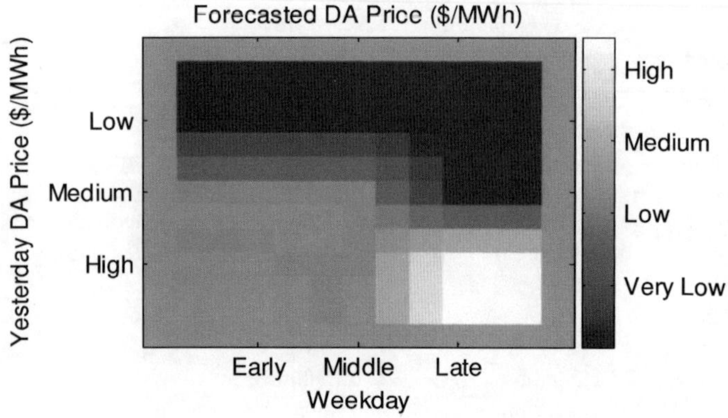

Fig. 4 A visible contour of nonlinear relationship in the rule base

The FIS can be manipulated for more accurate and efficient performance. Membership functions can be defined to respect distribution patterns of variable magnitudes and intervals. Time decay can be used to adjust the sample multiplication and assign higher priority to recent samples. The creation of fuzzy rules can also be facilitated. For example, fuzzy memberships of hours across midnight can be shifted together instead of following into separate partitions. Furthermore, the fine tune-up of each rule depends on the increased number of predefined membership functions. However, some finer rules may be infeasible due to insufficiency of training samples. As a result, the number of membership functions shall be balanced between fining and scattering (Li et al. 2007).

The proposed FIS is heuristic in data acquisition and knowledge interpretation, but its implementation is a systematic process. The correlations between the antecedent candidates and the consequents are investigated. Then antecedent candidates with significant correlations are selected and their membership functions initialized. The FIS architecture is manipulated by including additional antecedents, removing redundant antecedents, and adjusting the number and type of membership functions. The criterion of tune-up is the reduction in forecast errors. Notice that various types of membership functions can be configured for different antecedents (Li et al. 2007).

5.2 A FIS–LSE Mixture for Day-ahead Electricity Price Forecasting

This section demonstrates a hybrid forecasting system, which is implemented for day-ahead electricity price forecasting. This hybrid system integrates FIS and LSE as shown in Fig. 5 (Li et al. 2007). The LSE technique can be explained in both geometric and statistical domains. In the geometric context, the least-squares

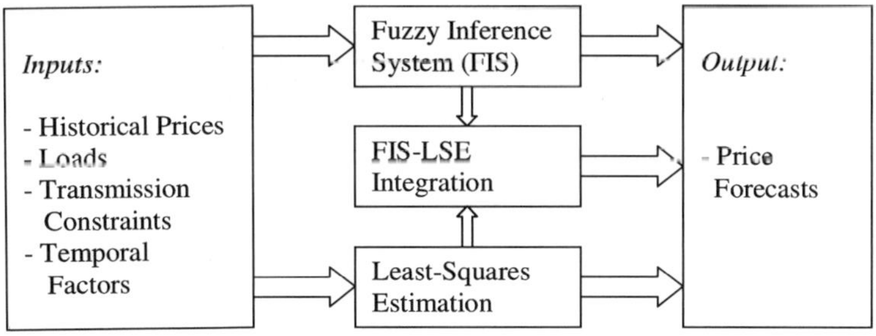

Fig. 5 A hybrid forecasting system integrates FIS and LSE

error is orthogonal to all linearly independent explanatory variables. In the statistical background, the least-squares estimate is unbiased and consistent (Astrom and Wittenmark 1995). Compared to elaborated AR models such as ARMA and GARCH, LSE is more reliable due to its simplified parameterization and unbiased consistency. However, LSE only performs accurately for normally distributed prices.

The forecasting system is composed of three modules: FIS module, LSE module, and FIS-LSE module. The outputs of the hybrid system are the day-ahead price forecasts of FIS, LSE, and FIS-LSE modules, respectively. The inputs of the FIS module contain temporal factors, prices of previous hours, and loads of current and previous hours. The inputs of LSE module contain historical prices, loads at current hour and previous hours, and transmission constraint information of study hour. The integrated FIS–LSE module is designed to be the same as FIS module with an additional input of LSE forecasts. The test results show that the least-squares estimate is an effective compensation to improve the forecasting accuracy of FIS (Li et al. 2007).

The output of the hybrid forecasting system contain the heuristic forecasts by FIS, the linear-regression forecasts by LSE, and the hybrid forecasts by FIS–LSE integration. LSE module shows the highest performance based on extensive testing. However, the linear regression of LSE only populates approximate weighting information, which indicates the contributions of explanatory variables to the forecasts. On the other hand, FIS forecasting and FIS–LSE mixture are also accurate and they provide a transparent knowledge set of fuzzy rules that interpret how prices are forecasted (Li et al. 2007).

5.2.1 Electricity Price Forecasting Using the FIS–LSE Hybrid System

The Pennsylvania–New Jersey–Maryland (PJM) electricity market is well recognized in North America and throughout the world. The hybrid forecasting system is studied with the public data of day-ahead energy market and system operations posted by the PJM Independent System Operator (ISO). The sample data has timestamps from January to December 2004 (see footnote 1).

The PJM market utilizes the LMP mechanism to appraise congestive conditions in the power grid (see footnote 1). The load-weighted-average day-ahead LMPs of the PECO load zone are studied for all 52 weeks in 2004. All load data are actual telemetered data in PJM due to the unavailability of load forecasts. The study results are creditable, given the average load forecasting error close to 2% (Li et al. 2005).

The antecedents of the FIS module contain weekday/weekend factor, peak-hour/off-peak-hour factor, zonal load-weighted-average day-ahead LMPs of previous 24, 48, and 168 h (same hours of one day, two days, and one week previously), and area-wide loads of study hour and same hour interval of the previous day. The consequent is the zonal load-weighted-average day-ahead LMP of study hour. The number of membership functions are set as 2 for temporal factors, 8 for historical prices and loads, and 40 for the price forecasts. The training data of FIS module includes samples available from 1st January 2003 to the beginning of the week for each specific study hour interval (Li et al. 2007).

The inputs of LSE module contain zonal load-weighted-average day-ahead LMPs of same hour in recent 2 weeks, and all zonal (PECO load zone and all other load zones in the PJM Mid-Atlantic area) and PJM Mid-Atlantic area-wide loads at study hour and same hour of previous day. The output is the zonal load-weighted-average day-ahead LMP forecasts at the study hour. The training period is 300 days previous to the beginning of the week for each specific study hour interval (Li et al. 2007).

In addition to above-mentioned antecedents of the FIS module, the integrated FIS–LSE module has one more antecedent, which is the output of LSE module. As an example, Fig. 6 illustrates the forecasts of PECO zonal load-weighted-average day-ahead LMPs by FIS, LSE, and FIS–LSE modules for the week of 12th-18th December 2004. The weekly average RMSE and MAPE of 52-week forecasts in 2004 are illustrated in Fig. 7 (Li et al. 2007).

Table 9 compares the performance of FIS, LSE, and FIS–LSE modules and those of prevailing statistical forecasting techniques such as ARMA, GARCH, FFNN, and SVM. The inputs and outputs of ARMA, GARCH, FFNN, and SVM are the same as those of FIS except of temporal factors. The training timestamps of ARMA, GARCH, FFNN, and SVM are recent 48 days previous to the beginning of the week of each specific study hour. The maximum number of iterations for ARMA and GARCH is 400. The volume of epochs for FFNN training is 50,000. All scenarios are computed by corresponding MATLAB Toolboxes in the MATLAB environment (see footnote 2), except that the SVM scenario is computed with the SVM-light software application.[3] Therefore, the comparison of CPU computation times between SVM and other scenarios is in a coarse context because the SVM-light application is compiled in C programming while all other scenarios are coded in MATLAB. Given same training data and test data, the specifications of each tested model have been manipulated to approach the best accuracy. For the ARMA model, both the autoregressive model order and the moving-average model order are set as 24. The maximum number of objective function evaluations is 52,000. For the GARCH

[3] SVM-Light. Available: http://svmlight.joachims.org/.

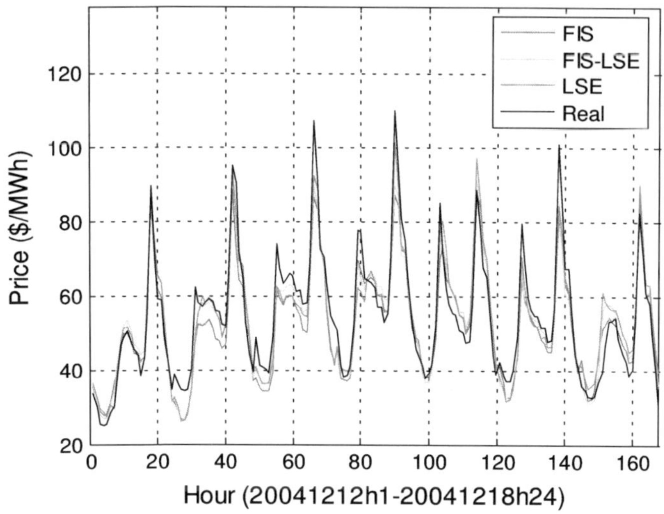

Fig. 6 A weekly forecasting example of PECO zonal load-weighted-average day-ahead LMPs using FIS, LSE, and FIS–LSE modules

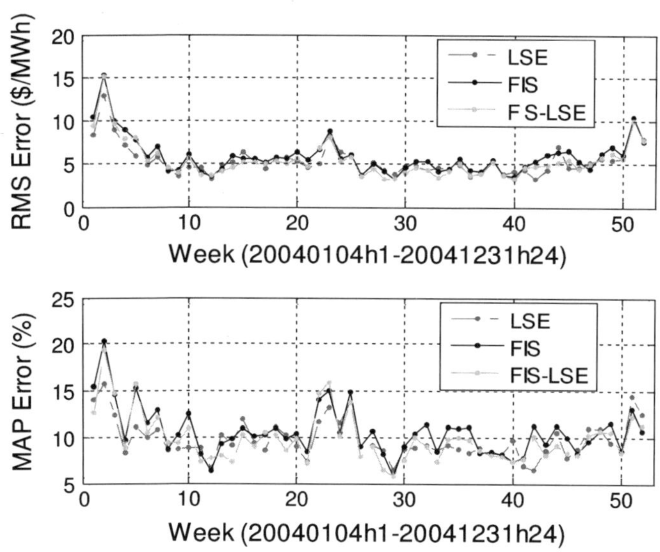

Fig. 7 Weekly average RMSE and MAPE of 52-week forecasts of PECO zonal load-weighted-average day-ahead LMPs in 2004

model, the autoregressive model order is set as 24, the moving-average model order is set as 24, the GARCH model order is set as 0, and the ARCH model order is set as 3. The maximum number of objective function evaluations is 53,500. For the FFNN model, a two-layer feed-forward back-propagation network was created.

Table 9 Performance comparison of various forecasting techniques

Technique	Yearly average RMSE ($/MWh)	Yearly average MAPE (%)	CPU computation time (min)
ARMA	16.49	22.54	416
GARCH	10.46	15.18	364
FFNN	6.22	10.36	546
SVM	6.16	10.72	4.33
FIS	6.28	10.56	1.51
FIS-LSE	5.85	9.83	2.37
LSE	5.68	9.70	0.24

The first layer has 20 neurons with hyperbolic tangent sigmoid transfer function, and the second layer has one neuron with linear transfer function. The network training method is gradient descent with momentum and adaptive learning rate back-propagation. The performance goal is 0.0001 (see footnote 2). All scenarios are computed on a Dell desktop PC of CPU speed of 3.2 GHz and physical memory of 1 GB (Li et al. 2007).

The process time is not a significant issue for forecasting a single time series of electricity prices with several inputs. However, forecasting all nodal prices in a LMP market requires the computation of price forecasts for hundreds of locations with dozens of inputs such as zonal loads and transmission constraints. In this situation, the requirement on efficiency becomes significant. The comparison indicates that FIS, LSE, and FIS–LSE are superior to ARMA, GARCH, FFNN, and SVM in terms of the computation time.

Given a specific identification, the configurations of ARMAX parameters and GARCH parameters have been carried out by the MATLAB GARCH Toolbox. A total of 13 out of 52 weekly ARMA simulations failed to regress due to nonstationary AR polynomials. The cause is the homoskedastic error specification of ARMA. In contrast, a total of 3 out of 52 weekly GARCH simulations failed to regress due to nonstationary AR polynomials. Notice that GARCH's formulation features heteroskedastic error specification. This observation indicates that GARCH's heteroskedastic error specification is more reliable than ARMA's homoskedastic one.

Table 9 compares the yearly average RMSE and MAPE of FIS, LSE, and FIS–LSE modules and those of prevailing statistical forecasting techniques such as ARMA, GARCH, FFNN, and SVM. In terms of yearly average RMSE and MAPE, ARMA and GARCH carry less accuracy than other techniques do. FIS, SVM, and FFNN are at the same level of accuracy, while LSE and FIS–LSE have less forecast errors than those of FIS, SVM, and FFNN.

Regarding the requirement on computation resources, FIS, LSE, and FIS–LSE are much more efficient than of that ARMA, GARCH, and FFNN. In terms of MAPE, the forecasting accuracy of FFNN and SVM are better than that in previous studies in the literature (Sansom et al. 2003; Szkuta et al. 1999). The forecasting accuracy of ARMA and GARCH are similar to those in previous studies in the literature (Garcia et al. 2005; Nogales et al. 2002).

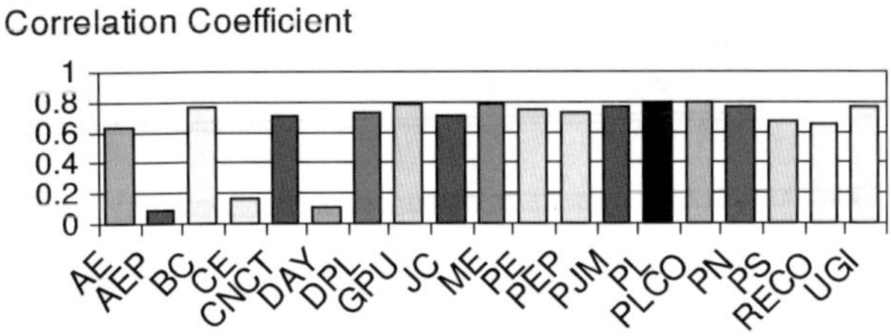

Fig. 8 Linear correlations between PJM PECO zonal LMPs and PJM zonal/area-wide loads in the years of 2003 and 2004

LSE provides the most accurate results as shown in Table 9. However, LSE's feature supports only coarse indication to explanatory variables' contributions. On the other hand, FIS performs at a significant level of accuracy with explicitly interpretable rules as shown in Table 8 and Fig. 4. Furthermore, the FIS–LSE mixture carries advantages of high accuracy and explicit interpretability.

5.2.2 Improvement by Considering Neighboring Loads and Transmission Congestions

The linear correlation between PECO zonal load-weighted-average day-ahead LMPs and zonal/area-wide loads in the years of 2003 and 2004 are illustrated in Fig. 8. Notice that LMPs are significantly correlated with local loads, most neighboring zonal loads, and area-wide loads. This locational correlation pattern is consistent with the LMP methodology. The LMPs appraise transmission congestions, which are caused by load variations in neighboring load zones of same area (Li et al. 2007).

The ten day-ahead transmission constraints with most binding occurrences in PJM are listed in Table 10 for the years of 2003 and 2004 (see footnote 1). The day-ahead transmission constraints are identified in terms of binding transmission elements or generic constraints with respect to contingency facilities.

The impact of day-ahead binding transmission constraints on LMPs is valuated by day-ahead shadow prices associated with these constraints. The linear correlations between PECO zonal load-weighted-average LMPs and the shadow prices of ten transmission constraints with most binding occurrences in PJM are summarized in Table 10. It is observed that a specific LMP sequence is only significantly correlated with some binding transmission constraints. This verifies that the impacts of day-ahead binding transmission constraints contribute at a different level to a specific LMP series (Li et al. 2007).

The following inputs are incrementally included to the initial LSE inputs of historical PECO zonal day-ahead load-weighted-average LMPs: Local zonal loads, area-wide loads, all available zonal/area-wide loads, and shadow prices of 8 day-ahead binding constraints, which are significantly correlated with PECO zonal load-weighted-average day-ahead LMPs. The training sample timestamps are 300 days previous to the beginning of the week of each specific study hour. For the

Table 10 Ten day-ahead transmission constraints with most binding occurrences in PJM for the years of 2003 and 2004

Binding thermal or generic constraint	Day-ahead contingency	Binding occurrence	Number of congested hours	Correlation between PECO zonal LMP and constraint shadow price
EAST	BASE CASE	357	3139	−0.25
CEDARGRO 230KV CED-CLIK	Roseland-Cedar Grove-Clifton-Athenia (B-2228) 230	245	1608	0.06
CENTRAL	BASE CASE	237	1415	−0.13
BED-BLA	PRUNTYTOWN MT STORM LINE	222	2021	−0.20
Cedar Interface	BASE CASE	216	2273	−0.16
LAUREL 69KV LAU-WOO	CUMBERLAND AE-CHURCHTOWN 230 LINE	216	1395	−0.10
LAUREL 69KV LAU-WOO	Cumberland (AE) 230/138kV transformer	166	937	−0.06
SHIELDAL 69KV SHI-VIN	CHAMBERS-CHURCHTOWN	156	1167	−0.16
NI to PJM PATHWAY	BASE CASE	153	3345	−0.02
SHIELDAL 69KV SHI-VIN	CUMBERLAND AE-CHURCHTOWN 230 LINE	151	1028	−0.07

Table 11 Ten day-ahead transmission constraints with most binding occurrences in PJM for the years of 2003 and 2004

Scenario	Incremental inclusion to inputs	RMSE ($/MWh)	MAPE (%)
A1	Historical PECO zonal LMPs	7.29	12.39
A2	Local zonal loads	6.00	10.12
A3	Area-wide loads	5.82	9.75
A4	All available zonal/area-wide loads	5.68	9.70
A5	Shadow prices of 8 day-ahead binding constraints	5.67	9.68

inputs, LMPs are of same hour in recent two weeks, loads are of current hour and same hour of previous day, and binding transmission constraints are of current hour. The current-hour shadow prices is not available at the moment of decision making; therefore, the current-hour shadow prices are substituted by the average of historical shadow prices of same constraints occurring at the same hour. Along with the extension to more correlated inputs, the enhancement of forecasting accuracy is summarized in Table 11. Notice that Table 9 demonstrates a horizontal comparison of various forecasting techniques. In contrast, Table 11 demonstrates a vertical comparison of the predictor selections. The higher deviation in Table 11 than that in Table 9 indicates that input selection is more significant than technique selection for the test scenarios (Li et al. 2007).

6 Conclusions

Various techniques have been reported in the literature for electricity price forecasting. The forecasting techniques can be categorized into statistical applications, simulation techniques, and equilibrium analysis. This study compares the categories of prevailing forecasting techniques based on their objective, time horizon, input–output specification, and level of accuracy. The summary facilitates the validation, comparison, and improvement of a specific technique or mixed technique for electricity price forecasting in a combined environment of competitive market and congestive grid.

A hybrid system integrating FIS, an intelligent system application, and LSE, a simplified regressive model, is studied for LMP forecasting. The forecasting system contains both verbal heuristics of FIS and linearly correlative factors of LSE. LSE generates the most accurate outputs and FIS supports transparency and interpretability. Therefore, this hybrid system carries the advantages of high accuracy and explicit interpretability. Moreover, it computes more efficiently compared to other intelligent systems such as neural networks and time series such as ARMA and GARCH. Since neighboring loads and day-ahead binding transmission constraints are significantly correlated with day-ahead LMPs, they are also included as inputs to improve the forecasting accuracy. Our future work will focus on improving short-term and medium-term electricity price forecasting by investigating more significantly correlated factors in the inputs, predicting volatility in the output, and extending the applications to market power measures.

Acknowledgements This work was supported in part by the National Science Foundation under Grant ECS-0424022 and in part by the Electric Power Research Center (EPRC) at the Iowa State University.

References

Alvarado FL, Rajaraman R (2000) Understanding price volatility in electricity markets. In: Proceedings of the 33rd Hawaii International Conference on System Sciences

Amjady N (2006) Day-ahead price forecasting of electricity markets by a new fuzzy neural network. IEEE Trans Power Syst 21:887–896

Anderson CL, Davison M (2008) A hybrid system-econometric model for electricity spot prices: considering spike sensitivity to forced outage distributions. IEEE Trans Power Syst 23:927–937

Astrom KJ, Wittenmark B (1995) Adaptive control, 2nd edn. Addison-Wesley, Reading, MA

Bastian J, Zhu J, Banunarayanan V, Mukerji R (1999) Forecasting energy prices in a competitive market. IEEE Comput Appl Power 12:40–45

Batlle C, Barqun J (2005) A strategic production costing model for electricity market price analysis. IEEE Trans Power Syst 20:67–74

Benini M, Marracci M, Pelacchi P, Venturini A (2002) Day-ahead market price volatility analysis in deregulated electricity markets. Proc IEEE Power Eng Soc Summer Meet 3:1354–1359

Botto C (1999) Price volatility-a natural consequence of electricity market deregulation and a trader's delight. Proc IEEE Power Eng Soc Summer Meet 2:1272

Bunn DW (2000) Forecasting loads and prices in competitive power markets. Proc IEEE 88: 163–169

Canazza V, Li G, Liu C-C, Lucarella D, Venturini A (2005) An intelligent system for price forecasting accuracy assessment. In: Proceedings of 13th International Conference on Intelligent Systems Application to Power Systems, Arlington, Virginia

Chan KF, Gray P (2006) Using extreme value theory to measure value-at-risk for daily electricity spot prices. Int J Forecast 22:283–300

Chen J, Deng S, Huo X (2008) Electricity price curve modeling and forecasting by manifold learning. IEEE Trans Power Syst 23:877–888

Conejo AJ, Plazas MA, Espinola R, Molina AB (2005) Day-Ahead electricity price forecasting using the wavelet transform and ARIMA models. IEEE Trans Power Syst 20:1035–1042

Contreras J, Espnola R, Nogales FJ, Conejo AJ (2003) ARIMA models to predict next-day electricity prices. IEEE Trans Power Syst 18:1014–1020

Dahlgren R, Liu C-C, Lawarree J (2001) Volatility in the California power market: source, methodology and recommendations. IEE Proc Generation Transm Distrib 148:189–193

Deb R, Albert R, Hsue L-L, Brown N (2000) How to incorporate volatility and risk in electricity price forecasting. Electricity J 13(4):65–75

Deng S (2000) Pricing electricity derivatives under alternative stochastic spot price models. In: Proceedings of the 33rd Hawaii International Conference on System Sciences

Deng S (1998) Stochastic models of energy commodity prices and their application: mean-reversion with jumps and spikes. PSERC report 98–28

Ethier R, Mount T (1998) Estimating the volatility of spot prices in restructured electricity markets and the implications for option values. PSERC Working Paper

Gao F, Guan X, Cao X, Papalexopoulos A (2000) Forecasting power market clearing price and quantity using a neural network method. In: Proceedings of the IEEE Power Engineering Society Summer Meeting, Seattle, WA, 2000, pp. 21832188

Garcia RC, Contreras J, van Akkeren M, Garcia JBC (2005) A GARCH forecasting model to predict day-ahead electricity prices. IEEE Trans Power Syst 20:867–874

Gonzalez AM, Roque AMS, Garcia-Gonzalez J (2005) Modeling and forecasting electricity prices with input/output hidden markov models. IEEE Trans Power Syst 20:13–24

Guillaume S (2001) Designing fuzzy inference systems from data: an interpretability-oriented review. IEEE Trans Fuzzy Syst 9:426–443

Guillaume S, Charnomordic B (2004) Generating an interpretable family of fuzzy partitions from data. IEEE Trans Fuzzy Syst 12:324–335

Guo J-J, Luh PB (2004) Improving market clearing price prediction by using a committee machine of neural networks. IEEE Trans Power Syst 19:1867–1876

Hong Y-Y, Hsiao C-Y (2002) Locational marginal price forecasting in deregulated electric markets using artificial intelligence. IEE Proc Gener Thansm Distrib 149:621–626

Kian A, Keyhani A (2001) Stochastic price modeling of electricity in deregulated energy markets. In: Proceedings of the 34th Hawaii International Conference on System Sciences, vol. 2

Kolos SP, Ronn EI (2008) Estimating the commodity market price of risk for energy prices. Energy Econ 30:621–641

Li G, Liu C-C, Mattson C, Lawarree J (2007) Day-ahead electricity price forecasting in a grid environment. IEEE Trans Power Syst 22:266–274

Li G, Liu C-C, Lawarree J, Gallanti M, Venturini A (2005) State-of-the-art of electricity price forecasting. In: Proceedings of the 2nd CIGRE/IEEE PES International Symposium 110–119

Lucia JJ, Schwartz E (2002) Electricity prices and power derivatives: evidence from the nordic power exchange. Rev Derivatives Res 5:5–50

Nauck D (2000) Data analysis with neuro-fuzzy methods. International ISSEK Workshop on Data Fusion and Perception, Udine, Italy

Ni E, Luh PB (2001) Forecasting Power market clearing price and its discrete PDF using a Bayesian-based classification method. IEEE Power Eng Soc Winter Meet 3:1518–1523

Nicolaisen JD, Richter CW Jr., Shebl GB (2000) Price signal analysis for competitive electric generation companies. In: Proceedings of the Conference on Electric Utility Deregulation and Restructuring and Power Technologies 66–71

Niimura T, Ko H-S, Ozawa K (2002) A day-ahead electricity price prediction based on a fuzzy-neuro autoregressive model in a deregulated electricity market. Proc Int Joint Conf Neural Netw 2:1362–1366

Nogales FJ, Contreras J, Conejo AJ, Espnola R (2002) Forecasting next-day electricity prices by time series models. IEEE Trans Power Syst 17:342–348

Olsson M, Soder L (2008) Modeling real-time balancing power market prices using combined SARIMA and Markov processes. IEEE Trans Power Syst 23:443–450

Pindoriya NM, Singh SN, Singh SK (2008) An adaptive wavelet neural network-based energy price forecasting in electricity markets. IEEE Trans Power Syst 23:1423–1432

Rodriguez CP, Anders GJ (2004) Energy price forecasting in the ontario competitive power system market. IEEE Trans Power Syst 19:366–374

Ruibal CM, Mazumdar M (2008) Forecasting the Mean and the Variance of electricity prices in deregulated markets. IEEE Trans Power Syst 23:25–32

Sansom DC, Downs T, Saha TK (2003) Evaluation of support vector machine based forecasting tool in electricity price forecasting for Australian national electricity market participants. J Electr Electron Eng Aust 22(3):227–234

Skantze P, Ilic M, Chapman J (2000) Stochastic modeling of electric power prices in a multi-market environment. In: Proceedings of Power Engineering Winter Meeting 1109–1114

Stevenson M (2001) Filtering and forecasting spot electricity prices in the increasingly deregulated Australian electricity market. In: International Institute of Forecasters Conference, Atlanta, June 2001

Szkuta BR, Sanavria LA, Dillon TS (1999) Electricity price short-term forecasting using artificial neural networks. IEEE Trans Power Syst 14:851–857

Valenzuela J, Mazumdar M (2001) On the computation of the probability distribution of the sport market price in a deregulated electricity market. In: Proceedings of the 21st Power Industry Computer Applications International Conference 268–271

Wang AJ, Ramsay B (1998) A Neural network based estimator for electricity spot-pricing with particular reference to weekends and public holidays. Neurocomputing 23:47–57

Wang H, Kwong S, Jin Y, Wei W, Man K-F (2005) Agent-based evolutionary approach for interpretable rule-based knowledge extraction. IEEE Trans Syst Man Cybern C Appl Rev 35:143–155

Wang L-X, Mendel JM (1992) Generating fuzzy rules by learning from examples. IEEE Trans Syst Man Cybern 22:1414–1427

Yen J, Wang L, Gillespie, CW (1998) Improving the interpretability of TSK fuzzy models by combining global learning and local learning. IEEE Trans Fuzzy Syst 6:530–537

Zareipour H, Canizares CA, Bhattacharya K, Thomson J (2006) Application of public-domain market information to forecast ontario's wholesale electricity prices. IEEE Trans Power Syst 21:1707–1717

Zhang L, Luh PB (2005) Neural Network-based market clearing price prediction and confidence interval estimation with an improved extended kalman filter method. IEEE Trans Power Syst 20:59–66

Zhang L, Luh PB, Kasiviswanathan K (2003) Energy clearing price prediction and confidence interval estimation with cascaded neural networks. IEEE Trans Power Syst 18:99–105

Zhao JH, Dong ZY, Li X, Wong KP (2007) A framework for electricity price spike analysis with advanced data mining methods. IEEE Trans Power Syst 22:376–385

Zhao JH, Dong ZY, Xu Z, Wong KP (2008) A statistical approach for interval forecasting of the electricity price. IEEE Trans Power Syst 23:267–276

Modelling the Structure of Long-Term Electricity Forward Prices at Nord Pool

Martin Povh, Robert Golob, and Stein-Erik Fleten

Abstract This chapter models long-term electricity forward prices with variables that influence the price of electricity. Long-term modelling requires consideration of expected changes in the demand and supply structure. The model combines high-resolution information on fuel costs from financial markets and low-resolution information on the demand/supply structure of the electricity market. We model the latter using consumption and supply capacity and the former with forward prices of fuels, emission allowances and imported electricity. The model is estimated using data from the Nordic electricity market and global long-term forward prices of energy. Owing to a lack of data on consumption and supply capacity, the estimated results provide only the broad influence of these variables on forward prices. Though extrapolation of the prices observed in Nord Pool may suffer from the influence of short-term variables, such as precipitation and temperature, the model yields robust forecasts of the prices of contracts that are not exchange traded.

Keywords Electricity markets · Electricity prices · Forward prices · Nord Pool · Regression models

1 Introduction

Since the beginning of the deregulation of electricity markets, a significant connection between electricity prices and different energy prices has prevailed. This is because the various energy sources serve either as a fuel for generating electricity or as a substitute for electricity use. Transparency in electricity markets has also improved significantly with the introduction of electricity exchanges. Although electricity markets struggle with low liquidity, they do provide information about

M. Povh (✉)
Faculty of Electrical Engineering, University of Ljubljana, Tržaška 25, SI-1000 Ljubljana, Slovenia
e-mail: martin.povh@fe.uni-lj.si

S. Rebennack et al. (eds.), *Handbook of Power Systems II*, Energy Systems,
DOI 10.1007/978-3-642-12686-4_7,

the competitive market price, often used as an indicator in over-the-counter (OTC) trading. Moreover, because the price of electricity is very volatile, both in the short and the long term, participants in electricity markets attempt to reduce price-related risk with forward trading.

A variety of forward contracts are traded on exchanges with times-to-delivery ranging from 1 week to a few years, and these give market participants a variety of choices for hedging against price risk. In liquid forward markets, forward trading also conveys the present information on the expected price in the future adjusted for the market price of risk. This information is especially important in the long term, not only for hedging purposes but also for different strategic purposes, including investment analysis, asset management, strategic decisions on mergers and acquisitions, state energy policies, etc. Excess capacity has gradually decreased since deregulation and investments in new capacity has not followed the growth in consumption. Apart from the increasing prices of other energy commodities, this has also had an important impact on the overall trend in increasing wholesale prices during this period.

In recent years, we have also witnessed a shift in electricity production technology from traditional coal and nuclear to natural gas and renewable sources. This shift in production requires investors to have relevant information about the electricity market in the future to support investment decisions today. Long-term information about electricity prices is one of the most important variables in these analyses. Investors in new production capacity need to estimate long-term expected spot prices. Alternatively, real option theory suggests the use of forward prices instead of expected spot prices, as forward prices already incorporate the appropriate market price of risk. Unfortunately, we do not have current information about prices in the distant future, for example 10 years ahead, because forward contracts with these delivery periods are not traded on the exchange. Typically, the term structure of forward contracts ends 5–6 years ahead.[1] This, however, does not mean that there is no forward trading beyond this horizon. For example, an OTC forward market with 10-year forward contracts existed for a few years in the Nordic electricity market. This provides evidence that investors seek long-term forward contracts to hedge long-term price related risks properly.

A simple way to estimate the value of the longer end of the term structure of electricity prices would be to extrapolate prices observable at electricity exchanges. Investors often use different rules of thumb to estimate the value of forward contracts beyond the traded horizon. These rules can be based on different historic data, intuition and experience. However, market forces other than the prices of observable exchange traded contracts may drive the prices of forward contracts beyond the traded horizon. This leads to potential errors in these estimates. Finally, the most accurate information about the value of electricity with a certain delivery period can only be provided by a liquid forward market. The absence of accurate information

[1] A forward price term structure is a set of prices of exchange traded forward contracts for various times-to-maturity.

indicates the need to extend the trading horizon on electricity exchanges into the future as far as possible. This would result in better market transparency, which is important not only for the investment purpose mentioned earlier but also for global benchmarking. Publicly available long-term price indicators in a particular electricity market would also be useful for investors and other institutions outside the market itself.

Long-term modelling of commodity forward prices is relatively new, as the availability of the long-term forward data remains low. Schwartz (1997) employs models based on short-term oil futures and tests their performance on available long-term oil futures. Pindyck (1999) models the long-term evolution of oil, coal and natural gas prices with mean reversion models using the theory of depletable resources. Schwartz and Smith (2000) use long-term commodity prices as an equilibrium to which short-term prices revert. Unfortunately, there is relatively little work on the modelling of long-term electricity forward prices.

However, an increasing number of studies focus on the short-term electricity forward market. For example, Byström (2003), Lucia and Schwartz (2002) and Solibakke (2006) focus on the valuation of short-term forward contracts and their hedging performance. Johnsen (2001) presents a short-term supply/demand model that we could potentially use in long-term modelling. Bessembinder and Lemon (2002) present a model in which the electricity forward price is the equilibrium of the supply and demand for forward contracts. Because electricity prices are influenced by the impossibility of storage and high uncertainty in underlying factors, some researchers suggest a combined approach to modelling electricity price. For instance, Eydeland and Wolyniec (2003), and Pirrong and Jermakyan (2008), combine the properties of financial models common in stock valuation with the properties of fundamental models where supply and demand are modelled using fundamental factors (Skantze et al. 2004). Here we follow a similar idea to model the prices of long-term electricity forwards with a regression approach. Our model depends on the assumption that the expected value of a commodity with a certain delivery period depends on expected supply and demand, as well as the supply and demand for risk hedging associated with uncertainty relating to its expected value. Modelling long-term supply and demand requires different fundamental factors from the short term, while structural changes in the market also require consideration. We identify which of these parameters influence the supply and demand for electricity in the long term and then seek long-term information that we can use to model these variables. We estimate the model parameters using the market data available for the Nordic electricity market.

The chapter is organized as follows. Section 2 identifies the long-term electricity forward price process with respect to supply, demand and the risk premium. We identify three groups of variables and present a simple model for each. At the end of this section, we present the model for long-term forward prices. Section 3 discusses the data and the parameter estimation process. In Section 4, we provide the estimation results and some indicators of model performance. Section 5 contains some conclusions and recommendations.

2 Long-term Forward Price Process

Modelling forward prices involves two time dimensions. The first is the observation time t, reflecting the time at which the price of a particular forward contract is observed or settled. The second is the delivery period T, which represent a period of time in which the forward contract matures or is delivered. The forward price is therefore denoted as $F_{t,T}$. We can forecast unobserved forward prices using either of these dimensions. The first strategy involves forecasting the future evolution of the price of an observed contract and using the n-step ahead forecast to obtain the value of a contract at time $t + n$. This model would typically involve autoregressive components of dependent and explanatory variables. The expected value of the forecast for the long-term forward price $E[F_{t,T}]$ would therefore be the average forecasted value during the delivery period T. This strategy is less convenient for a long-term forecast as the forecasting horizon could be very large when compared with the estimation sample.

An alternative strategy for forecasting the price of an unobserved forward contract would be forecasting with respect to the delivery period T. This strategy is based on the assumption that the relationships between the dependent and explanatory variables is dimension invariant; that is these relationships do not change with the observation time t or delivery period T. Under this assumption, the relationships estimated on observed contracts with delivery period T can be used to forecast the price of a contract with delivery period $T + N$. This strategy also has some advantages when it comes to modelling long-term prices. First, the forecasting horizon is very small if the delivery period T is long enough, for example 1 year. Second, we can use the available long-term information on the explanatory variables directly in the forecast as we assume the same relationship for the observed contracts. Third, we can design the underlying model to match the specifics of long-term prices, for example constructing a specific model for a specific type of contract.

2.1 Definition

We define long-term electricity forward prices as the prices of electricity forward contracts with a delivery period of 1 year and a time-to-delivery of more than 1 year ($T - t \geq 1$ year). Figure 1 depicts the prices of the yearly Nord Pool contracts ENOYR1, ENOYR2 and ENOYR3 with respective times-to-delivery of 1, 2 and 3 years.

Figure 1 demonstrates that the forward price dynamics depend very much on the time-to-delivery $T - t$. Short-term information, such as the level of water reservoirs, strongly influences the price of ENOYR1. Namely, a large proportion of hydropower plants in the Nordic market have the possibility of storing water up to 1 year or longer. There is also some degree of short-term information in ENOYR2 and ENOYR3; however, we assume that the prices of these contracts are mainly driven by the long-term information. Somewhat different laws therefore govern long-term

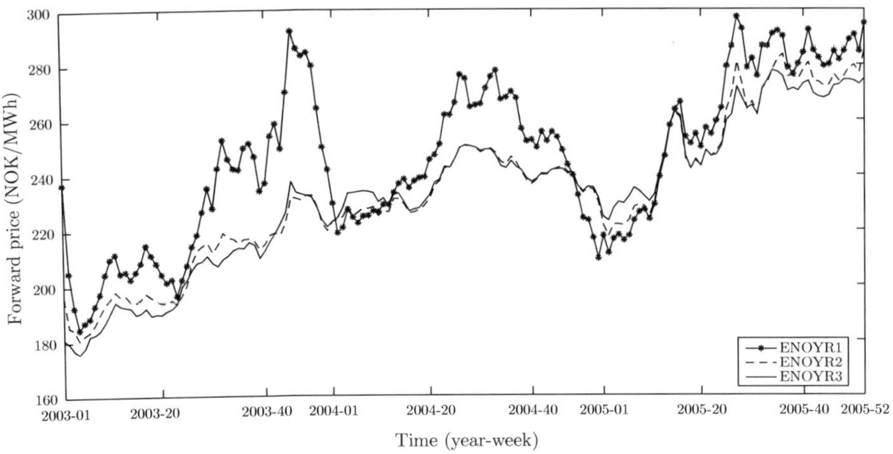

Fig. 1 Average weekly price of yearly forward contracts at Nord Pool

and short-term forward (or spot) prices, and this suggests the need for different modelling strategies. Using this definition, long-term forward prices in the Nordic electricity market are ENOYR2 and ENOYR3, whereas ENOYR1 is the mid-term and eventually short-term contract as time-to-delivery approaches zero.

2.2 Valuation of Forwards

Electricity is a very peculiar commodity because there is no economically efficient means of storage. This suggests that, to estimate the relation between the forward price and the spot price of electricity, we cannot use the standard *cost-of-carry* arbitrage:

$$F_{t,T} = (P_t + C_s)(1 + r - y)^{(T-t)}, \tag{1}$$

where the forward price $F_{t,T}$ is the sum of the spot price P_t and storage costs C_s discounted by the difference between the risk-free interest rate r and the convenience yield y. Putting counterparty risk aside, the arbitrage principle implies that there should be no difference between buying the forward contract and buying the commodity in the spot market, storing it and using it during the delivery period. Using the arbitrage principle to value electricity forward contracts is not advisable, because electricity cannot be stored today and consumed later in the future. One possible alternative for valuing electricity forward contracts is a risk-adjusted expected spot price:

$$F_{t,T} = E_t[P_T](1 + r - \lambda)^{(T-t)}, \tag{2}$$

where the forward price $F_{t,T}$ is the current expectation about the spot price during the delivery period $E_t[P_T]$ (or simply $P_{t,T}$) discounted by the difference between the risk-free interest rate r and the risk premium λ.

2.3 Risk Adjustment

Assuming that all investors hold the same expectation about the spot price, this approach (2) basically seeks the equilibrium between supply and demand for reducing risk by using an appropriate risk premium. Another interpretation of risk adjustment is the separation of the forward price data-generating process in two subprocesses. The underlying process of the expected supply and demand for electricity during the delivery period generates the expected spot price during the delivery period. Using this expectation, we obtain the probability distribution of the expected spot price, and this generates the risk premium based on the risk preferences of investors over this probability distribution. Transforming (2) to logs gives

$$\ln F_{t,T} = \ln P_{t,T} + \ln(1 + r - \lambda)(T - t). \tag{3}$$

While risk-free interest rates can be easily determined by looking at, for example the prices of government bonds, the assessment of λ, or the market price of risk, has always been a challenging task. A somewhat naive way to determine λ is to use the capital asset pricing model (CAPM), in which the risk premium in the electricity market depends on the risk premium in the overall capital market and the correlation between movements in the electricity market and movements in the overall capital market. Although CAPM is able to capture the price of risk, it is not well suited for pricing electricity derivatives because it assumes that (financial) electricity market is also used for diversification of general investors. In financial electricity markets, the participation of investors outside the industry is weak; hence, the dynamics of the risk premium is mainly driven by producers and consumers, who are motivated by hedging production and consumption. These investors are generally not diversified in the general capital market, and due to the ownership structure dominated by governmental influence, it is natural to consider them risk-averse. For instance, Bessembinder and Lemmon (2002) show that in the absence of outside speculators, different levels of risk aversion by producers and consumers lead to non-zero risk premiums in electricity forwards. These non-zero risk premiums may attract participants from outside the industry to include forwards in their portfolios, and this can gradually decrease the level of risk premium.

To estimate the risk premium, we use past empirical findings about the risk premium in the electricity market. First, Bessembinder and Lemmon (2002) and Longstaff and Wang (2004) argue that the risk premium in short-maturity electricity forwards is influenced by the probability of price spikes (load seasonality) and the level of prices. A few studies attempting to estimate the risk premium on far-maturity contracts have found that it is much lower than in the short-maturity contracts (Ollmar 2003; Diko et al. 2006). Ollmar (2003) also finds a connection between the risk premium and seasonal observation time in short- and far-maturity contracts from Nord Pool. This implies that there is a difference in supply and demand for long-term forwards at different times of the year. One of the reasons could be the yearly liquidity cycle, as the liquidity of yearly contracts from Nord Pool is typically lower at the beginning of the year and increases towards the end of the year when most of the yearly delivery contracts are settled. While this significantly influences the price of the first yearly contract, it also has some influence on the prices of all subsequent yearly contracts. Based on these findings, we use the following model for risk adjustment:

$$\ln F_{t,T} = \ln P_{t,T} + \kappa_0 r_{t,T} + \kappa_1 + \kappa_2 \tau_t, \quad (4)$$

where the risk premium in far-maturity contracts is a constant κ_1 plus a seasonal time-dependent term $\kappa_2 \tau_t$ in which τ_t is a seasonal time, defined as the time from January 1 in the observation year. Although the forward price in (4) is the price of a pure financial contract where there is no cash flow between the buyer and the seller at settlement time, there is a strong rationale for the prices of contracts beyond the traded horizon requiring at least some cash flow or financial guarantee at the time of the settlement. Therefore, long-term interest rates will have an influence on the value of the forward contract. We also believe that interest rates influence forward prices through their influence on the fundamental variables discussed in the following section. For this reason, the interest rate is included not only as a discounting variable but also as an explanatory variable for the expected spot price. The risk adjustment in (4) is no longer a function of time-to-delivery $T - t$. Because of the Samuelson effect (1965), the volatility of futures and forwards increases as time-to-delivery approaches. Hence, we assume that the volatility of far-maturity contracts does not vary significantly with time-to-delivery. The risk premium in far-maturity contracts is therefore a constant with respect to time-to-delivery.

2.4 *Modelling the Long-term Expected Spot Price*

The expected spot price during a particular period is a simple average of the expected spot prices during the delivery period. To model the long-term expected spot price, we model supply and demand using fundamental factors, whereas we obtain the price in equilibrium by matching supply and demand.

2.4.1 Long-Term Demand

Electricity demand is a process driven by short-term fundamental drivers, such as the daily and weekly cycle, temperature, the price elasticity of electricity and its substitutes, daylight hours, etc. Economic drivers (gross domestic product, household consumption expenditure) and demographic drivers (population, migration) influence electricity demand in the long term, typically causing demand to grow over time. Unfortunately, long-term information on short-term drivers does not exist. However, historical averages may be a reasonable estimate of their long-term expected value. This implies that these drivers do not influence the expected value of long-term demand, as their expected value is constant. However, we need to model the uncertainty in these variables as a short-term non-persistent error. Contrary to the situation for the short-term drivers, we can predict the long-term drivers to some extent, or alternatively, one can use existing forecasts produced by different institutions. We model long-term electricity demand as long-term electricity consumption adjusted by the price elasticity:

$$\ln q_t^d = \ln c_t + \eta \ln P_t \tag{5}$$

where q_t^d is the quantity of demand, P_t is the spot price, η is the demand elasticity and c_t is electricity consumption. Long-term consumption c_t is a well-understood process, driven by the long-term variables described earlier. However, because we quite often adequately model expected consumption growth in developed countries as a constant, we employ a simple linear model

$$\ln c_t = \alpha_0 + \alpha_1 t \tag{6}$$

and the long-term demand function is therefore

$$\ln q_t^d = \alpha_0 + \alpha_1 t + \eta \ln P_t + u_t^d. \tag{7}$$

In (7), a stochastic error term u_t^d is included to account for short-term non-persistent errors such as temperature, daylight hours, wind, etc.

2.4.2 Long-Term Supply

Electricity supply is more price elastic than electricity demand, though the underlying factors influencing supply have greater uncertainty. While many variables that influence long-term demand cannot be included in the model because we do not have reliable long-term information, this is not the case for long-term supply. Some important influencing variables are also traded commodities with transparent long-term price information. Nevertheless, long-term information about some variables that also play an important role in electricity supply does not exist. We can divide the variables that influence the shape of the supply function into two groups. The first group includes structural variables that influence the structure of the market. We capture these variables with supply capacity. The second group of supply variables consists of supply cost variables, and these cause changes in electricity supply costs.

We model long-term electricity supply as a function of supply capacity, supply costs and price elasticity:

$$\ln q_t^s = a_1 \ln C_t^s + a_2 \ln g_t + a_3 \ln P_t, \tag{8}$$

where P_t and q_t^s are the spot price and the supply quantity, respectively, g_t is the supply capacity and C_t^s are supply costs.

The supply function derives from the different units that supply the market. Each unit has limited installed capacity and therefore a limited amount of electricity it can produce in 1 year. Electricity producers also face other restrictions, including technical and environmental constraints and the weather. The level of precipitation or the available water for hydro production is a variable that has a significant impact on the supply capacity in some markets. However, we do not have any long-term information about this. Our expectation on hydro production for a particular year in the future instead derives from the historical average. We define supply capacity as the sum of total installed capacity in the market including net transfer capacity. Given that we cannot treat wind production the same as other types of capacity, we adjust the wind capacity with the average utilization factor of 0.25. Supply from neighbouring markets is a specific producer that participates in the market with a net transfer capacity.

Given the current market situation, it is reasonable to assume that until the start of the delivery period T a few years ahead, some new production units will enter the market and some old production units will be decommissioned. In theory, the expected change in supply capacity should depend on the spread between electricity prices and supply costs; however, in reality, new investments are to a large degree driven by political decisions, such as environmental pressures or the change in the level of reserve capacity associated with deregulation. We do not claim that the spread between electricity prices and supply costs provides no incentive for new investments. However, we believe that it is still not a prevailing factor influencing new investments. For this reason, we model the supply capacity g_t during time t with a simple linear model:

$$\ln g_t = \gamma_0 + \gamma_1 t + u_t^g, \tag{9}$$

where the error term u_t^g represent the uncertainty in the supply capacity. Alternatively, one can use long-term information about construction plans, which is often available, although in most cases it is still a rough expectation.

On the supply side, each unit has different production costs, which depend on the type of fuel, fuel costs and efficiency. While some units have very low or no fuel costs (wind, hydro, nuclear), other units have considerable and uncertain costs of fuel (coal, natural gas, oil). Similar to Eydeland and Wolyniec (2003), we divide the supply cost variables into three groups. The first group consists of non-tradable fuels such as water, uranium, wind and biomass. As there is no liquid market or long-term price information for these fuels, we assume that their expected costs do not change with t or T. The second group consists of tradable fuels, mostly coal, natural gas and oil derivatives. Long-term information about the prices of these fuels is available in the form of futures and forwards, sometimes traded on the exchange up to 6 years into the future. The third group of supply cost variables includes other costs of supply such as emission allowance prices and the prices of electricity imports. We assume that the introduction of the European Union Emission Trading Scheme (EU ETS) in 2005 increased production costs for fossil-fuel producers, thereby increasing electricity prices. Emission allowance prices should therefore have a significant influence on electricity prices. Imported electricity is also part of the supply function, and we model this as electricity from a specific type of producer. The price for this producer is the average price in the market from where the electricity is imported.

Supply costs C_t^s can then be expressed as a linear combination of the relevant supply cost variables:

$$\ln C_t^s = b_0 + b_1 \ln P_t^{oil} + b_2 \ln P_t^{coal} + b_3 \ln P_t^{ea} + b_4 \ln P_t^{import}, \qquad (10)$$

where P_t^{oil} and P_t^{coal} are the prices of crude oil and coal, P_t^{ea} is the emission allowance price and P_t^{import} is the price of electricity in the market from which the electricity is imported. In (10) we do not include the price of natural gas, as long-term information about the price of natural gas remains scarce and because the price of natural gas is highly correlated with the price of crude oil. We assume that the price of crude oil is a good proxy for the prices of all oil derivatives and natural gas.

Combining (8) and (10), we obtain the following linear supply model:

$$\begin{aligned}\ln P_t = {} & \beta_0 + \beta_1 \ln g_t + \beta_2 \ln q_t^s + \beta_3 \ln P_t^{oil} \\ & + \beta_4 \ln P_t^{coal} + \beta_5 \ln P_t^{ea} + \beta_6 \ln P_t^{import} + u_t^s. \end{aligned} \qquad (11)$$

To account for all uncertainties in the supply function, such as precipitation and wind, we also include a stochastic error u_t^s.

2.5 Long-Term Forward Price Model

The average electricity spot price during time t is obtained by matching the supply and demand function. In equilibrium, the demand quantity q_t^d equals the supply quantity q_t^s, giving

$$\begin{aligned}\ln P_t = {} & \beta_0 + \beta_1 \ln g_t + \beta_2 \ln q_t^d + \beta_3 \ln P_t^{oil} \\ & + \beta_4 \ln P_t^{coal} + \beta_5 \ln P_t^{ea} + \beta_6 \ln P_t^{import} + u_t^s. \end{aligned} \qquad (12)$$

Equation (12) is a model for expected spot price for any time t. By replacing spot price variables in (12) with forward-looking variables, we can express the expected spot price during delivery period T a few years ahead as follows:

$$\ln P_{t,T} = \beta_0 + \beta_1 \ln g_{t,T} + \beta_2 \ln q^d_{t,T} + \beta_3 \ln F^{oil}_{t,T}$$
$$+\beta_4 \ln F^{coal}_{t,T} + \beta_5 \ln F^{ea}_{t,T} + \beta_6 \ln F^{import}_{t,T} + u^s_{t,T}. \quad (13)$$

In (13), $F^{oil}_{t,T}$, $F^{coal}_{t,T}$ and $F^{ea}_{t,T}$ are the forward prices of oil, coal and emission allowances, respectively, whereas $F^{import}_{t,T}$ is the forward price of electricity in the market from where the electricity is imported.

The final step in modelling long-term forward prices is the risk adjustment of the expected spot price $P_{t,T}$, which can be obtained with (4). Combining (4), (7), (9) and (13) gives

$$\ln F_{t,T} = \delta_0 + \delta_1(\gamma_0 + \gamma_1 T) + \delta_2(\alpha_0 + \alpha_1 T) + \delta_3 \ln F^{oil}_{t,T}$$
$$+\delta_4 \ln F^{coal}_{t,T} + \delta_5 \ln F^{ea}_{t,T} + \delta_6 \ln F^{import}_{t,T} + \delta_7 r_{t,T} + \delta_8 \tau_t + u_{t,T}, \quad (14)$$

where $\delta_{1,\ldots,6}$ equals $\beta_{1,\ldots,6}$ divided by $1 - \beta_2\alpha_2$, whereas $\delta_{7,8}$ equals $\kappa_{0,2}$ divided by $1 - \beta_2\alpha_2$. The error term $u_{t,T}$ is a linear combination of $u^d_{t,T}$, $u^s_{t,T}$ and $u^g_{t,T}$. Expected supply capacity and expected consumption in (14) are modelled with two linear submodels; however, the model structure allows the use of any other information about expected consumption and supply capacity, and hence information that is more accurate can be used if available.

3 Model Estimation

The model (14) combines two important properties of long-term forward prices: changes in the structure of the market and changes in supply costs. We estimate the model parameters using data from the common Nordic electricity market, comprising Norway, Sweden, Finland and Denmark. The Nordic electricity market is one of the world's oldest electricity markets, with a high level of competition and with Nord Pool as the longest established power exchange. Nord Pool has high liquidity in the spot market and fair liquidity in the forward market. Because the Nordic electricity market is large (close to 400 TWh in yearly consumption) and the level of concentration on the production side is small, we assume that the market is close to efficient. Similar to most electricity markets, the Nordic electricity market still struggles with the lack of long-term information necessary to estimate the parameters of our model.

3.1 Data

All the data discussed in this section represent the average information for delivery period T. Most data are available at daily resolution, except for the data on structural variables that are available only on a monthly, quarterly or yearly basis. As we wish to minimize the influence of short term variations in price due to different short-term factors and, at the same time, produce an adequate data sample to obtain significant results, we use a weekly resolution ($t = 1$ week).

3.1.1 Electricity

Yearly electricity forward contracts at Nord Pool trade up to 5 years ahead, while contracts beyond this horizon trade on the OTC market. Because the liquidity and trading frequency of yearly forwards decreases significantly with time-to-delivery, we use only the prices of the 2-year ahead (ENOYR2) and 3-year ahead (ENOYR3) forward contracts to estimate the model parameters. To estimate the structural parameters, we use monthly average electricity spot prices from Nord Pool.

3.1.2 Electricity Consumption and Supply Capacity

The model for the long-term electricity forward price in (14) also draws on the expected long-term electricity consumption and supply capacity. Their influence, however, is difficult to quantify as we have 3 years of high-resolution electricity forward data and several years of low-resolution consumption and supply capacity data. Instead, we assume that the expected consumption and supply capacity influence the expected spot price in the same way that actual consumption and supply capacity influenced the spot price in the past. To estimate the structural parameters, we use historical data on yearly electricity consumption and supply capacity for all four countries. To produce a larger sample, we use monthly frequencies for the yearly data, where the individual observation is the sum or average of the previous 12 months. For supply capacity where only yearly data are available, we use linear interpolation to obtain monthly observations. These data are from Nordel (Organisation for Nordic Transmission System Operators). We modify the supply capacity data to represent the total average supply capacity in the market. We adjust the influence of wind capacity with an average utilization factor of 0.25 and include the net import capacity (available from Nordel). To estimate the demand variables in (7), we use annual data on temperature-adjusted consumption from Nordel. Temperature-adjusted consumption is consumption adjusted to normal temperature conditions using a heating-degree-day index (Johnsen and Spjeldnæs 2004).

3.1.3 Oil Prices

Most crude oil and oil derivatives are traded on two major international exchanges, London's Intercontinental Exchange (ICE) and the New York Mercantile Exchange (NYMEX). Because there is a small difference in the quoted price of crude oil in both exchanges, we choose NYMEX crude oil data. We also use NYMEX monthly averages of the crude oil spot prices for estimation of the structural parameters.

3.1.4 Coal Prices

The international coal market is based on coal prices offered at major shipping ports and freight prices. As a result, there are different prices for different delivery ports across Europe. Few institutions provide OTC coal prices based on reported deals from traders. We use McCloskey's North West European (NWE) Steam Coal Marker Price. This is a respected coal price marker and is particularly valid for the UK market. Given that long-term forward prices for coal are not available, we assume that the NWE coal marker price also provides adequate information about future coal prices.

3.1.5 Emission Allowance Prices

The EU ETS was implemented in 2005. Because Nord Pool began trading with CO_2 allowances in the beginning of 2005, we assume that the expected CO_2 allowance was constant and equalled approximately 8.7 EUR/ton, which is the price observed during the first days of trading. We base this assumption on the idea that the trading in the market usually starts at a price that is the average of different expectations of investors. We combine European Carbon Index data from the European Energy Exchange (EEX) and the Nord Pool data on European Emission Allowances (EUA) forward contracts. EEX started publishing the European Carbon Index in November 2004, and Nord Pool started trading EUA in March 2005. Given that emission allowances can be purchased and used anywhere in Europe, the differences in prices between the different markets are small.

3.1.6 Imported Electricity Prices

The Nordic electricity market has become a net importer of electricity in recent years. More than half of imported electricity comes from Russia and the remainder is imported from Germany and Poland. As there is no electricity price signal in Russia, we use only prices from EEX, which is a dominant price indicator for Germany and more or less all of Central Europe. We use data on electricity spot prices and yearly futures (traded up to 6 years ahead) in the estimation.

3.1.7 Interest Rates

Interest rates are needed not only as explanatory variables but also for estimating the forward exchange rates for converting US dollars and Euros to Norwegian kroner. For estimation of forward exchange rates, we use interest rate parity. Long-term government bond yields from the Bank of Norway provide the risk-free interest rate for estimation of the forward exchange rate. For estimation of the forward exchange rate for US dollars, we use Treasury bill long-term interest rates from the US Department of Treasury, and for conversion from the Euro, we use Eurozone government benchmark yields from Reuters.

3.2 Data Analysis

The data described in the previous section vary in terms of observation time and delivery period resolution. For long-term supply capacity, we use yearly data from 1990 to 2005, whereas for long-term consumption, we use yearly data from 1996 to 2005: this is because temperature-adjusted consumption is available only after 1996. For estimation of the structural parameters δ_1 and δ_2, we use monthly data from January 2001 to December 2005. Weekly parameters are estimated on weekly data from Week 1 in 2003 to Week 52 in 2005. Table 1 presents all the data used for the model estimation with their observation time, sample size and resolution.

3.2.1 Testing for Stationarity

A regression between non-stationary time series may result in a spurious regression, and we cannot rely upon the regression parameters and their confidence intervals. We assume that the structural data, such as the annual consumption and supply capacity, are stationary. Although the small data samples may not reveal stationarity, in theory, these variables have more or less constant growth, and shocks in these

Table 1 Data observation time and sample size

Variable	Observation time	Sample size	Resolution
$\ln F_{t,T}^{np}$	Week 1, 2003–Week 52, 2005	314	Weekly
$\ln F_{t,T}^{oil}$	Week 1, 2003–Week 52, 2005	314	Weekly
$\ln F_{t,T}^{coal}$	Week 1, 2003–Week 52, 2005	314	Weekly
$\ln F_{t,T}^{eua}$	Week 1, 2003–Week 52, 2005	314	Weekly
$\ln F_{t,T}^{eex}$	Week 1, 2003–Week 52, 2005	314	Weekly
$r_{t,T}$	Week 1, 2003–Week 52, 2005	314	Weekly
τ_t	Week 1, 2003–Week 52, 2005	314	Weekly
$\ln g_t$	Jan. 2001–Dec. 2005	60	Monthly
$\ln c_t$	Jan. 2001–Dec. 2005	60	Monthly
$\ln w_t$	Jan. 2001–Dec. 2005	60	Monthly
$\ln P_t^{np}$	Jan. 2001–Dec. 2005	60	Monthly
$\ln P_t^{oil}$	Jan. 2001–Dec. 2005	60	Monthly
$\ln P_t^{coal}$	Jan. 2001–Dec. 2005	60	Monthly
$\ln P_t^{eua}$	Jan. 2001–Dec. 2005	60	Monthly
$\ln P_t^{eex}$	Jan. 2001–Dec. 2005	60	Monthly
$\ln g_t^y$	1990–2005	16	Yearly
$\ln c_t^{ta,y}$	1996–2005	10	Yearly
$\ln P_t^{np,y}$	1996–2005	10	Yearly

Table 2 Unit root test results

Variable	t-value (constant)	t-value (constant + trend)
$\ln F_{t,T}^{np}$	-1.463^{a}	-3.103^{a}
$\ln F_{t,T}^{oil}$	0.729^{a}	-2.594^{a}
$\ln F_{t,T}^{coal}$	-2.666^{a}	-0.575^{a}
$\ln F_{t,T}^{eua}$	0.183^{a}	-1.750^{a}
$\ln F_{t,T}^{eex}$	-2.923^{b}	-3.211^{a}
$r_{t,T}$	-4.275	-1.922^{a}
τ_t	-3.502	-3.720^{b}

[a]Reject the null hypothesis at the 1% level of significance
[b]Reject the null hypothesis at the 5% level of significance

variables will eventually die out with respect to long-term growth. Weekly variables, on the other hand, include the forward prices of energy commodities, which in theory are not stationary, and so we perform a unit root test only on weekly variables. A standard way to test for stationarity is to run the Augmented Dickey–Fuller (ADF) test; however, as the weekly data series comprises two contracts, there is a significant shift in the mean where the contracts are rolled over. Here we employ an ADF test that allows for a known structural break (Perron 1989) at observation $T_b = 157$:

$$\Delta y_t = \alpha_0 + \alpha_1 DU_t + d(DT_b)_t + \beta t + \rho y_{t-1} + \sum_{j=1}^{p} c_j \Delta y_{t-j} + e_t, \qquad (15)$$

where y_t is the series under test, DU_t is a shift dummy variable $DU_t = 1$ if $t > T_b$ and 0 otherwise and DT_b is an impulse dummy $DT_b = 1$ for $T = T_b + 1$ and 0 otherwise. The optimal number of lags p in (15) minimizes Akaike's information criterion. Table 2 gives the results of the Perron-type ADF test for stationarity with intercept ($\beta = 0$) and trend stationarity. The critical value at the 5% level of significance for the constant ($\beta = 0$) is -2.871, and -3.424 for the constant plus trend. The results show that $F_{t,T}^{np}$, $F_{t,T}^{oil}$, $F_{t,T}^{coal}$, $F_{t,T}^{eua}$ and $F_{t,T}^{eex}$ are probably non-stationary, whereas $r_{t,T}$ and τ_t could be considered as stationary with intercept.

When forecasting the future evolution of a time series with respect to the observation time t, a regression between non-stationary data would require the use of first differences, providing that the series contain only one unit root $I(1)$ and after testing for cointegration when the series are driven by a common source of non-stationarity. Because stationarity is measured with respect to observation time, we believe that when it comes to forecasting with respect to delivery period T, non-stationarity is not an issue. Here relationships between contemporaneous values of variables with the same delivery period T are estimated, which are then used to forecast the dependent variable $\mathrm{E}[F_{t,T+N}]$. As these relationships are assumed to be independent of t or T, non-stationary data will not produce inconsistent forecasts. One may, for example, randomly rearrange the data sample with respect to t, which would significantly change the results of the stationarity test as well as $\mathrm{E}[F_{t+n,T}]$. However, the coefficient estimates, their confidence intervals and $\mathrm{E}[F_{t,T+N}]$ are unchanged. For this reason, we continue to employ a regression in levels, even in the presence of non-stationary data.

3.3 Parameter Estimation

As shown in Table 1, the estimation data are available at three different resolutions. For this reason, we estimate the model parameters sequentially. In the first step, regressions (7) and (11) are estimated to obtain the expected demand parameters α_0, α_1 and α_2 and the expected supply capacity parameters γ_0 and γ_1. The expected demand and supply capacity parameters are estimated using yearly data. In the second step, the structural parameters δ_1 and δ_2 are estimated. These describe how the expected spot prices change because of changes in the expected supply capacity and expected consumption. We estimate the structural parameters with monthly data. In the third step, the weekly parameters $\delta_0, \delta_3, \delta_4, \delta_5, \delta_6, \delta_7$ and δ_8 are estimated using weekly data.

In the first regression, we estimate the demand parameters α_0, α_1 and α_2 using (7). Since temperature is short-term non-persistent error, we use data on temperature-adjusted consumption to exclude variations in consumption due to temperature. This way, the demand model (7) yields the demand that would occur under expected normal temperature conditions. For the electricity price P_t^{np}, we use yearly averages of the Nord Pool spot price. The demand parameters in Table 3 show growth in electricity demand equivalent to approximately 1.3% growth in annual consumption. A negative and significant long-term demand elasticity indicates that an important proportion of electricity consumers in the Nordic electricity market adjust their consumption according to the electricity spot price.

In the second step, we estimate the supply capacity parameters γ_0 and γ_1 in (11). As shown in Fig. 2, we observe constant growth in supply capacity from the Nordic electricity market from 1990 to 2005, except in 1999 when Sweden and Denmark decommissioned some thermal and nuclear power plants.

To take this shift into account, we add a shift dummy DU_{1999}^{cap} with 1 at $t \geq 1999$ and 0 otherwise. The supply capacity parameters in Table 4 show that the logged supply capacity annual growth is approximately 1.2% per year.

Table 3 Demand parameters

Variable	Parameter value	Standard error	p-value
α_0	6.079	0.054	0.000
α_1	0.013	0.001	0.000
α_2	−0.052	0.012	0.003

$N = 10, \; R^2(\text{adj.}) = 0.895, \; SE = 0.012$

reg. eq. : $\ln c_t^{ta,y} = \alpha_0 + \alpha_1 t + \eta \ln P_t^{np,y} + u_t^d$

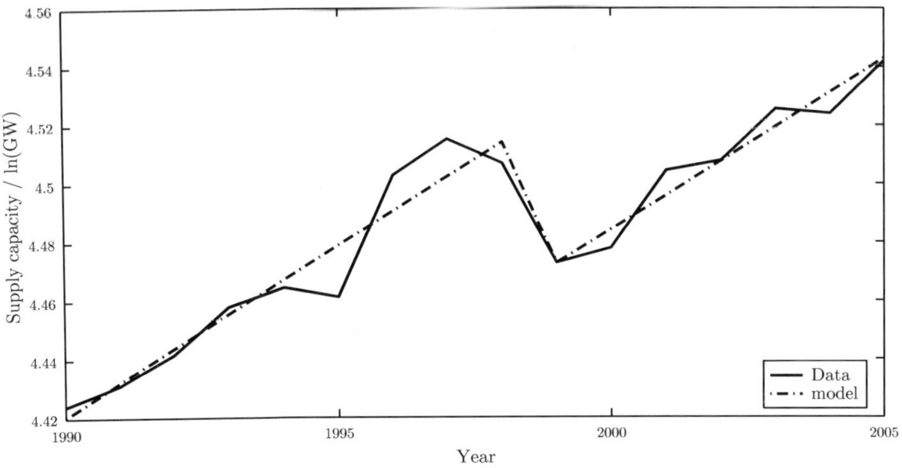

Fig. 2 Supply capacity in Nordic electricity market

Table 4 Supply capacity parameters

Variable	Parameter value	Standard error	p-value
γ_0	4.415	0.005	0.000
γ_1	0.012	0.001	0.000
γ_2	−0.053	0.008	0.000
$N = 16, \quad R^2(\text{adj.}) = 0.946, \quad SE = 0.008$			
reg. eq. : $\ln g_t^y = \gamma_0 + \gamma_1 t + \gamma_2 DU_{1999}^{cap} + u_t^g$			

In the third step, we estimate δ_1 and δ_2 with monthly parameters. The estimation of these parameters is the most difficult part of our model estimation, as we only possess annual data on supply capacity that we can only consistently use after 2000 with the formation of the final geographical scope of the Nordic market. We assume that expected supply capacity $g_{t,T}$ and expected consumption $c_{t,T}$ influence the expected spot price $P_{t,T}$ in the same way as actual supply capacity g_t and actual consumption c_t influenced the average spot price P_t in the past. In this manner, we can use historical average spot prices to estimate δ_1 and δ_2. First we estimate rough approximations of parameters β_0, β_3, β_4, β_5, β_6 in (12). These are estimated with an auxiliary regression (14) where $\delta_1 = \delta_2 = 0$ is assumed, hence $\beta_1 = \beta_2 = 0$ and $\delta_{0,3..6} = \beta_{0,3...6}$. The estimated approximations $\hat{\beta}_0$, $\hat{\beta}_3$, $\hat{\beta}_4$, $\hat{\beta}_5$, $\hat{\beta}_6$ are then used in the following regression:

$$\ln P_t^{np*} = \beta_1 \ln g_t + \beta_2 \ln q_t + \psi \ln w_t, \tag{16}$$

where $\ln P_t^{np*}$ is the electricity spot price adjusted for the influence of the spot prices of oil, coal, emission allowances and imports:

$$\ln P_t^{np*} = \ln P_t^{np} - \hat{\beta}_0 - \hat{\beta}_3 \ln P_t^{oil} - \hat{\beta}_4 \ln P_t^{coal} - \hat{\beta}_5 \ln P_t^{ea} - \hat{\beta}_6 \ln P_t^{eex}, \tag{17}$$

Table 5 Structural parameters

Variable	Parameter value	Standard error	p-value
β_1	−6.992	2.080	0.001
β_2	6.487	1.600	0.000
ψ	−1.680	0.146	0.000
$N = 60,\ R^2(\text{adj.}) = 0.689,\ SE = 0.152$			
reg. eq. : $\ln P_t^{np*} = \beta_1 \ln g_t + \beta_2 \ln q_t + \psi \ln w_t$			

Table 6 Weekly parameters

Variable	Parameter value	Standard error	p-value
δ_0	−4.543	0.117	0.000
δ_3	0.088	0.012	0.000
δ_4	0.131	0.011	0.000
δ_5	0.063	0.009	0.000
δ_6	0.343	0.024	0.000
δ_7	0.015	0.004	0.000
δ_8	0.046	0.005	0.000
$N = 314,\ R^2(\text{adj.}) = 0.962,\ SE = 0.022$			
reg. eq. : $\ln F_{t,T}^{np*} =$ $\delta_0 + \delta_3 \ln F_{t,T}^{oil} + \delta_4 \ln F_{t,T}^{coal} + \delta_5 \ln F_{t,T}^{ea} + \delta_6 \ln F_{t,T}^{eex} + \delta_7 r_{t,T} + \delta_8 \tau_t + u_{t,T}$			

and the two structural parameters are estimated with

$$\delta_1 = \frac{\hat{\beta}_1}{(1 - \hat{\beta}_2 \hat{\alpha}_2)}, \quad \delta_2 = \frac{\hat{\beta}_2}{(1 - \hat{\beta}_2 \hat{\alpha}_2)}. \tag{18}$$

In (16), the spot price is also adjusted for the influence of water reservoir levels w_t. In the Nordic electricity market, the levels of hydro power plant reservoirs heavily influence electricity spot prices. We use Nordel data on the average monthly reservoir potential in GWh.

Although two regressions are needed to estimate the structural parameters, only the results of regression (16) are presented in Table 5. In addition, although both structural parameters appear significant, they are approximations. Namely, we may question the supply capacity parameter, as it was estimated using few data. Nevertheless, the parameters have their expected sign and therefore can at least illustrate their influence. This is because expected forward prices decrease when expected supply capacity increases, and increase when expected consumption increases. The corresponding values of δ_1 and δ_2 when applying (18) are −5.3970 and 4.9554, respectively.

In the last step of the parameter estimation, we run regression (14) in which $F_{t,T}^{np}$ is adjusted for the influence of expected supply capacity and expected consumption:

$$F_{t,T}^{np*} = F_{t,T}^{np} - \hat{\delta}_1(\hat{\gamma}_0 + \hat{\gamma}_1 T) - \hat{\delta}_2(\hat{\alpha}_0 + \hat{\alpha}_1 T). \tag{19}$$

Table 6 provides the estimates of the weekly parameters with their corresponding standard errors and p-values. All parameters have their expected sign as all positively correlate with the electricity forward price. A negative constant parameter compensates for the high values of the structural parameters. The estimated parameters are all significant, with none of the p-values above the 1% level.

The parameters in Table 6 also indicate that the EEX price is the most significant parameter. This is somewhat unexpected as only a small share of electricity in the Nordic electricity market is from Central Europe. This implies that the EEX price may serve as a sort of marginal producer in the Nordic electricity market. The EEX price may also serve as one of the benchmarks for investors in the Nordic electricity market. It is also reasonable to assume that other information that we cannot quantify similarly influences the EEX and Nord Pool prices. Specific information, including construction plans for new generation and interconnection capacity or other political decisions that will influence prices in the future, is likely to influence prices in both markets. Because the same or similar variables as we use in the model may influence the price of EEX, there is the indication of a possible problem with multi-collinearity. We use variance inflation factors (VIF) to detect multi-collinearity in the explanatory variables in question. As a rule of thumb, if any VIF exceeds 10, the corresponding variable is said to be highly collinear (Myers 1986), in which case, it is inadvisable to regress one on another.

Although none of the VIF values in Table 7 exceeds the critical value, the model could still suffer from potential multi-collinearity. A detailed analysis of the regression results, however, indicates no significant sign of multi-collinearity. According to Brooks (2003), typical problems with multi-collinearity are as follows.

1. The regression is very sensitive to small changes in the specification, so that dropping or adding one variable in the regression leads to a large change in the level or significance of the other parameters.
2. The estimated parameters have high standard errors and wide confidence intervals.

Detailed analysis of the model specification and standard errors above shows no such problems. Finally, if the model parameters have appropriate sign and significance, and the selected variables have good theoretical background, Brooks suggests that the presence of multi-collinearity can be ignored. Although $F^{eex}_{t,T}$ indicates some degree of collinearity with $F^{np}_{t,T}$, we assume it also includes other valuable information that otherwise cannot be modelled, and so we choose not to remove it from the regression.

Table 7 Variance inflation factor

Variable	VIF value
$\ln F^{oil}_{t,T}$	8.1
$\ln F^{coal}_{t,T}$	5.1
$\ln F^{eua}_{t,T}$	7.5
$\ln F^{eex}_{t,T}$	9.2
$r_{t,T}$	6.0
τ_t	1.2

3.4 Model Testing

The main obstacle to long-term modelling and forecasting is insufficient market data to estimate the parameters and test model performance. We use the out-of-sample test to test the performance of the model on the available long-term contracts. We use an out-of-sample test to see how the model behaves using data that are not included in the estimation of the parameters. However, choosing the 2005 data for the out-of-sample tests would not provide a true picture of the model because the CO_2 allowance price is assumed to be constant during 2003 and most of 2004, and changes only in 2005. In this case, the model would not produce the true influence of CO_2 allowance prices, which were particularly important driver of the price increase in 2005. Given that the model does not include any autoregressive terms and is not intended to forecast prices with respect to t, it is in principle possible to choose any in-sample or out-of-sample period. Here we estimate the model parameters with in-sample data from 2004 to 2005 and test the model on the out-of-sample data from 2003. Only ENOYR3 is used to present the results of this test. As shown in Fig. 3, the model produces similar levels of prices for out-of-sample data. The standard error of the estimate on the in-sample data is 0.0219, and the standard error of the estimate on the out-of-sample data is 0.0258.

The problem of testing the performance of the model when predicting prices 4–10 years ahead offers no adequate solution, as there are no prices for us to benchmark the forecast. To estimate this set of prices, we use the variables with the same delivery period as the price in question. This means that, for estimation of the ENOYR5 contract, we use the forward prices of oil, imports, emission allowances and the EEX price with delivery 5 years ahead. Interest rates are available up to 10 years ahead, and the prices of oil, emission allowances and imports are available

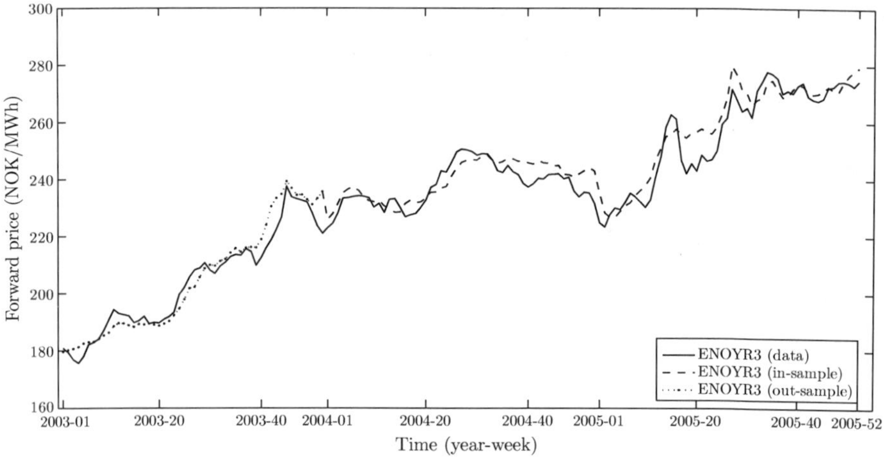

Fig. 3 Out-of sample test on ENOYR3 data

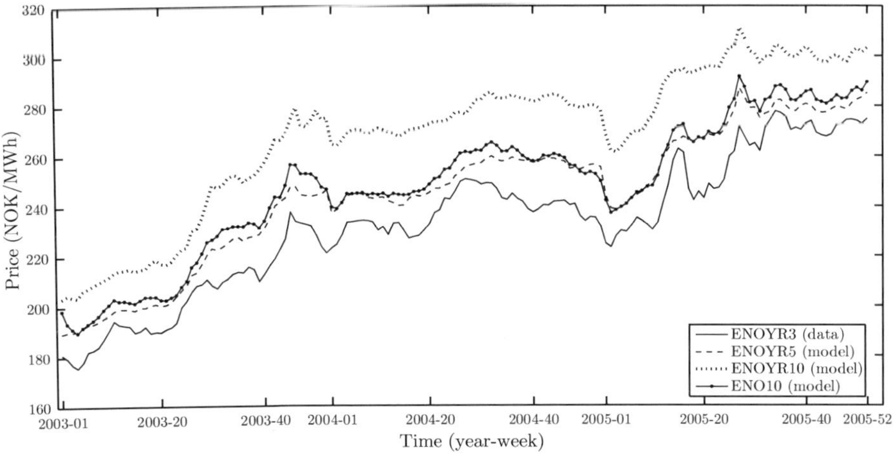

Fig. 4 The prices of ENOYR3, ENOYR5, ENOYR10 and ENO10

for up to 6 years ahead, whereas the coal price index we use in estimation is only relevant for spot prices. To estimate prices up to 10 years ahead, we forecast the missing variables needed for estimation. Here, for variables for which we have the information available up to 6 years ahead, we extrapolate the difference between the last two prices for the remainder of the forward curve. For the coal price, we assume zero growth in the term structure. Figure 4 provides the forward prices for ENOYR3, ENOYR5 and ENOYR10 and the price for a 10-year contract (ENO10), derived as the interest-rate-weighted average price of all yearly contracts.[2] Here we assume that settlement occurs only once a year.

While the price of ENO10 suggests reasonable dynamics compared with the ENOYR3 contract, the price level of the ENO10 is also important. ENO10 contracts are only traded OTC, and so there are no available data on their prices. We only obtained OTC data on ENO10 prices for the period from Week 48, 2000 to Week 33, 2002, but could not use this in the regression as data on the other variables are missing during this period. Even though these prices are based on known OTC deals and rough estimation, analysis shows that the ENO10 price has dynamics similar to the ENOYR2 and ENOYR3 prices, and is on average 6.9% higher than the ENOYR3 price during that period, whereas our forecast of the ENO10 price is on average 6.6% higher than the ENOYR3 price. While there is no strong reason to believe

[2] If r_{10} is a 10-year interest rate and k is the time to the first settlement, the value of ENO10 with the start of delivery in January next year is given by

$$\text{ENO10} = \frac{\text{ENOYR1}\frac{1}{(1+r_{10})^k} + \text{ENOYR2}\frac{1}{(1+r_{10})^{1+k}} + \ldots + \text{ENOYR10}\frac{1}{(1+r_{10})^{9+k}}}{\frac{1}{(1+r_{10})^k} + \frac{1}{(1+r_{10})^{1+k}} + \ldots + \frac{1}{(1+r_{10})^{9+k}}}. \quad (20)$$

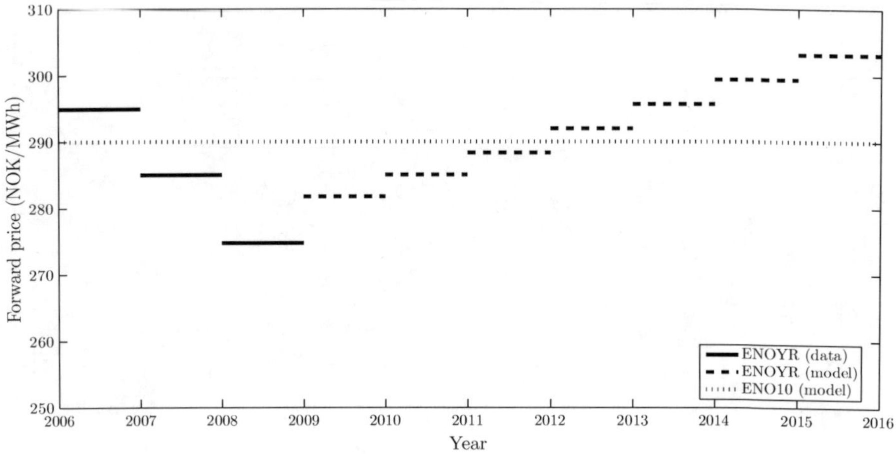

Fig. 5 Estimated term structure for week 52, 2005

that the ENO10 price in 2005 should behave similarly to the ENO10 price in 2000–2002, it is the only available indicator on how the model behaves when predicting the ENO10 year price. The slopes of the term structures of the explanatory variables influence the differences between the prices 4–10 years ahead. Among these, the slope of expected supply capacity and consumption are particularly important. The expected growth in supply capacity and consumption has not changed significantly since 2002, and therefore the difference between ENOYR3 and ENO10 should not have changed significantly since then. Figure 5 presents the estimated electricity forward price term structure for Week 52, 2005. Here the ENOYR1 and ENOYR2 prices are higher than the ENOYR3 price, indicating the higher prices in the spot and short-term forward market typical of low hydrological balances or temperatures. Extrapolation of the first three yearly prices up to 10 years ahead would in this case produce a term structure with a negative slope. As the slope of the far end of the term structure should not depend on short-term variations in weather, we should therefore extrapolate prices with care. Our model, although based on rough estimates of the structural parameters, produces a positive slope for the remainder of the term structure. Here the influence of the structural parameters plays an important role, as these parameters are independent of any short-term influences.

4 Conclusions

We present a regression model for long-term electricity forward prices. We construct the model for forecasting the prices of long-term forward contracts that we cannot observe on the exchange. Long-term forward prices mostly depend on variables that influence the supply costs of electricity. However, as the time horizon in long-term modelling is very long, we also consider expected changes in the structure of the supply and demand for electricity. We present ideas on how to combine high-resolution data on fuel prices from financial markets and low-resolution data on the expected structure of supply and demand, including expected electricity consumption and supply capacity. Combining both types of information yields a relationship for the expected long-term electricity spot price, which when adjusted for a risk premium provides the model for the long-term electricity forward price. Combining data with different observation time resolution and different delivery periods requires a work-around in parameter estimation. We estimate two submodels for expected supply capacity and consumption, whereas their influence on the expected spot price, which is one of the critical parts of the model, is estimated only roughly based on the historical influence of actual supply capacity and actual demand on actual spot price. The changes in supply costs are modelled with the crude oil price, coal prices, the emission allowance price and the price of imported electricity. Although stationarity tests indicate that most variables are likely non-stationary, we argue that this is not relevant in a model used to forecast prices with respect to the delivery period and not with respect to the observation time. We also detect the presence of multi-collinearity between variables. However, we observe none of the problems typically associated with multi-collinearity.

The estimated model provides the rough influence of expected supply capacity, expected consumption and long-term supply cost variables on long-term electricity forward prices at Nord Pool. The performance of the model is tested with out-of-sample data on ENOYR3 contracts from Nord Pool, and the results indicate stable parameter estimates. To test the performance of the model on contracts beyond 3 years is not possible as these contracts only trade OTC, and therefore their prices are not publicly available. Although the models suffer from a lack of data on structural variables, particularly expected supply capacity and expected consumption, the forecasts of the prices beyond the traded horizon provide robust and expected results independent of short-term variations (such as precipitation or temperature), unlike the simple extrapolation of current prices. The model also offers many possibilities for improvement, both in the choice of variables and the data underlying them, as well as for parameter estimation. We hope that the availability of these data in terms of accuracy, resolution and longer horizons will improve in the future.

Acknowledgements The authors thank Nord Pool ASA, Nordel and the McCloskey Group for access to data and Nico van der Wijst and Jens Wimschulte for useful comments. Fleten acknowledges the Centre for Sustainable Energy Studies (CenSES) at NTNU, and is grateful for financial support from the Research Council of Norway through project 178374/S30.

References

Bessembinder H, Lemmon ML (2002) Equilibrium pricing and optimal hedging in electricity forward markets. J Finance 57(3):1347–1382

Brooks C (2003) Introductory econometrics for finance. Cambridge University Press, Cambridge, UK

Byström HN (2003) The hedging performance of electricity futures on the nordic power exchange. Appl Econ 35:1–11

Diko P, Lawford S, Limpens V (2006) Risk premia in electricity forward prices. Stud Nonlinear Dyn Econom 10(3):1358–1358
Eydeland A, Wolyniec K (2003) Energy and power risk management. Wiley, Chichester, UK
Johnsen TA (2001) Demand, generation and price in the norwegian market for electric power. Energy Econ 23:227–251
Johnsen TA, Spjeldnæs N (2004) Electricity demand and temperature – an empirical methodology for temperature correction in Norway. Norwegian Water Resources and Energy Directorate
Longstaff F, Wang A (2004) Electricity forward prices: A high-frequency empirical analysis. J Finance 59(4):1877–1900
Lucia JJ, Schwartz ES (2002) Electricity prices and power derivatives: Evidence from the nordic power exchange. Rev Deriv Res 5(1):5–50
Myers RH (1986) Classical and modern regression with applications. Duxbury Press, Boston
Ollmar F (2003) Empirical study of the risk premium in an electricity market. Working paper. Norwegian School of Economics and Business Administration
Perron P (1989) The great crash, the oil price shock, and the unit root hypothesis. Econometrica 57:1361–1401
Pindyck RS (1999) The long-run evolution of energy prices. Energy J 20:1–27
Pirrong C, Jermakyan M (2008) The price of power: The valuation of power and weather derivatives. J Bank Finance 32(12):2520–2529
Samuelson PA (1965) Proof that properly anticipated futures prices fluctuate randomly. Ind Manag Rev 6: 41–49
Schwartz ES (1997) The stochastic behavior of commodity prices: Implications for valuation and hedging. J Finance 52:923–973
Schwartz E, Smith JE (2000) Short-term variations and long-term dynamics in commodity prices. Manag Sci 46:893–911
Skantze PL, Ilic M, Gubina A (2004) Modeling locational price spreads in competitive electricity markets: Applications for transmission rights valuation and replication. IMA J Manag Math 15:291–319
Solibakke PB (2006) Describing the nordic forward electric-power market: A stochastic model approach. Int J Bus 11(4):345–366

Hybrid Bottom-Up/Top-Down Modeling of Prices in Deregulated Wholesale Power Markets

James Tipping and E. Grant Read

Abstract "Top-down" models, based on observation of market price patterns, may be used to forecast prices in competitive electricity markets, once a reasonable track record is available and provided the market structure is stable. But many studies relate to potential changes in market structure, while prices in hydro-dominated markets are driven by inflow fluctuations and reservoir management strategies, operating over such a long timescale that an adequate track record may not be available for decades, by which time the system itself will be very different. "Bottom-up" analysis can readily model structural change and hydro variation, but must make assumptions about fundamental system data, commercial drivers, and rational optimizing behavior that leave significant unexplained price volatility. Here we describe a technique for fitting a hybrid model, in which a "top-down" approach is used to estimate parameters for a simplified "bottom-up" model of participant behavior, from market data, along with a stochastic process describing residual price volatility. This fitted model is then used to simulate market behavior as fundamental parameters vary. We briefly survey actual and potential applications in other markets, with differing characteristics, but mainly illustrate the application of this hybrid approach to the hydro-dominated New Zealand Electricity Market, where participant behavior can be largely explained by fitted "marginal water value curves." A second application of a hybrid model, to the Australian National Electricity Market, is also provided.

Keywords Cournot gaming · Electricity price · Estimation · Forecasting · Hybrid · Hydro · Marginal Water Value · Mean reversion · Reservoir management · Simulation · Stochastic model · Time series.

E Grant Read (✉)
Energy Modelling Research Group, University of Canterbury, Private Bag 4800
Christchurch 8140, New Zealand
e-mail: grant.read@canterbury.ac.nz

S. Rebennack et al. (eds.), *Handbook of Power Systems II*, Energy Systems,
DOI 10.1007/978-3-642-12686-4_8,

1 Introduction

Since the advent of electricity sector deregulation, spot prices for wholesale electricity have largely been set by market forces rather than central planners. Realistic modeling of the behavior of these prices is crucial for many purposes. Forecasts of future prices are important for current market participants and potential investors, for electricity retailers setting tariffs and forecasting profitability, and for pricing financial derivatives, such as forward contracts. Policy makers have a particular interest in determining the extent to which observed price patterns reflect underlying fundamentals versus exercise of market power, and the impact that structural or policy changes might have on the latter.

Because of the relatively recent development of competitive electricity markets around the world, modeling of electricity prices using historic market data has only been undertaken in the past decade. Green (2003) summarizes the questions commonly asked with regard to how best to model spot prices:[1]

> Should we be following the approach of the finance literature, which treats the price of electricity as a stochastic variable and concentrates on studying its properties in terms of volatility, jumps, and mean reversion? Or should we concentrate on the fact that the short-term price for every period is set by some intersection of demand and supply, and study the interaction of these factors with the market rules? I would be uneasy if the stochastic approach is taken to imply that we cannot understand the out-turn values for each period's price. However, it may be that explaining every price in turn is too cumbersome, and randomness should be taken as a shortcut for "things we could explain, but don't have time for in this application."

The tools currently developed for the purpose of modeling and forecasting electricity prices can be broadly categorized into Green's two groups of models, which Davison et al. (2002) classify as "bottom-up" and "top-down" models.

Top-down models (also referred to as econometric, time series, or statistical models) do not attempt to model the electricity systems themselves, but instead look at series of historic prices, or other data, and attempt to infer aspects such as price trends and volatility directly from those series. Bottom-up models, in contrast, may include detailed modeling of factors such as the marginal costs of generation, generating capacities, transmission constraints, and load, and calculate prices from this information in much the same way as do the actual market dispatch algorithms.

Recent top-down models of spot market prices aim to account for all the characteristics exhibited by electricity price time series: seasonality, mean reversion, price jumps/spikes, and time- and price-dependent volatility. However, they do this not by explaining the fundamental changes in demand and supply that cause these characteristics, but in terms of technical features in time series models usually developed for financial applications. The parameters for some of these models are estimated only from the series of prices themselves, and so, while the formulation and estimation of the models is often complex, only one series of data is required. Other models

[1] These views are echoed by Anderson and Davison (2008) and Karakatsani and Bunn (2008).

draw on other relevant data to help "explain" electricity price patterns, but the kind of explanation offered is statistical and is based on inferences drawn from correlations, rather than attempting to build up a physical, organizational, or economic model of how decisions are actually made in the sector.

In contrast, bottom-up models estimate prices in much the same way as the market-clearing process of an electricity market. In their simplest form, they construct a market offer stack, using information on the generating capacities and marginal costs of each generating company, and then calculate the spot price by finding the intersection of the market demand curve with the market offer stack. More sophisticated models typically employ optimization techniques such as LP or MILP to model "rational" participant decision-making, given dynamic and locational constraints, etc. Many studies assume that the market is perfectly competitive, so that prices can be predicted using an optimization identical to that which would be performed by a central planner, given the same assumed information set. Other models employ some particular form of gaming analysis, such as Cournot or Supply Function Equilibrium. Such models can be difficult to solve and are data-intensive. Importantly, they require commercially sensitive data, which may not be available to the analyst. They must also make assumptions about factors such as risk aversion, gaming strategies, and discount rates, which are essentially subjective and may differ widely among participants.

Thus while they are rich in terms of system specification, bottom-up price models are unlikely to replicate market price behavior, due to the complexity and subjectivity of the real-time decision-making process and the impact of random factors external to the model. Much simpler top-down models can actually perform better, particularly with respect to modeling price volatility, which has a crucial impact on many commercial valuations. Bottom-up models do, however, have the distinct advantage of being able to estimate the impact of changes to underlying factors that have not yet been observed in the market data, including major investments in generation or transmission infrastructure, shifts in technology or relative fuel prices or, in the case of gaming models, changes to industry structure or market rules.

So the goal of hybrid top-down/bottom-up modeling is twofold. First, we wish to make better estimates of price behavior for the market, as it is. Second, though, we wish to create a more robust model that can produce realistic "market-like" simulations of price behavior for a market operating under conjectured changes. We do this by separating the model to be estimated into two components: a behavioral model that can be modeled as changing in response to conjectured changes in market structure, policy, or environment; and a random component that may or may not be assumed to be independent of those factors.

We recognize that such models can never be accurate, and that this methodology can be criticized for its bold assumptions. But the fact is that such speculative analysis is a routine activity, guiding major investment and policy decisions, in markets worldwide. Analysts must, of pragmatic necessity, make bold assumptions about the modeling of situations that have not yet been observed, typically without any framework at all to guide assumptions about the extent to which various characteristics of observed market behavior might be carried over into, or modified by, these conjectured situations. In this context, the concept and study presented here may be regarded as a first step toward providing such a framework.

We assume that most readers are familiar with the great variety of bottom-up models used to calculate electricity price, but the following section describes the characteristics of top-down models in more detail, before detailing our methodology for applying such top-down techniques to estimate a bottom-up model, then combining the two into a hybrid simulation model. The paper concludes by describing the estimation, from readily available market data, of a hybrid model for the hydro-dominated New Zealand Electricity Market (NZEM), where the approach has proved surprisingly effective.[2] A second application is also included, this time calibrating a more complex bottom-up model to the Australian National Electricity Market (NEM) and combining that model with a state-of-the-art top-down model to account for residual volatility in prices.

2 Top-Down Models for Electricity Price Forecasting

The top-down models most commonly applied are time series models, which aim to account for the specific characteristics exhibited by time series of asset prices. Until recently, these models were not applied to electricity prices, because there was no "market" data to study. Recently though, econometric models have been developed specifically for application to electricity price time series. Some models are based on observation of electricity prices alone, while others also take account of exogenous factors such as fuel prices, system load, and available capacity.

The most common feature of such models is a mean-reverting stochastic process, as proposed by Lucia and Schwartz (2002) in their study of Norwegian electricity derivative prices. Thus prices are decomposed as the sum of two components: A deterministic component, $f(t)$, reflecting underlying seasonality etc.; a stochastic component modeling the impact of random shocks to the price series, which have a lasting but decaying effect over time. Mean-reverting jump diffusion (MRJD) models have most often been applied to electricity prices in recent years, with several types of jump process being proposed. Escribano et al. (2002) present one such model, decomposing the stochastic component into three separate terms: An autoregressive (AR) component; Plus "normal" volatility, Plus "jumps," that,

$$P_t = f(t) + X_t = f(t) + \theta X_{t-1} + \text{"normal"} + \text{"jumps."} \tag{1}$$

The AR factor, θ, models persistence in the effect of shocks to electricity price series, determining the speed at which prices revert back to their mean level after a shock. Escribano et al. model "normal" volatility as a General Autoregressive

[2] This paper describes the original NZEM study by Tipping et al. (2005a), but the technique has since been applied by Westergaard et al. (2006), Castalia Strategic Advisors (2007), and Peat (2008).

Conditional Heteroskedastic (GARCH) process. Heteroskedasticity, or non-constant variability in the residuals of a model, is appropriate for electricity price time series, because there are periods both of prolonged high variance and prolonged low variance. GARCH processes include their own AR component to model persistence in volatility. In a GARCH (1,1) process, the "usual" volatility, e_t, is distributed normally, with mean zero and variance h_t, modeled by

$$h_t = \omega + \alpha e_{t-1}^2 + \beta h_{t-1}. \tag{2}$$

That is, this period's forecast variance in the residuals is a function of an underlying (or unconditional) variance, ω, last period's forecast conditional variance (h_{t-1}), and the square of last period's actual movement from the expected value (e_{t-1}^2). The ARCH term (α) is included to account for short-term clustering and persistence in the volatility of the residuals due, for example, to temporary transmission outages or experimental strategies. The GARCH term (β) contributes more to long-term volatility persistence due, for example, to plant maintenance or seasonal factors. For the jump component, Escribano et al. use a Poisson process, with the jump intensity λ_j varying by season, j. If a jump occurs in a given time period, then its size is drawn randomly from a normal distribution with mean μ_j and standard deviation σ_j. The stochastic component of the price model can be represented as follows:

$$X_t = \begin{cases} \theta X_{t-1} + e_t & \textit{with rate } (1-\lambda_j) \quad \textit{No Jump} \\ \theta X_{t-1} + e_t + \mu_j + \sigma_j \varepsilon_{2t} & \textit{with rate } \lambda_j \quad \textit{Jump} \end{cases},$$

where

$$e_t = \sqrt{h_t}\varepsilon_{1t} \text{ and } \varepsilon_{1t}, \varepsilon_{2t} \sim \mathrm{N}(0,1). \tag{3}$$

Fitting this model requires estimation of all the parameters discussed earlier. Relatively straightforward estimation procedures are available for less complex models. A MRJD model will suffice for the representation of many stochastic processes. Clewlow and Strickland (2000) describe a method using a recursive filter to estimate jump parameters and linear regression to estimate the mean-reversion rate, without the need for specialist software. But Escribano et al. employ an optimization technique called constrained maximum likelihood estimation (CMLE) to find the set of parameters, Θ, for the model that is *most likely* to have produced the actual series of data observed, via the following steps:

1. Define the log-likelihood function for the observations of price. The concept of the likelihood function (LF) is similar to that of the probability density function (PDF) for a single observation; while the PDF provides the probability of realizing an observation x, given a set of parameters Θ, the LF calculates the probability of the set of parameters being Θ given an observation x. The LF is the joint PDF of the sample data and is constructed by multiplying together the probabilities of each of the individual observations of price.[3]

---

[3] The LF for the Escribano et al. model is provided in their paper.

2. Find the set of parameters Θ for which the LF is maximized or, in other words, the *most likely* values of Θ. The first step in this procedure is to take the log of the LF (the LLF), taking advantage of the fact that the logarithm is an increasing function, as maximizing the LLF is a more straightforward procedure. For a very simple LLF, the set Θ can be computed analytically, though a more complex LLF will require the use of line search algorithms.[4] Often the computational effort required in maximizing the LLF will dictate the form and complexity of the underlying model.

The top-down models referred to above are based solely on analysis of price series, without any reference to other data. But a great many other factors may also be considered relevant, and many "top-down" electricity price models now involve estimation based on other variables representing aspects of supply and demand. The most common explanatory variables include load and temperature-related variables (e.g., Longstaff and Wang 2004); supply-side factors, such as fuel prices (e.g., Guirguis and Felder 2004) and measures of available capacity; or combinations of both demand- and supply-side variables (e.g., Karakatsani and Bunn 2008). Aggarwal et al. (2009) survey many of the recent spot price models and give further examples of their extension to account for physical data series.

Because the underlying level of electricity prices is driven by the combination of supply and demand, inclusion of even a simplistic representation of the factors driving electricity prices is likely to improve the overall performance of a model. The equilibrium models of Pirrong and Jermakyan are two of the most widely referenced. Their first model (1999) specifies price as the product of the fuel price g_t raised to a power and an exponential function of load q_t, as shown below in (4). Both the fuel price and load are modeled as stochastic processes:

$$P\left(q_t, g_t, t\right) = g_t^{\gamma} \exp\left[\alpha q_t^2 + c(t)\right]. \tag{4}$$

In this model, $c(t)$ is a deterministic seasonal function that accounts for seasonality in both demand and supply, which "shifts the pricing function up or down over time." Their second model (2001) is similar, except the demand-side term is a function of the natural log of load rather than an exponential function of load. They use these functions as a result of their (intuitive) suggestion that "the price function is increasing and convex in (load)."

This example illustrates that representations of the electricity market do not have to be sophisticated to improve modeling performance over a top-down model that ignores such factors in explaining shifts in the price level and price volatility. Such models could be described as "hybrid," inasmuch as they use nonprice time series

[4] Several statistical computation packages contain procedures for CMLE. The analysis in this paper was undertaken using the CMLE procedure in GAUSS 6.0. Constraints are required for the Escribano et al. model, as the variance parameters ω and σ^2 and jump intensity λ must be non-negative. Parameters for some other models may be calculated using conventional maximum likelihood estimation.

to help "explain" electricity price behavior.[5] But they do not involve any significant "bottom-up" modeling component. That is, they do not attempt to build up a model of how decisions are actually made in the industry.[6] Thus, in our view, they do not make full use of the large body of knowledge about electricity sector behavior developed by bottom-up modeling over the years. Nor do they provide a direct way to calibrate and improve bottom-up models or make them more suitable for modeling market behavior, which is our goal. Many of these models still rely on observation of "physical" data (e.g., the available generation capacity or the marginal fuel type) as exogenous inputs to the model, rather than generating a forecast of such data, endogenously, as part of a market simulation.

Thus we will largely ignore this type of modeling in what follows. But there is no reason, in principle, why the factors accounted for in these models could not also be accounted for in the type of hybrid model discussed here, either as additional explanatory factors driving the stochastic component, or as inputs to the bottom-up models estimated. Indeed the way in which inflow data, in particular, is used in constructing our hybrid model is not too different from the kind of estimation performed by some of the models listed earlier.

3 Hybrid Bottom-Up/Top-Down Modeling

Motivated by concerns about the inadequacies of bottom-up and top-down models alone, the general concept of hybrid modeling has received attention recently for a variety of purposes. Often the bottom-up/top-down terminology merely refers to methods that produce system costs or load forecasts, say, by combining and reconciling detailed bottom-up assessments of specific components, with broader top-down assessments of aggregate levels. But a recent special edition of *The Energy Journal* was devoted entirely to hybrid bottom-up/top-down models, in the context of energy/economy modeling, over relatively long time scales. Hourcade et al. (2006) note, in the introduction to that edition, that while bottom-up energy economy models are technology rich, they have been criticized for their inability to model both microeconomic decision-making and macroeconomic feedback from different technology outcomes. Top-down energy models are able to capture these features, but, like top-down pricing models, require calibration with observed data and thus lack the flexibility to deal with structural and policy changes. Hence experimentation with hybrid models such as the combination of the top-down energy policy analysis tool EPPA with the bottom-up systems engineering tool MARKAL (Schäfer and Jacoby 2006) has become more common.

---

[5] Pure top-down models such as that of Conejo et al. (2005) have been referred to as "hybrid" because they blend more than one type of top-down model into a single modeling framework.

[6] Artificial intelligence models including Artificial Neural Networks and other similar techniques (see Szkuta et al. 1999 and Georgilakis 2007) would also fall into this category, unless they attempt to develop a model of supply/demand interaction within their forecasting framework.

In many ways this literature is reminiscent of the energy modeling literature of the 1970s and 1980s, as surveyed by Manne et al. (1979), where several modeling systems combined top-down econometric and bottom-up optimization models of consumer choice in some way. One approach was the "pseudo-data" technique of Griffin (1977), whereby top-down econometric representations were fitted to data generated by detailed bottom-up optimization models, as if it were actual market data. This takes advantage of the accuracy of bottom-up models in predicting prices in hypothetical market situations and the computational efficiency of using functional representations similar to those from top-down models, in simulating prices. In a sense, our methodology here reverses this by fitting bottom-up models to actual market data. Here, though, we ignore wider energy/economy interactions and focus only on models for relatively short-term electricity price forecasting and on the hybrid methodology already outlined in the introduction.

Our focus is on hybrid models that allow industry behavior to be simulated using an endogenous bottom-up model and its complementary stochastic process, both estimated by top-down methods from available market data and driven by fundamental physical or organizational drivers. This enables us to model the way in which system behavior, including at least some of the variables treated as inputs in the original estimation process, can be expected to respond to changes in those fundamental drivers. At a conceptual level, simulation requires us to be able to describe how the "state" of the system, S_t, evolves from one period to the next. Here S_t need not be the whole vector of system variables that may be of interest to the analyst, but a subset of variables that is sufficient to define how the system evolves to the next stage. We envisage evolution as involving a three-step process, as described by the following conceptual equation:

$$S_{t+1} = S_{t+1}(S_t, A_t(S_t), D_t(S_t, A_t(S_t)), X_t(S_t, A_t(S_t), D_t(S_t, A_t(S_t)))). \quad (5)$$

Working from left to right, this conceptual equation describes the evolutionary sequence in order of time and hence of dependency. Starting from an initial state, S_t, a set of fundamental drivers, A_t, is first observed. These are drawn from some random set and may (or may not) depend on S_t. Then a decision, D_t, is made and it may (or may not) depend on S_t or A_t, in accordance with our fitted bottom-up model. Then the system experiences further uncertainty, X_t, which may (or may not) depend on S_t or A_t or D_t, as determined by our fitted complementary process. This then determines the market outcome for the period, and that may include a number of factors that are simply recorded, and have no influence on subsequent periods. But among those outcomes is the vector S_{t+1}, from which the simulation may proceed in the next period.

Simplifying this equation in various ways yields various classes of model. If we eliminate both A_t and D_t, it describes a simulation model in which the price in one period depends only on (a system state vector determined by) prices in previous periods, plus a new random component, as in the top-down "price only" models of the previous section. Adding in A_t allows us to expand the description to include top-down models in which price is driven by exogenous factors other than previous prices. It also requires us to develop models to forecast, or simulate, the evolution of those exogenous factors, but standard techniques exist for this, and here we assume that such forecasting models already exist.

The distinctive feature added by the hybrid models discussed here is the decision phase, producing D_t. As noted, D_t could depend on S_t or A_t, and it could also play a role in determining X_t. The idea is that, as it becomes more sophisticated, the bottom-up model will explain more of the observed volatility, thus progressively reducing the role of the complementary stochastic process. Few models of this type have appeared in the literature on electricity price forecasting, particularly for higher-frequency data, and none has fully exploited the possibilities inherent in the concept. Our own model is described in the final section, but here we canvas some alternative approaches that have been taken, or could be taken, to produce such a hybrid model, in accordance with a four step methodology:

1. Determine the form of the conceptual bottom-up model and how it will be represented in a simplified form, with a set of parameters limited enough to facilitate estimation from market data.
2. Determine the form of the complementary stochastic (top-down) process to account for residual variation in prices.
3. Estimate simultaneously the parameters for the empirical bottom-up and top-down models from market data.
4. Combine both into a hybrid simulation model to test the performance of the model and to perform market studies.

3.1 *Form of the Bottom up Model*

In terms of (5), the bottom-up model is what determines the decision, at each stage of the simulation, $D_t(S_t, A_t(S_t))$. Pragmatically, no matter how complex this model is, we require its overall behavior to be governed by a small enough set of parameters that they can be estimated from historical data. Thus an LP, for example, may be an excellent bottom-up model, but we cannot expect to estimate all of its constraint or objective function coefficients. But we could assume, for instance, that capacities are known, with fuel costs estimated from market data, but that participant offers add a markup to those fuel costs. We could then estimate the markup, or markup structure, from market data, treating all other deviations from our assumptions as "noise" to be described by the complementary stochastic process.

Neuhoff et al. (2008) use a simple stack model, rather than LP, to forecast price levels and volatility in the British electricity market. The marginal generating unit in each hour is found at the intersection of the stochastic market supply stack and demand, but a stochastic "uplift" component is added to the market-clearing SRMC, depending on system "tightness". To parameterize the uplift component, the same stack model is used to back-cast estimates of the marginal SRMC from historic half-hourly data, using observed load, capacities, availability, and fuel prices as inputs, with the uplift in each period calculated as the difference between price and calculated SRMC. The observed relationship between observed capacity margins and calculated uplift is then parameterized with a functional form and stochastic parameter to account for residual volatility. In another approach, Bessembinder and Lemmon (2002) assume that all companies in the market act as perfectly competitive profit-maximizers, and have identical cost functions, of a simple form, whose parameters can be estimated via linear regression from observed price and load.

Cournot models are another option. Tipping et al. (2005b) note that such models are largely driven by assumptions about demand elasticity and contracting levels. But those parameters may not be known, not least because we do not know over what time period participants consider the effect of their actions. Thus Tipping et al. estimate the parameter values implicit in participant offers from data for peak, shoulder, and off-peak periods in the thermal-dominated Australian electricity market. For quite different reasons, Wolak (2003) also estimates residual elasticities and implied net contract positions, assuming Cournot gaming by Australian market participants; the difference being that Wolak had historical offer data for that case, whereas Tipping et al. do not assume that such data will be available.

Vehviläinen and Pyykkönen (2005) focus on monthly average spot prices in the Nordic market. They estimate a simplified Marginal Water Value (MWV) curve from market data, and use that curve, along with a piecewise linear supply curve for thermal power (also estimated from market data), to calculate the level, and price, of both hydro and thermal power required to meet system load.[7] Our New Zealand model uses a different, but similar, MWV curve representation as its bottom-up model, as described in Sect. 4.

3.2 Form of the Complementary Stochastic Process

The choice of stochastic process to complement the bottom-up model depends on which features of electricity price time series that the bottom-up model is not expected to be able to account for well, and may be constrained by computational requirements. A simplified stochastic process may suffice if capacity scarcity and fuel price fluctuations account well for price spikes and seasonality, for example. But complexity may be increased if stochastic elements of the top-down process are linked to variables in the simulation or other exogenous variables. Modeling the top-down jump intensity parameter λ as a function of system variables such as load or scarcity is common practice, but other elements such as the unconditional volatility ω may also be driven by such factors.

Some of the bottom-up models cited earlier were not intended to form part of a simulation model, and do not explicitly fit a complementary stochastic process, thus implicitly treating the residual fitting errors as unstructured random noise.

[7] MWV curves are well known and frequently used in reservoir management. See Read and Hindsberger (2009) for discussion and examples.

Vehviläinen and Pyykkönen do form a simulation model, but do not employ a complementary stochastic process, primarily because, in a market with significant storage, monthly average prices are largely explained by their bottom-up model. Tipping et al. (2005b) complement their fitted Cournot model with a stochastic MRJD process, while Neuhoff et al. account for deviations of price away from modeled SRMC with their stochastic representation of "uplift." As discussed in Sect. 4, our New Zealand model uses a stochastic process similar to the Escribano et al. model described in Sect. 2.

3.3 Estimation

Ideally, the estimation of all parameters will be performed simultaneously, but this depends on the structure of the bottom-up and top-down models selected. Bessembinder and Lemmon, and Vehviläinen and Pyykkönen, used simple bottom-up models, for which the parameters could be estimated via ordinary least squares regression, and neither required the estimation of complex complementary stochastic processes such as those described in Sect. 2. For the latter model, parameter estimation proceeded in three sequential steps: First, parameters for functions of climate processes were estimated directly from observed climate data; second, functions for physical demand and inflexible supply were estimated directly from climate and production data (approximately covering A in (5)); and finally, functions relating to the pricing of flexible generation, including the marginal water value function, were estimated from observed data and the calculated outcomes of the functions estimated in the first two steps corresponding approximately to estimation of a model to determine D in (5).

For hybrids incorporating more complex bottom-up models, a multi-step empirical estimation procedure may be required, solving first for the set of bottom-up model parameters that provides the best fit to empirical data, and second for the set of top-down parameters that provides the best fit to the residuals. One way to do this is to apply standard econometric techniques, such as those described in Sect. 2 above, to determine a set of parameters for the bottom-up model that best fit market data. This process itself may be accomplished in multiple steps, as per Vehviläinen and Pyykkönen, to enable independent auditing of the parameters estimated in each step. The complementary stochastic process can then be fitted to a series of residuals generated by simulation using the fitted bottom-up model, again using the methods of Sect. 2. The model of Neuhoff et al. is estimated using this procedure, in which the parameters for the bottom-up stack model are estimated prior to the parameters for the "uplift" component.

An alternative approach to estimating the parameters of the bottom-up model essentially reverses the pseudo-data approach. As in that technique, the bottom-up model is first run for a specific historic period multiple times, assuming multiple combinations of the parameters being estimated and thus generating whole series of "pseudo-data" for each parameter combination. Then, rather than fitting a function to describe the output surface as a function of the parameters, we choose the parameter combination for which the results of interest best fit the historical data. Tipping et al. estimated elasticity and contract parameters for their Cournot model using this method, with the criterion being the match of simulated to market prices.[8] As per the first approach, they then fit a stochastic process to the residuals of the simulated series, yielding the best fit to market prices.

3.4 Simulation

Having estimated the parameters for our hypothesized bottom-up model of decision-making and its complementary stochastic process, we wish to combine the two models within a Monte Carlo simulation framework to forecast prices. If the goal were to examine the performance of the model in replicating market behavior already observed, we may use as input an historic series of observed variables from which the models were estimated. In many cases, though, those observations are outputs from the electricity system, rather than fundamental drivers of electricity system behavior. Thus we cannot conduct a meaningful simulation of hypothetical future behavior unless we have some means of generating consistent sets of such data for hypothetical future situations.

In other words, forming a self-contained simulation model requires us to either produce a process to generate random series for the underlying drivers, A, as part of the simulation process, or if A does not depend on S, form a set of series for A, for example, from an extended set of historical data. Our New Zealand model, which is ultimately driven by hydrology data, takes the latter approach, as discussed in Sect. 4, and so do Bessembinder and Lemmon (2002). Vehviläinen and Pyykkönen take the former approach, however, simulating deviation away from long-run average temperature and precipitation levels, the two primary drivers of their model. Neuhoff et al. (2008) also take the former approach, simulating underlying data stochastically rather than drawing repeatedly from a larger set of observed data.

4 A Hybrid Model for the New Zealand Electricity Market

In this section we use the framework outlined earlier to develop a hybrid model for electricity prices in a hydro-dominated market. The model is fitted to historic data from the New Zealand electricity market (NZEM), and the section concludes with illustrations of the performance of the model in back-casting and forecasting over a range of hydrological conditions.

The NZEM began trading in October 1996, and had a 7 year data history at the time of this study, although with some changes in market structure over that period.

[8] A simple criterion such as the root mean squared error may suffice, but a criterion based on median errors may be more suitable if the distributions are likely to be skewed.

Hydro is the primary form of electricity generation (approximately 65% of average annual output) but, unlike the Nordic market, for example, reservoir storage is quite limited, with storage capacity for only around 12% of annual energy. Thus security of supply is dependent on the management of the hydro storage capacity, on which this model focuses. The NZEM operates across the North and South Islands, linked by an HVDC cable. The market is cleared half-hourly, but the data modeled in this paper were aggregated at the daily-average level.[9]

Recent top-down models of electricity prices are particularly well-suited for thermal-dominated systems, in which extreme short-term price volatility and strong mean-reversion are dominant characteristics. However, these models can be shown to fit poorly to price behavior in hydro-dominated markets (Tipping et al. 2004), in which prices vary greatly in response to changing hydrological conditions. This therefore motivates the incorporation into such models of some form of bottom-up representation of the hydrology underpinning the price-setting process. The model described in the following sections follows (1), in that prices are explained by the addition of deterministic and stochastic components. But rather than the deterministic component being simply a function of time, $f(t)$, we take it to be the MWV, which we estimate as a simplified function of the water storage level. Several parameters in the complementary stochastic process, X_t, are also linked to MWV.

4.1 Bottom-up Representation

The MWV plays a critical role in determining reservoir management behavior, and there are well-established techniques for bottom-up optimization of MWV,[10] or of an MWV curve/surface to be used in a simulation context.[11] Such models have been used extensively to simulate market behavior, and hence to forecast prices and investment behavior. But the actual MWV curves used by market participants are assessed internally, and are not public knowledge. We do not know the extent to which they may be adjusted to account for gaming, for instance, or risk aversion, and nor do we necessarily have data on the underlying factors, such as contract levels, that would affect such adjustments. In any case, different participants will have different perspectives on likely future conditions, and hence differing assessments of the MWV if water stored for future use.

Therefore, any MWV model that is to be used to forecast actual market behavior will need some form of market calibration. Pragmatically, our goal here is not to gather as much detailed data as we can or to build the most sophisticated bottom-up model possible. While future studies may find value in more sophisticated

[9] Daily-average prices are presented from the Benmore node, measured in NZD per MWh, and storage level observations are aggregated across the country, measured in GWh. The dataset analysed in this paper, provided by M-co, covers August 1999 – June 2003.

[10] See Pereira (1989) and Pereira and Pinto (1991), for example.

[11] See Read and Hindsberger (2009), for example.

bottom-up models, our aim here was to see how much of the market price volatility can be explained using a bottom-up model that is as simple as possible, both to construct and to estimate. Accordingly, to form the "bottom-up" component of our hybrid model, we calibrated a very simple MWV model based on the concept of the "relative storage level" (RSL), explained below.

Storage levels in hydro reservoirs are inherently seasonal due to seasonality in the major drivers of those levels, namely load and inflows. Capturing this seasonality is crucial to the calculation of prices in any hydro-dominated market, but the single biggest factor, determining whether prices are high or low relative to the mean for any time of year, is where the storage level sits, relative to the range of possible storage levels for that time of year. Somewhat arbitrarily, we defined the RSL on day t as the difference between the actual storage level on day t and the historic tenth percentile of all observed storage levels on the same day of the year, d:[12]

$$RSL_t = Storage\ level_t - Historic\ tenth\ percentile_d \tag{6}$$

We effectively replace the deterministic component $f(t)$ in the price model with the functional form shown in (7), for the MWV on day t, as a function of RSL_t. Parameter c is a constant lower bound price assumed to apply for very large values of RSL, while x, y, and w are to be fitted. The parameter y is required to ensure that the function is defined for negative values of RSL, x is required to scale the value of the bracketed terms down to a number that can be raised sensibly to an exponential power, and w is the maximum marginal water value obtainable when the RSL is at the lowest level allowed by the function.

$$f(t) = MWV_t = c + we^{x(y+RSL_{t-1})^z} \tag{7}$$

Hydro generation levels, or releases from hydro reservoirs, are a key part of any bottom-up model of a hydro-dominated system and form the decision component D in (5). We model releases from the reservoir, R_t, with a dynamic regression model, hypothesizing that R_t is a function of the marginal water value MWV_t, inflows into the reservoir I_t, price P_t, and a stochastic component N_t. The stochastic component accounts for variation in R_t attributable to other significant drivers of release, including generators' contract levels and system load, which were not publicly available at the time this analysis was undertaken.

4.2 Top-down Stochastic Process

Incorporating a function for the MWV curve increases the ability of the price model to track the underlying level of prices in the NZEM. However, as mentioned

[12] This implies that while the absolute storage level must be greater than zero, the RSL will be negative about 10% of the time. Since any storage level guideline for reservoir management would be a smooth curve, a 45-day centered moving average of the tenth percentile was used.

previously, a cost-based model cannot account for a significant amount of volatility in prices and persistent deviation away from the underlying level. Other physical factors such as transmission constraints will lead to short-term spikes in prices, and particular participant behavior leads to deviations away from the underlying level for days, if not weeks.

As in the Escribano et al. model, we complement the hydrology-driven bottom-up component of the price model with a stochastic process including mean-reversion, jumps, and a GARCH process. Two modifications are made, however. First, the jump intensity at time t is modeled as a linear function of MWV_t, rather than varying simply by season j, and the mean jump size is modeled similarly.[13] This reflects the fact that a higher MWV may be considered a proxy for overall system tightness, and jumps in price are more likely and often larger at times when the RSL is particularly low.

As mentioned in the previous section, several factors that influence hydro generation levels were unable to be accounted for in the model for releases, thereby necessitating the inclusion of a stochastic error term in the model. Releases may deviate away from the expected levels for days, if not weeks, due to factors that are not directly related to hydrology. We account for these deviations in the release model with the stochastic component N_t, described by an ARIMA model,[14] which is able to account for serial correlation in the residuals.

4.3 Parameter Estimation

In combination with (1)–(3) and (6)–(7), the following equations map out the combined price and release model:

$$P_t = MWV_t + X(t, MWV) \tag{8}$$

$$R_t = aMWV_t + bI_t + c\ P_t + N_t \tag{9}$$

The data series required for the estimation of parameters in the price and release models are the following:

P_t: daily average NZEM spot prices

RSL_t: daily aggregate NZEM relative storage levels

---

[13] Similar functions were tested for the unconditional variance ω and the jump size variance σ^2, however these proved not to improve the explanatory power of the model.

[14] ARIMA (Autoregressive Integrated Moving Average) models were formalised by Box and Jenkins (1976, as cited in Makridakas et al. 1998, p. 312). These models include terms to account for multi-period autoregression in both the dependent variable and the model residuals. Some series must be differenced one or more times in order to ensure stationarity. Standard notation for ARIMA models is ARIMA(p,d,q), where p is the order of autoregression in the dependent variable, q is the order of autoregression in the residuals of the model, and d is the order of integration of the series (or the number of times the time series must be differenced before it becomes stationary). An ARIMA(p,0,q) or ARMA(p,q) model is fitted to series that do not need any order of differencing.

R_t: daily aggregate NZEM reservoir releases

I_t: daily aggregate NZEM reservoir inflows

The parameters of the price and release models were estimated in two separate steps to enable independent analysis of each model; however, it would have been possible to estimate them simultaneously. The parameters of the price model, including the MWV function and MWV-driven stochasticity, were estimated via CMLE, as per the original models of Escribano et al. The release model was also estimated using CMLE, although several iterations of the procedure were required to finalize the parameters of the error process, which yielded the best fit to the overall model while accounting for the serial correlation in the error.[15] The estimation process is summarized in the steps below:

1. Calculate RSL, the series of daily relative storage levels, from historic daily aggregate NZEM storage levels. The series should extend far enough into the past to allow a reasonably smooth tenth percentile benchmark to be calculated, although this can be smoothed further by calculating a moving average.
2. Estimate the parameters of the price model using CMLE.
3. Calculate a series of MWV_t using the parameters estimated in step 2.
4. Estimate the parameters of the release model using CMLE.

Figure 1 shows a scatter plot of observed value of RSL vs. daily-average spot prices from the Benmore node, and reveals a reasonably strong (negative) relationship between the RSL and daily average spot prices, suggesting the inclusion of the RSL in the price model. As expected, prices rise rapidly when storage falls towards historically low levels, but have a slower rate of change at higher levels of relative

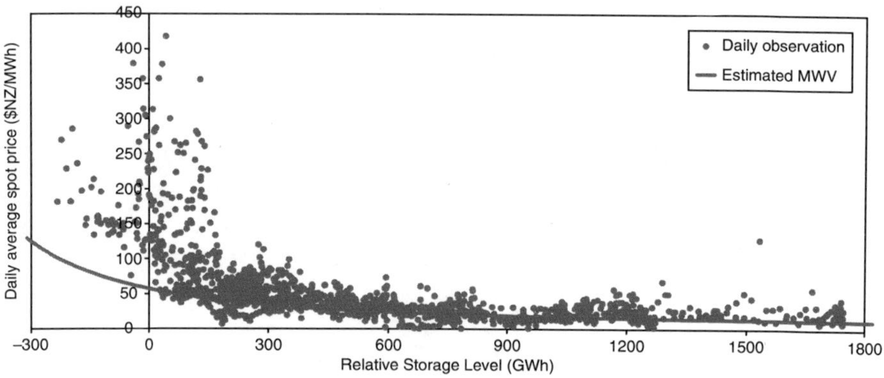

Fig. 1 Daily average Benmore spot prices vs. Waitaki RSL, August 1999–June 2003

[15] The final specification of the ARIMA (p,d,q) model was $p = 3$, $d = 0$, $q = 1$.

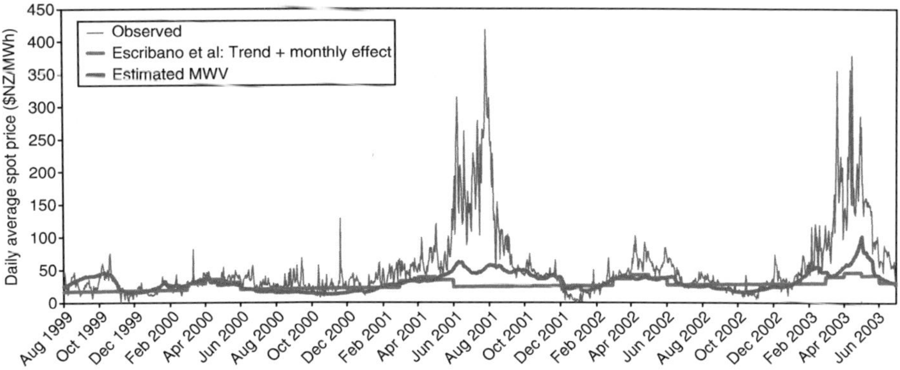

Fig. 2 Daily average spot prices from the Benmore node, August 1999–June 2003, observed prices vs. fitted deterministic components

storage. The red line in the chart represents the MWV curve fitted as part of the price model.[16]

When the estimated MWV is plotted against the actual time series of spot prices, as in Fig. 2, it compares favourably with the deterministic component of the original Escribano et al. model, estimated over the same time period. While the seasonal pattern in prices attributable to load is captured by the original model, it takes no account of the influence of low relative storage levels, which can occur at any time of year. The estimated MWV accounts for far more variation in the underlying price series than a deterministic function based solely on the time of year, as discussed in Tipping et al. (2004). Although there is still a significant stochastic component "explaining" extreme price levels, the fit to observed prices is improved significantly. However, to take advantage of this improved performance, the price forecasting model requires accurate predictions of storage level, as discussed in the following section.

Further analysis of the price model for NZEM data may be found in Tipping et al. (2004). A key conclusion in that paper was that the speed of mean reversion increased markedly once MWV was included in the model, which may be interpreted as proving that the MWV tracks the underlying price level more accurately than a standard time-based deterministic component.

Turning to the release model, the variable that accounted for the majority of the variation in releases was MWV_t, with, as expected, a statistically significant negative coefficient. Inflows accounted for the second greatest proportion of variation,

[16] Note that the fitted curve does not follow the scattered points to the highest levels of prices; because of the variability in prices at low levels of RSL, the CMLE procedure finds a better fit to the overall model by attributing those high prices to jumps, rather than to movements in the underlying MWV. This is probably because there were so few high price episodes in the sample. Much better-looking fits were produced manually, and might have produced a better hybrid simulation. However, the scatter plot does not show the time dimension, and thus it is hard to tell how well the stochastic component in the model is accounting for deviation away from the estimated MWV.

but surprisingly, while positive and statistically significant, the coefficient on P_t (effectively X_t, the stochastic component of prices) was relatively small. This suggests that releases in the NZEM are driven primarily by the RSL and inflows, with higher prices having only a minor impact, largely because prices rise to match MWV, leaving release much the same. There was also a positive constant in the model, which could be interpreted as representing factors such as minimum release constraints, and, perhaps, contracted generation levels.

4.4 Hybrid Simulation Model

Our hybrid simulation model follows a similar structure to that of Vehviläinen and Pyykkönen. The storage level on each day, S_t, is the sum of the previous day's storage level, S_{t-1}, and today's inflows, I_t,[17] less today's releases, R_t. Our fitted MWV curve defines MWV_t as a function of the previous day's storage level, S_{t-1}, while the daily average price is the sum of MWV_t, and the fitted complementary stochastic process, X_t. Releases, R_t, are a function of MWV_t; inflows, I_t; price, P_t; and a further stochastic process, N_t.

Back-casting performance of the model is particularly strong. Figure 3 below shows the simulated median storage levels and simulation intervals over the period

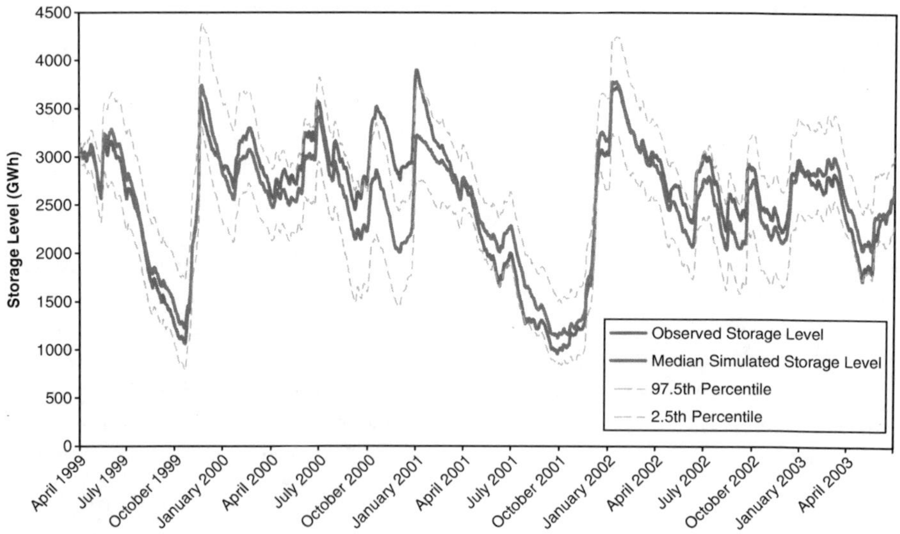

Fig. 3 Historic storage and 95% back-casting simulation intervals, 1999–2003

[17] Our model uses historical inflow data, but incorporating an inflow model would be a simple extension.

in which the model was fitted, assuming the same (observed) sequence of inflows in each simulation. It is interesting to note that the width of the 95% simulation interval is fairly constant throughout the back-casting period. This comes as a result of the self-correcting nature of the MWV process. As the RSL falls due to the rate of inflows being less than the rate of release, the MWV increases, reducing releases and slowing the rate at which storage declines. Similarly, as the RSL rises through increased inflows, MWV decreases, increasing the rate of release. Ultimately, it is the series of inflows in the past that determines storage levels in the present.

4.5 Forecasting Results

Inflows are the key underlying driver of the hydrological system in the NZEM and are independent of the other variables that we model. Rather than model and simulate inflows independently in the hybrid simulation model, we have chosen to collate annual sequences of inflows dating back over 70 years, and sample from those series within the Monte Carlo simulation framework of the hybrid model. The random errors in both the price and release models are assumed to be independent of each other and of the inflow series.

For many applications in the areas of policy development, regulation, and investment analysis, the exact timing of specific prices is less important than the overall shape and level of prices in aggregate. These prices are often represented in a price duration curve (PDC), which is effectively an inverse cumulative distribution function for prices. Thus for many purposes, the ultimate test of our hybrid simulation model is how well it simulates the PDC. Figure 4 shows the observed PDC for the 1999–2003 sample period along with two simulated PDCs.

Comparing the observed PDC with the PDC simulated using inflows from 1999–2003, we can see that the fit is, in fact, very good. What really matters, though, is the PDC simulated using a wider set of inflows from 1980–2003, because this is more representative of "long run" market performance. It may be seen that the model, which has been fitted to observed market behavior, predicts a long run PDC significantly lower than that observed to date, reflecting the fact that dry years have been relatively common over the time the market has been running. Still, the simulated PDC also suggests the possibility that even higher prices could be observed, in the top 3% of the PDC

This example illustrates the benefit of using a hybrid model to fit a simple bottom-up model, allowing industry behavior to be simulated over a wider set of inflows than those actually observed during the market era. Further forecasting results using a similar hybrid simulation model may be found in Peat (2008), who provides illustrations of the back-casting performance of the hybrid model in several key periods in the history of the NZEM. The results of his application, and the results illustrated above, provide evidence for the validity of the model. Note that, once fitted, the MWV surface can be used just like an MWV surface produced by a bottom-up optimization such as that in Read and Hindsberger, and embedded into

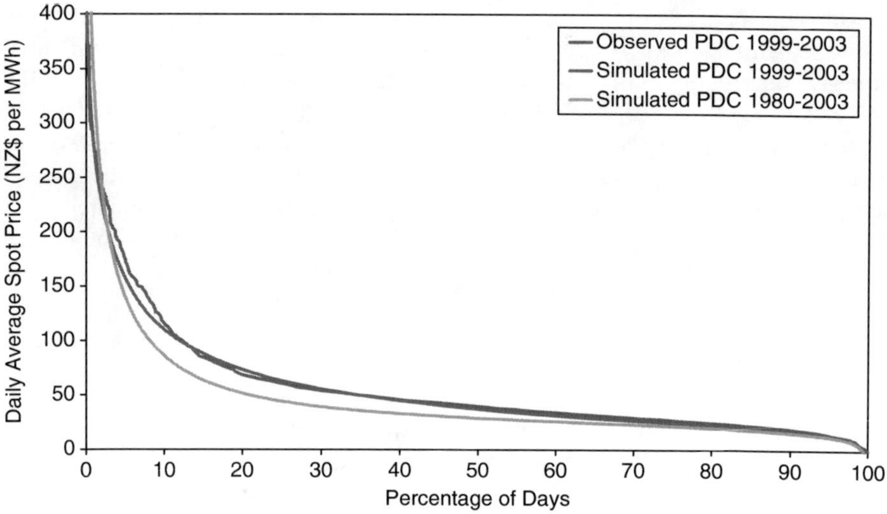

Fig. 4 Simulated NZEM price duration curves: 1999–2003 and 1980–2003

more detailed models for study of particular aspects of system operation. Similarly, the complementary stochastic process can be used to model the additional market price volatility, which may be expected, over and above that due to hydrological variations, as managed by a bottom-up reservoir optimization model.

5 A Hybrid Model for the Australian Electricity Market

Arguably, the hybrid NZEM model described earlier did not, in fact, incorporate a specific bottom-up reservoir optimization model per se, but rather estimated parameters for the output of such a model, the MWV curve. The crucial aspect of the hybridization process is that it assumes that, while a model producing such curves may be used by market participants, the details of that model are unknown to the analyst modeling market prices. Tipping et al. (2005b) apply a similar methodology, but calibrate a bottom-up Cournot model for the purpose of modeling prices in the Australian National Electricity Market (NEM). They complement this with the top-down model of Escribano et al. to account for residual volatility in prices not captured by the Cournot model.

The particular transmission-constrained Cournot model they use is described by Chattopadhyay (2004). Two of the key input parameters in this model (as in any Cournot model) are the elasticity of the market demand curve and the extent to which participants are contracted. Let the intercept and slope of the (linear) inverse market demand curve be A and b, respectively, the marginal cost function for firm i be $\mathrm{d}_i + \mathrm{c}_i G_i$, and its contract level be K_i. As in Chattopadhyay, the profit-maximizing output level of firm i in a Cournot market, G_i^*, is given by (10)

$$G_i^* = \frac{A - b(G_{i'} - K_i) - d_i}{2b + c_i} \tag{10}$$

Thus, without knowing (or at least estimating) values for the slope (or the elasticity) of the demand curve or the extent to which each firm is contracted, it is impossible to calculate realistic generation levels and market prices using such a model. But neither of these parameters is publicly available in many markets. Indeed, neither parameter is particularly well-defined. The market rules do not really define a Cournot game and participants are not really focused on the effective demand elasticity in a single period. They must consider the market response to a repeated bidding strategy over at least the medium term. Thus the "elasticity of demand" must really represent demand- and supply-side responses over an extended period. The contract level in the current period is not the only relevant measure of contractual commitments, either. Contract levels are often only implicit in the need to meet retail load requirements, and the way in which contractual commitments typically fall over time is also important. As participants look forward, contractual commitments typically fall, implying greater incentives to push prices up, but elasticity also rises, implying less ability to do so. And participants must also factor in risk aversion, and the threat of regulation, among other things. Even if participants were to use a formal Cournot model, which they probably do not, each must find their own balance between these factors, introducing a significant subjective element into the choice of Cournot parameters.

Thus Tipping et al. do not speculate about the actual elasticity or contract levels in any period, but attempt to estimate a combination of pseudo elasticity of demand (PED) and pseudo contract level (PCL), which best explains observed behavior across the entire sample period. They model peak, shoulder, and off-peak prices for each day in 2003–2004, in each region of the NEM, utilizing the following publicly available data:

1. Observed half-hourly load for each state
2. Information on each generating plant, including capacity, technology characteristics, and ownership
3. Daily fuel prices
4. Transmission capacities between regions

The data are used to calculate company cost functions for each day of the sample period and, using the Cournot framework, Nash Equilibrium market prices over the same time horizon. No assumptions are made around plant unavailability; effectively each plant is available at all times in the model, despite many units having been taken offline for planned and unplanned outages over the course of the sample period. Ex ante, information of this kind is rarely available to the market analyst.

The fitting procedure involved drawing up a grid of values of PCL and PED and simulating price series over that grid, then selecting the combination which yielded

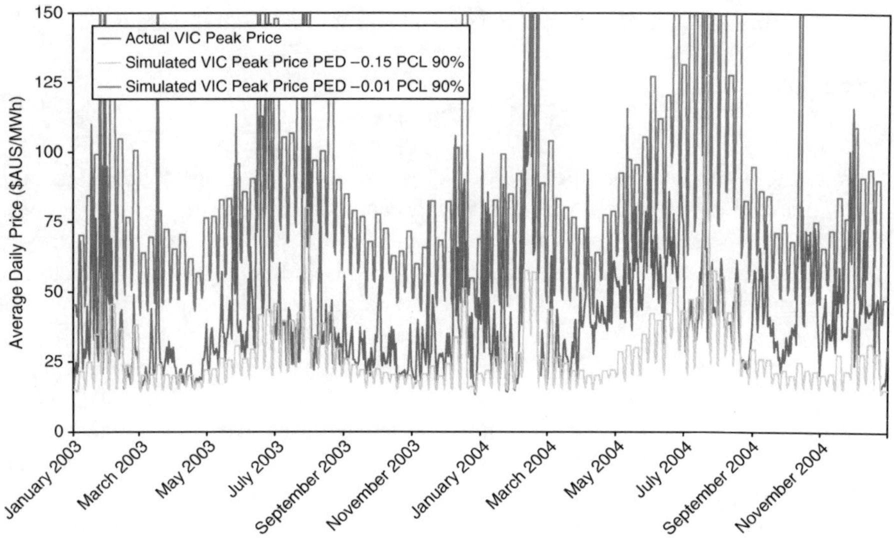

Fig. 5 Average daily spot price vs. simulated series for two different elasticities (PED)

the lowest median absolute deviation (MAD) between the observed and simulated prices.[18]

Figure 5 shows observed daily average peak spot prices for Victoria over the period 2003–2004, as well as two simulated price series using different combinations of PED and PCL. While both levels of PED are perfectly plausible, the differences between the two simulated series and the observed prices illustrate the importance of using accurate values for the PED and PCL within the model. The modeling showed that similarly accurate fits to observed prices can be obtained with a combination of a relatively inelastic demand curve and high contract levels, and vice versa.

As an illustration of the fitting performance of the model, modeled peak prices for Victoria produced with the final combinations of PED and PCL are shown below in Fig. 6. It can be observed that the underlying level of the observed price series is tracked fairly well by the modeled prices. However, there is still some residual volatility in prices, and several periods in which modeled prices deviate significantly from observed prices for weeks, if not months. While the fit could obviously be improved through the incorporation of extra information such as maintenance schedules or by varying the PED and PCL across time, the hybrid modeling framework allows this residual variation to be accounted for by the top-down Escribano et al. model, which includes jumps and mean reversion. Tipping et al. show that the

[18] The MAD is a less common measure than the mean absolute deviation, but the mean can be skewed by price spikes. Spikes could be expected to be accounted for by the stochastic price component.

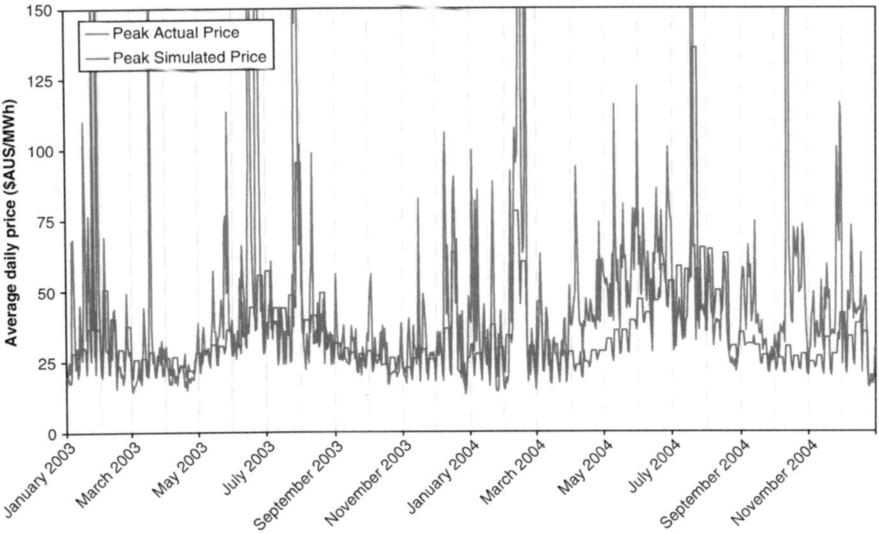

Fig. 6 Average peak spot price vs. simulated price series

fit to prices of the combined model is greater than the fit achieved by the top-down model on its own.

5.1 *Application of a Hybrid Simulation Model*

It could be argued that a similar fit to the prices in Fig. 6 could be obtained by including load and fuel prices in a top-down model, in which prices are modeled via correlations rather than by explicit calculation of the intersection between demand and supply. Once the Cournot parameters have been fitted, though, we can use the hybrid model to simulate not only prices, but also other metrics such as generation levels, company profits, transmission flows, or emissions, which the top-down price model cannot. We can also simulate situations in which underlying market conditions differ from those in the periods over which model parameters can be estimated, including hypothetical situations such as asset divestment or structural changes in the market. A calibrated top-down model is unable to do this, however well estimated it may be, while a bottom-up model without calibration and/or without a complementary stochastic process will be unable to replicate realistic market outcomes.

6 Conclusions

As electricity prices are driven by supply and demand, forecasting performance of models for prices can be maximized by incorporating information relating to one or both of these factors. Pure bottom-up models incorporate as much information as possible relating to both sides of the demand-supply equation. But such models are data- and computationally-intensive, and still do not capture a range of factors driving real-world market volatility. Pure top-down models, which can be estimated from market data and mimic market price volatility, may be much simpler. But they say nothing about why prices move as they do, and provide no mechanism to model the impact of a wide range of possible changes to market structure, etc.

There are pros and cons associated with each type of model. However, forecasting performance and modeling application can be improved by combining both types of model within a hybrid framework. This approach seems particularly suited to situations where studies must be performed, often within a limited time frame, to analyze market situations that are not completely observable or may yet be hypothetical. Of course, accuracy will diminish if we try to model situations that are very different from those observed historically. But the point is that such imperfect analyses are performed routinely in the real world, as a matter of pragmatic necessity. Our goal has been to illustrate the potential performance of a hybridization approach, providing a framework for the development of models, which allow observations of real market behavior to shape and complement the kind of simplified bottom-up models commonly employed in such situations. The NZEM and Australian NEM models outlined here were deliberately constructed so as to discover how much information might be gained from the simplest possible model, rather than to create the best possible model for the purpose. Thus they may be seen as exploratory studies, in this regard, while also providing improved fits to models that are currently applied extensively in many markets around the world.

Acknowledgements The authors acknowledge the assistance of Miguel Herce and John Parsons of CRA International in developing the original GAUSS model described in this paper, and Stephen Batstone, Deb Chattopadhyay, and Don McNickle for their assistance in model development for the New Zealand and Australian electricity markets.

References

Aggarwal SK, Saini LM, Kumar A (2009) Electricity price forecasting in deregulated markets: a review and evaluation. Int J Electr Power Energy Syst 31(1):13–22

Anderson CL, Davison M (2008) A hybrid system-econometric model for electricity spot prices: considering spike sensitivity to forced outage distributions. IEEE Trans Power Syst 23(3): 927–937

Bessembinder H, Lemmon ML (2002) Equilibrium pricing and optimal hedging in electricity forward markets. J Finance LVII(3):1347–1382

Castalia (2007) Electricity security of supply policy review, consultation paper for the electricity commission. Wellington, New Zealand, Available at: http://www.castalia.fr/files/22631.pdf

Chattopadhyay D (2004) Multicommodity spatial cournot model for generator bidding analysis. IEEE Trans Power Syst 19(1):267–275

Clewlow L, Strickland C (2000) Energy derivatives: pricing and risk management. Lacima Group, London

Conejo AJ, Plazas MA, Espínola R, Molina AB (2005) Day-ahead electricity price forecasting using the wavelet transform and ARIMA models. IEEE Trans Power Syst 20(2):1035–1042

Davison M, Anderson CL, Marcus B, Anderson K (2002) Development of a hybrid model for electrical power spot prices. IEEE Trans Power Syst 17(2):257–264

Escribano A, Peña JI, Villaplana P (2002) Modeling electricity prices: International evidence. Working Paper 02–27, Economic Series 08, Universidad Carlos III de Madrid, Spain

Georgilakis PS (2007) Artificial intelligence solution to electricity price forecasting problem. Appl Artif Intell 21(8):707–727

Green RJ (2003) Electricity markets: challenges for economic research. Proceedings of the Research Symposium on European Electricity Markets, The Hague, (26 September)

Griffin JM (1977) Long-run production modeling with pseudo data: electric power generation. Bell J Econ 8(1):112–127

Guirguis HS, Felder FA (2004) Further advances in forecasting day-ahead electricity prices using time series models. KIEE Int Trans Power Eng 4-A(3):159–166

Hourcade JC, Jaccard M, Bataille C, Ghersi F (2006) Hybrid modeling: New answers to old challenges. Introduction to the Special Issue of the Energy Journal. Energy J 2:1–12

Karakatsani NV, Bunn DW (2008) Forecasting electricity prices: The impact of fundamentals and time-varying coefficients. Int J Forecas 24(4):764–785

Longstaff FA, Wang AW (2004) Electricity forward prices: A high frequency empirical analysis. J Finance 59(4):1877–1900

Lucia JJ, Schwartz ES (2002) Electricity prices and power derivatives: evidence from the nordic power exchange. Rev Derivatives Res 5(1):5–50

Manne AS, Richels RG, Weyant JP (1979) Energy Policy Modeling: A Survey. Oper Res 27(1): 1–36

Makridakas SG, Wheelwright SC, Hyndman RJ (1998) Forecasting: methods and applications, 3rd ed. Wiley, New York

Neuhoff K, Skillings S, Rix O, Sinclair D, Screen N, Tipping J (2008) Implementation of EU 2020 Renewable Target in the UK Electricity Sector: Renewable Support Schemes. A report for the Department of Business, Enterprise and Regulatory Reform by Redpoint Energy Limited, University of Cambridge, and Trilemma UK. Available at http://www.decc.gov.uk/en/content/cms/consultations/con_res/

Peat M (2008) STEPS: A stochastic top-down electricity price simulator. Presented at *EPOC Winter Workshop*, Auckland. Available at http://www.esc.auckland.ac.nz/epoc/

Pereira MVF (1989) Stochastic operation scheduling of large hydroelectric systems. Electric Power Energy Syst 11(3):161–169

Pereira MVF, Pinto LMG (1991) Multi-stage stochastic optimization applied to energy planning. Math Program 52:359–375

Pirrong C, Jermakyan M (1999) Valuing power and weather derivatives on a mesh using finite difference methods. In: Jameson R (ed.). Energy modeling and the management of uncertainty. Risk Publications, London

Pirrong C, Jermakyan M (2001) The price of power: The valuation of power and weather derivatives. Working paper, Oklahoma State University

Read EG, Hindsberger M (2010) Constructive dual DP for reservoir optimisation. In Rebennack S, Pardalos PM, Pereira MFV, Iliadis NA (ed.) Handbook of power systems, vol. I. Springer, Heidelberg, pp. 3–32

Schäfer A, Jacoby HD (2006) Experiments with a hybrid CGE-MARKAL Model. Hybrid modeling of energy-environment policies. Reconciling bottom-up and top-down – Special Issue of the Energy Journal. Energy J 2:171–178

Szkuta BR, Sanabria LA, Dillon TS (1999) Electricity price short-term forecasting using artificial neural networks. IEEE Trans Power Syst 14(3):851–857

Tipping J, Read EG, McNickle D (2004) The incorporation of hydro storage into a spot price model for the New Zealand electricity market. Presented at the Sixth European Energy Conference: Modeling in Energy Economics and Policy. Zurich. Available at http://www.mang.canterbury.ac.nz/research/emrg/

Tipping J, McNickle DC, Read EG, Chattopadhyay D (2005a) A model for New Zealand hydro storage levels and spot prices. Presented to *EPOC* Workshop, Auckland. Available at http://www.mang.canterbury.ac.nz/research/emrg/

Tipping J, Read EG, Chattopadhyay D, McNickle DC (2005b) Can the shoe be made to fit? - Cournot modeling of Australian electricity prices. ORSNZ Proceedings. Available at: http://www.mang.canterbury.ac.nz/research/emrg/

Vehviläinen I, Pyykkönen T (2005) Stochastic factor model for electricity spot price – the case of the nordic market. Energy Econ 27(2):351–367

Westergaard E, Chattopadhyay D, McCall K, Thomas M (2006) Analysis of Transpower's Proposed 400kV Project and Alternatives. CRA report to Transpower NZ Ltd, Wellington. Available at http://www.electricitycommission.govt.nz/pdfs/submissions/

Wolak FA (2003). Identification and estimation of cost functions using observed bid data: an application to electricity markets. In: Dewatripont M, Hansen LP, Turnovsky SJ (eds.) Advances in economics and econometrics: theory and applications, vol. II. Cambridge University Press, New York, pp. 115–149

Part III
Energy Auctions and Markets

Agent-based Modeling and Simulation of Competitive Wholesale Electricity Markets

Eric Guerci, Mohammad Ali Rastegar, and Silvano Cincotti

Abstract This paper sheds light on a promising and very active research area for electricity market modeling, that is, agent-based computational economics. The intriguing perspective of such research methodology is to succeed in tackling the complexity of the electricity market structure, thus the fast-growing literature appeared in the last decade on this field. This paper aims to present the state-of-the-art in this field by studying the evolution and by characterizing the heterogeneity of the research issues, of the modeling assumptions and of the computational techniques adopted by the several research publications reviewed.

Keywords Agent-based computational economics · Agent-based modeling and simulation · Electricity markets · Power system economics

1 Introduction

In the last decade, several countries in the world have been obliged to intensively regulate the electric sector, either for starting and supporting the liberalization process (e.g., Directives 96/92/EC and 2003/54/EC of the European Commission recommended all European countries to switch to market-based prices) or for amending and improving proposed regulations because of shortcomings in the market design or market failures (e.g., the 2001 California crisis has motivated the US Federal Energy Regulatory Commission to propose a common wholesale power market for adoption by all United States). In all cases, common restructuring proposals are to adopt complex market structures, where several interrelated market places are envisaged and where the integration of the national markets towards an inter-regional/continental perspective is encouraged (e.g., European Union and USA).

S. Cincotti (✉)
Department of Biophysical and Electriconic Engineering, University of Genoa
Via Opera Pia 11a, 16146, Genoa, Italy
e-mail: cincotti@dibe.unige.it

S. Rebennack et al. (eds.), *Handbook of Power Systems II*, Energy Systems,
DOI 10.1007/978-3-642-12686-4_9,

A generic wholesale power market is composed by energy markets such as spot and forward markets for trading electricity, ancillary service markets for guaranteeing security in the provision, and even market couplings with foreign markets. Furthermore, wholesale electricity markets are vertically integrated with other commodity markets, mainly fuels markets, such as natural gas, oil, and coal markets, and moreover climate policy measures are imposing market-based mechanisms for climate change regulation to foster a carbon-free electricity production. The rationale for this complex structure relies upon the nature of the traded good and of the technologies of the production. Electricity is essentially not storable in large quantities, thus a real-time balancing in the production and consumption of electrical power must be carefully guaranteed. Furthermore, the electric sector is a highly capital-intensive industry due to economies of scale occurring mainly in the production sector. This favors market concentration in the supply side and increases the opportunity for the exertion of market power by a reduced number of market players. Furthermore, the producers' long-term investment plans must be supported by a clear, adequate, and also stable regulation. Thus regulatory interventions must be accurately devised and implemented.

Such fast-changing economical sector together with its complexity from both an economical and a technological viewpoint have fostered practitioners and power system/economics researchers to acquire new competencies and to adopt innovative research approaches to study realistic market scenarios and propose appropriate regulations for this original market environment. The several high-quality scientific contributions gathered in this book witness the broadness and multidisciplinarity of the research domain and the great interest around this topic. Several approaches have been proposed in the electricity markets literature. Generally speaking, classical analytical power system models are no longer valid because they are a poor fit to the new market-based context. Now all operation decisions are decentralized and strategically taken by each market operator so as to maximize individual trading profit. Similarly, standard analytical economic approaches, based on game-theoretic models, are usually limited to stylized market situations among few actors neglecting technological and economical constraints such as transmission networks. Furthermore, modeling aspects such as repeated interactions among market operators, transmission network constraints, incomplete information, multi-settlement market mechanisms, or market coupling are seldom, if never, addressed by the theoretical literature. On the other side, human-subject experiments have been proposed to complement theoretical approaches (Staropoli and Jullien 2006); however, realistically replicating the bidding behavior of a power generator in laboratory experiments requires strong expertise by the participants. Moreover, the relevant literature mainly focuses on market design issues concerning single stylized marketplaces, such as the day-ahead market session, thus neglecting the structural complexity of wholesale electricity markets. Conversely, computational methodologies offer an appealing approach to cope with the complexity of power market mechanisms by enabling researchers to realistically simulate such market environments.

This paper focuses attention to the computational literature and in particular with the promising approach of agent-based modeling and simulation (ABMS), which has now became a widespread methodology in the study of power systems and electricity markets. Current strands of research in power systems ABMS are distributed electricity generation models, retail and wholesale electricity market models. Space reasons as well as the fact that last topic has been the major strand of research in electricity market ACE have motivated the choice to focus on the literature about wholesale electricity market models.

It is worth mentioning that two literature reviews have recently appeared,[1] that is, Ventosa et al. (2005) and Weidlich and Veit (2008b). The first paper presents electricity market modeling trends enclosing both contributions from the theoretical and the computational literature. In particular, the authors identify three major trends: optimization models, equilibrium models, and simulation models. The second one is a critical review about agent-based models for the wholesale electricity market. Their focus is on the behavioral modeling aspect, with a discussion about methodological issues to raise awareness about critical aspects such as appropriate agent architecture and validation procedures.

This review paper offers an updated outlook on the vast and fast-growing literature, thus enhancing previous surveys (49 papers are reviewed in details). Moreover, this paper differs from previous reviews both for the chronological perspective and for the agent-based modeler perspective adopted to organize the contents of the writing. As far as concerns the former aspect, this paper reports three distinct summary tables where all papers are listed and compared in chronological order. This enables to highlight modeling trends with respect to different viewpoints, that is, research issues, behavioral adaptive models, and market models. As far as concerns the latter aspect, this paper discusses in separate sections the modeling solutions for the major agents populating an electricity market models, that is, market operators and market institutions, thus allowing the proper identification of the solutions proposed by each reviewed paper. As a final remark, this paper provides an overview of the challenging ABMS research methodology in the domain of electricity markets to facilitate newcomers and practitioners in the attempt to better disentangle in the vast literature and to present the state-of-the-art of the several computational modeling approaches.

The paper is structured as follows: Sect. 2 introduces notions about the agent-based modeling and simulation approach, from the perspective of agent-based computational economics. A specific focus is then devoted to the electricity market literature. Section 3 focuses on the behavioral modeling aspect by reviewing several papers and the variety of approaches adopted. On the other side, Sect. 4 lists and discusses the market models proposed. Large-scale models as decision support systems are also presented. For each of the three previous sections, a table is reported

[1] A recent working paper by Sensfuß et al. (2007) proposes a further literature review mainly focusing on the research activity led by some research groups active in the field. The discussion is grouped around three major themes, that is, the analysis of market power and design, the modeling agent decisions, and the coupling of long-term and short-term decisions. The authors also review papers dealing with electricity retail markets.

to summarize and highlight the major modeling features of each reviewed paper. Finally, Sect. 5 draws conclusions.

2 Agent-based Modeling and Simulation

ABMS refers to a very active research paradigm, which has gained more and more popularity among a wide community of scientist, ranging from natural science like biologist up to human and social science, that is, economists and sociologist, and engineers. The rationale is twofold: a scientific reason referring to the notion of complexity science and a technological one, that is, the increased availability of cheap and powerful computing facilities.

The development of ABMS in social sciences, and in particular in Economics, is closely linked with the pioneering work conducted at the Santa Fe Institute (SFI). SFI grouped 20 worldwide renowned experts from different disciplines, that is, economists, physicists, biologists, and computer scientists, to study the economy as an evolving complex system. Their authoritative work greatly influenced researchers by adopting agent-based models in the study of social systems.[2] Their philosophical approach can be summarily expressed by "If you didn't grow it, you didn't explain it," supporting the view of ABMS approach as a "generative social science" (Epstein 1999). The basic units of an agent-based model (ABM) are agents, which are considered entities characterized by autonomy, that is, there is no central or "top down" control over their behavior, and by adaption, that is, they are reactive to the environment. The notion of agent is better understood within the framework of complex adaptive systems (CAS) (Blume and Easley 1992). A CAS is a complex system, that is, a system of interacting units, that includes goal-directed units, that is, units that are reactive and that direct at least some of their reactions towards the achievement of built-in or evolved goals (Tesfatsion and Judd 2006). This definition enables agents to be entities ranging from active data-gathering decision-makers with sophisticated learning capabilities to passive world features with no cognitive functioning.

2.1 ACE: A New Paradigm for Economics

A modern market-based economy is an example of a CAS, consisting of a decentralized collection of autonomous agents interacting in various market contexts. Agent-based computational economics (ACE) (Tesfatsion and Judd 2006; Richiardi 2009) is the computational study of economic processes modeled as dynamic

[2] At SFI, Swarm (2009), the ancestor of agent-based software programming tools, was developed, and the Santa Fe Artificial Stock Market, one of the first agent-based financial market platforms, was realized (Ehrentreich 2002).

systems of interacting agents.[3] ACE researchers and practitioners are intrigued by the appealing perspective of taking into account in their economic models of several aspects of the procurement processes, the "dark side" of the economic modeling, that is, all economic events occurring in reality among customers and suppliers during negotiation and trading processes, which are constrained by economic and social institutions. These aspects are greatly simplified or even neglected by standard analytical economic theories; indeed, substituting equilibrium assumptions for procurement processes is one way to achieve a drastically simplified representation of an economic system.[4] The drawback but also the opportunity of this challenging methodological approach is to consider aspects such as asymmetric information, strategic interaction, expectation formation, transaction costs, externalities, multiple equilibria, and out-of equilibrium dynamics.

However, this fairly young methodological approach in spite of an apparently ease-of-use in modeling and simulations requires great expertise in implementing the models and in obtaining significant simulation results. To succeed with the modeling goals, ACE models must be carefully devised in their attempt to increase realism. A correct identification of irrelevant modeling aspects is still required, and in any case theoretical models are effective starting-points for the modeling activity. Moreover, ACE models must be adequately implemented and an object-oriented programming approach is strongly recommended for scalability and modularity. In particular, agent-oriented programming offers a suitable implementing framework for building artificial worlds populated by interacting agents (Jennings 2000). This is a strong point of the methodology that both the modeler and the software programmer perspectives usefully coincide. Several agent-based software tools[5] have appeared in the last 10 years and are fostering the adoption of such methodology.

2.2 ACE in Electricity Markets: Research Issues

The great versatility of ACE methodology has motivated researchers to adopt it for the study of several electricity market issues. Some researchers have applied agent-based models for examining electricity consumer behavior at the retail level, for example, Hämäläinen et al. (2000); Roop and Fathelrahman (2003); Yu et al. (2004); and Müller et al. (2007) or for studying distributed generation models, for example, Newman et al. (2001); Rumley et al. (2008); and Kok et al. (2008). This paper focuses on a third strand of research, that is, the study of competitive

[3] The "mother" of the expression agent-based computational economics (ACE) is Professor Leigh Tesfation at Iowa State University, who has been one of the leader of agent-based modeling and simulation applied to economics and also to electricity markets (Nicolaisen et al. 2000).

[4] For a detailed discussion on procurement process and neoclassical approaches to economic modeling refers to Tesfatsion and Judd (2006).

[5] A description of software tools for agent-oriented programming is available at Tesfatsion (2007).

wholesale electricity markets, which are mainly characterized by centralized market mechanisms such as double-auction markets.

Table 1 reports a broad selection of papers written in the last decade about ACE applied to the study of wholesale electricity markets.[6] These 49 works are the ones that are considered and reviewed in the rest of this paper. Table 1 lists the papers in a chronological order[7] to show how the researchers' focus of attention in the field has evolved throughout the years. Eight major research issues are highlighted from column 2 to column 9 from the most present to the less present in the considered literature. These aspects, which are not mutually exclusive, have been selected because they are the ones of major concern in the literature. In particular, these correspond to (1) market performances and efficiency, (2) market mechanisms comparison, (3) market power, (4) tacit collusion, (5) multi-settlement markets, (6) technological aspects affecting market performances, (7) diversification or specialization, and (8) divestiture or merging of generation assets. Finally, the last column of Table 1 reports a concise explanation of the objectives for every paper, and the last row reports the sum of the occurrences of the specific research issue in the literature.

The chronological listing points out the progressive increase in the number of contributions in this field throughout the last decade. The tendency pinpoints the still rising interest in the topic and in the computational approach. Table 1 further shows that there has not been a clear evolution in the study of the different research themes. The first three reported issues, that is, columns numbered one to three, have been and are the most addressed research areas since the early computational studies. Indeed, market performances and efficiency, market mechanisms comparison, and market power have been and are the major research issues also of the theoretical and experimental literature. The great emphasis placed on these issues denote the confidence that ACE researchers have on this approach to provide guidance for market design issues. The remaining columns, and in particular columns numbered 6–8, highlight further research areas, which have been currently investigated only by few research groups, which certainly deserve more research efforts. As a final remark, it is important to note that the majority of these papers are purely computational studies, that is, empirical validation is seldom addressed. This is a critical aspect that needs to be addressed by researcher to assess the effectiveness of their modeling assumptions. Furthermore, it is often difficult to compare simulation results of papers dealing with the same issue because of the great heterogeneity

[6] In this paper, all major ACE publications dealing with wholesale electricity markets known to the authors have been considered. The selection has been based on the surveys of previous review papers, most recent papers published by the major groups active in this field of research and their relevant cites. Moreover, the long-standing authors' knowledge in the field helped selecting further original contributions. It is worth remarking that due to the fast-growing and broad literature in this research area, authors may have probably, but involuntarily, missed citing some contributions that appeared in the literature. However, the authors believe that the paper provides a vast and representative outlook on the ACE literature on wholesale electricity markets.

[7] The papers are ordered in descending order with respect to the year of publication and if two or more papers have appeared in the same year, then an alphabetical descending order has been considered.

Table 1 Description of the research issues for each paper of the considered literature

Paper	*1*	*2*	*3*	*4*	*5*	*6*	*7*	*8*	*Explanation*
Curzon Price (1997)	✓								Co-evolutionary programming technique to study Bertrand and Cournot competition, a vertical chain of monopolies and a simple model of electricity pool
Richter and Sheblé (1998)	✓								Study of bidding strategies with GA, with sellers endowed with price forecasting techniques
Visudhiphan and Ilic (1999)	✓		✓						Comparison of hour- and day-ahead spot market performances with different demand elasticities
Bagnall and Smith (2000)	✓	✓		✓					Study on UK wholesale electricity market with a complex agent architecture based on learning classifier systems
Bower and Bunn (2000)		✓							ACE investigation of bilateral and pool market system under UA and DA clearing price mechanisms
Lane et al. (2000)	✓		✓						Study of market power exertion by sellers in different oligopolistic scenarios with intelligent buyers
Nicolaisen et al. (2000)	✓	✓	✓						Market power analysis under different concentration and capacity scenarios
Bower and Bunn (2001b)		✓							ACE investigation of bilateral and pool market system under UA and DA clearing price mechanisms. The paper is similar to Bower and Bunn (2000)
Bower et al. (2001)								✓	Evaluation of mergers operations in the German electricity market
Bunn and Oliveira (2001)	✓				✓				Study about market performances for the NETA of UK electricity market
Day and Bunn (2001)		✓	✓					✓	Impact of divestiture proposals in the price market
Nicolaisen et al. (2001)	✓	✓	✓						Market power and efficiency analysis in the discriminatory double-auction
Visudhiphan and Ilic (2001)	✓	✓	✓						Investigation of market power abuse with generators strategically withholding capacity
Cau and Anderson (2002)	✓			✓					Study of bidding behavior of a simple duopolistic model of the Australian wholesale electricity market

(*Continued*)

Table 1 (Continued)

Paper	*1*	*2*	*3*	*4*	*5*	*6*	*7*	*8*	*Explanation*
Koesrindartoto (2002)	✓	✓	✓						Study on DA efficiency with respect to Roth and Erev learning model specification
Bunn and Oliveira (2003)	✓		✓		✓				Study about the opportunity of exertion of market power by two companies in the UK electricity market
Cau (2003)	✓			✓					Investigation of tacit collusion behavior in an oligopolistic market based on the Australian wholesale power market
Visudhiphan (2003)	✓	✓	✓						Comparing several learning algorithms to study how simulated price dynamic is affected
Atkins et al. (2004)		✓			✓				Large scale model for studying electricity market
Bin et al. (2004)		✓	✓						Pricing mechanisms comparison for pool-based electricity market (UA, DA, EVE)
Minoia et al. (2004)	✓					✓			Proposal of an original model of pricing settelment rule for wholesale electricity markets providing incentives for transmission investments
Xiong et al. (2004)	✓	✓							Comparison of discriminatory and uniform auction markets in different demand scenarios
Bagnall and Smith (2005)	✓	✓		✓					Study on UK wholesale electricity market with a complex agent architecture based on learning classifier systems
Bunn and Martoccia (2005)			✓	✓					Market power and tacit collusion analysis of the electricity pool of England and Wales
Bakirtzis and Tellidou (2006)		✓	✓						Analysis of the effect of the market power under uniform and discriminatory settlement rule
Botterud et al. (2006)		✓	✓			✓			Analysis of the effect of congestion management on market price
Chen et al. (2006)				✓					Analysis of Cournot models of electricity pool

Krause et al. (2006)	✓					Study of market dynamics for network-constrained pool markets
Krause and Andersson (2006)		✓	✓			Market power analysis under different congestion management schemes
Ma et al. (2006)	✓			✓		Study of the bidding behavior of suppliers in network-constrained electricity market
Naghibi-Sistani et al. (2006)	✓					Study of simulation convergence properties with Nash equilibrium
Veit et al. (2006)	✓			✓		Impact of forward market on the spot price volatility
Weidlich and Veit (2006)	✓	✓		✓		Analysis of day ahead and real time market under discriminatory and uniform auctions
Banal-Estanol and Rupérez-Micola (2007)					✓	Study on how the diversification of generation portfolio influences market prices
Bunn and Oliveira (2007)		✓			✓	Investigation about impact of market design on technological diversification among portfolio generators
Guerci et al. (2007)	✓	✓				Detailed analysis of learning bidding behavior in UA and DA market mechanisms
Hailu and Thoyer (2007)	✓	✓				Study the performance of uniform, discriminatory, and Vickrey auctions
Ikeda and Tokinaga (2007)	✓					Analysis of electricity market performances with a genetic programming approach
Nanduri and Das (2007)		✓	✓			Market power analysis under different auction mechanisms
Sueyoshi and Tadiparthi (2007)	✓					Assessment of adaptive agent-based model as good estimator of electricity price dynamics

(Continued)

Table 1 (Continued)

Paper	*1*	*2*	*3*	*4*	*5*	*6*	*7*	*8*	*Explanation*
Sun and Tesfatsion (2007)	✓								Presentation of the AMES simulator with a test bed case
Tellidou and Bakirtzis (2007)	✓		✓	✓					Analysis of market power and tacit collusion among producers
Guerci et al. (2008a)	✓	✓		✓					Study of tacit collusive behavior with two RL algorithm in UA and DA market mechanisms
Guerci et al. (2008b)	✓		✓						Study of market power in network-constrained electricity market with different learning algorithms
Li et al. (2008)	✓	✓							Analysis of LMP mechanisms in different price-cap and price-sensitive demand scenarios
Somani and Tesfatsion (2008)	✓		✓						Comparisons of market efficiency indexes for the detection of market power in wholesale power markets
Sueyoshi and Tadiparthi (2008a)		✓				✓			Study on how transmission limits affect the market price and its volatility in different market mechanism
Weidlich and Veit (2008a)	✓				✓				Analysis of German electricity market performances by an ACE model considering day-ahead, real-time balancing and CO2 emission allowances markets
Bunn and Day (2009)	✓	✓		✓					Comparison between real UK market performances and a detailed computational model of price formation
Total occurrences	34	25	18	9	7	3	2	2	

From column 2 to column 9, the following research issues are considered: (1) market performances and efficiency, (2) market mechanisms comparison, (3) market power, (4) tacit collusion, (5) multi-settlement markets, (6) technological aspects affecting market performances, (7) diversification or specialization, and (8) divestiture or merging of generation assets performances. Last row reports the sum of the occurrences of the specific research issue (column) in the behavioral modeling assumptions. Next section presents the heterogeneity of the behavioral modeling assumptions adopted for modeling market participants in wholesale electricity markets.

3 Behavioral Modeling

Key activity of the agent-based modeler concerns the ability to choose appropriate behavioral models for each type of agent. Important sources of heterogeneity, among agents populating an artificial economy, are not only economical or technological endowments such as marginal costs or revenues, generating capacity, but also behavioral aspects such as available information, backward- or forward-looking horizon, risk-aversion, and generically reasoning capabilities. This heterogeneity is exogenously set by the modeler and should be adequately justified by correct modeling assumptions.

In general, the ABSM literature has adopted several approaches for modeling agent's behavior ranging from zero-intelligence to complex black-box data-driven machine learning approaches such as neural networks, from social learning such as genetic algorithms to individual learning approaches such as reinforcement learning algorithms.[8] This great variability is present also in the ACE literature about electricity markets, even if the model choice is seldom adequately justified.

The summary Table 2 is reported to facilitate the comparison of the reviewed papers, for easily identifying key behavioral modeling aspects for each paper. This table lists in a chronological order papers to pinpoint the modeling trends in the last decade. Three major aspects are examined, that is, the type of adaptive behavioral model (columns 2–6), some features of the generation companies model (columns 7 and 8), and of the demand model (columns 9 and 10). In particular, columns 2–6 refer to the adaptive behavioral model chosen for suppliers, buyers, or even transmission operators. We have categorized the papers under five distinct classes in an attempt to gather all computational approaches in nonoverlapping groups. However, it is always difficult to exactly categorize each element, and so in some cases approximate solutions may have been proposed. The five classes are genetic algorithms (GA), co-evolutionary genetic algorithms (CGA), learning classifier systems (LCS), algorithm models tailored on specific market considerations (Ad-hoc), and Reinforcement Learning algorithms (RL). In the rest of the section, all such techniques will be adequately illustrated. As far as concerns columns 7 and 8, wholesale electricity market models usually focus on suppliers behavior, and so the table reports the bidding format of the spot market (Actions) and the suppliers characteristic of being generating units portfolio (Portf.). On the contrary, the demand-side is often

---

[8] A major distinction in ACE models regards the so-called social/population learning and individual learning models (Vriend 2000). The former refers to a form of learning where learned experiences are shared among the players, whereas the latter involves learning exclusively on the basis of his own experience.

Table 2 Categorization of agents' adaptive models and supply- and demand-side models for each paper

Paper	Adaptive model					Supply		Demand	
	GA	*CGA*	*LCS*	*Ad-hoc*	*RL*	*Action*	*Portf.*	*Elas.*	*Adapt.*
Curzon Price (1997)		✓				P			
Richter and Sheblé (1998)	✓					P,Q			
Visudhiphan and Ilic (1999)				✓		SF		✓	
Bagnall and Smith (2000)			✓			P			
Bower and Bunn (2000)					✓	P	✓	✓	
Lane et al. (2000)	✓					P		✓	✓
Nicolaisen et al. (2000)	✓					P		✓	✓
Bower et al. (2001)					✓	P	✓	✓	
Bower and Bunn (2001b)					✓	P	✓	✓	
Bunn and Oliveira (2001)				✓		P	✓	✓	✓
Day and Bunn (2001)				✓		SF	✓	✓	
Nicolaisen et al. (2001)					✓	P		✓	✓
Visudhiphan and Ilic (2001)				✓		P,Q			
Cau and Anderson (2002)		✓				SF			
Koesrindartoto (2002)					✓	P		✓	✓
Bunn and Oliveira (2003)				✓		P	✓	✓	✓
Cau (2003)		✓				SF	✓		
Visudhiphan (2003)				✓	✓	P,Q		✓	✓
Atkins et al. (2004)				✓		P,Q	✓	✓	✓
Bin et al. (2004)					✓	P	✓		
Minoia et al. (2004)				✓		P			
Xiong et al. (2004)					✓	P		✓	
Bagnall and Smith (2005)			✓			P			
Bunn and Martoccia (2005)				✓		P	✓	✓	
Bakirtzis and Tellidou (2006)					✓	P	✓		

Botterud et al. (2006)				✓		P	✓		
Chen et al. (2006)		✓				Q		✓	
Krause et al. (2006)					✓	SF			
Krause and Andersson (2006)					✓	SF		✓	✓
Ma et al. (2006)					✓	SF		✓	
Naghibi-Sistani et al. (2006)					✓	SF			
Veit et al. (2006)					✓	SF	✓	✓	
Weidlich and Veit (2006)					✓	P,Q			
Banal-Estanol and Rupérez-Micola (2007)					✓	P	✓		
Bunn and Oliveira (2007)				✓		Q		✓	
Guerci et al. (2007)					✓	P,Q			
Hailu and Thoyer (2007)					✓	SF			
Ikeda and Tokinaga (2007)		✓				P,Q	✓		
Nanduri and Das (2007)					✓	SF			
Sueyoshi and Tadiparthi (2007)					✓	P,Q		✓	✓
Sun and Tesfatsion (2007)					✓	SF			
Tellidou and Bakirtzis (2007)					✓	P,Q			
Guerci et al. (2008a)					✓	P,Q			
Guerci et al. (2008b)					✓	SF	✓		
Li et al. (2008)					✓	SF		✓	
Somani and Tesfatsion (2008)					✓	SF		✓	
Sueyoshi and Tadiparthi (2008a)					✓	P,Q		✓	✓
Weidlich and Veit (2008a)					✓	P,Q	✓		
Bunn and Day (2009)				✓		SF	✓	✓	
Total occurrences	3	5	2	12	28		18	24	11

The five classes for the adaptive models are genetic algorithms (GA), co-evolutionary genetic algorithms (CGA), learning classifier systems (LCS), algorithm models tailored on specific market considerations (ad-hoc), and reinforcement learning algorithms (RL). For the suppliers model, the bidding strategy (Actions) and the suppliers characteristic of being generating units portfolio (Portf.) are highlighted. Finally, the demand side model is described by the properties of being price-elastic (Elas.) or adaptive (Adapt.). A ✓notation is adopted for paper presenting the column feature; otherwise, an empty field means that the negation is valid.

represented by an aggregate price-inelastic or -elastic demand mainly without adaptive behavior, thus these two aspects are highlighted in columns 9 and 10 of the table (Elas. and Adapt., respectively).

It happens that some papers describe the great flexibility of a computational model by illustrating an extended software framework, but reports simulation studies in a simplified case. In this case, table's values refer to the simulation test case proposed.

3.1 Evolutionary Computation

Evolutionary computation (EC) is the use of computer programs that not only self-replicate but also evolve over time in an attempt to increase their fitness for some purpose. Evolutionary algorithms have been already widely adopted in economics and engineering for studying socio-economic systems. One major and most famous algorithm for EC is the genetic algorithm (GA).[9] GA are commonly adopted as adaptive heuristic search algorithm, but they have also been applied as a social learning (mimicry) rule across agents (Vriend 2000). They are premised on genetic and natural selection principles and on the evolutionary ideas of survival of the fittest. GAs are conceived on four analogies with evolutionary biology. First, they consider a genetic code (chromosomes/genotypes). Second, they express a solution (agents/phenotypes) from the code. Third, they consider a solution selection process, the survival of the fittest; the selection process extracts the genotypes that deserve to be reproduced. Finally, they implement the so-called genetic operators, such as elitism, recombination, and mutation operations, which are used to introduce some variations in the genotypes.

Key element for the implementation of the genetic algorithm is the definition of a utility function, or fitness function, which assigns a value to each candidate solution (potential phenotype). This evolving function is thus used to evaluate the interest of a phenotype with regard to a given problem. The survival of the corresponding solution or equivalently its number of offspring in the next generation depends on this evaluation.

A summarily description of the algorithm flow is: Initialization. Random generation of an initial population P_1 of N agents

for iteration t $= 1, \ldots,$ T:

1. Evaluation of the fitness $F_t(n)$ for each agent $n = 1, \ldots, N$ of the current population P_t.
2. Selection of the best-ranking phenotypes in P_t according to selection probabilities $\sigma(n)$. $\sigma(n)$ are proportional to $F_t(n)$.

[9] First pioneered by John Holland in the 1960s, genetic algorithms (GA) has been widely adopted by agent-based modelers. Introduction to GA learning can be found in Holland (1992) and Mitchell (1998).

3. Adoption of genetic operators to generate offspring. The offspring of an individual are built from copies of its genotype to which genetic operators are applied.
4. Evaluation of the individual fitnesses $F_t^*(n)$ of the offspring.
5. Generation of P_{t+1} by probabilistically selecting individuals from P_t according to $F_t(n)$ and $F_t^*(n)$.

end

This approach has been adopted by ACE researchers to mimic social learning rule. However, modifications to the standard single-population GA implementation have been proposed. Multi-population GAs have been introduced as coevolutionary computational techniques to extend conventional evolutionary algorithms. A multi-population GA models an ecosystem consisting of two or more species. Multiple species in the ecosystem coevolve, that is, they interact in the environment with each other without exchanging genetic material and thus without the possibility of hybridization, and they independently evolve by implementing a GA. This latter approach is commonly adopted by ACE researchers for implementing individual learning models, where each species is a specific agent endowed with a population of actions, for example, bidding-strategies, evolving according to GA rules.

3.1.1 Social Learning: Single-population Genetic Algorithms (GA)

In the early literature, the adoption of standard GA methods for evolving the economic systems concerned mainly single-population GA. These early attempts (Richter and Sheblé 1998; Lane et al. 2000; Nicolaisen et al. 2000) consider simple electricity market implementing double-auctions with a discriminatory price mechanism based on mid-point pricing. Lane et al. (2000) and Nicolaisen et al. (2000) implement in their market model also grid transmission constraints. The authors consider generation companies (GenCos) with single production-unit.

Richter and Sheblé (1998) is one of the first attempt to adopt GA for studying the bidding behavior of electricity suppliers. The GenCos forecast equilibrium prices according to four distinct techniques, that is, moving average, weighted moving average, exponentially weighted moving average, and linear regression. Each GenCo/chromosome is composed of three parts or genes for determining the offer price and the offer quantity and for choosing the price forecast method. Simulation results suggest that the followed GA approach furnishes reasonable findings.

The papers of Nicolaisen et al. (2000) and Lane et al. (2000) are closely related in their aims and approaches. They both adopt the same EPRI market simulator and they address the same research issue, that is, to study the exertion of market power in different market conditions. They both consider adaptive sellers and buyers. Price is the unique decision variable of market operators, being always quantity in their maximum production or purchase capacity, respectively. Different market scenarios are considered to test for market power. In particular, Lane et al. (2000) consider three market scenarios where sellers differ with respect to market concentrations and marginal costs and where the three buyers are identical. Nicolaisen et al. (2000) consider nine scenarios where both buyers and sellers differ in number and purchase or production capacities, being marginal revenues and costs always equal. In both papers, results are contradictory also with economic theory. The authors suggests that this approach is not suitable, in particular when few agents are considered. They envisage the adoption of individual learning techniques.

3.1.2 Individual Learning: Multipopulation or Coevolutionary Genetic Algorithms (CGA)

Curzon Price (1997) studies Bertrand and Cournot competition, a vertical chain of monopolies, and a simple model of an electricity pool identical to the one considered by von der Fehr and Harbord (1993), by means of co-evolutionary programming. This paper is the first attempt in the electricity market ACE literature to implement a multi-population GA (CGA). In particular, the author considers a simplified market scenario where only two sellers/producers, that is, two-population GA, are competing on prices or quantities and each producer autonomously adapts to the changing environment by implementing a GA where the solutions are his own strategies. Results show that this evolutionary approach is suitable to converge to Nash equilibria in pure strategies in this rather simple economic setting. The author suggests to extend this research by adopting richer evolutionary approaches such as learning classifier systems.

Cau and Anderson (2002) and Cau (2003) propose a wholesale electricity market model similar to the Australian National Electricity Market. Cau (2003) extend the model of Cau and Anderson (2002) by considering both step-wise and piece-wise linear functions as bidding strategies and by studying several market settings with different number of sellers (two and three), with asymmetric and symmetric cost structures and production capacities, and with producers with a portfolio of generating units. An important aspect is the fact that they consider condition-action rules, that is, they consider a mapping from a state (defined by the previous spot market price, the past market demand, and the forecast market demand) to a common set of possible actions (step-wise and piece-wise linear functions). The demand is assumed price-inelastic and uniformly randomly varying between a high and a low level. The aim of the authors is to study the agents ability to achieve tacit collusion in several market settings. Summarily, the author finds that tacit collusive bidding strategies can be learned by the suppliers in the repeated-game framework. This is true for both market bidding formats even with randomness in the environment. In particular, on the demand side, high overall demand, high uncertainty, and low price elasticity facilitate tacit collusion. On the supply side, situations where tacit collusion is easier to achieve are characterized by symmetry in cost and capacity and small hedging contract quantity. Furthermore, the author experimentally confirms that an increasing number of agents makes more difficult, but still possible, tacit collusion to occur in a sustainable way. Finally, the author states that the results are quite analogous to classical supergame theory.

Chen et al. (2006) address the issue of the emergence of tacit collusion in the repeated-game framework by means of a co-evolutionary computation approach. In particular, they consider a simplified model of an electricity market based on the classical Cournot model with three generating units. The authors propose to test their CGA models for convergence towards game-theoretical solutions, that are, the Cournot–Nash equilibrium and the solutions set Pareto-dominating the Cournot–Nash equilibrium to consider also equilibria in infinitely repeated games. In particular, they adopt a standard GA to study the bidding behavior with respect to Cournot–Nash equilibrium, whereas for studying tacit collusion they introduce a more complicated GA technique used for multiobjective optimization problems. Furthermore, the authors adopt two type of demand functions, the inverse linear and the constant elasticity demand functions. Under this respect, the authors aim to demonstrate the ability of their coevolutionary approach for finding Cournot–Nash equilibrium and for solving nonlinear Cournot models (the case with constant elasticity demand function).

Ikeda and Tokinaga (2007) investigate the properties of multi-unit multi-period double-auction mechanisms by a genetic programming (GP) approach, which is an extension of the coevolutionary GA. Each solution/agent in the population is a function (or computer program) composed of arithmetic operations, standard mathematical expressions, if–then type condition, and variables. Standard mutation, crossover, recombination operations are then applied to improve the fitness of the population. Producers are backward-looking adaptive agents selling electricity if they think that the cost to generate electricity is costly than buying electricity. The sellers are endowed with a Cobb–Douglas type production function and a multi-objective fitness function evaluating both profits and capacity utilization rate of their units. Two different demand market settings are then considered, a constant price-inelastic and a randomly time-varying price-inelastic demand scenarios. The authors compare their computational results in both market settings with a perfect competition scenario and furthermore they discuss about the price fluctuations of the simulation results, that is, spikes and volatility, with respect to real price performances. Finally, the authors propose two simplified market cases as control schemes to stabilize the price fluctuations. In particular, in the second case they introduce an exogenous force to control the auction price, and they assume that buyers additionally pay (get) some amount of money from sellers if market prices falls higher (lower) than a reference level, that is, the theoretical equilibrium value. The additional payment schemes work as factors to remove the instability, but simultaneously they produce unavoidable costs for untraded electricity.

3.2 *Individual Learning "Ad-hoc" Models*

3.2.1 Microsimulation Model with No Adaption

Bunn and Martoccia (2005) present a static microsimulation analysis based on an agent-based model of the electricity pool of England and Wales for studying unilateral and collusive market power. They do not consider a multiagent coevolutionary market setting, and the authors realize static scenario simulations by exogenously imposing bidding strategies to each market agent. This paper is not a proper ACE model in the sense that agents do not present autonomy in the decision-making; however, this paper illustrates how agent-based modeling framework can be adopted. The simulations consider a demand profile based on the NGC typical winter day and own estimates of real marginal cost functions. Starting from an empirical observation on generators supply functions that the mark-up is an increasing function of their units' marginal costs, they develop a behavioral model where generating portfolio companies consider to bid-up some of their plants beyond the competitive levels of marginal costs by a particular percentage mark-up value. The simulation results report the change in profits due to different mark-up levels. Then, the authors attempt to estimate the degree of tacit collusion present during the two divesture programs required by the UK Office for Electricity Regulation (OFFER). Results support the thesis that National Power in the early years (owning 50% of total capacity production) exercised unilateral market power and that in the later years, as the market concentration declined, a regime of tacit collusion occurred. The authors further conclude that market concentration measures, such as HHI, do not give a reliable diagnostic aid to the potential exercise of market power in the wholesale electricity sector.

3.2.2 Simple Behavioral Rules

Botterud et al. (2006) study the effect of transmission congestion management on the exertion of market power in electricity markets, in particular focusing on the ability of portfolio generating companies to raise prices beyond competitive levels. They adopt EMCAS software framework, which is a powerful and flexible agent-based simulation platform for electricity market (see Sect. 4.1). However, they do not exploit the complex agent architecture, but they consider simple adaptive behavioral rules to replicate the bidding strategy of generation companies. Only three strategies are considered. The first is the production cost bidding strategy, so that generation companies act as pure price-taker bidding the marginal production cost of their plants. The second regards physical withholding strategies where generation companies attempt to raise the market price by withholding units during hours when the expected system capacity is low. Random outages are also considered in the simulations. Finally, the third strategy implements an economic withholding strategy, that is, generation companies try to probe their influence on market prices by increasing (decreasing) the bid price with a specific percentage for the same unit in the same hour for the following day if the bid for a unit in a certain hour was accepted (not accepted). The authors test two pricing mechanisms a locational marginal pricing and a system marginal pricing (uniform price mechanisms) for two sequential markets comprising day-ahead (DA) and real-time (RT) sessions. A case study for an 11-node test power system is presented. An aggregate price-inelastic demand and eight generation companies are considered. Each GenCo owns three identical thermal plants, that is, one base load coal plant (CO), one combined cycle plant (CC) to cover intermediary load, and one gas turbine (GT) peaking unit. Three scenarios are then simulated: a base case, a physical withholding case, and an economic withholding case. Results show that the dispatch of the system does not depend on the market pricing mechanism. Furthermore, the results show that unilateral market power is exercised under both pricing mechanisms, even if locational marginal pricing scheme exhibits the largest changes in consumer costs and generation company profits.

3.2.3 Comparisons Among Standard Multiagent Learning and Ad-hoc Learning Models

A group of two researchers at the Massachusetts Institute of Technology repeatedly investigated (Visudhiphan and Ilic 1999, 2001, 2002; Visudhiphan 2003) the opportunity to adopt different agent-based models for studying market efficiency and market power issues in wholesale electricity markets. In their research activity, these researchers have compared own learning models based on electricity market considerations with learning models inspired by the multiagent learning literature.

In their first paper, Visudhiphan and Ilic (1999) consider a repeated day-ahead and hour-ahead spot markets adopting a uniform double-auction. Three generators submit bids as linear supply or single step supply functions. Pursuing their profit maximization strategy, they implement a derivative follower strategy (Greenwald et al. 1999) to learn how to adjust to the market conditions in near-real time. The agents exploit the available information about the past market clearing prices and their own marginal-cost functions to adjust the chosen strategy. On the demand side, time varying price-inelastic and -elastic loads are analyzed. The authors conclude that the market clearing price in the hour-ahead market model is lower than the day-ahead model and confirm that generators exert more market power facing price-inelastic demand, that is, prices are significantly higher. In Visudhiphan and Ilic (2001), they propose an own learning model for studying market power where agents can strategically withhold capacity. The algorithm reproduces a two-step decision process for setting bid quantity and price, respectively. The agents strategically withhold capacity if their expected profit increases. The bidding price strategy is richer and is decided on the basis of past market performances. The agent stores information about past market prices with respect to predefined load ranges and consider if market outcomes are a result of strategic or competitive behavior. Six possible price strategies for setting the marginal-unit bidding price, conditioned on such information, are available. Each strategy corresponds to a different statistical estimation performed on historic prices, for example, maximum, minimum, or mean price, or the sum of weighted prices. The authors conclude that the strategic outcomes exhibit prices higher than competitive ones for different level of demand and for both market settings of available capacity considered. Finally, Visudhiphan (2003), in is PhD thesis, extends previous works by considering further adaptive learning models and provides a more detailed explanation of the agent bidding formulation. His aim is to test market efficiency for two price mechanisms commonly adopted in the spot electricity market, that is, discriminatory and uniform double-auction. In particular, the author considers three learning algorithms tested with different parameters to investigate the generators' bidding behaviors. In addition to an own learning model similar to the one proposed in Visudhiphan and Ilic (2001), the author proposes two learning models derived from the multi-agent learning literature. In particular, the first model is inspired by the work of Auer et al. (2002), and three distinct implementations of such learning model are considered. These algorithms are based on the assumption that the agent knows the number of actions and the rewards of selected actions in previous trials. The last learning model considered is a simple reinforcement learning algorithm 3.3 that maintains the balancing between exploration and extraction in the action selection by implementing a softmax action selection using a Boltzmann distribution (Sutton and Barto 1998). They conclude that learning algorithms significantly affect simulated market results, in any case discriminatory price mechanism tends to exhibit higher market price, confirming other studies such as Bower and Bunn (2001a). Finally, discussing about the empirical validation issue, they highlight that, to carry out a correct validation, many important data such as marginal cost functions, bilateral contract obligations, plant outages are not publicly available and thus render difficult the validation procedure.

3.2.4 Model Based Upon Supply Function Optimization

Professor D.W. Bunn at the London Business School (LBS) has been very active in this research domain. He has proposed several ACE studies where models tailored on specific market assumptions have been adopted for studying several research issues about the England and Wales wholesale electricity market. In the following, we consider two papers (Day and Bunn 2001; Bunn and Day 2009), which share the same behavioral modeling approach based on a supply-function optimization heuristic. Both papers consider adaptive portfolio generating companies which, as profit maximizers, seek to optimize both their daily profits and their long-term financial contracts, typically contracts for differences (CFD). In the daily market session, they are modeled to submit piecewise linear supply functions. All such strategies are built on the same discrete price grid and thus differ only on quantity value assignments. The short-run marginal costs have been estimated on real data. Their strategic reasoning is based upon the conjecture that their competitors will submit the same supply functions as they did in the previous day. The optimization routine is conceived to modify iteration-after-iteration just one supply-function relative to only one generating unit, that is, the unit that increase the most objective function. The authors then study several market settings with a price-elastic demand. Both papers are theoretically validated in the sense that a simplified scenario where three symmetric firms are considered is run and the theoretical continuous supply-function equilibria is evaluated. Simulation results show, on average, a good fit with the theoretical solution. However, each simulation exhibits no convergence to a stationary supply function, instead a repeated cycle behavior is highlighted. These findings motivate the authors to extend such models to study more realistic market settings where asymmetric firms hedge market risk with contract cover. In particular, Day and Bunn (2001) study on the second divestiture plan of generating assets imposed by the electricity regulatory authority in 1999. The authors show that also this second round of divesture (40%) might result insufficient because results show prices substantially (about 20%) above short-run marginal costs for the proposed divestiture plan. In general, for the different divestiture plan scenarios tested (25% and 50%), the average bid significantly decreases. So the authors conclude that probably a further period of regulatory price management will be required.

Bunn and Day (2009) focus on market power issues. The authors remark that the England and Wales electricity market is not characterized by perfect competition, but on the contrary is a daily profit-maximizing oligopoly. Thus, they suggest and computationally prove that it is not correct to assess market power exertion by adopting marginal cost as a benchmark, that is, perfect competition. The results of their several tested market settings show that the simulated supply-functions lie always above marginal cost functions. The authors conclude that this is evidence of the market structures and thereby of the strategic opportunities offered to the market participants, and no collusive aspects are modeled. Thus, the authors propose the outcomes of their simulations as realistic baselines for the identification of market power abuse. These results compared to real market supply functions according to such hypothesis show that a clear market power abuse occur in the England and Wales wholesale electricity market.

Minoia et al. (2004) propose an original model for wholesale electricity markets providing incentives for transmission investments. The transmission owner is supposed to strategically compete in the spot market, thus ISO objective function takes into account not only the power producers' bids but also the transmission owners' bids. In this model, the authors consider a simplified two node grid, two constant price-inelastic loads, and asymmetric single-unit generators, which have fixed marginal cost. Each generator submits the minimum price willing to be produced and the transmission owner submits the price she asks per power quantity (MW) flowing through the line. The total expenditure is a function of generators' produced power quantity and the power quantity flowing into the line. The market clearing mechanism is built with the aim to minimize the total expenditure function subject to some technological constraints, that is, the balancing between supply and demand, generators' maximum capacities, and line constraints. Throughout the simulation, each agent maximizes its profits conjecturing that all other agents keep fixed their previous strategies. The authors discuss four different market settings with two and five generators and two different transmission capacities. They compare simulation results with a standard locational marginal price settlement rule. The results show that the market settlement rules remarkably affect generators, transmission owner, loads, and ISO rewards. However, the authors are not able to establish which settlement rule is more efficient.

3.3 Reinforcement Learning (RL) Models

The majority of wholesale electricity market ACE models adopt standard RL algorithms, such as classical Roth and Erev (Roth and Erev 1995; Erev and Roth 1998) and Q-learning (Watkins and Dayan 1992) algorithms or modifications (see Table 2).[10] These algorithms, and mainly all reinforcement learning models based on model-free approaches (Shoham et al. 2007), share some common modeling features. First, these algorithms assume that individual agents are endowed with minimal information about the evolution of the game, that is, they record only their own past history of plays and their associate instantaneous rewards. Moreover, each agent has no prior knowledge about the game structure or other players. Second, they assume no communication capabilities among the agents. The only permitted communication is from the environment/system to the agent. Third, these algorithms implement backward-looking stimulus and response learning, that is, they learn from past experience how to best react to the occurrence of a specific event without relying on forward-looking reasoning. Finally, these algorithms do not take into account opponents' strategies, and agents learn a strategy that does well against the opponents without considering the path of plays of their opponents.

In general, a standard feature of all RL algorithms consists on implementing a sort of exploitation and exploration mechanism by relying on a specific action selection policy, which map the strenght/Q-value/propensity values $S_t^i(a^i)$ at time t to a probability distribution function over actions $\pi_t^i(a^i)$. Standard models are a proportional rule (see (1)) and an exponential rule based on Gibbs–Boltzmann distribution (see (2)).

$$\pi_t^i(a^i) = \frac{S_t^i(a^i)}{\sum_{a^i} S_t^i(a^i))}. \tag{1}$$

$$\pi_t^i(a^i) = \frac{e^{\lambda S_t^i(a^i)}}{\sum_{a^i} e^{\lambda S_t^i(a^i)})}. \tag{2}$$

The idea is to increase the probability of selecting a specific action after probing the economic environment, that is, changing the relative values among the strenghts/

[10] An instructive special issue about "Foundations of multi-agent learning"(Vohra and Wellman 2007) has been published by the journal *Artificial Intelligence*. The special issue has been devoted to open a debate on the multiagent learning (MAL) agenda by bringing joint contributions of "machine learners" and "economists" to highlight different viewpoints and experiences in the field. The starting point of the discussion is the paper by Shoham et al. (2007), where they attempt to pinpoint the goal of the research on MAL and the properties of the online learning problem. The authoritative contribution of Fudenberg and Levine (2007) underlines that the theory of mechanism design can well benefit from development of computational techniques and with respect to learning models envisages: "these models may be useful for giving people advice about how to play in games, and they may also help us make better predictions. That is because learning rules for games have evolved over a long period of time, and there is some reason to think that rules that are good rules from a prescriptive point of view may in fact be good from a descriptive point of view" (Fudenberg and Levine 2007).

Q-values/propensities values. A widely adopted feature of the exponential model is that the λ parameter can be defined to increase with time. Thereby, as the learning phase proceeds, the action selection model becomes more responsive to propensity values differences, and so agents are more and more likely to select better than worse choices. At the final simulations stages, the very high values reached by the λ parameter lead to a peaked probability distribution function on the strategy with the highest propensity, that is, the best strategy.

A further method, commonly adopted for QL implementation, is the ϵ-greedy policy. It selects with probability $1-\epsilon$ the action that maximizes its expected reward and it chooses with probability ϵ one of the alternative actions. The ϵ parameter may vary with time.

3.3.1 Naive Reinforcement Learning Models

Some early models adopted own learning approaches based on elementary RL mechanisms. In the following we report some of them.

A second research activity at LBS has concerned the adoption of learning models to study electricity spot market. A common ACE modeling framework has been adopted for a series of papers (Bower and Bunn 2000, 2001b; Bower et al. 2001). In the first two papers, the authors address the issue of market efficiency by comparing different market mechanisms: pool day-ahead daily uniform price, pool day-ahead daily discriminatory (pay as bid) , bilateral hourly uniform price, and bilateral hourly discriminatory. Pool market determines a single price for each plant for a whole day, bilateral market 24 separate hourly prices for each plant. On the other hand, in the last paper, the authors adopt the same ACE model for studying the German electricity market focusing on the impact of different mergers operations proposed at that time. All papers consider a fixed and price-elastic aggregate demand and portfolio generating companies. The objectives of a generating company are twofold: first, to obtain at least the target rate of utilization for their whole plant portfolio, and second, to achieve a higher profit on their own plant portfolio, than for the previous trading day. Each GenCo has a set of four rules/actions to choose from conditioned to past market performances, that is, the achievement of target rate of utilization and the increase or decrease of profits. These actions implement the decision on diminishing, keeping or increasing prices for all or part of the power plants of the GenCo.

Bower and Bunn (2000) conclude that prices are higher under discriminatory mechanism for both pool and bilateral models. In particular, uniform settlement simulations show that most low cost plants bid at close to zero, whereas for discriminatory settlement they learn to bid prices close to the market clearing prices, intuitively for maximizing their profits. Bilateral models exhibit higher prices than pool ones. Another finding is that GenCo with few power plants perform better in the uniform than in the discriminatory settlement. The authors argue that under the uniform setting all GenCos receive the same information, that is, the unique market clearing price, whereas in the pay as bid mechanism there is also informational disadvantage among GenCos on the basis of their market share, that is, GenCos owning more power-plants receive more price signals from the market. Bower and Bunn (2001b) enrich this previous paper with a theoretical validation of the simulation model by means of classical model of perfect competition, monopoly, and duopoly. On average, the simulation results of perfect competition and monopoly are in great accordance with the unique equilibrium solutions, that is, same price for both settlement mechanisms. On the contrary, in the duopolistic case, the mean simulated market clearing price for the uniform settlement mechanism is equal with the theoretical value, but for the discriminatory settlement mechanisms is significantly lower than the theoretical value.

Finally, Bower et al. (2001) consider only the pay as bid mechanisms to study the German electricity market. In particular, they focus on potential mergers solutions. In particular, six market settings are studied for different demand level and different plant utilization rate objectives: marginal cost, no mergers, two mergers, four mergers, two mergers, and the closing of peaking oil plants by the two mergers, two mergers, and the closing of all nuclear plants. The authors conclude that prices raise considerably as an effect of mergers for all scenarios. In particular, for the two scenarios, considering the closing of power plants, the raise in prices is more significant.

A third ACE model has been proposed at LBS. Bunn and Oliveira (2001) and Bunn and Oliveira (2003) describe the model and adopt for studying pricing and strategic behavior of the proposed NETA (new electricity trading arrangements) for the UK electricity market and the presence of market power abuse by two specific generation companies, respectively. The ACE model is an extension of previous model in several directions. First, it actively models the demand side, by considering suppliers, that is, agents purchasing from wholesale market to "supply" end-use customers. Second, it models the interactions between two different markets the bilateral market (PX) and the balancing mechanism (BM) as sequential one-shot markets. Both are modeled in a simplified way, as single call markets, that is, sealed-bid multiple-unit auctions. As far as concerns market participants, both suppliers and portfolio generating companies are modeled. Suppliers are characterized by the following parameters: a retail market share, a BM exposure (percentage of forecast demand that they intend to purchase in the PX), retail price, mean average prediction error in forecasting the contracted load, and a search propensity parameter (how fast or slow the agents change their strategy with experience). Generators are characterized by the following parameters: plants portfolio, plant cycles, installed capacity, plant availability corresponding to outage rates for each plants, BM exposure, and search propensity parameter. Both types of agent seek to maximize total daily profits and to minimize the difference between its fixed objective for the BM exposure and the actual BM exposure. For PX, agents learn to set mark-ups relative to the PX price in the previous day, and for BM they learn to set mark-ups relative to the PX price in the same day. They update the propensities for each mark-up level considered according to a reinforcement learning model. Finally, some lower bounds of rationality through operational rules are considered, for example, peak plants never offer prices below their marginal costs.

In Bunn and Oliveira (2001), the authors model only a typical day, and they consider a realistic market scenario based on power plant data of the UK electricity market. Simulation results confirm the intuitive behavior that system buy price and sell price in the balancing mechanism are considerably distant and the average bilateral price is centrally located between them. In Bunn and Oliveira (2003), the authors study whether two specific generation companies in the UK electricity market can exert market power. They first propose a simplified model of a one-shot Bertrand oligopoly model with constraints to assess basic model performances. Then a realistic scenario of the UK electricity market in 2000 is studied. The authors find that only one generator can raise PX prices unilaterally, whereas BM prices are quite robust against manipulation by the two generating companies. The authors conclude that learning in repeated games is thus a suitable modeling framework for studying market behavior such as tacit collusion.

A fourth ACE model is developed at LBS to investigate how market performance depends on the different technological types of plant owned by the generators, and whether, through the strategic adaptation of their power plant portfolios, there is a tendency for the market to evolve into concentrations of specialized or diversified companies. Bunn and Oliveira (2007) consider a sequential market game. The first stage is the plant trading game where generators trade power generation assets among themselves. The second stage is the electricity market where generating companies trade electricity and is modeled as a Cournot game. In particular, two electricity market mechanisms are compared: a single-clearing (power pool) and a multi-clearing (bilateral market) mechanism. In the latter also baseload, shoulder, and peakload plants are traded in three different markets. The authors simulate two realistic scenarios using data from the England and the Wales electricity market and study both under the two market clearing mechanisms. In the first market setting (specialization scenario), three players own baseload, shoulder, and peakload plants, respectively. The pool model evolves towards a monopoly (maximum concentration), whereas in the bilateral market two companies (baseload and shoulder generating companies) at the end own equally most of the capacity. In the second market setting (diversification scenario), the three players are similar in their portfolio. In this case, the market structure does not converge on the monopolistic solution for both market types, and in the pool mechanism the structure converged to a more concentrated configuration than in the multi-clearing mechanism. The authors conclude that if the industry is at a state of great diversification it will tend to remain so, independently of the market-clearing mechanism.

3.3.2 Classical Reinforcement Learning Models

In the following, two reinforcement learning algorithms are described, which are far the most adopted by the wholesale electricity market literature. A list of agent-based model implementing such algorithms are also discussed.

Roth and Erev Reinforcement Learning Algorithm (RE). The original algorithm formulation is in Roth and Erev (1995) and Erev and Roth (1998). In this model, three psychological aspects of human learning in the context of decision-making behavior are considered: the power law of practice, that is, learning curves are initially steep and tend to progressively flatten out, the recency effect, that is, forgetting effect, and an experimentation effect, that is, not only experimented action but also similar strategies are reinforced. Nicolaisen et al. (2001) studied market power occurrence in the context of discriminatory auction, proposing some modifications to the original algorithm to play a game with zero and negative payoffs. In particular, they compare simulation results with the previous paper (Nicolaisen et al. 2000) adopting GA. The modified formulation is: for each strategy $a_i \in \mathcal{A}_i$, a propensity $S_{i,t}(a_i)$ is defined. At every round t, propensities $S_{i,t-1}(a_i)$ are updated according to a new set of propensities $S_{i,t}(a_i)$. The following formula holds:

$$S_{i,t}(a_i) = (1 - r) \cdot S_{i,t-1}(a_i) + E_{i,t}(a_i) \tag{3}$$

where $r \in [0, 1]$ is the recency parameters, which contributes to exponentially decrease the effect of past results. The second term of (3) is called experimentation function.

$$E_{i,t}(a_i) = \begin{cases} \Pi_{i,t}(\hat{a}_i) \cdot (1 - e) & a_i = \hat{a}_i \\ S_{i,t-1}(a_i) \cdot \frac{e}{n-1} & a_i \neq \hat{a}_i; \end{cases} \tag{4}$$

where $e \in [0, 1]$ is an experimentation parameter, which assigns different weight between played action and nonplayed actions and n is the number of sellers.

Propensities are then normalized to determine the mixed strategy or action selection policy $\pi_{i,t+1}(a_i)$ for next auction round $t + 1$. The original formulation (Roth and Erev 1995) considered a proportional action selection model 1, but it has also been adopted for the exponential model 2 (Nicolaisen et al. 2001).

The group led by Prof. Tesfatsion at Iowa state has repeatedly considered this algorithm in its works (Nicolaisen et al. 2001; Koesrindartoto 2002; Sun and Tesfatsion 2007; Li et al. 2008; Somani and Tesfatsion 2008). The first two papers (Nicolaisen et al. 2001; Koesrindartoto 2002) aim to assess the effectiveness of the agent-based model proposed to study market efficiency of a wholesale electricity market with a discriminatory double-auction. The authors conclude that the approach provide useful insights into the study of market efficiency; in particular, the second paper shows that changes in the learning parameter have a substantial systematic effect on market efficiency. The last three papers are related to the large scale agent-based simulator project AMES (see 4.1), which is still in its development phase. They all describe the software framework and present simulation results for the same test case with a five node transmission grid and five learning single-unit generation companies. The market model considers a day-ahead market with locational marginal pricing. Two different scenarios are always simulated and compared, that is, a competitive scenario where the five generators submit their true linear marginal costs and a strategic scenario where the generators submit "reported" linear marginal cost functions. In the work of Sun and Tesfatsion (2007), the most detailed description of the features of AMES simulator is provided and simulation results are provided to assess the effectiveness of the agent-based model. Li et al. (2008) investigate how demand-bid price sensitivity, supply offer price caps, and generator learning affect dynamic wholesale power market performance. They discuss simulation results with respect to six different performance measures, for example, average locational marginal price and average Lerner index. They conclude that generators' learning affects significantly market outcomes and that the presence of a binding price cap level affects market price spiking and volatility, either increasing or decreasing the effect. Finally, Somani and Tesfatsion (2008) aim to study market efficiency under transmission congestions and to detect the exertion of market power by producers. They adopt five market performance indexes: the Herfindahl–Hirschman Index, the Lerner Index, the Residual Supply Index, the Relative Market Advantage Index, and the Operational Efficiency Index. A detailed analysis of simulation results is provided for each of the previous indexes to highlight shortcomings or virtues.

Another group of research very active in this research domain is leaded by Prof. Veit at the University of Mannheim in Germany. In two papers (Veit et al. 2006; Weidlich and Veit 2006), they study wholesale electricity markets focusing on two-settlement market mechanisms, a day-ahead market and a real-time balancing market. Generators act strategically on the day-ahead electricity market and on the balancing power by bidding both prices and quantities to maximize their individual profits. In Weidlich and Veit (2006), the day-ahead market is a uniform double-auction receiving bids by each seller as a couple of price and quantity, whereas the balancing market for managing minute reserve is played in two stages. In the first stage, occurring one day ahead, TSO selects the power plants that are held in reserve, in the second stage, both pay as bid and uniform double-auctions are considered. The learning model has been subdivided into two separate learning task, one for each market, but the reinforcement of each chosen action in both markets comprises the profit achieved on the associated market, but it also includes opportunity costs, which take into account profits potentially achieved in the other market. Four market scenarios are studied for different market sequence combinations: first day-ahead, second balancing with uniform; first day-ahead, second balancing with pay as bid; first balancing with uniform, second day-ahead; and first balancing with pay as bid, second day-ahead. Results show that prices attain a higher (lower) level if the day-ahead market is cleared after (before) the balancing power market, and average prices are higher under uniform price than under pay-as-bid, although agents bid at higher prices under pay-as-bid. In Veit et al. (2006), the two-settlement electricity market is modeled differently, based on the Allaz (1992) work. Thus, they define a day-ahead market as the forward market and they model it as a classical Cournot model. The real-time market is called spot market and is modeled with a locational marginal pricing. The generators are portfolio traders and they learn to strategically bid on both markets separately. In particular, the propensities for forward bid quantities are updated on the basis of the generators total profit, that is, these actions are likely updated enclosing potential future profits earned in the spot market. Action propensities for the spot market are updated only on the basis of their achieved spot market profits. The authors simulate the Belgian electricity network with five different demand scenarios for the spot market. The strategic behavior of two portfolio generating companies is studied. Simulation results show that the access to the forward market leads to more competitive behaviors of the suppliers in the spot market, and thus to lower spot energy prices. Furthermore, the authors computationally prove that the introduction of a forward market creates incentives for suppliers to engage in forward contracts. Such incentives determine lower energy prices and smaller price volatility as compared to a system without the forward market.

Bin et al. (2004) investigate pool-based electricity market characteristic under different settlement mechanisms, uniform, discriminatory (pay as bid), and the Electricity Value Equivalent (adopted in China) methods with portfolio generation companies bidding prices and demand in price-inelastic. The authors conclude that the EVE pricing method has many market characteristics better than other pricing methods, because it exists little room for a power supplier to raise the market price, whereas uniform and discriminatory settlements are not able to restrain generators' market power.

Banal-Estanol and Rupérez-Micola (2007) build an agent-based model to study how the diversification of electricity generation portfolios affects wholesale prices. The authors study a duopolistic competition with a uniform price double-auction, where the demand is price-inelastic and the portfolio generators submit single-price bids for each unit. The learning model is the classical Roth and Erev except for the inclusion of a mechanism called "extinction in finite time," where actions whose probability value falls below a certain threshold are removed from the action space. Simulation results show that technological diversification often leads to lower market prices. They test for different demand to supply ratio and they identify for each ratio value a diversification breaking point. Thus, the authors conclude that, up to the breaking point, more intense competition due to higher diversification always leads to lower prices. However, in high-demand cases, further diversification leads to higher prices, but prices remain lower or equal to those under perfect specialization.

Finally, Hailu and Thoyer (2007) aim to compare the performances of three common auction formats, that is, uniform, discriminatory (pay as bid), and generalized Vickrey auctions by means of an agent-based model of an electricity market. Sellers submit linear continuous bid supply functions and they employ the classical Roth and Erev algorithm to update their bidding strategies. Four different market scenarios are studied, where eight suppliers varies from a homogenous to a completely differentiated setting with respect to maximum capacity of production and marginal-cost functions. Simulation results enable to conclude that bidding behavior cannot be completely characterized by auction format because heterogeneity in cost structure and capacity may play an important role. In general, the authors conclude that in most cases the discriminatory auction is the most expensive format and Vickrey is the least expensive.

Q-Learning (QL). The QL algorithm was originally formulated by Watkins (1989) in his PhD thesis. Since then, QL algorithm is a popular algorithm in computer science. It presents a temporal-difference mechanism (Sutton and Barto 1998), which derives from considering the intertemporal discounted sum of expected rewards as utility measure, originally conceived to solve model-free dynamic programming problems in the context of single-agent. Unlike Roth and Erev RL algorithm, QL updates only the Q-value (propensity value for RE algorithm) of the last played action and implements state-action decision rules. In the standard modeling framework, states are identical and common knowledge among the agents.

The ith agent takes an action a_i in state s at time t and obtains a reward $R_{i,t}(a_i, a_{-i}, s)$, depending also on the actions played by the opponents a_{-i}. Both action and state spaces must be discrete sets. Then, it performs an update of the Q-function $Q_i(a_i, s)$ according to the following recursive formula:

$$Q_{i,t}(a_i, s) = \begin{cases} (1-\alpha_t)Q_{i,t-1}(a_i, s) + \alpha_t[R_{i,t} + \gamma \max_a Q_{i,t-1}(a, s')] \\ \qquad\qquad\qquad\qquad\qquad\qquad\qquad \text{if } a_i = a_{i,t}, \\ Q_{i,t-1}(a_i, s) \qquad\qquad\qquad\qquad\qquad\quad \text{otherwise;} \end{cases} \tag{5}$$

where $0 \leq \gamma < 1$ is the discount factor of the discounted sum of expected rewards and $0 < \alpha_t \leq 1$ is the time-varying learning rate. γ determines the importance of future rewards, whereas α_t determines to what extent the newly acquired information will override the old information. The most common action selection policy is the ϵ-greedy selection rule (see Par. 3.3).

Several papers have adopted QL algorithm or extensions (Xiong et al. 2004; Krause et al. 2006; Krause and Andersson 2006; Naghibi-Sistani et al. 2006; Nanduri and Das 2007; Guerci et al. 2008a,b; Weidlich and Veit 2008a).

In particular, some papers (Krause et al. 2006; Krause and Andersson 2006; Guerci et al. 2008a,b) adopt the formulation with no state representation or equivalently with a single state.

Krause and Andersson (2006) compare different congestion management schemes, that is, locational marginal pricing (LMP), market splitting, and flow-based market coupling. In particular, the authors investigate two market regimes, a perfect competition and a strategic case, where single-unit producers bid linear cost functions and they face a price-elastic demand. To compare simulation results, the authors focus on some performance measures such as producer, consumer and overall surplus, congestion cost, and maximum and minimum nodal prices. The authors conclude that in both perfect competition and oligopolistic cases, strategic locational marginal pricing exhibits the highest overall welfare, followed by market splitting. In a second paper, Krause et al. (2006) adopt the same learning model to study network-constrained electricity market. Generation companies' action space is a discrete grid of intercept value of a linear cost function that is very shallow, that is, markup values. The demand is assumed price-inelastic. They aim of the work is to validate theoretically learning models by comparing the convergence properties of the coevolving market environment to Nash equilibria. Two case studies are considered, the first one where a unique equilibria is present and the second one where two equilibria are present. In the former situation, simulation results show that, in a robust way with respect to different parameters, there is high likelihood for the Q-learning algorithm to converge to the strategic solution, whereas in the latter a cyclic behaviors is observed.

Guerci et al. (2008a) address a similar issue in an extension of a previous paper (Guerci et al. 2007). The authors compare the discriminatory (pay as bid) and uniform market settlement rules with respect to two game-theoretical solution concepts, that is, Nash equilibria in pure strategies and Pareto optima. The authors aim to investigate the convergence to tacit collusive behavior by considering the QL algorithm, which implements the discounted sum of expected rewards. On this purpose, they also compare QL algorithm simulation results with a further RL algorithm proposed originally by Marimon and McGrattan (1995), which is supposed to converge to Nash equilibria. Three market settings are studied for each pricing mechanism, two duopolistic competition cases with low and high price-inelastic demand levels and one tripolistic case with the same low demand value. All these market settings are characterized by a multiplicity of Nash equilibria (up to 177 in the tripoly uniform case). The difference between payments to suppliers and total generation costs are estimated so as to measure the degree of market inefficiency. Results point out that collusive behaviors are penalized by the discriminatory auction mechanism in low demand scenarios, whereas in a high demand scenario the difference appears to be negligible.

In the following papers, a state space is defined and consequently agents learn action-state decision rules. In particular, Xiong et al. (2004) compare market performances of the two standard discriminatory (pay as bid) and uniform auctions under a price-inelastic and -elastic demand scenarios, respectively. Single-unit producers submit a price bid and they learn from experience with a classical Q-learning adopting as discrete state space a grid of market prices. Simulation results show that discriminatory prices are lower and less volatile than those in the uniform price auction, and furthermore discriminatory is less affected by demand side response. Naghibi-Sistani et al. (2006) consider a rather simplified pool-based electricity market settings where two single-unit producers compete in a uniform double-auction facing a price-inelastic demand. The state space is composed by two states relative to two different reference value of fuel costs. Simulation results show to converge to the unique strategic solution and to be robust to parameter variation. Nanduri and Das (2007) aim to investigate market power and market concentration issues in a network-constrained day-ahead wholesale electricity markets. Three different pricing settlement rules are considered discriminatory, uniform, and second-price uniform. The Herfindahl-Hirschmann and Lerner indexes and two own indexes, quantity modulated price index (QMPI), and revenue-based market power index (RMPI) are adopted to measure the exertion of market power. A state of the system is defined as the vector of loads and prices at each bus realized in the last auction round. A 12-bus electric power network with eight generators and price-inelastic loads is simulated. High and low demand scenarios are studied. The authors conclude that in most of the cases, discriminatory auction determines highest average prices followed by the uniform and second price uniform auctions, confirming computational results of other paper (Visudhiphan and Ilic 2001; Bower and Bunn 2000). In any case, generators in a discriminatory auction tend to bid higher as the network load increases; consequently, the Lerner Index and the QMPI values show that a consistently high bid markup is exhibited under this settlement mechanism. In the uniform and second price uniform auctions, the bids do not change appreciably.

Weidlich and Veit (2008a) is an original paper studying three interrelated markets reproducing the German electricity market: a day-ahead electricity market with uniform price, a market for balancing power, which is cleared before the day-ahead market, and a carbon exchange for CO_2 emission allowances modeled as a uniform double-auction. Agents learn to bid separately their generating capacities for the day-ahead and for the balancing power market. The prices for carbon dioxide emission allowances are also included into the reinforcement process as opportunity costs. The state space for each agent is represented by three levels of bid prices, that is, low, intermediate, and high with respect to maximum admissible bid price, and three levels of trading success defined as marginal, intra-marginal, and extra-marginal. The reward function for each market session used to reinforce specific actions considers the consequence of actions played on other markets as opportunity costs. A classical QL algorithm is thus implemented with ϵ-greedy action selection rule. The authors empirically validate at macrolevel simulation results by comparing prices with real market prices observed in the German electricity market in 2006. Simulated prices on both day-ahead and balancing market are in accordance to real-world prices in many months of the year 2006. Besides, influence of CO_2 emissions trading on electricity prices is comparable to observations from real-world markets.

Some papers have adopted variants of the original QL algorithm. Bakirtzis and Tellidou (2006) and Tellidou and Bakirtzis (2007) adopt the SA-Q learning formulation, which apply the Metropolis criterion, used in the Simulated Annealing (SA) algorithm, to determine the action-selection strategy instead of the classical ϵ-greedy rule. The former paper considers uniform and pay as bid pricing mechanisms to study the bidding behaviors of single-unit suppliers. The state space for each agent is represented by the previous market price and the action space by a grid of bid prices. Four market scenarios are studied under different market concentration conditions and number of players. The simulation results confirm the fact that the prices under uniform pricing are lower compared to the ones under pay-as-bid. This occurs if no supplier is market dominant; otherwise, the difference in prices is negligible due to the exertion of market power in both pricing settlement rules. Tellidou and Bakirtzis (2007) investigate tacit collusion in electricity spot markets using location marginal pricing. Unlike previous paper, they use no space representation and the action space is two-dimensional, that is, offer quantity and price. The authors aim to consider capacity withholding strategies. A simple two-node system is used for studying two market settings: a duopolistic test case and a case with eight asymmetric generators. Simulation results show that under high market concentration suppliers may learn to adopt capacity withholding strategies and that also in competitive settings, that is, eight generators, tacit collusion behaviors may arise.

3.4 Learning Classifier Systems (LCS)

The last class of agent-based adaptive behavioral models considered refers to Learning Classifier Systems (LCSs).[11] LCSs are complex adaptive[12] behavioral models linking GA and RL models. The basic element of such systems are classifiers defined as condition-action rules or state-action rules in a Markov-decision process framework. The algorithm's goal is to select the best classifier within a ruleset or action-space in those circumstances and in the future by altering the likelihood of taking that action according to past experience. They can be categorized as backward-looking learning models, even if anticipatory LCSs have been devised (Butz 2002). The standard formulation is to adopt GA to select the best condition-action rules by exploiting mutation and crossover operators. Moreover, fitness is based on future performance according to a reinforcement learning mechanism, that is, the rule that receives the most reward from the system's environment will be more reinforced, thus resulting in a higher fitness and consequently his "offspring" will likely be larger.

Bagnall and Smith (2000) and Bagnall and Smith (2005) aim to increase market realism by enhancing the modeling of the agent architecture. Their work has been widely presented and discussed in several papers (Bagnall 1999, 2000a,b, 2004; Bagnall and Smith 2005). In any case, Bagnall and Smith (2005) present aims and results of all previous works and extends them, enclosing also a detailed description of the common agent and market architectures. Their long-term project objectives which are clearly stated in their last paper are as follows: first to study if the agents are capable of learning real-world behavior, second to test for alternative market structures, and finally to find evidences for the evolution of cooperation. The market model focuses on a simplified pre-NETA UK electricity market and in particular on studying the supply side where 21 generator agents are considered, classified as one of the four types: nuclear, coal, gas, or oil/gas turbine. Generators differ also for fixed cost, start-up cost, and generation cost. Agents are grouped into three different groups: constrained on (required to run), or constrained off (forced not to generate), and unconstrained. In particular, three fixed groups of seven units are considered. The agents implement a multiobjective adaptive agent architecture based on two related objectives, that is, maximize profits avoiding losses. Both goals are represented by an own learning classifier system within the agent. The learning task is to set the bid price. An important part of the agent architecture is the state representation. Every day, each agent receives a common message from the environment to properly identify the state of the economic system. The environment state is randomly drawn every day. The state is represented by three

[11] Invented in 1975 by John Holland (Holland 1975), LCSs are less famous than GA though; GAs were originally invented as a sub-part of LCSs. At the origin of Hollands work, LCSs were seen as a model of the emergence of cognitive abilities, thanks to adaptive mechanisms, particularly evolutionary processes.

[12] In the sense that their ability to choose the best action improves with experience

variables, related to electricity demand level (four levels are considered, summer weekend, summer weekday, winter weekend, and winter weekday), the nature of the constraints, and information concerning capacity premium payments (an additional payment mechanism used in the UK market as an incentive to bid low in times of high demand).

In particular, Bagnall (2000b) examine how the agents react under alternative market mechanisms (uniform and discriminatory price mechanisms) and under what conditions cooperation is more likely to occur. The author also reports and discusses results obtained within previous works (Bagnall 1999, 2000a). Findings shows that the uniform mechanism determines higher total cost of generation, but agents tend to bid higher under the discriminatory one. Finally, they show that cooperation can emerge even if it is difficult to sustain the cooperation pattern for longer period. They impute this to the greedy behavior of agents always striving to find new ways of acting in the environment.

Bagnall and Smith (2005) address all three original research issues. As regards the first question of interest, the authors show that nuclear units bid at or close to zero and are fairly unresponsive to demand, gas and coal units bid close to the level required for profitability by increasing generally their bid-prices in times of high demand, and finally oil/GT units are bidding high to capture peak generation. They conclude that the agents' behavior broadly corresponds to real-word strategies. As far as concerns the comparison of alternatives market structures, the authors confirm previous findings that, on average, pay-as-bid price mechanism determines for each type of agent to bid higher than in the uniform one. Furthermore, they show that uniform settlement method is more expensive overall than payment at bid price even if the nuclear units perform worse while coal units perform better. Finally, they discuss about the third research issue, that is, evolution of cooperation. They consider cooperation as the situation in which two or more players make the same high bid. They find only situations where for a limited number of days this cooperation criterion is met. They ascribe this difficulty in maintaining for long-term cooperation patterns to the large number of available actions, the exploration/exploitation policy, and the potential incorrect generalization over environments.

4 Market Modeling

The wholesale electricity market agent-based models widely discussed in previous section have shown the great variability of modeling approaches adopted for both the agent architecture and the market model. Several market models have been proposed from rather simplified market settings, such as duopolistic competitions in network-unconstrained market settings, to very complex market mechanisms aiming to replicate real market features. In particular, as far as concerns latter models, large-scale agent-based models comprising multiple sequential markets and considering different time scales have been recently developed with the aim to fully exploit the flexibility and descriptive power of the ACE approach to set up decision support systems for the practitioners of this industrial sector. In this section, an overview of the major features of some of these large-scale ACE models is presented. Furthermore, Table 3 reports a schematic representation of the market modeling approach for the considered literature. Highlighted aspects are the considered market environment, that is, single or multi-settlement markets. In particular, day-ahead market (DAM), forward market (FM) corresponding to purely financial markets, and real-time balancing market for minute reserve (RT) are considered. However, it is often difficult to classify exactly all contributions within one of the considered classes. Some proposed market mechanisms may differ with respect to the standard one. Moreover, a list of pricing settlement rules are highlighted from column 2 to 7: discriminatory settlement (DA), such as midpoint or pay as bid pricing; uniform settlement (UA); classical theoretical model (Class.), such as Cournot and Bertrand oligopoly models; locational marginal pricing (LMP), such as DCOPF or ACOPF; and finally other less common market mechanisms (Other), such as EVE (Bin et al. 2004). Finally, a further column is added to indicate if network constraints are considered in the market settlement. Two adopted notations are the following: first, if one paper adopts both network-constrained and -unconstrained market mechanisms, the table reports network-constrained, and second, if two or more market clearing mechanisms or market architectures are considered, these are highlighted in all relevant columns.

4.1 Large Scale

Several large scale agent-based models have been developed. The common philosophy of these projects is to replicate national electricity market by considering multiple markets and time scales. ACE simulators for the US are EMCAS (Conzelmann et al. 2005), Marketecture (Atkins et al. 2004), N-ABLE (Ehlen and Scholand 2005), MAIS (Sueyoshi and Tadiparthi 2008b), AMES (Sun and Tesfatsion 2007); for Australia NEMSIM (Batten et al. 2005); and for Germany PowerACE (Weidlich et al. 2005; Genoese et al. 2005). Some of them are commercial, for example, EMCAS and MAA, whereas AMES results to be the first open-source software project. In the following four of them are discussed to introduce some modeling trends.

The Electricity Market Complex Adaptive System (EMCAS) developed at Argonne National Laboratory is an outgrowth of SMART II+, which originally was a project for integrating long-term-model of the electric power and natural gas markets North (2001). EMCAS is a large scale agent-based modeling simulator aiming to be used as a decision support system tool for concrete policy making as well as to simulate decisions on six different time scales from real time to multi-year planning. It comprises several kind of agents such as generating, transmission, and distribution companies to capture the heterogeneity of real markets (Conzelmann et al. 2005). Different markets are implemented, such as spot markets either adopting uniform or discriminatory mechanism, bilateral markets or four markets for the grid regulation

Table 3 Categorization of market models for each paper

Paper	Market clearing rule						Single markets	Multi-settlement	
	DA	*UA*	*Class.*	*LMP*	*Other*	*N-const.*	*DAM*	*DAM & FW*	*DAM & RT*
Curzon Price (1997)	✓		✓				✓		
Richter and Sheblé (1998)	✓						✓		
Visudhiphan and Ilic (1999)		✓					✓		
Bagnall and Smith (2000)	✓	✓					✓		
Bower and Bunn (2000)	✓	✓						✓	
Lane et al. (2000)	✓					✓	✓		
Nicolaisen et al. (2000)	✓					✓	✓		
Bower et al. (2001)	✓							✓	
Bower and Bunn (2001b)	✓	✓						✓	
Bunn and Oliveira (2001)	✓								✓
Day and Bunn (2001)		✓						✓	
Nicolaisen et al. (2001)	✓						✓		
Visudhiphan and Ilic (2001)		✓					✓		
Cau and Anderson (2002)				✓[13]			✓		
Koesrindartoto (2002)	✓						✓		
Bunn and Oliveira (2003)	✓								✓
Cau (2003)				✓			✓	✓	
Visudhiphan (2003)	✓	✓					✓		
Atkins et al. (2004)		✓	✓			✓		✓	
Bin et al. (2004)	✓	✓			✓		✓		
Minoia et al. (2004)				✓		✓	✓		
Xiong et al. (2004)	✓	✓					✓		
Bagnall and Smith (2005)	✓	✓					✓		
Bunn and Martoccia (2005)		✓					✓		
Bakirtzis and Tellidou (2006)			✓				✓		

(Continued)

Table 3 (Continued)

Paper	Market clearing rule						Single markets	Multi-settlement	
	DA	*UA*	*Class.*	*LMP*	*Other*	*N-const.*	*DAM*	*DAM & FW*	*DAM & RT*
Botterud et al. (2006)	✓	✓					✓		
Chen et al. (2006)		✓		✓		✓	✓		✓
Krause et al. (2006)				✓		✓	✓		
Krause and Andersson (2006)				✓	✓	✓	✓		
Ma et al. (2006)				✓		✓	✓		
Naghibi-Sistani et al. (2006)		✓					✓		
Veit et al. (2006)				✓		✓			✓
Weidlich and Veit (2006)	✓	✓							✓
Banal-Estanol and Rupérez-Micola (2007)		✓					✓		
Bunn and Oliveira (2007)	✓	✓	✓				✓[14]		
Guerci et al. (2007)	✓	✓					✓		
Hailu and Thoyer (2007)	✓	✓	✓				✓		
Ikeda and Tokinaga (2007)	✓						✓		
Nanduri and Das (2007)				✓	✓	✓	✓		
Sueyoshi and Tadiparthi (2007)				✓		✓			✓
Sun and Tesfatsion (2007)				✓		✓	✓		
Tellidou and Bakirtzis (2007)				✓		✓	✓		
Guerci et al. (2008a)	✓	✓					✓		
Guerci et al. (2008b)				✓		✓	✓		

Li et al. (2008)				✓		✓	✓		✓
Somani and Tesfatsion (2008)				✓		✓	✓		
Sueyoshi and Tadiparthi (2008a)				✓		✓			✓
Weidlich and Veit (2008a)		✓							✓[15]
Bunn and Day (2009)		✓						✓	
Total occurrences	22	23	5	15	3	17	35	6	8

[13] They do not consider transmission constraints, so it is equivalent to a uniform pricing settlement rule.

[14] A second market is considered where generators trade power generation assets among themselves.

[15] A third interrelated market is also considered for CO_2 emission allowances.

From columns 2–6 market, clearing price mechanisms are reported: discriminatory settlement (DA), such as midpoint or pay as bid pricing; uniform settlement (UA); classical theoretical model (Class.), such as Cournot and Bertrand oligopoly models; locational marginal pricing (LMP), such as DCOPF or ACOPF; and finally other less common market mechanisms (Other). Column 7 lists the paper where transmission network constraints are considered in the market clearing rule (N-const.). Finally, single or multi-settlement markets are highlighted in columns 8–10. In particular, day-ahead market (DAM), forward market (FM) corresponding to purely financial markets, and real-time balancing market for minute reserve (RT) are considered. A ✓notation is adopted for paper presenting the column feature, otherwise an empty field means that the negation is valid.

used for managing reserves. The demand side is populated of retail consumers supplied by demand companies, and on the supply side portfolio generating companies are considered. A complex agent architecture is developed to enable researchers to reproduce complicated decision-making patterns. Agents are endowed with a multiobjective utility function, where configurable objectives may be selected such as minimum profit or minimum market. Furthermore, they have risk-preferences and they are also able to forecast market prices. An application of the EMCAS platform is described in Conzelmann et al. (2005), which shows the great modeling power of the software platform. The future power market configuration for a part of the Midwestern United States is studied. This work reproduces a market setting including about 240 generators, about 2,000 buses, over 30 generation companies, and multiple transmission and distribution companies. Batten et al. (2005) developed the first large scale agent-based simulation model (NEMSIM), which represents Australia's National Electricity Market (NEM). This simulator enables researchers to automatically retrieve a huge amount of historical data and other information to simulate market structure from long time decisions to real time dispatch. Unlike EMCAS, NEMSIM comprises also a greenhouse gas emissions market module. Batten et al. (2005) describe the adaptive agent architecture, which implements both backward- and forward-looking learning models, and enables agents to learn from other participants strategies. Agents may have different goals at different timescales, and a specific multi-objective utility function characterizes each type of agent. Since 2004, at the University of Karlsruhe and at the Fraunhofer Institute of Karlsruhe PowerACE, an agent-based model of electric power and CO2-certificate markets has been developed (Weidlich et al. 2005; Genoese et al. 2005). The main objectives of the simulator is to promote renewable energy and to study CO_2-emission trading schemes. The impacts of renewable power energy sources in the wholesale market and their CO_2 savings can be simulated as well as long term analysis, such as investment in power plant capacities development. Suppliers can participate in the integrated market environment (comprising spot, forward, balancing, and CO_2 markets) as plant dispatchers or traders, whereas, in the demand side, consumers negotiate with the supplier agents the purchase of electric power and suppliers in their turn purchase the required electricity on the power markets. Several papers adopt PowerACE to study the German liberalized electricity market (Sensfuß and Genoese 2006; Sensfuß et al. 2008; Genoese et al. 2008). In particular, Sensfuß and Genoese (2006) and Sensfuß et al. (2008) investigate the price effects of renewable electricity generation on the CO_2 emissions, and Genoese et al. (2008) study how several emission allocation schemes affect investment decision based on both expected electricity prices and CO_2 certificate prices. Koesrindartoto and Tesfatsion (2004) report of a collaborative project between Iowa State University and the Los Alamos National Laboratory for the development of an agent-based modeling of electricity systems simulator (AMES) for testing the economic reliability of the US wholesale power market platform. The leader of the project is Prof. Leigh Tesfatsion at Iowa State University who has already published several interesting research works (see Koesrindartoto et al. (2005), Sun and Tesfatsion (2007), Li et al. (2008), and Somani and Tesfatsion (2008)). One remarkable feature of this project is that it is the first noncommercial open source software project. It is entirely developed in JAVA to be machine-independent and it enables the computational study of the US Wholesale Power Market Platform design. The AMES market architecture includes several sequential markets such as forward and financial transmission rights markets, a day-ahead and realtime market session, and finally a re-bid period. DC optimal power flow and AC optimal power flow locational pricing mechanisms for day-ahead and realtime market settlement are implemented. The AMES software package comprises a scalable software module of reinforcement learning algorithms Gieseler (2005).

5 Conclusions

Agent-based computational economics applied to wholesale electricity markets is an active and fast-growing branch of the power system literature. This paper has documented the great interest of the international scientific community in this research domain by reviewing a large number of papers that appeared throughout the last decade in computer science, power system engineering, and economics journals. Reviewed papers deal exclusively with wholesale electricity markets models. In any case, space reasons as well as the fact that this topic has been the major strand of research in electricity market ACE motivated our choice to focus on it. Nonetheless, several research issues can be highlighted within this scientific literature and a comprehensive outlook of them is reported in Table 1. For the sake of comparison, three summary tables (Tables 1, 2, and 3) are proposed, providing a complete outlook on all considered publications from a chronological viewpoint. The reported tables enable to assess the evolution of the research in the specific area during more or less one decade of research in the field.

As far as concerns research issues (Table 1), it is not possible to highlight a clear tendency by the ACE research community to progressively focus on some topics. Since the early works, the addressed issues have regarded mainly the study of market performance and efficiency and of market mechanisms comparison, which were, and still are, among the major issues also for theoretical and experimental economics. This fact probably denotes the great confidence placed by ACE researchers in providing useful and complementary insights in the market functioning by a "more realistic" modeling approach. Conversely, it is possible to highlight a clear evolution in the selection of modeling assumptions and solutions for both agent's architecture and market properties. In particular, as far as concerns the former (Table 2), the early works experienced several adaptive behavioral models ranging from social to individual learning models and from naive rules of thumbs to very complex agent's architecture based on learning classifier systems. However, progressively researchers have focused their research activity on reinforcement learning models, such as Roth and Erev and Q-learning algorithms. Currently, these are by far the most frequently adopted models. As far as concerns wholesale market models (Table 3), classical discriminatory and uniform double-auction clearing mechanisms have been intensively studied and compared throughout all these years. Recently, network-constrained market models, such as locational marginal pricing, have been studied to better model some real wholesale power markets.

The heterogeneity of agent-based modeling approaches witnesses the great flexibility of such methodology, but it also suggests the risk of neglecting to assess the quality and relevance of all these diverse approaches. Moreover, it is often difficult to compare them because of different modeling assumptions and also because detailed and clear explanations of the agent architecture are not always provided. On this purpose, this survey of ACE works aims to suggest or raise awareness in ACE researchers about critical aspects of modeling approaches and unexplored or partially explored research topics. One of the mostly evaded research approach concerns the validation of simulation results. Two approaches, which have been rarely pursued, can be adopted. The first one regards a theoretical validation based on analytical benchmark models. The drawback is that the validation occurs with analytical models, which are commonly oversimplified models of wholesale electricity markets. This approach usually involves that behavioral models successfully validated on simplified market settings are automatically adopted to investigate more realistic market scenarios where the complexity of the market scenario may obviously arise. This aspect is rarely discussed or studied. A second fruitful approach is to address the task of an empirical validation. This latter approach is seldom adopted either at a macro or a micro level. The rationale is the unavailability of real datasets comprising historical market outcomes, and real features of market operators and transmission network. The few researchers who have performed an empirical validation at a macro-level, that is, they have compared simulated prices with market prices, have often limited their comparisons to verbal or graphic considerations. No paper has tackled a statistical analysis at an aggregate level to prove the statistical significance of the computational model results, to the best of the authors' knowledge. A micro-behavioral statistical validation is an intriguing perspective, which, however, is more demanding in terms of availability of real data and in terms of modeling assumptions. The building of a model to be validated at micro-level requires careful generalization of real world markets and agent's features to enable researchers to test for computational models of market operators' behavior. Indeed, the development of effective agent architecture is a major task also without aiming to empirically validate the model. This consideration underlines a further disregarded modeling issue, that is, the relevance of the model for the decision-making process of market operators. It is worth remembering that the literature shows that simulation results are considerably different by comparing performances of several learning algorithms in an identical market setting (e.g., Koesrindartoto (2002); Visudhiphan (2003); Krause et al. (2006); Guerci et al. (2008a)). Thus, some questions need an answer: is there a suitable minimum degree of rationality? to what extent very complicated agent architectures, such as the one based on learning classifier systems or coevolutionary genetic programming, enhance the description of real market participants? In general, a high degree of bounded rationality characterizes the selected behavioral models such as the Roth and Erev classical reinforcement learning model. The choice of selecting this behavioral model is implicitly justified by the fact that this adaptive model has been studied and then calibrated on human-laboratory experiments. However, market operators do not take their decisions uniquely on the basis of private information, but also they take advantage of a more or less public information shared among all or the majority of market operators. This common knowledge may comprise technological features and historical behavior of opponents.

Furthermore, all market operators are described with the same behavioral features, for example, reasoning and learning capabilities or risk aversion, irrespective of a great heterogeneity in real world market participants. Some market actors may correspond to departments of quantitative analysis and/or operations management dedicated to the optimization scheduling of a portfolio of power-plants on a long-term horizon, whereas others may correspond to purely financial traders speculating on a short-term horizon.

Last but not least, a challenging issue is to adopt adaptive models with the capability of enhancing the action and/or state space representation to continuous variables. In the current literature, often discrete action and state spaces with low cardinality are considered, whereas the bidding format, for example, continuous, step-, or piece-wise supply functions, may be reproduced by divers parameters. Thus, this simplified settings may preclude the relevance of simulation results.

Finally, the research efforts should be further directed in the study of markets interdependence. Indeed, the strong point of ACE methodology is to provide a research tool for addressing the complexity of these market scenarios by modeling highly interconnected economic environments. Electricity markets are very complex economic environments, where a highly interdependence exists among several markets. In particular, from a modeling point of view, multi-settlement markets and market coupling (with foreign countries) issues must be adequately investigated, because they may be prominent factors in determining market outcomes. Some papers have started to investigate the interdependence of day-ahead market sessions with forward or real-time market sessions, analogously other works have started to study how CO_2 emission allowances, transmission rights or other commodities markets, such as gas, may affect market outcomes. However, this great opportunity offered by ACE methodology to model the complexity of electricity markets must be taken not to the detriment of an empirical validation of simulation results. This is obviously the challenging issue and certainly the final goal of a successful ACE agenda in the study of wholesale electricity markets.

Nonetheless, the ACE methodology has been a fertile approach for successfully studying electricity market. The several ACE works presented in this paper witness the richness and validity of contributions brought to electricity market studies. The agent-based computational approach confirms to be a very promising approach, which can greatly complement theoretical and human-subject experiments research studies.

Acknowledgements This work has been partially supported by the University of Genoa, by the Italian Ministry of Education, University and Research (MIUR) under grant PRIN 2007 "Long-term and short-term models for the liberalized electricity sector" and by the European Union under NEST PATHFINDER STREP Project COMPLEXMARKETS.

References

Allaz B (1992) Oligopoly, uncertainty and strategic forward transactions. Int J Ind Organ 10:297–308

Atkins K, Barret CL, Homan CM, Marathe A, Marathe M, Thite S (2004) Marketecture: A simulation-based framework for studying experimental deregulated power markets citeseer. ist.psu.edu/atkins04marketecture.html

Auer P, Cesa-Bianchi N, Freund Y, Schapire R (2002) The nonstochastic multiarmed bandit problem. Siam J Comput 32(1):48–77

Bagnall A (2000a) Modelling the UK market in electricity generation with autonomous adaptive agents. PhD thesis, School of Information Systems, University of East Anglia

Bagnall A (2000b) A multi-adaptive agent model for generator company bidding in the UK market in electricity. In: Whitely D, Goldberg D, Cant-Paz E, Spector L, Parmee I, Beyer H-G (eds) Genetic and evolutionary computation conference (GECCO-2000), Morgan Kaufmann, San Francisco, USA, pp. 605–612

Bagnall A (2004) A multi-agent model of the UK market in electricity generation. In: Applications of learning classifier systems. Studies in fuzziness and soft computing. Springer, Berlin Heidelberg New York, pp. 167–181

Bagnall A, Smith G (2000) Game playing with autonomous adaptive agents in a simplified economic model of the UK market in electricity generation. In: IEEE International Conference on Power System Technology, pp. 891–896

Bagnall A, Smith G (2005) A multi-agent model of the UK market in electricity generation. IEEE Trans Evol Comput 9(5):522–536

Bagnall SG A (1999) An adaptive agent model for generator company bidding in the UK power pool. In: Proceedings of artificial evolution. Lecture notes in computer science, vol. 1829. Springer, Berlin Heidelberg New York, pp. 191–203

Bakirtzis AG, Tellidou AC (2006) Agent-based simulation of power markets under uniform and pay-as-bid pricing rules using reinforcement learning. In: Proceedings of the IEEE Power Systems Conference and Exposition PSCE 06, Atlanta, USA, pp. 1168–1173

Banal-Estanol A, Rupérez-Micola A (2007) Composition of electricity generation portfolios, pivotal dynamics and market prices. Working Paper

Batten D, Grozev G, Anderson M, Lewis G (2005) Nemsim: Finding ways to reduce greenhouse gas emissions using multi-agent electricity modelling. Draft Working Paper http://smaget.lyon.cemagref.fr/contenu/SMAGET20proc/PAPERS/Batten.pdf

Bin Z, Maosong Y, Xianya X (2004) The comparisons between pricing methods on pool-based electricity market using agent-based simulation. In: Proceedings of the IEEE International Conference on Electric Utility Deregulation, Restructuring and Power Technologies (DRPT2004), Hong Kong

Blume L, Easley D (1992) Evolution and market behaviour. J Econ Theor 58:9–40

Botterud A, Thimmapuram PR, Yamakado M (2006) Simulating GenCo bidding strategies in electricity markets with an agent-based model. Proceedings of the Seventh Annual IAEE (International Association for Energy Economics) European Energy Conference (IAEE-05), Bergen, Norway

Bower J, Bunn D (2001a) Experimental analysis of the efficiency of uniform-price versus discrimatory auctions in the england and wales electricity market. J Econ Dyn Control 25(3–4):561–592

Bower J, Bunn DW (2000) Model-based comparisons of pool and bilateral markets for electricity. Energy J 21(3):1–29

Bower J, Bunn DW (2001b) Experimental analysis of the efficiency of uniform-price versus discriminatory auctions in the england and wales electricity market. J Econ Dynam Control 25:561–592

Bower J, Bunn DW, Wattendrup C (2001) A model-based analysis of strategic consolidation in the german electricity industry. Energy Policy 29(12):987–1005

Bunn DW, Day C (2009) Computational modelling of price formationin the electricity pool of englandandwales. J Econ Dynam Control 33:363–376, www.elsevier.com/locate/jedc
Bunn DW, Martoccia M (2005) Unilateral and collusive market power in the electricity pool of england and wales. Energy Econ 27:305–315
Bunn DW, Oliveira F (2001) Agent-based simulation: An application to the new electricity trading arrangements of england and wales. IEEE Trans Evol Comput 5(5):493–503; special issue: Agent Based Computational Economics
Bunn DW, Oliveira F (2003) Evaluating individual market power in electricity markets via agent-based simulation. Ann Oper Res 121(1-4):57–77
Bunn DW, Oliveira F (2007) Agent-based analysis of technological diversification and specialization in electricity markets. Eur J Oper Res 181:1265–1278
Butz MV (2002) Anticipatory learning classifier systems, genetic algorithms and evolutionary computation, vol. 4. 1st edn. Springer Berlin Heidelberg New York
Cau TDH (2003) Analysing tacit collusion in oligopolistic electricity markets using a co-evolutionary approach. PhD Thesis, Australian Graduate School of Management
Cau TDH, Anderson EJ (2002) A co-evolutionary approach to modelling the behaviour of participants in competitive electricity markets. IEEE Power Eng Soc Summer Meeting 3:1534–1540
Chen H, Wong KP, Nguyen DHM, Chung CY (2006) Analyzing oligopolistic electricity market using coevolutionary computation. IEEE Trans Power Syst 21(1):143–152
Conzelmann G, Boyd G, Koritarov V, Veselka T (2005) Multi-agent power market simulation using emcas. IEEE Power Eng Soc General Meeting 3:2829–2834
Curzon Price T (1997) Using co-evolutionary programming to simulate strategic behaviour in markets. J Evol Econ 7(3):219–254, http://ideas.repec.org/a/spr/joevec/v7y1997i3p219-254.html
Day CJ, Bunn DW (2001) Divestiture of generation assets in the electricity pool of england andwales: A computational approach to analyzing market power. J Regulat Econ 19:123–141
Ehlen M, Scholand A (2005) Modeling interdependencies between power and economic sectors using the n-able agent-based model. Proceedings of the IEEE Power Engineering Society General Meeting, pp. 2842–2846
Ehrentreich N (2002) The santa fe artificial stock market re-examined – suggested corrections. Computational Economics 0209001, EconWPA, http://ideas.repec.org/p/wpa/wuwpco/0209001.html
Epstein JM (1999) Agent-based computational models and generative social science. Complexity 4(5):41–60
Erev I, Roth AK (1998) Predicting how people play games: Reinforcement learning in experimental games with unique, mixed strategy equilibria. Am Econ Rev 88(4):848–881
von der Fehr N, Harbord D (1993) Spot market competition in the UK electricity industry. Econ J 103:531–546
Fudenberg D, Levine D (2007) An economist's perspective on multi-agent learning. Artif Intelligence 171(7):378–381
Genoese M, Sensfuß F, Weidlich A, Mst D, Rentz O (2005) Development of an agent-based model to analyse the effect of renewable energy on electricity markets. In: Proceedings of the 19th Conference Informatics for Environmental Protection (EnviroInfo). Brno
Genoese M, Most D, Gardyan P, Rentz O (2008) Impact of emmiossion allocation schemes on power plant investments. In: Proceeding Of The First European Workshop On Energy Market Modeling Using Agent-Based Computational Economics, pp. 29–47
Gieseler CJ (2005) A java reinforcement learning module for the repast toolkit: Facilitating study and experimentation with reinforcement learning in social science multi-agent simulations. Master's thesis, Computer Science, Iowa State University.
Greenwald A, Kephart J, Tesauro G (1999) Strategic pricebot dynamics. EC 99: Proceedings of the 1st ACM conference on Electronic commerce, USA ACM Press, New York, NY, pp. 58–67
Guerci E, Ivaldi S, Raberto M, Cincotti S (2007) Learning oligopolistic competition in electricity auctions. J Comput Intelligence 23(2):197–220

Guerci E, Ivaldi S, Cincotti S (2008a) Learning agents in an artificial power exchange: Tacit collusion, market power and efficiency of two double-auction mechanisms. Comput Econ 32:73–98

Guerci E, Rastegar MA, Cincotti S, Delfino F, Procopio R, Ruga M (2008b) Supply-side gaming on electricity markets with physical constrained transmission network. In: Proceedings of Eem08 - 5th International Conference On The European Electricity Market

Hailu A, Thoyer S (2007) Designing multi-unit multiple bid auctions: An agentbased computational model of uniform, discriminatory and generalised vickrey auctions. The Economic Record (Special Issue) 83:S57–S72

Hämäläinen R, Mäntysaari J, Ruusunen J, Pineau P (2000) Cooperative consumers in a deregulated electricity market - dynamic consumption strategies and price coordination. Energy 25:857–875

Holland JH (1975) Adaptation in natural and artificial systems. University of Michigan Press, Ann Arbor, MI

Holland JH (1992) Adaptation in natural and artificial systems. MIT, Cambridge, MA, USA, http://portal.acm.org/citation.cfm?id=129194

Ikeda Y, Tokinaga S (2007) Analysis of price changes in artificial double auction markets consisting of multi-agents using genetic programming for learning and its applications. Ieice Trans Fundamentals E90A(10):2203–2211

Jennings NR (2000) On agent-based software engineering. Artif Intelligence 117:277–296

Koesrindartoto D (2002) Discrete double auctions with artificial adaptive agents: A case study of an electricity market using a double auction simulator. Department of Economics Working Papers Series, Working Paper, 02005

Koesrindartoto D, Tesfatsion L (2004) Testing the reliability of ferc's wholesale power market platform: an agent-based computational economics approach. Proceedings of the 24th USAEE/IAEE North American Conference, Washington, DC, USA

Koesrindartoto D, Sun J, Tesfatsion L (2005) An agent-based computational laboratory for testing the economic reliability of wholesale power market desings. Proc IEEE Power Eng Soc General Meeting 3:2818–2823

Kok K, Derzsi Z, Gordijn J, Hommelberg M, Warmer C, Kamphuis R, Akkermans H (2008) Agent-based electricity balancing with distributed energy resources, a multiperspective case study. In: HICSS '08: Proceedings of the 41st Annual Hawaii International Conference on System Sciences, IEEE Computer Society, Washington, DC, USA, http://dx.doi.org/10.1109/HICSS.2008.46, http://dx.doi.org/10.1109/HICSS.2008.46

Krause T, Andersson G (2006) Evaluating congestion management schemes in liberalized electricity markets using an agent-based simulator. In: Proceedings of the Power Engineering Society General Meeting, Montreal.

Krause T, Beck E, Cherkaoui R, Germond A, Andersson G, Ernst D (2006) A comparison of nash equilibria analysis and agent-based modelling for power markets. Electrical Power and Energy Systems 28:599–607

Lane D, Kroujiline A, Petrov V, Shebl G (2000) Electricity market power: marginal cost and relative capacity effects. In: Proceedings of the 2000 Congress on Evolutionary Computation, vol. 2, pp. 1048–1055

Li H, Sun J, Tesfatsion L (2008) Dynamic lmp response under alternative price-cap and price-sensitive demand scenarios. In: Proceedings, IEEE Power Engineering Society General Meetings, Carnegie-Mellon University, Pittsburgh

Marimon R, McGrattan E (1995) On adaptive learning in strategic games. In: Kirman A, Salmon M (eds) Learning and rationality in economics. Blackwell, Oxford, pp. 63–101

Minoia A, Ernst D, Ilic M (2004) Market dynamics driven by the decision-making of both power producers and decision makers. http://www.montefiore.ulg.ac.be/services/stochastic/pubs/2004/MEI04

Mitchell M (1998) An Introduction to genetic algorithms. MIT, Cambridge, MA, USA

Müller M, Sensfuß F, Wietschel M (2007) Simulation of current pricing-tendencies in the german electricity market for private consumption. Energy Policy 35(8):4283–4294

Naghibi-Sistani M, Akbarzadeh Tootoonchi M, Javidi-Dashte Bayaz M, Rajabi-Mashhadi H (2006) Application of q-learning with temperature variation for bidding strategies in market-based power systems. Energy Conversion and Management 47:1529–1538

Nanduri V, Das TK (2007) A reinforcement learning model to assess market power under auction-based energy pricing. IEEE Trans Power Syst 22(1):85–95

Newman HB, Legr IC, Bunn JJ (2001) A distributed agent-based architecture for dynamic services. In: CHEP 2001, Beijing, Sept 2001, http://clegrand.home.cern.ch/clegrand/CHEP01/che, pp. 01–10

Nicolaisen J, Petrov V, Tesfatsion L (2001) Market power and efficiency in a computational electricity market with discriminatory double-auction pricing. IEEE Trans Evol Comput 5(5):504–523

Nicolaisen M Jand Smith, Petrov V, Tesfatsion L (2000) Concentration and capacity effects on electricity market power. In: Proceedings of the 2000 Congress on Evolutionary Computation, La Jolla, USA, vol. 2, pp. 1041–1047

North M (2001) Multi-agent social and organizational modeling of electric power and natural gas markets. Comput Math Organ Theor 7:331–337

Richiardi M (2007) Agent-based Computational Economics. A Short Introduction. Working Papers 69 of the LABORatorio Riccardo Revelli - Centre for Employment Studies, Collegio Carlo Alberto Moncalieri (To), Italy. http://www.laboratoriorevelli.it/_pdf/wp69.pdf

Richter CW, Sheblé GB (1998) Genetic algorithm evolution of utility bidding strategies for the competitive marketplace. IEEE Trans Power Syst 13:256–261

Roop J, Fathelrahman E (2003) Modeling electricity contract choice: An agent-based approach. Proceedings of the ACEEE Summer Study Meeting, Rye Brook, New York

Roth AE, Erev I (1995) Learning in extensive form games: Experimental data and simple dynamic models in the inteermediate term. Games Econ Behav 8(1):164–212

Rumley S, Kägi E, Rudnick H, Germond A (2008) Multi-agent approach to electrical distribution networks control. In: COMPSAC '08: Proceedings of the 2008 32nd Annual IEEE International Computer Software and Applications Conference, IEEE Computer Society, Washington, DC, USA, pp. 575–580, http://dx.doi.org/10.1109/COMPSAC.2008.224

Sensfuß F, Genoese M (2006) Agent-based simulation of the german electricity markets-an analysis of the german spot market prices in the year 2001. Proceedings of the 9 Symposium Energieinnovationen, 15-17022006 Graz, Austria

Sensfuß F, Ragwitz M, M G, D M (2007) Agent-based simulation of electricity markets - a literature review. Working Paper Sustainability and Innovation

Sensfuß F, Ragwitz M, M G, D M (2008) An agent-based simulation platform as support tool for the analysis of the interactions of renewable electricity generation with electricity and co2 market. In: Proceeding Of The First European Workshop On Energy Market Modeling Using Agent-Based Computational Economics

Shoham Y, Powers R, Grenager T (2007) If multi-agent learning is the answer, what is the question? J Artif Intelligence 171(7):365–377

Somani A, Tesfatsion L (2008) An agent-based test bed study of wholesale power market performance measures. IEEE Computational Intelligence Magazine, pp. 56–72

Staropoli C, Jullien C (2006) Using laboratory experiments to design efficient market institutions: The case of wholesale electricity markets. Annals of Public and Cooperative Economics 77(4):555–577, http://ideas.repec.org/a/bla/annpce/v77y2006i4p555-577.html

Sueyoshi T, Tadiparthi G (2007) Agent-based approach to handle business complexity in U.S. wholesale power trading. IEEE Trans Power Syst 22(2):532–543

Sueyoshi T, Tadiparthi G (2008a) Wholesale power price dynamics under transmission line limits: A use of an agent-based intelligent simulator. IEEE Transactions On Systems, Man, And Cybernetics – Part C: Applications And Reviews 38(2):229–241

Sueyoshi T, Tadiparthi GR (2008b) An agent-based decision support system for wholesale electricity market. Decision Support Systems 44:425–446

Sun J, Tesfatsion L (2007) Dynamic testing of wholesale power market designs: An open-source agent-based framework. Comput Econ 30:291–327

Sutton RS, Barto AG (1998) Reinforcement Learning: An introduction. MIT, Cambridge
Swarm (2009) Website available at: http://www.swarm.org/index.php/Swarm_main_page
Tellidou AC, Bakirtzis AG (2007) Agent-based analysis of capacity withholding and tacit collusion in electricity markets. IEEE Trans Power Syst 22(4):1735–1742
Tesfatsion L (2007) General software and toolkits: Agent-based computational economics. Website available at http://www.econ.iastate.edu/tesfatsi/acecode.htm
Tesfatsion L, Judd K (2006) Handbook of computational economics: Agent-based computational economics, Handbook in Economics Series, vol. 2. North Holland, Amsterdam
Veit DJ, Weidlich A, Yao J, Oren SS (2006) Simulating the dynamics in two-settlement electricity markets via an agent-based approach. Int J Manag Sci Eng Manag 1(2):83–97
Ventosa M, Baillo A, Ramos A, Rivier M (2005) Electricity market modeling trends. Energy Policy 33:897–913
Visudhiphan P (2003) An agent-based approach to modeling electricity spot markets. PhD thesis, Massachusetts Institute of Technology
Visudhiphan P, Ilic MD (1999) Dynamic games-based modeling of electricity markets. Power Engineering Society 1999 Winter Meeting IEEE 1:274–281
Visudhiphan P, Ilic MD (2001) An agent-based approach to modeling electricity spot markets. Proceedings of IFAC Modeling and Control of Economics Systems 2001 Meeting, Klagenfurt, Austria
Visudhiphan P, Ilic MD (2002) On the necessity of an agent-based approach to assessing market power in the electricity markets. In: Proceedings of the International Symposium on Dynamic Games and Applications, Sankt-Petersburg, Russia
Vohra R, Wellman M (2007) Special issue on foundations of multi-agent learning. Artif Intelligence 171(7):363–364
Vriend NJ (2000) An illustration of the essential difference between individual and social learning, and its consequences for computational analyses. J Econ Dynam Control 24(1):1–19, http://ideas.repec.org/a/eee/dyncon/v24y2000i1p1-19.html
Watkins C (1989) Learning from delayed rewards. PhD thesis, King's College, Sunderland University, UK
Watkins C, Dayan P (1992) Q-learning. Machine Learn 8(3-4):279–292
Weidlich A, Veit DJ (2006) Bidding in interrelated day-ahead electricity markets: Insights from an agent-based simulation model. In: Proceedings of the 29th IAEE International Conference, Potsdam
Weidlich A, Veit DJ (2008a) Analyzing interrelated markets in the electricity sector – the case of wholesale power trading in germany. IEEE Power Engineering Society General Meeting, pp. 1–8
Weidlich A, Veit DJ (2008b) A critical survey of agent-based wholesale electricity market models. Energy Econ 30:1728–1759
Weidlich A, Sensfuß F, Genoese M, Veit D (2005) Studying the effects of co2 emissions trading on the electricity market - a multi-agent-based approach. In: Proceedings of the 2nd Joint Research Workshop Business and Emissions Trading, Wittenberg.
Xiong G, Okuma S, Fujita H (2004) Multi-agent based experiments on uniform price and pay-as-bid electricity auction markets. In: Proceedings of the IEEE International Conference on Electric Utility Deregulation, Restructuring and Power Technologies (DRPT2004), Hong Kong.
Yu J, Zhou J, Yang J, Wu W, Fu B, Liao R (2004) Agent-based retail electricity market: modeling and analysis. In: Proceedings of the Third International Conference on Machine Learning and Cybernetics, Shanghai
Yuchao M, Bompard EF, Napoli R, Jiang C (2006) Modeling the Strategic Bidding of the Producers in Competitive Electricity Markets with the Watkins' Q (lambda) Reinforcement Learning. International Journal of Emerging Electric Power Systems. vol. 6, Iss.2, Article 7. http://www.bepress.com/ijeeps/vol6/iss2/art7

Futures Market Trading for Electricity Producers and Retailers

A.J. Conejo, R. García-Bertrand, M. Carrión, and S. Pineda

Abstract Within a yearly time framework this chapter describes stochastic programming models to derive the electricity market strategies of producers and retailers. Both a financial futures market and a day-ahead pool are considered. Uncertainties on hourly pool prices and on end-user demands are represented modeling these factors as stochastic processes. Decisions pertaining to the futures market are made at monthly/quarterly intervals while decisions involving the pool are made throughout the year. Risk on profit variability is modeled through the CVaR. The resulting decision-making problems are formulated and characterized as large-scale mixed-integer linear programming problems, which can be solved using commercially available software.

Keywords CVaR · Futures market · Power producer · Power retailer · Risk · Stochastic programming

1 Introduction: Futures Market Trading

This chapter provides stochastic programming decision-making models to identify the best trading strategies for producers and retailers within a yearly time horizon in an electricity market (Birge and Louveaux 1997; Conejo and Prieto 2001; Ilic et al. 1998; Kirschen and Strbac 2004; Shahidehpour et al. 2002; Sheblé 1999).

We consider that electric energy can be traded in two markets, a pool and a futures market. The pool consists in a day-ahead market while the futures market allows trading electricity up to several years ahead. The futures market may present a lower/higher average price for the seller/buyer than the pool but involves a reduced volatility. Thus, it allows hedging against the financial risk inherent to pool price volatility.

A.J. Conejo (✉)
University of Castilla-La Mancha, Campus Universitario sn, Ciudad Real, Spain
e-mail: antonio.conejo@uclm.es

S. Rebennack et al. (eds.), *Handbook of Power Systems II*, Energy Systems,
DOI 10.1007/978-3-642-12686-4_10,

Within a decision horizon of 1 year or longer, an appropriate bundling of hourly pool-price values and demand values is a requirement to achieve computational tractability. Therefore, hourly pool prices and demands are aggregated into a smaller number of values providing an appropriate trade-off between accuracy and computational tractability.

Pool price and end-user demand uncertainty are described through a set of *scenarios*. Each scenario contains a plausible realization of pool prices for a producer, and a plausible realization of pool prices and end-user demands for a retailer, with an occurrence probability. Scenarios are organized using scenario trees. Procedures on how to build efficiently scenario trees are explained, for instance, in Høyland and Wallace (2001) and Høyland et al. (2003).

Care should be exercised in constructing the scenario tree and in the actual generation of scenarios to achieve a comprehensive description of the diverse realizations of the involved stochastic processes.

Figure 1 depicts a typical two-stage scenario tree. A total number of n scenarios are represented by this tree. The tree includes branches corresponding to different realizations of uncertain parameters. We have only a futures market trading decision vector for the single black node of the tree. Pool trading decisions, made at the white nodes, depend on the scenario realizations.

For simplicity, a two-stage decision framework is considered throughout this chapter. However, note that a multi-stage decision framework is both possible and desirable for some applications. Nevertheless, care should be exercised to build multi-stage models since these models become easily intractable from a computational viewpoint.

Note that pool prices and forward contract prices are, in general, dependent stochastic processes. The joint treatment of these two stochastic processes to build scenario trees that embody the stochastic dependencies of these processes is a subject of open research. However, in two-stage stochastic programming models, forward contract prices are known to the decision maker and an appropriate forecasting procedure can be used to generate pool price scenarios.

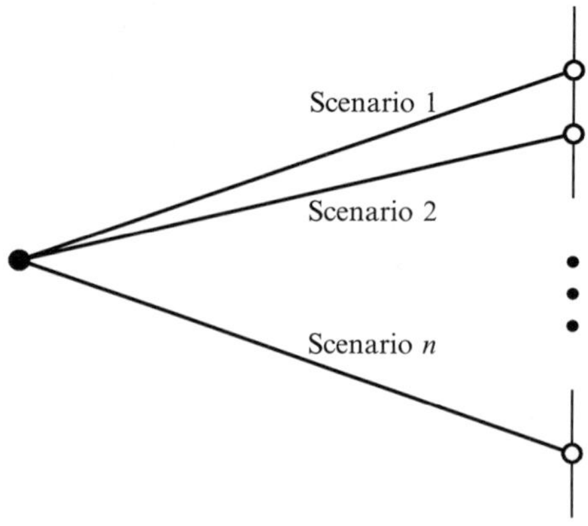

Fig. 1 Scenario tree example

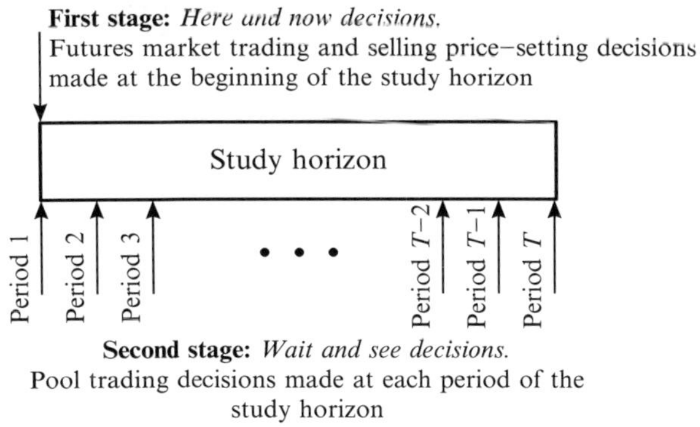

Fig. 2 Decision framework

Decisions pertaining to the futures market and to selling price determination, the latter only by a retailer, are made at monthly or quarterly intervals, while pool decisions are made throughout the year. Thus, futures market selling and price-setting decisions are made before knowing the realization of the stochastic processes and they are denoted as *here-and-now* decisions. In contrast, pool decisions are deferred in time with respect to the futures market decisions and we consider that they are made with perfect information and are referred to as *wait-and-see* decisions. This two-stage decision framework is illustrated in Fig. 2.

Specifically, a producer decides at monthly or quarterly intervals the energy to sell (and eventually to buy) in the futures market, and throughout the year the energy to sell (and eventually to buy) in the pool. A retailer decides at monthly or quarterly intervals the energy to buy (and eventually to sell) in the futures market and the selling price to be charged to its clients, and throughout the year the energy to buy (and eventually to sell) in the pool. The target of both the producer and the retailer is to maximize their respective expected profit subject to a level of risk on profit variability. The risk on profit variability refers to the volatility of that profit. The tradeoff expected profit vs. profit standard deviation is illustrated in Fig. 3. This figure represents two different probability distribution functions of the profit. The flatter one (right-hand side) has a high expected profit but a high volatility. The pointy one (left-hand side) has a low volatility as a result of a smaller expected profit.

The models presented in this chapter are to be used on a monthly basis within a rolling window information framework. At the beginning of each month decisions pertaining to the following 12 months involving futures markets and selling price-setting are made. During the considered month, trade is carried out in the pool until the beginning of the following month when futures market and selling price determination decisions spanning the following 12 months are made again, and so on. As a result of the procedure above, forward contracts and selling prices can be modified once a month. Observe that the above decision framework should be tailored to suit the trading preferences of any given producer or retailer. Note also that a

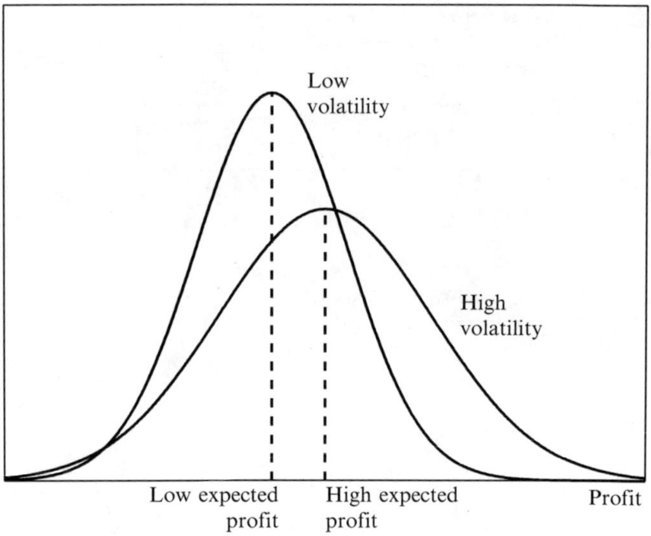

Fig. 3 Trade-off expected profit vs. profit standard deviation

multi-stage stochastic programming model suits properly the decision framework above, but the computational burden of such multi-stage approach leads more often than not to intractability.

The rest of this chapter is organized as follows. Sections 2 and 3 describe and characterize the decision-making problems of a producer and a retailer, respectively. Examples and case studies are presented and discussed. Section 4 provides some relevant conclusions. In the Appendix, the conditional value at risk (CVaR) metric is described.

2 Producer Trading

2.1 Producer: Trading

At monthly/quarterly intervals, the producer decides which forward contracts to sign in order to sell (or to buy) electric energy in the futures markets, and throughout the year it decides how much energy to sell (or to buy) in the pool.

In the derivations below, $\lambda(\xi)$ represents the different scenario realizations of the vector of electricity pool prices (stochastic process). Variable x represents the vector defining the energy traded (mostly sold) in the futures market, and $y(\xi)$ represents the variable vector defining the energy traded (mostly sold) in the pool. Note that pool decisions depend on price scenario realizations while futures market decisions do not. $R^F(\cdot)$ is the revenue associated with selling energy in the futures market, while $R^P(\cdot)$ is the revenue associate with selling in the pool. $C^O(\cdot)$ is the total production cost of the producer. $\mathcal{E}_\xi\{\cdot\}$ is the expectation operator over the stochastic process vector represented using scenarios ξ, while $\mathcal{R}_\xi\{\cdot\}$ is a risk measure over ξ, for example, the minus CVaR (Rockafellar and Uryasev 2000, 2002). If uncertainty is described via scenarios, then a linear expression for CVaR is provided in the Appendix. Sets Ω^F, Ω^P, Ω^O, and Ω^R represent, respectively, the feasibility region of futures market trading, the feasibility region of pool operation, the operating constraints of the producer, and the set of constraints needed to model the risk of profit variability.

The two-stage stochastic programming problem pertaining to a risk-neutral producer can be formulated as

$$\underset{x}{\text{maximize}} \quad R^F(x) + S(x) \tag{1}$$

subject to

$$x \in \Omega^F, \tag{2}$$

where

$$S(x) = \mathcal{E}_\xi \left\{ \underset{y(\xi)}{\text{maximize}} \quad \left(R^P(\lambda(\xi), y(\xi)) - C^O(x, y(\xi))\right) \right\} \tag{3}$$

subject to

$$y(\xi) \in \Omega^P, \forall\xi; \quad (x, y(\xi)) \in \Omega^O, \forall\xi. \tag{4}$$

Objective function (1) represents the expected profit of the producer, which is the sum of the revenue from selling in the futures market and, as expressed by (3), the expected revenue from selling in the pool minus the expected production cost. Constraint (2) imposes the feasibility conditions pertaining to the futures market, and constraints (4) enforce the feasibility conditions pertaining to the pool and the operating constraints of the producer.

Under rather general assumptions (Birge and Louveaux 1997), the maximization and expectation operators can be swapped in (3). Then, the two-stage stochastic programming problem (1)–(4) is conveniently formulated as the deterministic mathematical programming problem stated below,

$$\underset{x,\, y(\xi)}{\text{maximize}} \quad R^F(x) + \mathcal{E}_\xi \left\{R^P(\lambda(\xi), y(\xi)) - C^O(x, y(\xi))\right\} \tag{5}$$

subject to

$$x \in \Omega^F; \quad y(\xi) \in \Omega^P, \forall\xi; \quad (x, y(\xi)) \in \Omega^O, \forall\xi. \tag{6}$$

If risk is considered, the producer two-stage stochastic programming problem is formulated as the deterministic mathematical programming problem stated below,

$$\underset{x,\,y(\xi)}{\text{maximize}} \quad R^F(x) + \mathcal{E}_\xi\left\{R^P(\lambda(\xi), y(\xi)) - C^O(x, y(\xi))\right\} \\ -\beta\,\mathcal{R}_\xi\left\{R^F(x) + R^P(\lambda(\xi), y(\xi)) - C^O(x, y(\xi))\right\} \tag{7}$$

subject to

$$x \in \Omega^F;\;\; y(\xi) \in \Omega^P, \forall \xi;\;\; (x, y(\xi)) \in \Omega^O, \forall \xi;\;\; (x, y(\xi)) \in \Omega^R, \forall \xi. \tag{8}$$

Objective function (7) expresses the expected profit of the producer minus a risk measure of its profit. The expected profit is the sum of the revenue from selling in the futures market and the expected revenue from selling in the pool minus the expected production cost. Note that β is a weighting factor to realize the tradeoff expected profit vs. profit variability. Last constraints in (8) enforce conditions pertaining to the risk term. It should be noted that the risk term depends on both futures market and pool trading.

Relevant references related to the medium-term decision-making problem faced by an electricity producer include Collins (2002), Conejo et al. (2007), Conejo et al. (2008), Gedra (1994), Kaye et al. (1990), Niu et al. (2005), Shrestha et al. (2005), Tanlapco et al. (2002).

Results obtained solving the model described earlier for different values of the risk weighting factor (β) include the efficient frontier for the producer. Figure 4 illustrates how the expected profit of a producer increases as its standard deviation (risk) increases. This information allows the decision-maker to make an informed decision on both futures market and pool involvement. Note that this curve is generally concave; that is, if the risk assumed by a producer is small, an increase in risk results in a significant increase in expected profit, but if the level of risk assumed by a producer is high, an increase in the level of risk results in a rather small increase in the expected profit.

For the risk-neutral case ($\beta = 0$), an appropriate measure of the advantage of a stochastic programming approach with respect to a deterministic one is the value of the stochastic solution (VSS) (Birge and Louveaux 1997).

The objective function value of the stochastic programming problem at its optimal solution is denoted by z^{SP} and represents the average value of profit over all scenarios for the optimal solution obtained solving the stochastic programming problem.

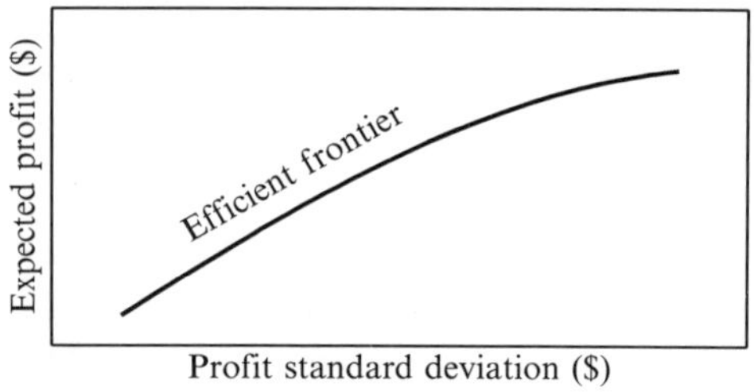

Fig. 4 Producer: expected profit vs. profit standard deviation

On the other hand, let us consider a deterministic problem in which the stochastic processes are replaced by their respective expected values. The solution of this deterministic problem provides optimal values for the futures-market decisions. The original stochastic programming problem is then solved fixing the values of the futures-market decisions to those obtained by solving the deterministic problem. The objective function value of the modified stochastic problem is denoted by z^{DP}.

The value of the stochastic solution is then computed as

$$\mathrm{VSS} = z^{\mathrm{SP}} - z^{\mathrm{DP}} \tag{9}$$

and provides a measure of the gain obtained from modeling stochastic processes as such, avoiding substituting them by average values.

2.1.1 Alternative Formulation

If the CVaR is used as the risk measure (see the Appendix), the risk can be considered in the producer problem using an alternative formulation to (7)–(8). Describing uncertainty via scenarios, the profit of the producer is a discrete random variable characterized by its probability density function (pdf). The value at risk (VaR) corresponds to the $(1-\alpha)$ percentile of this pdf, α being a parameter representing the confidence level. The CVaR represents the quantile of this pdf corresponding to the average profit conditioned by not exceeding the VaR. Therefore, the VaR at the α confidence level is the greatest value of profit such that, with probability α, the actual profit will not be below this value; and CVaR at the α confidence level is the expected profit below the profit represented by VaR.

Considering the definition of CVaR, an alternative formulation for problems (7)–(8) is the problem below whose objective is maximizing the quantile represented by the CVaR, that is,

$$\underset{x,\, y(\xi)}{\text{maximize}} \quad -\mathcal{R}_{\xi}\left\{R^{F}(x) + R^{P}(\lambda(\xi), y(\xi)) - C^{O}(x, y(\xi))\right\} \tag{10}$$

subject to

$$x \in \Omega^{F};\;\; y(\xi) \in \Omega^{P}, \forall \xi;\;\; (x, y(\xi)) \in \Omega^{O}, \forall \xi;\;\; (x, y(\xi)) \in \Omega^{R}, \forall \xi, \tag{11}$$

where the risk measure used, $\mathcal{R}_{\xi}\{\cdot\}$, is the minus CVaR.

By definition of CVaR, a confidence level $\alpha = 0$ in problem (10)–(11) corresponds to maximizing the expected profit of the producer. Increasing the confidence level α entails that the producer is willing to assume a lower risk of having low profits, which results in a decrease in the expected profit of the producer. A confidence level $\alpha = 1$ corresponds to maximizing the lowest possible profit.

The main difference between problems (7)–(8) and (10)–(11) is the number of required parameters. While weighting factor β and confidence level α need to be specified for solving problem (7)–(8), only the confidence level α is needed for solving problem (10)–(11). However, both formulations provide the same efficient frontier for the case study analyzed.

2.2 Producer: Model Characterization

Model (7)–(8) can be formulated as a large-scale linear programming problem of the form (Castillo et al. 2002)

$$\underset{v}{\text{maximize}} \quad z = d^{\mathrm{T}} v \tag{12}$$

subject to

$$Av = b, \tag{13}$$

where $v \in \mathcal{R}^n$ is the vector of optimization variables comprising first and second-stage variables, z is the objective function value, and A, b, and d are a matrix and two vectors, respectively, of appropriate dimensions.

The above problem involving several hundreds of thousands of both real variables and constraints can be readily solved using commercially available software (The FICO XPRESS Website 2010; The ILOG CPLEX Website 2010). However, a large number of scenarios may result in intractability. Note that the continuous variables and the constraints increase with the number of scenarios.

For example, consider a producer owning six thermal units, a decision horizon involving 72 price values describing price behavior throughout 1 year, 12 forward contracts, and 1,000 price scenarios. This model involves 505,013 continuous variables and 937,012 constraints.

To reduce the dimension of this problem while keeping as much as possible the stochastic information embodied in the scenario structure, scenario-reduction techniques can be used (Dupačova et al. 2000; Gröwe-Kuska 2003; Pflug 2001). These techniques trim down the scenario tree, reducing its size while keeping as intact as possible the stochastic information that the tree embodies. Appropriate decomposition techniques are also available to attack particularly large problems. These techniques include Benders decomposition, Lagrangian relaxation, and other specialized techniques (Conejo et al. 2006; Higle and Sen 1996).

2.3 Producer: Example

To clarify and illustrate the derivations above, a simple example is provided below. This simple example illustrates the decision-making process of a producer who can sell its energy either through forward contracts at stable prices or in the pool at uncertain prices. The objective of this producer is to maximize its profit while

Table 1 Producer example: pool price scenarios ($/MWh)

Period	Scenario				
	1	2	3	4	5
1	56	40	62	52	58
2	58	41	65	56	59

Table 2 Producer example: generator data

Capacity (MW)	Minimum power output (MW)	Linear production cost ($/MWh)
500	0	45

Table 3 Producer example: forward contract data

Contract	Price ($/MWh)	Power (MW)
1	52	100
2	50	50
3	46	25

Table 4 Producer example: power sold through forward contracts (MW)

Period	$\beta = 0$	$\beta = 1$	$\beta = 5$	$\beta = 100$
1	0	100	150	175
2	0	100	150	175

controlling the volatility risk of this profit. Forward contracts allow selling energy at stable prices, but they prevent selling the energy already committed in forward contracts to the pool during periods of high prices. For a given level of risk aversion, the producer should determine how much of its energy production should be sold through forward contract and how much in the pool. The example below illustrates this decision-making process.

A planning horizon of two periods is considered. The pool price is treated as a stochastic process using a tree of five equiprobable scenarios. Table 1 provides the pool prices for each period and scenario.

We consider a producer that owns a generator whose technical characteristics are provided in Table 2. This producer can sign the three forward contracts detailed in Table 3. Forward contracts are defined as a single block of energy at fixed price spanning the two periods. For instance, signing contract 1 implies selling 100 MW during the two considered time periods. For the sake of clarity, note that arbitrage between the pool and the futures market is explicitly excluded.

The risk measure used is the CVaR at 0.95 confidence level (see the Appendix).

This problem is solved for four different values of the weighting factor, $\beta = \{0, 1, 5, 100\}$. The power sold through forward contracts in each period is provided in Table 4. Table 5 provides the power generated and the power sold in the pool by the producer for each period and scenario, respectively.

Table 5 Producer example: power generated/sold in the pool (MW)

Scenario	$\beta = 0$		$\beta = 1$	
	t = 1	t = 2	t = 1	t = 2
1	500/500	500/500	500/400	500/400
2	0/0	0/0	100/0	100/0
3	500/500	500/500	500/400	500/400
4	500/500	500/500	500/400	500/400
5	500/500	500/500	500/400	500/400

Scenario	$\beta = 5$		$\beta = 100$	
	t = 1	t = 2	t = 1	t = 2
1	500/350	500/350	500/325	500/325
2	150/0	150/0	175/0	175/0
3	500/350	500/350	500/325	500/325
4	500/350	500/350	500/325	500/325
5	500/350	500/350	500/325	500/325

Table 6 Producer example: expected profit and profit standard deviation

	$\beta = 0$	$\beta = 1$	$\beta = 5$	$\beta = 100$
Expected profit ($)	10,600.0	9,880.0	9,320.0	8,840.0
Profit standard deviation ($)	6,850.2	5,480.1	4,795.1	4,452.6

Both the power generation and the power sold in the pool are dependent on the pool price scenarios. For example, the pool price in both periods of scenario 2 is lower than the production cost of the generator (see Tables 1 and 2, respectively). Thus, as Table 5 shows, the producer decides not to sell power in the pool in scenario 2 and only to produce the energy contracted by forward contracts (cases $\beta = 1$, $\beta = 5$ and $\beta = 100$).

The expected profit and the profit standard deviation obtained for the four values of β are provided in Table 6. Figure 5 depicts the efficient frontier, that is, the expected profit vs. the profit standard deviation for different values of β. For the risk-neutral case ($\beta = 0$), the expected profit is \$10,600 with a standard deviation of \$6,850.2. On the other hand, the risk-averse case ($\beta = 100$) results in a profit of \$8,840 with a smaller standard deviation of \$4,452.6.

2.4 *Producer Case Study*

The model corresponding to the producer is further illustrated through a realistic case study based on the electricity market of mainland Spain.

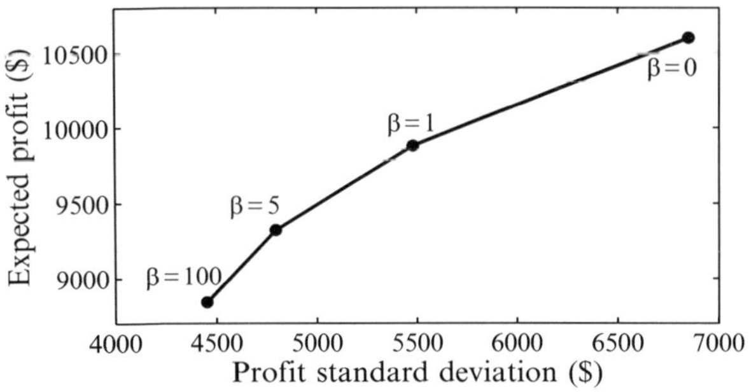

Fig. 5 Producer example: expected profit vs. profit standard deviation

Table 7 Producer case study: characteristics of the thermal units

Unit	Type	Cost ($/MWh)	Capacity (MW)	Peak/off-peak (%)
1	$Coal_1$	57	255	30
2	$Coal_2$	61	255	30
3	Oil_1	80	295	25
4	Oil_2	74	295	25
5	$CCGT_1$	50	300	25
6	$CCGT_2$	45	300	25

2.4.1 Data

The considered producer owns six thermal units whose characteristics are provided in Table 7. The last column of this table indicates the minimum percentage of the energy produced by each unit during peak periods that must be produced during off-peak periods. The minimum power output of all units is 0 MW. A decision horizon of 1 year is considered and hourly pool prices of the whole year are aggregated in 72 price values. Within this framework, 12 forward contracts are considered, one per month. Each forward contract consists of two selling blocks of 80 MW each. Table 8 lists the data of prices for both blocks of each forward contract. Initially 200 price scenarios are generated, and the fast forward scenario reduction algorithm explained in Gröwe-Kuska (2003) is used to reduce this number to 100, considering the profit obtained for each scenario as the scenario reduction metric. Figure 6 depicts the evolution of the price scenarios, the evolution of the average pool price and the forward contract prices throughout the year.

2.4.2 Results

The risk measure used is the CVaR at 0.95 confidence level. The resulting problem is solved for different values of the weighting factor β.

Figure 7 illustrates how the optimal objective function value changes as the number of scenarios increases for the risk-neutral case. The information provided by this figure allows selecting 100 as an appropriate number of scenarios, which yields an adequate tradeoff between tractability and accuracy.

Table 8 Producer case study: monthly forward contracts

Contract	Price ($/MWh)		Contract	Price ($/MWh)	
	First block	Second block		First block	Second block
1	76	73	7	64	57
2	68	66	8	64	62
3	59	56	9	61	58
4	59	57	10	61	58
5	63	58	11	64	61
6	60	57	12	62	60

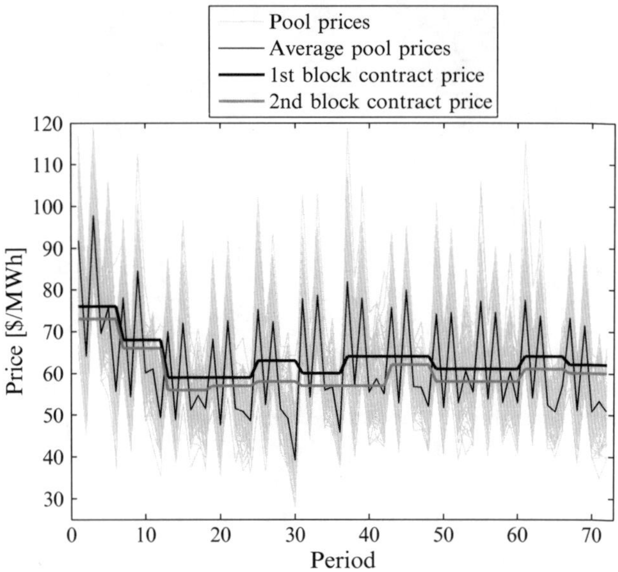

Fig. 6 Producer case study: pool prices and forward contract prices

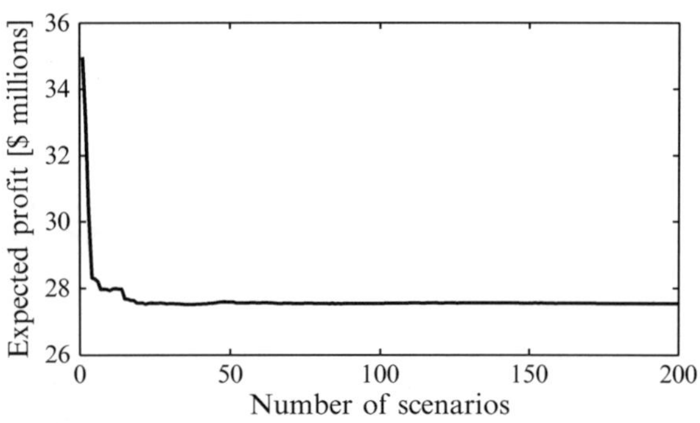

Fig. 7 Producer case study: expected profit vs. number of scenarios

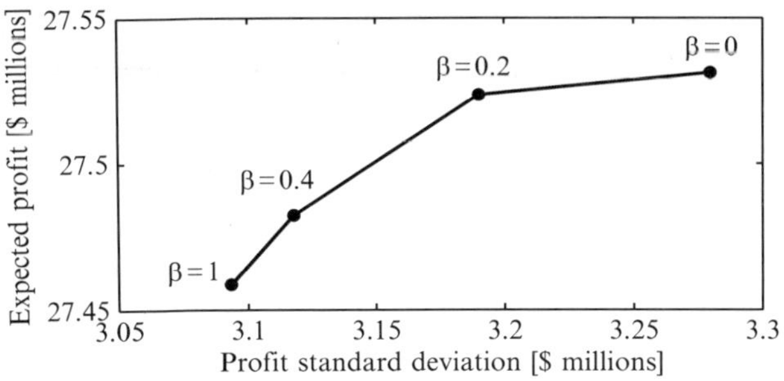

Fig. 8 Producer case study: evolution of expected profit vs. profit standard deviation

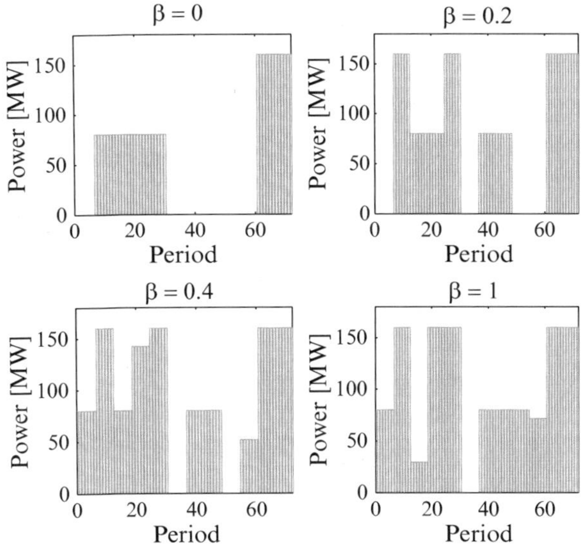

Fig. 9 Producer case study: evolution of power involved in futures market

Figure 8 shows the efficient frontier, that is, the expected profit as a function of the standard deviation. A risk-neutral producer ($\beta = 0$) expects to achieve a profit of \$27.531 million with a standard deviation of \$3.280 million. For a risk-averse producer ($\beta = 1$), the expected profit is \$27.463 million with a standard deviation of \$3.082 million. Observe that the expected profit increases as its corresponding standard deviation, which is related to risk, increases.

Figure 9 depicts thc power traded in the futures market for different values of β. The y-axis represents power, not energy; since time periods include different number of hours, the above remark is necessary. We observe that as the concern on risk increases, the selling power in the futures market increases. This behavior is justified by the fact that forward contracts involve more stable prices than the pool. Therefore, selling energy in futures market generally results in lower risk and lower profit.

The value of the stochastic solution for the risk-neutral case is VSS $= \$27.531 - \$27.491 = \$0.04$ million, that is, 0.15%. Note that \$27.491 million is the average profit obtained solving the stochastic programming problem once forward contracting decisions are fixed to the optimal values obtained from the deterministic problem.

The considered problem, characterized by 108,124 constraints and 50,525 continuous variables, has been solved using CPLEX 10.0 (The ILOG CPLEX Website 2010) under GAMS (Rosenthal 2008) on a Linux-based server with two processors clocking at 2.4 GHz and 8 GB of RAM. CPU time required to solve it is less than 10 s.

3 Retailer Trading

3.1 Retailer: Trading

A retailer must procure energy in both the futures market and the pool to sell that energy to its clients. On a monthly/quarterly basis, the retailer must decide which forward contracts to sign and the optimal selling prices to be charged to its clients, and throughout the year it must determine the energy to be traded (mostly bought) in the pool.

Note that $\lambda(\xi)$ and $d(\lambda^C, \xi)$ represent different scenario realizations of the pool prices and demands (stochastic processes), respectively, and that client demands depend on the selling price offered by the retailer, λ^C.

The retailer decision-making problem is conveniently formulated as the deterministic mathematical programming problem stated below,

$$\begin{aligned} \underset{x,\, y(\xi),\, \lambda^C}{\text{maximize}} \quad & \mathcal{E}_\xi \left\{ R^C(\lambda^C, d(\lambda^C, \xi)) - C^P(\lambda(\xi), y(\xi)) \right\} - C^F(x) \\ & -\beta \mathcal{R}_\xi \left\{ R^C(\lambda^C, d(\lambda^C, \xi)) - C^P(\lambda(\xi), y(\xi)) - C^F(x) \right\} \end{aligned} \tag{14}$$

subject to

$$x + y(\xi) = d(\lambda^C, \xi), \forall \xi, \tag{15}$$

$$x \in \Omega^F; \;\; y(\xi) \in \Omega^P, \forall \xi; \;\; (x, y(\xi)) \in \Omega^R, \forall \xi. \tag{16}$$

where x represents the variable vector defining the energy traded (mostly bought) in the futures market, while $y(\xi)$ represents the variable vector defining the energy traded (mostly bought) in the pool. Pool decisions depend on scenario realizations while futures market decisions do not. λ^C is the variable vector of selling prices to clients, $R^C(\cdot)$ the revenue from selling energy to clients, $C^F(\cdot)$ the cost associated with buying energy in the futures market, and $C^P(\cdot)$ the cost associated with buying in the pool. Sets Ω^F, Ω^P, and Ω^R represent, respectively, the feasibility region of forward contracting, the feasibility region of pool operation, and the set of constraints needed to model the risk of profit variability.

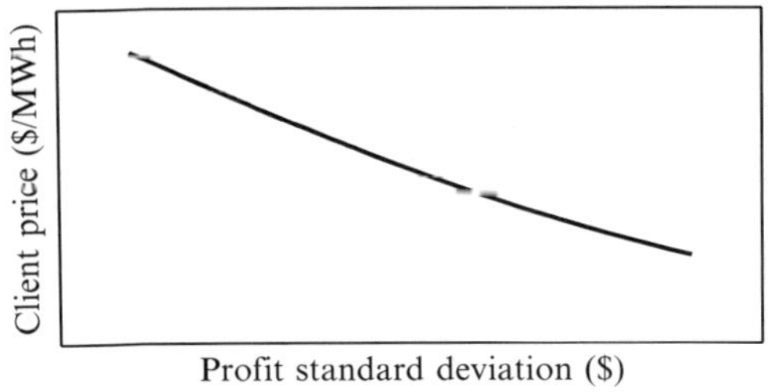

Fig. 10 Retailer: client price vs. profit standard deviation

Objective function (14) expresses the expected profit of the producer minus a risk measure of it. The expected profit is obtained as the expected revenue from selling to the clients minus the expected cost from buying from the pool minus the cost of buying through forward contracts. Constraints (15) impose that the demand of the clients should be supplied. Constraints (16) enforce the feasibility conditions pertaining to the futures market and the pool and the conditions pertaining to the risk term. Note that the risk term depends on both futures market and pool trading decisions.

Results obtained solving this model for different values of the risk weighting factor (β) include the efficient frontier for the retailer, which is similar to the one obtained for a producer (Fig. 4), and how the prices set by a retailer to its clients change with the standard deviation of the profit of the retailer. Figure 10 shows how the selling price to clients (fixed by the retailer) decreases with the standard deviation (risk) of the profit of the retailer. As the retailer relies more on its purchases from the pool (higher profit standard deviation), the selling price offered to its clients decreases to attract as much consumption as possible. Conversely, as the retailer relies more on its purchases from forward contracts (lower profit standard deviation), the selling price offered to its clients increases as the energy from forward contracts is more expensive than that from the pool.

Relevant references on the medium-term decision-making problem faced by an electric energy retailer include Carrión et al. (2007), Conejo et al. (2007), Fleten and Pettersen (2005), Gabriel et al. (2002), Gabriel et al. (2006).

The value of the stochastic solution can also be obtained as explained in Sect. 2.1. This measure is an indication of the quality of the stochastic solution vs. the corresponding deterministic one.

3.1.1 Alternative Formulation

As explained in Sect. 2.1.1 for the producer case, the retailer problem (14)–(16) can also be formulated as

$$\underset{x,\, y(\xi),\, \lambda^C}{\text{maximize}} \quad -\mathcal{R}_\xi \left\{ R^C(\lambda^C, d(\lambda^C, \xi)) - C^P(\lambda(\xi), y(\xi)) - C^F(x) \right\} \tag{17}$$

subject to

$$x + y(\xi) = d(\lambda^C, \xi), \forall \xi, \tag{18}$$

$$x \in \Omega^F; \;\; y(\xi) \in \Omega^P, \forall \xi; \;\; (x, y(\xi)) \in \Omega^R, \forall \xi. \tag{19}$$

where the risk measured used, $\mathcal{R}_\xi\{\cdot\}$, is the minus CVaR.

Problems (14)–(16) and (17)–(19) provide the same efficient frontier for the case study analyzed.

3.2 Retailer: Model Characterization

The model presented for the retailers (14)–(16) can be formulated as large-scale mixed-integer linear programming problem of the form (Castillo et al. 2002)

$$\underset{v,\,u}{\text{maximize}} \quad z = d^{\mathrm{T}} v + e^{\mathrm{T}} u \tag{20}$$

subject to

$$Av + Bu = b, \tag{21}$$

where $v \in \mathcal{R}^n$ is a vector of real optimization variables comprising first- and second-stage decisions; $u \in \{0, 1\}^m$ is a vector of binary optimization variables pertaining to both first- and second-stage decisions; and A, B, d, e, and b are two matrices and three vectors, respectively, of appropriate dimensions.

If the selling price is modeled through a stepwise price-quota curve, binary variables are needed to identify the interval of the curve corresponding to the selling price. The price-quota curve provides the relationship between the retail price and the end-user demand supplied by the retailer.

The above problem involving a few hundreds of binary variables and several hundreds of thousands of both real variables and constraints can be readily solved using commercially available software (The FICO XPRESS Website 2010; The ILOG CPLEX Website 2010). However, a very large number of scenarios may result in intractability.

For example, consider a retailer that provides the electricity demand of three end-user groups throughout 1 year, which is divided into 72 time periods. The relationship between the selling price and the demand provided by the retailer is modeled by price-quota curves with 100 steps. The retailer has the possibility of signing 12 contracts in the futures market. A 1,000-scenario tree is used to take into account the uncertainty of pool prices and end-user demands. The formulation of this problem requires 650,930 constraints, 577,508 continuous variables, and 300 binary variables.

As in the producer case, if a very large number of scenarios results in intractability, scenario-reduction and decomposition techniques can be used to achieve tractability.

3.3 Retailer: Example

This simple example illustrates the decision-making process of a retailer that, to supply its contracting energy obligations can buy energy either through forward contacts at stable prices or in the pool at uncertain prices. Forward contracts allow buying energy at stable prices, but they prevent buying the energy already committed in these contracts from the pool during periods of low prices. The objective of the retailer is to maximize its profits from selling energy to its clients while controlling the volatility risk of this profit. Note that the amount of energy to be supplied to clients is uncertain and, additionally, depends on the selling price offered to these clients by the retailer. For a given level of risk aversion, the retailer should determine the energy selling price to its clients, as well as how much of the energy to be supplied should be bought from forward contacts and how much from the pool. The example below illustrates this decision-making process.

In this example we consider a retailer that buys energy in the futures market through forward contracts and in the pool to provide electricity to its clients. The demand of the clients is modeled as a stochastic process with three equiprobable scenarios. Table 9 provides the demand of the clients for each period and scenario. Note that Table 9 provides the maximum demand values that the retailer may supply to its clients and correspond to 100% of demand supplied in Fig. 11. These maximum values are supplied only for selling prices not above 60 \$/MWh, as illustrated in Fig. 11.

The response of the clients to the price offered by the retailer is represented through the price-quota curve shown in Fig. 11, which expresses how the demand decreases as the price increases. A single price-quota curve is used for all periods and scenarios.

The retailer can purchase energy in the futures market through two forward contracts. The characteristics of these forward contracts are provided in Table 10.

The problem is solved for three different values of the weighting factor, $\beta = \{0, 0.5, 5\}$. The power purchased through forward contracts in each period is provided in Table 11. The price offered by the retailer to its clients is given in Table 12 for the different values of β. The price offered increases from 70 \$/MWh ($\beta = \{0, 0.5\}$) to 82 \$/MWh ($\beta = 5$). A higher price originates a lower demand to be supplied but also a smaller risk of incurring a financial loss. Table 13 provides the power purchased in the pool for each period and scenario.

Table 9 Retailer example: client demands (MW)

Scenario	Period	
	1	2
1	450	412
2	482	395
3	421	381

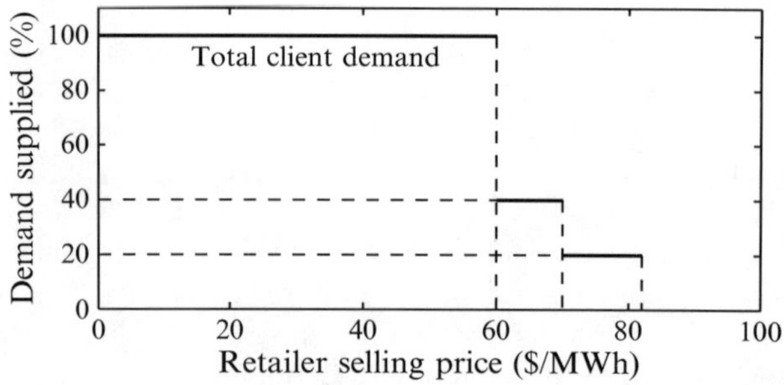

Fig. 11 Retailer example: price-quota curve

Table 10 Retailer example: forward contracts

Contract	Price ($/MWh)	Power (MW)
1	57	50
2	59	25

Table 11 Retailer example: power purchased through forward contracts (MW)

Period	$\beta = 0$	$\beta = 0.5$	$\beta = 5$
1	0	50	75
2	0	50	75

Table 12 Retailer example: price offered by the retailer to its clients ($/MWh)

$\beta = 0$	$\beta = 0.5$	$\beta = 5$
70	70	82

Table 13 Retailer example: power purchased in the pool (MW)

Scenario	$\beta = 0$		$\beta = 0.5$		$\beta = 5$	
	t = 1	t = 2	t = 1	t = 2	t = 1	t = 2
1	180.00	164.80	130.00	114.80	15.00	7.40
2	192.80	158.00	142.80	108.00	21.40	4.00
3	168.40	152.40	118.40	102.40	9.20	1.20

The expected profit and the profit standard deviation obtained for the three values of β are provided in Table 14, while Fig. 12 depicts the efficient frontier. The risk-neutral case ($\beta = 0$) results in a high expected profit of $4,507.3 involving a high profit standard deviation of $1,359.8. On the other hand, a highly risk-averse case ($\beta = 5$) results in a low profit, $4,136.5, involving a low profit standard deviation, $249.7. Finally, Fig. 13 depicts the selling price offered to its clients by the retailer vs. the profit standard deviation. Observe that as the standard deviation of the profit increases, the selling price decreases.

Table 15 provides the power sold by the retailer to its client in each scenario and each value of the weighting factor β.

Table 14 Retailer example: expected profit and profit standard deviation

	$\beta = 0$	$\beta = 0.5$	$\beta = 5$
Expected profit ($)	4,507.3	4,474.0	4,136.5
Profit standard deviation ($)	1,359.8	1,009.2	249.7

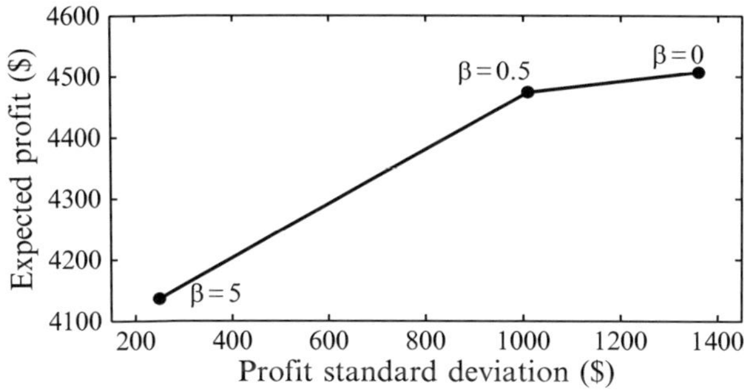

Fig. 12 Retailer example: expected profit vs. profit standard deviation

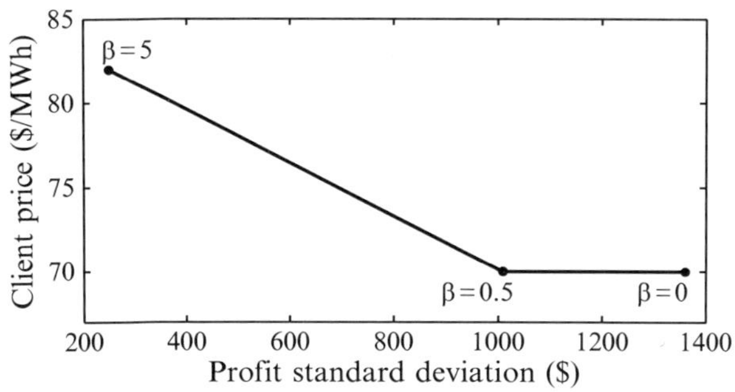

Fig. 13 Retailer example: client price vs. profit standard deviation

Table 15 Retailer example: power sold to the clients (MW)

Scenario	$\beta = 0$		$\beta = 0.5$		$\beta = 5$	
	t = 1	t = 2	t = 1	t = 2	t = 1	t = 2
1	180.00	164.80	180.00	164.80	90.00	82.40
2	192.80	158.00	192.80	158.00	96.40	79.00
3	168.40	152.40	168.40	152.40	84.20	76.20

3.4 Retailer Case Study

Below, the model of the retailer is analyzed and tested using a realistic case study based on the electricity market of mainland Spain.

Table 16 Retailer case study: forward contracts

Contract	Price ($/MWh)	
	First block	Second block
1	76	82
2	68	71
3	59	62
4	59	61
5	63	68
6	60	67
7	64	71
8	64	68
9	61	65
10	61	64
11	64	67
12	62	64

3.4.1 Data

A planning horizon of 1 year is considered. The year is divided into 72 periods. The retailer participates in a futures market where 12 forward contracts are available, one per month. Each forward contract consists of two selling blocks of 80 MW each. Table 16 lists the data of prices for both blocks of each forward contract. Three groups of clients are considered according to different consumption patterns, namely residential, commercial, and industrial. The response of each group of client to the price offered by the retailer is modeled through the 100-block price-quota curves depicted in Fig. 14. Note that a single price-quota curve is used for each client group in all periods and scenarios. An initial scenario tree of 200 pool price scenarios and client demands is generated. Taking into account the profit associated to each scenario, this scenario tree is trimmed down using the fast forward scenario reduction algorithm provided in Gröwe-Kuska (2003), considering the profit obtained for each scenario as the scenario reduction metric, yielding a final tree comprising 100 scenarios. The evolution of the pool price scenarios, the average pool price and the forward contract prices are plotted in Fig. 15. The client demand for each group is depicted in Fig. 16.

3.4.2 Results

The risk measure used is the CVaR at 0.95 confidence level. The resulting problem is solved for different values of the weighting factor β.

The evolution of the expected profit with the number of scenarios in provided in Fig. 17. This figure shows that the selection of 100 scenarios is adequate in terms of tractability and accuracy.

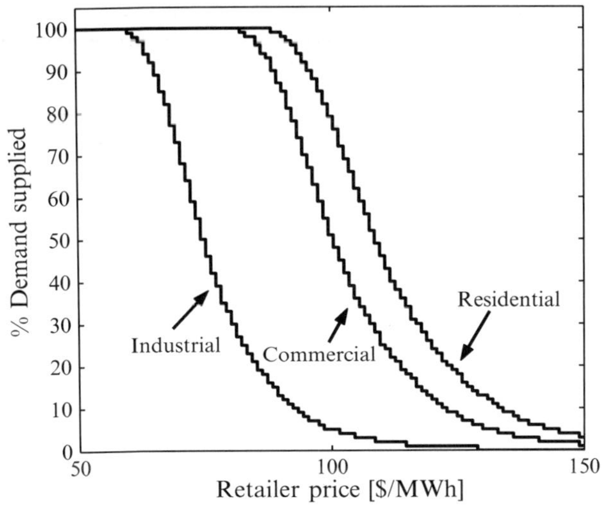

Fig. 14 Retailer case study: price-quota curve

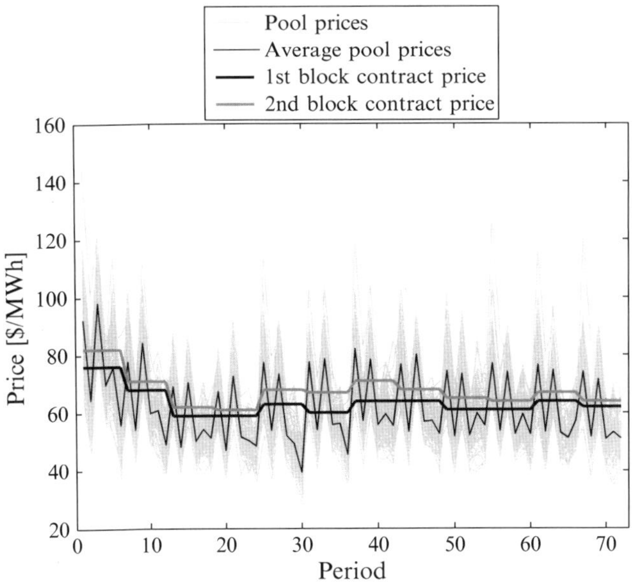

Fig. 15 Retailer case study: pool prices and forward contract prices

Figure 18 provides the efficient frontier, that is, the expected profit vs. the profit standard deviation for different values of the weighting factor β. This figure shows that the expected profit of a risk-neutral retailer is equal to \$595.99 million with a standard deviation of \$52.29 million. A risk-averse retailer ($\beta = 100$) expects to achieve a profit of \$577.70 million with a standard deviation equal to \$37.62 million.

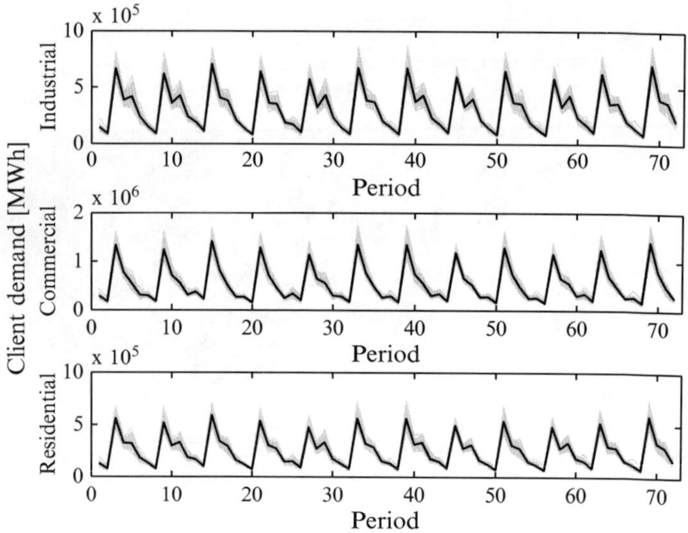

Fig. 16 Retailer case study: client demand

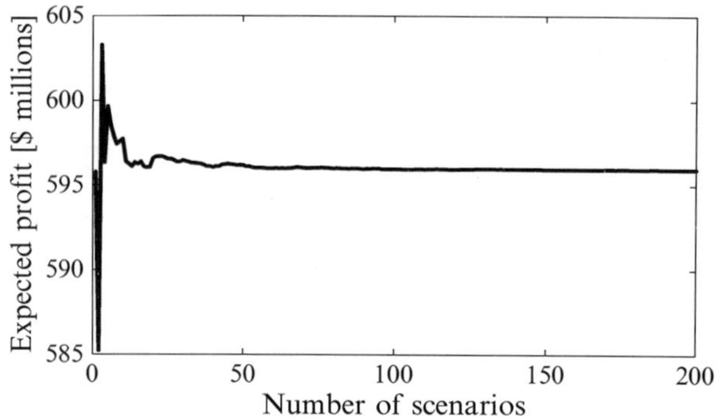

Fig. 17 Retailer case study: expected profit vs. number of scenarios

Then, we can conclude that a retailer who desires to achieve large expected profits must assume high risk.

The value of the stochastic solution (risk-neutral case) is VSS = \$591.991 – \$591.961 = \$0.03 million, that is, 0.005%. Note that \$591.961 million is the expected profit obtained if the optimal forward contract decisions derived using the deterministic problem are fixed for solving the stochastic problem.

Figure 19 shows the resulting power purchased in the futures market for different values of β. Note that the y-axis represents the power purchased from each contract for all hours within the delivery period of the contract. It can be observed that the quantity of power purchased increases as the risk-aversion of the retailer becomes

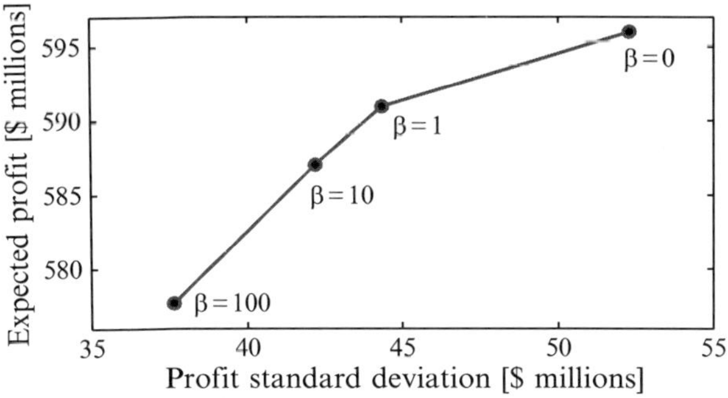

Fig. 18 Retailer case study: evolution of expected profit vs. profit standard deviation

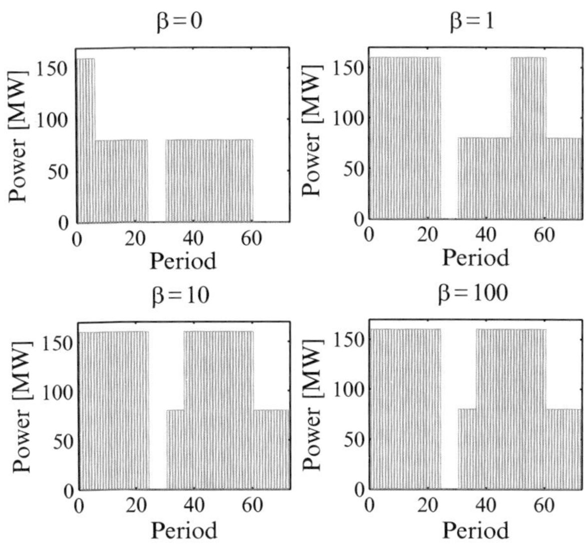

Fig. 19 Retailer case study: evolution of power involved in futures market

higher (smaller standard deviation of profit). This result is coherent since the prices of the forward contracts are more stable than the pool prices.

The prices offered to each clients group for different values of β are depicted in Fig. 20. As the risk-aversion increases (decreasing standard deviation of profit), observe that the price offered also increases. An increase in the selling price causes a reduction of both quantity and volatility of the demand supplied to the clients. In addition, the rise of the selling price is motivated by an increase of the forward contract purchases, because forward contracting is a more expensive electricity source than the pool.

The above problem, characterized by 7,925 constraints, 7,926 continuous variables, and 300 binary variables is solved using CPLEX 10.0 (The ILOG CPLEX Website 2010) under GAMS (Rosenthal 2008) on a Linux-based server with two

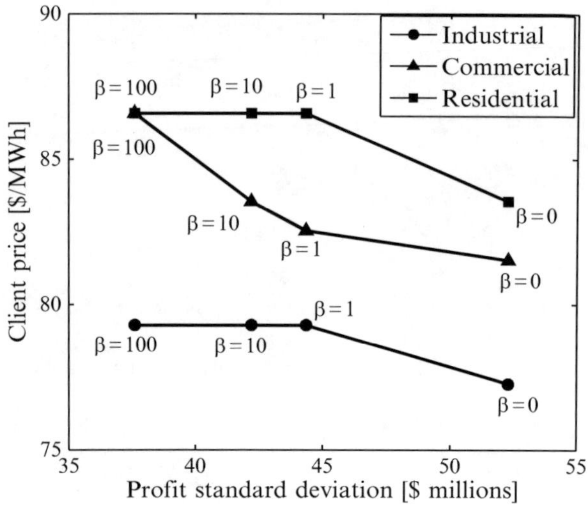

Fig. 20 Retailer case study: client price vs. profit standard deviation

processors clocking at 2.4 GHz and 8 GB of RAM. All cases has been solved in less than 13 s.

4 Conclusions

Stochastic programming is an appropriate methodology to formulate the medium-term (e.g., 1 year) trading problems faced by producers and retailers in an electricity market. This chapter provides models that allow a power producer/retailer engaging in forward contracting to maximize its expected profit for a given risk-level.

A power producer facing the possibility of signing forward contracts has to deal with uncertain pool prices. For the case of a power retailer, it copes with not only uncertain pool prices, but also with end-user demands dependent on the selling price offered by the retailer. Pool prices and demands are stochastic processes that can be characterized through a scenario set that properly represents the diverse realization of these stochastic processes. Care should be exercised in the generation of scenarios so that the solution attained is independent of any specific scenario set.

The resulting mathematical programming models are generally large-scale mixed-integer linear programming problems that can be solved using commercially available software. If these problems become intractable, then scenario-reduction or decomposition techniques make them tractable.

The results provided by these models include the efficient frontier giving expected profit vs. profit standard deviation (risk). Efficient frontier curves are used by decision-makers to achieve informed decisions on futures market, pool involvement, and client prices.

Acknowledgements We are thankful to A. Canoyra, Á. Caballero, and A. de Andrés from UNION FENOSA Generación for insightful comments and relevant observations linking our models to the actual world.

Appendix: Conditional Value at Risk

The conditional value at risk (CVaR) is a measure of risk of particular importance in applications where uncertainty is modeled through scenarios (Rockafellar and Uryasev 2002).

Given the probability distribution function of the profit (see Fig. 21), the CVaR is the expected profit not exceeding a value called value at risk (VaR), that is,

$$\text{CVaR} = \mathcal{E}\{\text{profit}|\text{profit} \leq \text{VaR}\}, \tag{22}$$

where $\mathcal{E}$ is the expectation operator.

The VaR is a measure computed as

$$\text{VaR} = \max\{\text{x}|\text{p}\{\text{profit} \leq \text{x}\} \leq 1-\alpha\}, \tag{23}$$

where confidence level α usually lies between 0.9 and 0.99.

If profit is described via scenarios, CVaR can be computed as the solution of Rockafellar and Uryasev (2000):

$$\underset{\zeta,\,\eta(\omega)}{\text{maximize}} \quad \zeta - \frac{1}{(1-\alpha)} \sum_{\omega=1}^{N_\omega} p(\omega)\eta(\omega) \tag{24}$$

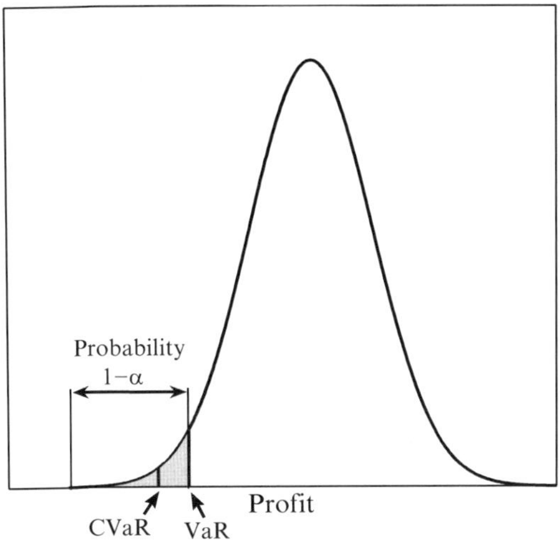

Fig. 21 Conditional value at risk (CVaR)

subject to

$$-\text{profit}(\omega) + \zeta - \eta(\omega) \leq 0 \quad \forall \omega \,, \tag{25}$$

$$\eta(\omega) \geq 0 \quad \forall \omega \,, \tag{26}$$

where ω is the scenario index, ζ is the VaR at the optimum, $p(\omega)$ is the probability of scenario ω, profit(ω) is the profit for scenario ω, and $\eta(\omega)$ is a variable whose value at the optimum is equal to zero if the scenario ω has a profit greater than VaR. For the rest of scenarios, $\eta(\omega)$ is equal to the difference of VaR and the corresponding profit. N_ω is the total number of considered scenarios.

To consider risk, the objective function (24) is added to the original objective function of the considered producer/retailer problem using the weighting factor β (Rockafellar and Uryasev 2000, 2002), while constraints (25)–(26) are incorporated as additional constraints to that problem. Note that constraints (25) couple together the variables related to risk and the variables pertaining to the producer/retailer problem through the "profit(ω)" term, which is the profit obtained by the producer/retailer in scenario ω.

References

Birge JR, Louveaux F (1997) Introduction to stochastic programming. Springer, New York

Carrión M, Conejo AJ, Arroyo JM (2007) Forward contracting and selling price determination for a retailer. IEEE Trans Power Syst 22:2105–2114

Castillo E, Conejo AJ, Pedregal P, García R, Alguacil N (2002) Building and solving mathematical programming models in engineering and science. Wiley, New York

Collins RA (2002) The economics of electricity hedging and a proposed modification for the futures contract for electricity. IEEE Trans Power Syst 17:100–107

Conejo AJ, Prieto FJ (2001) Mathematical programming and electricity markets. TOP 9:1–54

Conejo AJ, Castillo E, Mínguez R, García-Bertrand R (2006) Decomposition techniques in mathematical programming. Engineering and science applications. Springer, Heidelberg

Conejo AJ, Carrión M, García-Bertrand R (2007) Medium-term electricity trading strategies for producers, consumers and retailers. Int J Electron Bus Manage 5:239–252

Conejo AJ, García-Bertrand R, Carrión M, Caballero A, de Andrés A (2008) Optimal involvement in futures markets of a power producer. IEEE Trans Power Syst 23:703–711

The FICO XPRESS Website (2010). `http://www.fico.com/en/Products/DMTools/Pages/FICO-Xpress-Optimization-Suite.aspx`

Dupačova J, Consigli G, Wallace SW (2000) Scenarios for multistage stochastic programs. Ann Oper Res 100:25–53

Fleten S-E, Pettersen E (2005) Constructing bidding curves for a price-taking retailer in the Norwegian electricity market. IEEE Trans Power Syst 20:701–708

Gabriel SA, Genc MF, Balakrishnan S (2002) A simulation approach to balancing annual risk and reward in retail electrical power markets. IEEE Trans Power Syst 17:1050–1057

Gabriel SA, Conejo AJ, Plazas MA, Balakrishnan S (2006) Optimal price and quantity determination for retail electric power contracts. IEEE Trans Power Syst 21:180–187

Gedra TW (1994) Optional forward contracts for electric power markets. IEEE Trans Power Syst 9:1766–1773

Gröwe-Kuska N, Heitsch H, Römisch W (2003) Scenario reduction and scenario tree construction for power management problems. Proc IEEE Bologna Power Tech, Bologna, Italy

Higle JL, Sen S (1996) Stochastic decomposition. A statistical method for large scale stochastic linear programming. Kluwer, Dordrecht

Høyland K, Wallace SW (2001) Generating scenario trees for multistage decision problems. Manage Sci 47:295–307

Høyland K, Kaut M, Wallace SW (2003) A heuristic for moment-matching scenario generation. Comput Optim Appl 24:169–185

Ilic M, Galiana F, Fink L (1998) Power systems restructuring: engineering and economics. Kluwer, Boston

The ILOG CPLEX Website (2010). http://www.ilog.com/products/cplex/

Kaye RJ, Outhred HR, Bannister CH (1990) Forward contracts for the operation of an electricity industry under spot pricing. IEEE Trans Power Syst 5:46–52

Kirschen DS, Strbac G (2004) Fundamentals of power system economics. Wiley, Chichester

Niu N, Baldick R, Guidong Z (2005) Supply function equilibrium bidding strategies with fixed forward contracts. IEEE Trans Power Syst 20:1859–1867

Pflug GC (2001) Scenario tree generation for multiperiod financial optimization by optimal discretization. Math Program Series B 89:251–271

Rockafellar RT, Uryasev S (2000) Optimization of conditional value-at-risk. J Risk 2:21–41

Rockafellar RT, Uryasev S (2002) Conditional value-at-risk for general loss distributions. J Bank Financ 26:1443–1471

Rosenthal RE (2008) GAMS, a user's guide. GAMS Development Corporation, Washington

Shahidehpour M, Yamin H, Li Z (2002) Market operations in electric power systems: forecasting, scheduling and risk management. Wiley, New York

Sheblé GB (1999) Computational auction mechanisms for restructured power industry operation. Kluwer, Boston

Shrestha GB, Pokharel BK, Lie TT, Fleten S-E (2005) Medium term power planning with bilateral contracts. IEEE Trans Power Syst 20:627–633

Tanlapco E, Lawarree J, Liu CC (2002) Hedging with futures contracts in a deregulated electricity industry. IEEE Trans Power Syst 17:577–582

A Decision Support System for Generation Planning and Operation in Electricity Markets

Andres Ramos, Santiago Cerisola, and Jesus M. Latorre

Abstract This chapter presents a comprehensive decision support system for addressing the generation planning and operation. It is hierarchically divided into three planning horizons: long, medium, and short term. This functional hierarchy requires that decisions taken by the upper level model will be internalized by the lower level model. With this approach, the position of the company is globally optimized. This set of models presented is specially suited for hydrothermal systems. The models described correspond to long-term stochastic market planning, medium-term stochastic hydrothermal coordination, medium-term stochastic hydro simulation, and short-term unit commitment and bidding. In the chapter it is provided a condensed description of each model formulation and their main characteristics regarding modeling detail of each subsystem. The mathematical methods used by these models are mixed complementarity problem, multistage stochastic linear programming, Monte Carlo simulation, and multistage stochastic mixed integer programming. The algorithms used to solve them are Benders decomposition for mixed complementarity problems, stochastic dual dynamic programming, and Benders decomposition for SMIP problems.

Keywords Electricity competition · Market models · Planning tools · Power generation scheduling

1 Introduction

Since market deregulation was introduced in the electric industry, the generation companies have shifted from a cost minimization decision framework to a new one where the objective is the maximization of their expected profit (revenues minus costs). For this reason, electric companies manage their own generation resources

A. Ramos (✉)
Santiago Cerisola, and Jesus M. Latorre, Universidad Pontificia Comillas
Alberto Aguilera 23, 28015 Madrid, Spain
e-mail: andres.ramos@upcomillas.es

S. Rebennack et al. (eds.), *Handbook of Power Systems II*, Energy Systems,
DOI 10.1007/978-3-642-12686-4_11, © Springer-Verlag Berlin Heidelberg 2010

and need detailed operation planning tools. Nowadays, planning and operation of generation units rely upon mathematical models whose complexity depends on the detail of the model that needs to be solved.

These operation planning functions and decisions that companies address are complex and are usually split into *very-long-term*, *long-term*, *medium-term*, and *short-term* horizons (Wood and Wollenberg 1996). The very long-term horizon decisions are mainly investment decisions and their description and solution exceed the decision support system (DSS) that we are presenting in this chapter. The long-term horizon deals with risk management decisions, such as electricity contracts and fuel acquisition, and the level of risk that the company is willing to assume. The medium-term horizon decisions comprise those economic decisions such as market share or price targets and budget planning. Also operational planning decisions like fuel, storage hydro, and maintenance scheduling must be determined. In the short term, the final objective is to bid to the different markets, based on energy, power reserve, and other ancillary services, for the various clearing sessions. As a result, and from the system operation point of view, the company determines first the unit commitment (UC) and economic dispatch of its generating units, the water releases, and the storage and pumped-storage hydro system operation.

In hydrothermal systems and in particular in this DSS, special emphasis is paid to the representation of the hydro system operation and to the hydro scheduling for several reasons (Wood and Wollenberg 1996):

- Hydro plants constitute an energy source with very low variable costs and this is the main reason for using them. Operating costs of hydro plants are due to operation and maintenance and, in many models, they are neglected with respect to thermal units' variable costs.
- Hydro plants provide a greater regulation capability than other generating technologies, because they can quickly change their power output. Consequently, they are suitable to guarantee the system stability against contingencies.
- Electricity is difficult to store, even more when considering the amount needed in electric energy systems. However, hydro reservoirs and pumped-storage hydro plants give the possibility of accumulating energy, as a volume of water. Although in some cases the low efficiency of pumped-storage hydro units may be a disadvantage, usually this water storage increases the flexibility of the daily operation and guarantees the long-term electricity supply.

In this chapter we present a proposal for covering these three hierarchical horizons with some planning models, which constitute a DSS for planning and operating a company in an electricity market. The functional hierarchy requires that decisions taken by the long-term level will be internalized by the medium-term level and that decisions taken by the medium-term level will be internalized by the short-term level. With this approach, the position of the company is globally optimized. At an upper level, a stochastic market equilibrium model (Cabero et al. 2005) with monthly periods is run to determine the hydro basin production, as well as fuel and electricity contracts, while satisfying a certain risk level. At an intermediate level, a medium-term hydrothermal coordination problem obtains weekly water release tables for large reservoirs and weekly decisions for thermal units. At a lower level, a stochastic simulation model (Latorre et al. 2007a) with daily periods incorporates those water release tables and details each hydro unit power output. Finally, a detailed strategic UC and bidding model defines the commitment and bids to be offered to the energy market (Baillo et al. 2004; Cerisola et al. 2009). In Fig. 1 the hierarchy of these four models is represented. Different mathematical methods are used for modeling the electric system: mixed complementarity problem (MCP) (Cottle et al. 1992), multistage stochastic programming (Birge and Louveaux 1997), Monte Carlo simulation (Law and Kelton 2000), and stochastic mixed integer programming. With the purpose of solving realistic-sized problems, those models are combined with special purpose algorithms such as Benders decomposition and stochastic dual dynamic programming to achieve the solution of the proposed models.

For using the DSS in real applications, it is important to validate the consistency of the system results. As it can be observed from Fig. 1, several feedback loops are included to check the coherence of the DSS results. As the upper level models pass target productions to lower level models, a check is introduced to adjust the system results to those targets with some tolerance.

Although the DSS conception is general, it has been applied to the Spanish electric system, as can be seen in the references mentioned for each model. In the following Table 1 we present the summary of the demand balance and the installed capacity by technologies of the mainland Spanish system in 2008, taken from Red Electrica de Espana (2008).

2 Long-term Stochastic Market Planning Model

In a liberalized framework, market risk management is a relevant function for generating companies (GENCOs). In the long term they have to determine a risk management strategy in an oligopolistic environment. Risk is defined as the probability of a certain event times the impact of the event in the company's objective of an expected return. Some of the risks that the GENCOs face are operational risk, market risk, credit risk, liquidity risk, and regulatory risk. Market risk accounts for the risk that the value of an investment will decrease due to market factors' movements. It can be further divided into equity risk, interest rate risk, currency risk, and commodity risk. For a GENCO the commodity risk is mainly due to the volatility in electricity and fuel prices, in unregulated hydro inflows and in the demand level.

The purpose of the long-term model of the DSS is to represent the generation operation by a market equilibrium model based on a conjectural variation approach, which represents the implicit elasticity of the residual demand function. The model decides the total production for the considered periods (months) and the position in futures so as to achieve the acceptable risk for its profit distribution function. Stochasticity of random variables are represented by a scenario tree that is computed by clustering techniques (Latorre et al. 2007b). Traditionally, models that represent

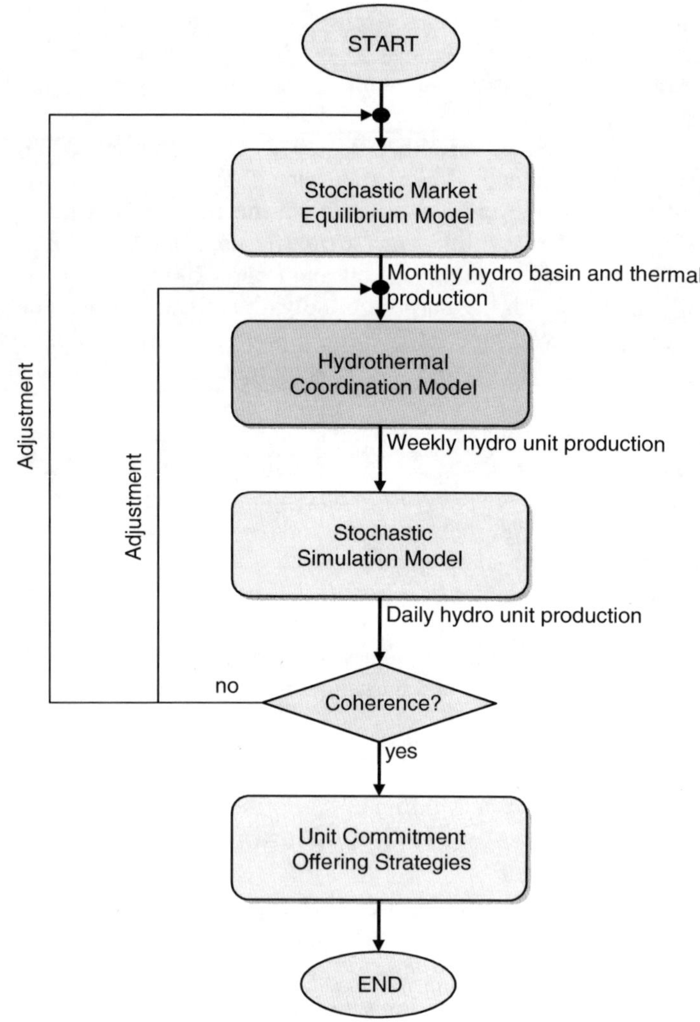

Fig. 1 Hierarchy of operation planning models

market equilibrium problems are based on linear or mixed complementarity problem (Hobbs et al. 2001a; Rivier et al. 2001), equivalent quadratic problem (Barquin et al. 2004), variational inequalities (Hobbs and Pang 2007), and equilibrium problem with equilibrium constraints (EPEC) (Yao et al. 2007). The formulation of this problem is based on mixed complementarity problem (MCP), that is, the combination of Karush–Kuhn–Tucker (KKT) optimality conditions and complementary slackness conditions, and extends those techniques that traditionally are used to represent the market equilibrium problem to a combined situation that simultaneously

Table 1 Demand balance and installed capacity of mainland Spanish electric system

	GWh	MW
Nuclear	58,756	7,716
Coal	46,346	11,359
Combined cycle	91,821	21,667
Oil/Gas	2,454	4,418
Hydro	21,175	16,657
Wind	31,102	15,576
Other renewable generation	35,434	12,552
Pumped storage hydro consumption	3,494	
International export	11,221	
Demand	263,961	

considers the market equilibrium and the risk management decisions, in a so-called integrated risk management approach.

2.1 Model Description

The market equilibrium model is stated as the profit maximization problem of each GENCO subject to the constraint that determines the electricity price as a function of the demand, which is the sum of all the power produced by the companies. Each company profit maximization problem includes all the operational constraints that the generating units must satisfy. The objective function is schematically represented by (Fig. 2).

In the long term the demand is represented by a load duration curve divided into peak, shoulder, and off-peak levels by period, the period being a month. For each load level the price p is a linear function of the demand d,

$$p = p_0 - p_0' \sum_c q_c \tag{1}$$

$$d = \sum_c q_c = q_c + q_c^- \tag{2}$$

p_0, p_0' being the intercept and slope of the inverse demand function, q_c the production of company c, and q_c^- the production of the remaining companies different from c. Then, the profit of each company becomes quadratic with respect to the quantities offered by companies. It accounts for those revenues that depend on the spot price and those that depend on long-term electricity contracts or take-or-pay fuel contracts. Schematically it can be stated as

$$\max \sum_\omega pr^\omega \sum_{t \in c} \left[p^\omega q_t^\omega - c^\omega (q_t^\omega) \right] \tag{3}$$

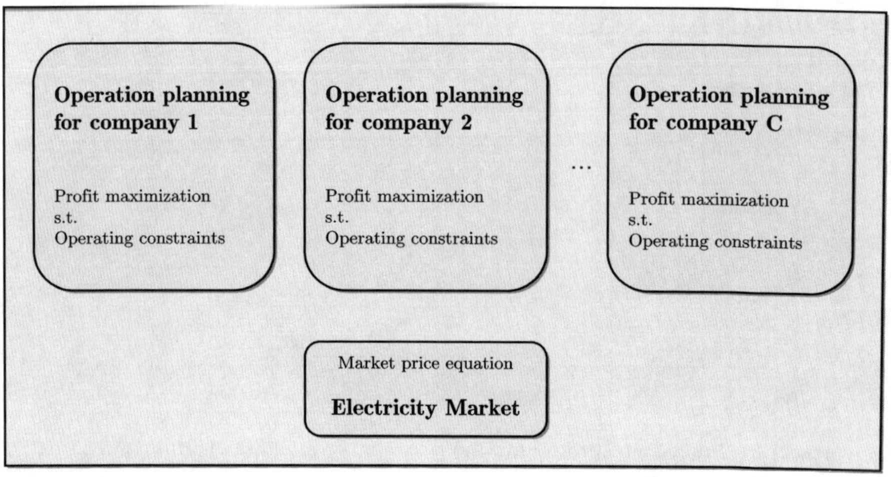

Fig. 2 Market equilibrium problem

ω being any scenario of the random variates, pr^{ω} the corresponding probability, p^{ω} the price, q_t^{ω} the energy produced by thermal unit t belonging to company c in scenario ω, and $c^{\omega}(q_t^{\omega})$ the thermal variable costs depending quadratically on the output of thermal units.

When considering the Cournot's approach, the decision variable for each company is its total output, while the output from competitors is considered constant. In the conjectural variation approach the reaction from competitors is included into the model by a function that defines the sensitivity of the electricity price with respect to the output of the company. This function may be different for each company:

$$\frac{\partial p}{\partial q_c} = -p_0' \left(1 + \frac{\partial q_{\bar{c}}}{\partial q_c}\right). \tag{4}$$

Operating constraints include fuel scheduling of the power plants, hydro reservoir management for storage and pumped-storage hydro plants, run-of-the-river hydro plants, and operation limits of all the generating units.

We incorporate in the model several sources of uncertainty that are relevant in the long term, such as water inflows, fuel prices, demand, electricity prices, and output of each company sold to the market. We do this by classifying historical data into a multivariate scenario tree. The introduction of uncertainty extends the model to a stochastic equilibrium problem and gives the company the possibility of finding a hedging strategy to manage its market risk. With this intention, we force currently future prices to coincide with the expected value of future spot prices that the equilibrium returns for each node of the scenario tree. Future's revenues are calculated as gain and losses of future contracts that are canceled at the difference between future and spot price at maturity. Transition costs are associated to contracts and computed when signed.

The risk measure used is the *Conditional Value at Risk* (CVaR), which computes the expected value of losses for all the scenarios in which the loss exceeds the *Value at Risk* (VaR) with a certain probability α, see (5).

$$CVaR_X(\alpha) = \mathbb{E}\left[X|X \geq VaR_X(\alpha)\right], \quad (5)$$

where X are the losses, $CVaR_X(\alpha)$ and $VaR_X(\alpha)$ are the CVaR and VaR of α quantile.

All these components set up the mathematical programming problem for each company, which maximizes the expected revenues from the spot and the futures market minus the expected thermal variable costs and minus the expected contract transaction costs. The operating constraints deal with fuel scheduling, hydro reservoir management, operating limits of the units for each scenario, while the financial constraints compute the CVaR for the company for the set of scenarios. Linking constraints for the optimization problems of the companies are the spot price equation and the relation of future price as the expectation of future spot prices.

The KKT optimality conditions of the profit maximization problem of each company together with the linear function for the price define a *mixed linear complementarity problem*. Thus the market equilibrium problem is created with the set of KKT conditions of each GENCO plus the price equation of the system, see Rivier et al. (2001). The problem is linear if the terms of the original profit maximization problem are quadratic and, therefore, the derivatives of the KKT conditions become linear.

The results of this model are the output of each production technology for each period and each scenario, the market share of each company, and the resulting electricity spot price for each load level in each period and each scenario. Monthly hydro system and thermal plant production are the magnitudes passed to the medium-term hydrothermal coordination model, explained below.

3 Medium-term Stochastic Hydrothermal Coordination Model

By nature, the medium-term stochastic hydrothermal coordination models are high-dimensional, dynamic, nonlinear, stochastic, and multiobjective. Solving these models is still a challenging task for large-scale systems (Labadie 2004). One key question for them is to obtain a feasible operation for each hydro plant, which is very difficult because models require a huge amount of data, due to complexity of hydro systems and by the need to evaluate multiple hydrological scenarios. A recent review of the state of the art of hydro scheduling models is done in Labadie (2004).

According to the treatment of stochasticity hydrothermal coordination models are classified into deterministic and stochastic ones.

- *Deterministic models* are based on network flows, linear programming (LP), nonlinear programming (NLP), or mixed integer linear programming (MILP), where binary variables come from commitment decisions of thermal or hydro units or from piecewise linear approximation of nonlinear and nonconvex water head effects. For taking into account these nonlinear effects, successive LP solutions are often used. This process does not converge necessarily to the optimal solution, see Bazaraa et al. (1993).

- *Stochastic models* are represented by stochastic dynamic programming (SDP), stochastic linear programming (SLP) (Seifi and Hipel 2001), and stochastic nonlinear programming (SNLP). For SLP problems decomposition techniques like Benders (Jacobs et al. 1995), Lagrangian relaxation, or stochastic dual dynamic programming (SDDP) (Pereira and Pinto 1991) can be used.

In this medium-term model, the aggregation of all the hydro plants of the same basin in an equivalent hydro unit (as done for the long-term model) is no longer kept. We deal with hydro plants and reservoir represented individually, as well as we include a cascade representation of their physical connections. Besides, thermal power units are considered individually. Thus, rich marginal cost information is used for guiding hydro scheduling.

The hydrothermal model determines the optimal yearly operation of all the thermal and hydro power plants for a complete year divided into decision periods of 1 week. The objective function is usually based on cost minimization because the main goal is the medium-term hydro operation, and the hydro releases have been determined by the upper level market equilibrium model. Nevertheless, the objective function can be easily modified to consider profit maximization if marginal prices are known (Stein-Erik and Trine Krogh 2008), which is a common assumption for fringe companies.

This model has a 1 year long scope beginning in October and ending in September, which models a hydraulic year, with special emphasis in large reservoirs (usually with annual or even hyperannual management capability). Final reserve levels for large reservoirs are given to the model to avoid the initial and terminal effects on the planning horizon. Uncertainty is introduced in natural inflows and the model is used for obtaining optimal and "feasible" water release tables for different stochastic inflows and reservoir volumes (Fig. 3).

The demand is modeled in a weekly basis with constant load levels (peak and off-peak hours, e.g.). Thermal units are treated individually and commitment decisions are considered as continuous variables, given that the model is used for medium-term analysis. For hydro reservoirs a different modeling approach is followed depending on the following:

- *Owner company*: Own reservoirs are modeled in water units (volume in hm^3 and inflow in $m^3\,s^{-1}$) while reservoirs belonging to other companies are modeled in energy units as equivalent and independent power plants with one reservoir each, given that the reservoir characteristics of the competitors are generally ignored.
- *Relevance of the reservoir*: Important large reservoirs are modeled with nonlinear water head effects while smaller reservoirs are represented with a linear dependency; therefore, the model does not become complex unnecessarily.

Unregulated hydro inflows are assumed to be the dominant source of uncertainty in a hydrothermal electric system. Temporal changes in reservoir reserves

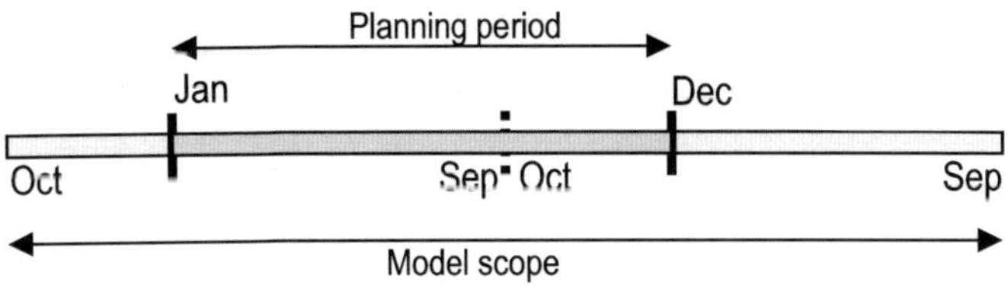

Fig. 3 Model scope for yearly operation planning

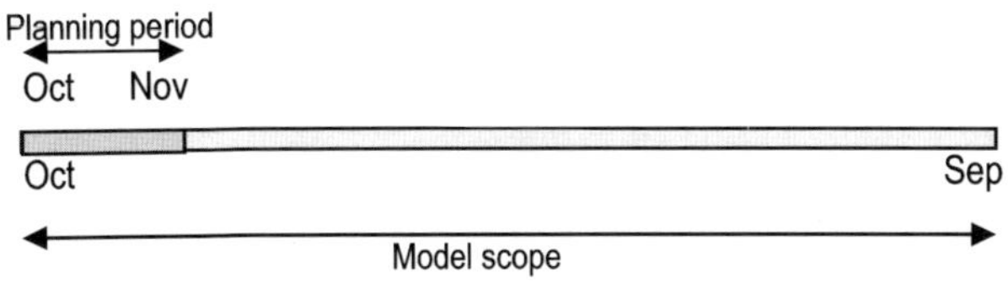

Fig. 4 Model scope for next future decisions under uncertain inflows

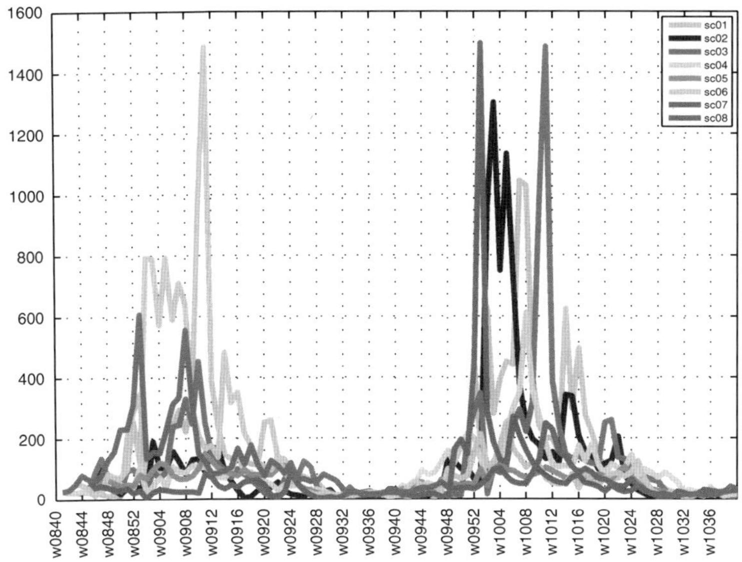

Fig. 5 Scenario tree with eight hydro inflows' scenarios from week 40 of year 2008 to week 39 of year 2010 expressed in ($m^3\,s^{-1}$)

are significant because of stochasticity in hydro inflows, highly seasonal pattern of inflows, and capacity of each reservoir with respect to its own inflow (Fig. 4).

Stochasticity in hydro inflows is represented for the optimization problem by means of a multivariate scenario tree, see Fig. 5 as a real case corresponding to a specific location. This tree is generated by a neural gas clustering technique (Latorre et al. 2007b) that simultaneously takes into account the main stochastic inflow series and their spatial and temporal dependencies. The algorithm can take historical or synthetic series of hydro inflows as input data. Very extreme scenarios can be artificially introduced with a very low probability. The number of scenarios generated is enough for medium-term hydrothermal operation planning.

3.1 Model Description

The *constraints* introduced into the model are the following:

- Balance between generation and demand including pumping: Generation of thermal and storage hydro units minus consumption of pumped-storage hydro units is equal to the demand for each scenario, period (week), and subperiod (load level).
- Minimum and maximum yearly operating hours for each thermal unit for each scenario: These constraints are relaxed by introducing slack and surplus variables that are penalized in the objective function. Those variables can be strictly necessary in the case of many scenarios of stochasticity. This type of constraints is introduced to account for some aspects that are not explicitly modeled into this model like unavailability of thermal units, domestic coal subsidies, CO_2 emission allowances, capacity payments, etc.
- Minimum and maximum yearly operating hours for each thermal unit for the set of scenarios.
- Monthly production by thermal technology and hydro basin: These constraints establish the long-term objectives to achieve by this medium-term model.
- Water inventory balance for large reservoirs modeled in water units: Reservoir volume at the beginning of the period plus unregulated inflows plus spillage from upstream reservoirs minus spillage from this reservoir plus turbined water from upstream storage hydro plants plus pumped water from downstream pumped-storage hydro plants minus turbined and pumped water from this reservoir is equal to reservoir volume at the end of the period. An artificial inflow is allowed and penalized in the objective function. Hydro plant takes water from a reservoir and releases it to another reservoir. The initial value of reservoir volume is assumed known. No lags are considered in water releases because 1 week is the time period unit.
- Energy inventory balance for reservoirs modeled in energy: Reservoir volume at the beginning of the period plus unregulated inflows minus spillage from this reservoir minus turbined water from this reservoir is equal to reservoir volume at the end of the period. An artificial inflow is allowed and penalized in the objective function. The initial value of reservoir volume is assumed known.
- Hydro plant generation is the product of the water release and the production function variable (also called efficiency): This is a nonlinear nonconvex constraint that considers the long-term effects of reservoir management.
- Total reservoir release is equal to the sum of reservoir releases from each downstream hydro plant.

- Pumping from a reservoir: Pumped water is equal to the pumped-storage hydro plant consumption divided by the production function.
- Achievement of a given final reservoir volume with slack and surplus variables: This final reserve is determined by the upper level long-term stochastic market equilibrium model of the DSS. The reserve levels at the end of each month of the problem are also forced to coincide with those levels proposed by the stochastic market equilibrium model.
- Minimum and maximum reservoir volume per period with slack and surplus variables: Those bounds are included to consider flood control curve, dead storage, and other plan operation concerns. The slack variables will be strictly necessary in the case of many scenarios.
- Computation of the plant water head and the production function variable as a linear function: Production function variable is a linear function of the water head of the plant that is determined as the forebay height of the reservoir minus the tailrace height of the plant. Tailrace height of the plant is the maximum of the forebay height of downstream reservoir and the tailrace height of the plant.
- Computation of the reservoir water head and the reservoir volume as a nonlinear function: Reservoir water head is determined as the forebay height minus the reference height. Reserve volume is a quadratic function of the reservoir water head.
- Variable bounds, that is, reservoir volumes between limits for each hydro reservoir and power operation between limits for each unit.

The *multiobjective function* minimizes:

- Thermal variable costs plus

$$\min \sum_{t\omega} pr^{\omega} c_t q_t^{\omega} \tag{6}$$

  q_t^{ω} the energy produced by thermal unit t in scenario ω and c_t the variable cost of the unit.
- Penalty terms for deviations from the proposed equilibrium model reservoir levels, that is, slack or surplus of final reservoir volumes, exceeding minimum and maximum operational rule curves, artificial inflows, etc.
- Penalty terms for relaxing constraints like minimum and maximum yearly operation hours of thermal units.

It is important for this model to obtain not only optimal solutions but also feasible solutions that can be implemented. Different solutions and trade-offs can be obtained by changing these penalties.

The main results for each load level of each period and scenario are storage hydro, pumped-storage hydro and thermal plant operation, reservoir management, basin and river production, and marginal costs. As a byproduct the optimal water release tables for different stochastic inflows and reservoir volumes are obtained. They are computed by stochastic nested Benders' decomposition technique (Birge and Louveaux 1997) of a linear approximation of the stochastic nonlinear optimization problem. These release tables are also used by the lower level daily stochastic simulation model, as explained in the next section.

4 Medium-term Stochastic Simulation Model

Simulation is the most suitable technique when the main objective is to analyze complex management strategies of hydro plants and reservoirs and their stochastic behavior. Simulation of hydrothermal systems has been used in the past for two main purposes:

- *Reliability analysis of electric power systems.* An example of this is given in Roman and Allan (1994), where a complete hydrothermal system is simulated. The merit order among all the reservoirs to supply the demand is determined as a function of their reserve level. Simulated natural hydro inflows and transmission network are considered. The goal is to determine the service reliability in thermal, hydro, or hydrothermal systems.
- *Hydrothermal operation.* In De Cuadra (1998) a simulation scheme for hydrothermal systems is proposed, where medium and long-term goals (maintenance, yearly hydro scheduling) are established. The system is simulated with stochastic demand and hydro inflows. For each day an optimization problem is solved to achieve the goals obtained from long-term models.

The hydro simulation model presented in this section takes into account the detailed topology of each basin and the stochasticity in hydro inflows, see Latorre et al. (2007a). It is directly related to the previous medium-term hydrothermal model, based on optimization. The stochastic optimization model guides the simulation model through a collection of hydro weekly production objectives that should be attained in different weeks for the largest hydro reservoirs. Once these guidelines are provided, the simulation model checks the feasibility of these goals, may test the simplifications made by the optimization model, and determines the energy output of hydro plants, the reserve evolution of the reservoirs and, therefore, a much more detailed daily operation. This double hierarchical relation among different planning models to determine the detailed hydro plants operation has also been found in Turgeon and Charbonneau (1998). It is a dynamic model, whose input data are series of historical inflows in certain basins' spots. Historical or synthetic series (obtained by forecasting methods) can be used for simulation. For this reason, it is also a stochastic model. Finally, system state changes take place once a day. These events can be changes of inflows, scheduled outages, etc. Consequently, the model is discrete with 1 day as time step. This is a reasonable time step because the usual model scope is 1 year and no hourly information is needed. The simulation model deals with plausible inflow scenarios and generates statistics of the hydro operation. Its main applications are the following:

- Comparison of several different reservoir management strategies
- Anticipation of the impact of hydro plant unavailabilities for preventing and diminishing the influence of floods
- Increment of hydro production by reducing the spillage

The model is based on the object-oriented paradigm and defines five classes that are able to represent any element of a hydro basin. The object oriented programming (OOP) paradigm (Wirfs-Brock et al. 1990) becomes very attractive for simulation because it allows to encapsulate the basic behavior of the elements and permits the independent simulation of each system element, which simply needs to gather information from incoming water flows from the closest elements to it. The model incorporates a simulation algorithm in three phases that decides the production of the hydro plants following several strategies about reservoir management. These management strategies and a description of the five classes are presented next.

4.1 Data Representation

A natural way to represent the hydro basin topology is by means of a graph of nodes, each one symbolizing a basin element. Those nodes represent reservoirs, plants, inflow spots, and river junctions. Nodes are connected among them by arcs representing water streams (rivers, channels, etc). Each node is independently managed, although it may require information about the state of other upstream basin elements. As a result of this data structure, object-oriented programming is a suitable approach to solve the problem of the simulation of a hydro basin (Fig. 6).

Analyzing real hydro basin patterns, we have concluded that five classes are enough to represent adequately every possible case. These object types are described in the next section. Additionally, different reserve management strategies can be pursued in a reservoir element. These nodes represent reservoirs, channels, plants, inflows, and river junctions, which are now described.

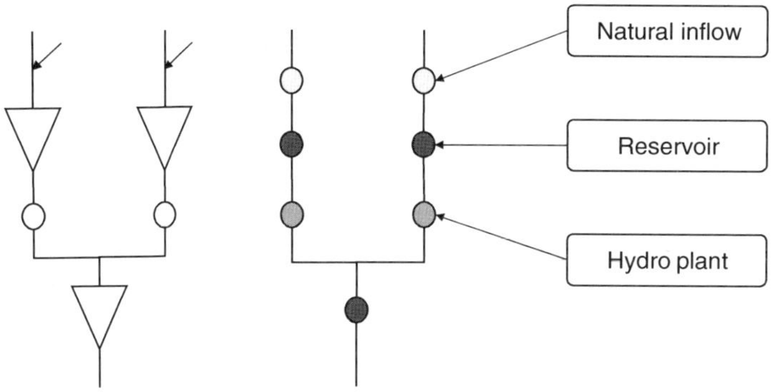

Fig. 6 Basin topology represented by a graph of nodes

4.1.1 Reservoirs

The objects representing reservoirs have one or more incoming water streams and only one outgoing. Apart from other technical limitations, they may have a minimum outflow release, regarding irrigation needs, sporting activities, or other environmental issues. Besides, they may have rule volume curves guiding their management. Examples are minimum and maximum rule curves, which avoid running out of water for irrigation or spillways risk.

Reservoirs are the key elements where water management is done. The chosen strategy decides its outflow, taking into account minimum and maximum guiding curves, absolute minimum and maximum volume levels, and water release tables. The different management strategies are described in Sect. 4.2.

4.1.2 Channels

These elements carry water between other basin elements, like real water streams do. They do not perform any water management: they just transport water from their origin to their end. However, they impose an upper limit to the transported water flow, which is the reason to consider them.

4.1.3 Plants

Kinetic energy from the water flow that goes through the turbine is transformed into electricity in the plant. In electric energy systems, hydro plants are important elements to consider due to their flexibility and low production costs. However, in this simulation model water management is decided by the reservoir located upstream. Hence, from this point of view they are managed in the same fashion as channels: they impose an upper limit to the transported flow.

As a simulation result, electric output is a function of the water flow through the plant. This conversion is done by a production function depending on the water head, which is approximated linearly. Water head is the height between the reservoir level and the maximum between the drain level and the level of the downstream element. In hydro plants, once the water flow has been decided, daily production is divided between peak and off-peak hours, trying to allocate as much energy as possible in peak hours where expensive thermal units are producing.

In addition, some plants may have pumped-storage hydro units, which may take water from downstream elements and store it in upstream elements (generally, both elements will be reservoirs). It is important to emphasize that, in this simulation model, pumping is not carried out with an economic criterion, as it does not consider the thermal units, but with the purpose of avoiding spillage.

4.1.4 Natural Inflows

These objects introduce water into the system. They represent streamflow records where water flow is measured, coming from precipitation, snowmelt, or tributaries. These elements have no other upstream elements. The outflow corresponds to the day been simulated in the series. These series may come from historic measures or from synthetic series obtained from forecasting models based on time series analysis.

4.1.5 River Junctions

This object groups other basin elements where several rivers meet. An upper bound limits the simultaneous flow of all the junction elements. An example of this object appears when two reservoirs drain to the same hydro plant. As both reservoirs share the penstock, this element has to coordinate both reservoirs' behavior.

4.2 Reservoir Management Strategies

Reservoir management is the main purpose of the simulation; the rest of the process is an automatic consequence of this. Different strategies represent all possible real alternatives to manage reservoirs with diverse characteristics. These strategies combine the implicit optimization of the upper level models with operational rules imposed by the river regulatory bodies to the electric companies. They are discussed in the following paragraphs.

4.2.1 Water Release Table

This strategy is used for large reservoirs that control the overall basin operation. Typically, those reservoirs are located at the basin head. The water release table is determined by the long-term optimization model and gives the optimal reservoir outflow as a multidimensional function of the week of the year of the simulated day, the inflows of the reservoir, the volume of the reservoir being simulated, and the volume of another reservoir of the same basin, if that exists, that may serve as a reference of the basin hydrological situation. The reservoir outflow is computed by performing a multidimensional interpolation among the corner values read from the table.

4.2.2 Production of the Incoming Inflow

This strategy is specially indicated for small reservoirs. Given that they do not have much manageable volume, they must behave as run-of-the-river plants and drain the incoming inflow.

4.2.3 Minimum Operational Level

In this strategy, the objective is to produce as much flow as possible. Of course, when the reservoir level is below this operational level no production is done. When the volume is above this minimum operational level, the maximum available flow must be turbined to produce electricity. This strategy can be suitable for medium-sized reservoirs when little energy is available in the basin.

4.2.4 Maximum Operational Level

With this strategy, the volume is guided to the maximum operational level curve. This is a curve that prevents flooding or spillage at the reservoirs. The reason behind this operation is that when the water head is larger, the production will be higher. However, in case of extreme heavy rain it can be dangerous to keep the reservoir at this level. This strategy can be suitable for medium-sized reservoirs when enough energy is available in the basin.

4.3 Simulation Method

Simulating a hydro basin allows to observe its evolution for different possible hydro inflows. Operation of hydro units follows the management goals of the reservoirs and the limitations of the other river basin objects. However, other factors can force changes in previous decisions, for example, avoiding superfluous spillage and assurance of minimum outflows. To achieve this purpose we propose a three-phase simulation method consisting in these phases:

1. Decide an initial reservoir management, considering each reservoir independently. It also computes the ability of each reservoir to modify its outflow without reaching too low or high water volumes.
2. Modify the previous management to avoid spillage or to supply irrigation and ecological needs. This uses the modification limits computed in the previous step.
3. Determine the hydro units' output with the final outflows decided in the previous step.

Results are obtained for each series, both in the form of detailed daily series and mean values, and mean and quantiles of the weekly values are also calculated. This permits general inspection of the results for each reservoir as well as a more thorough analysis of the daily evolution of each element of the river basin.

5 Short-term Unit Commitment and Bidding Strategies

The last step in the decision process is faced once the weekly production decisions for the thermal units and the daily hydro production are obtained with the medium-term stochastic optimization and simulation models, respectively. The optimal unit commitment schedule for the next day will comply with those decisions taken by upper level models of the DSS.

We assume an electricity market where agents have to indicate, by means of an offering curve, the amount of energy they are willing to sell for different prices for each of the 24 h of the next day. In the same manner, agents willing to purchase electricity may indicate the quantities they are willing to buy for different prices. An independent system operator (ISO) intersects both curves and sets the hourly price for the very next day. Those prices are denoted as market clearing prices. We focus our attention on a marginal pricing scheme where those offers with prices less than the market clearing price are accepted, while those whose price is higher than the market clearing price are rejected.

In this framework, companies are responsible of their offer and suffer the uncertainty of the disclosure of the market price, which is mainly induced by the uncertain behavior of the agents. Even more, if they have a large market share, their own offer may affect the final price. We model this uncertainty of the market clearing price by means of the residual demand function. The residual demand function is created for each company eliminating from the purchase offer curve the sell offer curves of the remaining agents. By doing this we obtain a function that relates the final price to the total amount of energy that the company may sell. Once this relation is available, the company may optimize its benefit determining the amount of energy that maximizes their profit, defined as the difference between the revenues earned in the market and the production costs.

The residual demand function so far commented is clearly unavailable before the market clearing process is done, and thus the company has to estimate it based on historical data (e.g., from the market operator). Nevertheless to say, for a relative small company that can be considered as a price taker, it is just enough to estimate the market prices for the next day.

With the purpose of deciding the committed units for the next day and with the intention of including the weekly thermal and hydro production decision taken by the DSS, we consider the problem of operating a diversified portfolio of generation units with a 1 week time horizon. This problem decides the hourly power output of each generation unit during the week of study, which implies choosing the generating units that must be operating at each hour. We introduce uncertainty by means of a weekly scenario tree of residual demand curves. The scenario tree branches at the beginning of each day and serial correlation is considered for the residual demand curves of the same day. This residual demand curves considered are neither convex nor concave, and we model them as well as the profit function by introducing binary variables.

So, the weekly unit commitment of the DSS is formulated as a large scale stochastic mixed integer programming problem. The problem decides the commitment schedule for the 7 days of the upcoming week, although just the solution of the very first day is typically the accepted solution. For the next day a new weekly unit commitment problem ought to be solved. For realistic large-scale problems a decomposition procedure may be used. The reader is referenced to Cerisola et al.

(2009), where a Benders-type algorithm for integer programming is applied for the resolution of a large-scale weekly unit commitment problem.

The objective function of the unit commitment problem maximizes the revenues of the company for the time scope:

$$\max \sum_h \Pi_h(q_h) = \sum_h p_h q_h = \sum_h R_h^{-1}(q_h) q_h, \tag{7}$$

where $\Pi_h(q_h)$ is the revenue of company in hour h, q_h the company output, p_h the spot price, $R_h(p_h)$ the residual demand faced by the company, and $R_h^{-1}(q_h)$ the inverse residual demand function.

In general, this revenue function is not concave but can be expressed as a piecewise-linear function by using auxiliary binary variables Cerisola et al. (2009).

We complete this section with a brief description of the *constraints* that constitute the mathematical problem of the unit commitment model:

- Market price is a function of total output and revenue is also a function of total output; both are modeled as piecewise linear equations using the δ-form as in Williams (1999).
- The production cost of each thermal unit is a linear function of its power output and its commitment state, which is modeled with a binary variable.
- The company profit is defined as the difference between the market revenue and the total operating cost.
- Maximum capacity and minimum output of thermal units are modeled together with the commitment state variable as usual. If the commitment state is off, the unit output will be zero.
- Ramp limits between hours are modeled linearly.
- Start-up and shut-down decisions are modeled as continuous variables, and their values decided by a dynamic relation between commitment states of consecutive hours.
- The power output for each hydro unit is decided by the higher level model of the DSS.
- The model forces the weekly thermal production of each plant to be equal to the decision of the medium-term hydrothermal coordination problem.

6 Conclusions

In this chapter we have presented a complete DSS that optimizes the decisions of a generation company by a hierarchy of models covering the long-term, medium-term, and short-term planning functions, see Fig. 7. The decisions taken from the highest model are passed to the lower level model and so on. These models are specially suited for representing a hydrothermal system with the complexities derived from

Table 2 Summary of characteristics of the models

	Long-term stochastic market planning	Medium-term stochastic hydrothermal coordination	Medium-term stochastic hydro simulation	Short-term unit commitment and bidding strategies
Scope	1 year	1 year	1 year	1 week
Time step	1 month	1 week	1 day	1 h
Stochastic variables	Fuel prices Hydro inflows aggregated monthly Demand	Hydro inflows aggregated weekly	Daily inflows	Residual demand curves
Stochastic representation	Monthly scenario tree	Weekly scenario tree	Series with given initial point	Daily scenario tree
Aggregation level	Hydro system aggregated by basins Thermal system aggregated by technologies	Hydro and thermal systems represented by units	Hydro units represented separately	Hydro units aggregated by offering unit Thermal units represented separately
Input data from higher level model	Risk management level (CVaR)	Hydro production for each month and basin Thermal production for each month and technology	Weekly water release tables for large reservoirs	Weekly production for thermal units Daily production for hydro units
Economic and operational decisions	Long-term fuel and electricity contracts Hydro production for each month and basin Thermal production for each month and technology	Budget planning Weekly water release tables for large reservoirs Weekly production for thermal plants	Daily production for hydro units	Bidding strategy for the next 24 h
Mathematical method	Mixed complementarity problem	Multistage stochastic linear programming	Monte Carlo simulation	Multistage stochastic mixed integer programming
Algorithm	Benders decomposition for mixed complementarity problems	Stochastic dual dynamic programming		Benders decomposition for MIP problems

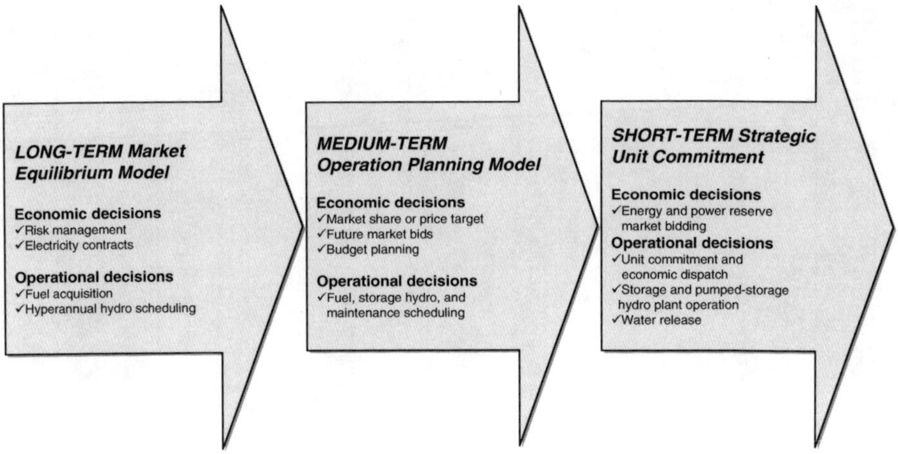

Fig. 7 Hierarchy of operation planning functions

hydro topology and stochastic hydro inflows that are conveniently incorporated into the models.

Table 2 summarizes the main characteristics of the models that are solved hierarchically passing the operation decisions to achieve the optimality of the planning process.

References

Baillo A, Ventosa M, Rivier M, Ramos A (2004) Optimal offering strategies for generation companies operating in electricity spot markets. IEEE Trans Power Syst 19(2):745–753

Barquin J, Centeno E, Reneses J (2004) Medium-term generation programming in competitive environments: a new optimisation approach for market equilibrium computing. IEE Proc Generat Transm Distrib 151(1):119–126

Bazaraa MS, Sherali HD, Shetty CM (1993) Nonlinear programming. Theory and algorithms, 2nd edn. Wiley, New York

Birge JR, Louveaux F (1997) Introduction to stochastic programming. Springer, New York

Cabero J, Baillo A, Cerisola S, Ventosa M, Garca-Alcalde A, Peran F, Relano G (2005) A medium-term integrated risk management model for a hydrothermal generation company. IEEE Trans Power Syst 20(3):1379–1388

Cerisola S, Baillo A, Fernandez-Lopez JM, Ramos A, Gollmer R (2009) Stochastic power generation unit commitment in electricity markets: A novel formulation and a comparison of solution methods. Oper Res 57(1):32–46

Cottle RW, Pang J-S, Stone RE (1992) The linear complementarity problem. Kluwer, Dordrecht

De Cuadra G (1998) Operation models for optimization and stochastic simulation of electric energy systems (original in Spanish), PhD Thesis, Universidad Pontificia Comillas

Garcia-Gonzalez J, Roman J, Barquin J, Gonzalez A (1999) Strategic bidding in deregulated power systems. 13th Power Systems Computation Conference, Trondheim, Norway

Hobbs BF (2001) Linear complementarity models of nash-cournot competition in bilateral and POOLCO power markets. IEEE Trans Power Syst 16(2):194–202

Hobbs BF, Pang JS (2007) Nash-cournot equilibria in electric power markets with piecewise linear demand functions and joint constraints. Oper Res 55(1):113–127

Hobbs BF, Rothkopf MH, O'Neill RP, Hung-po C (2001) The next generation of electric power unit commitment models. Kluwer, Boston

Jacobs J, Freeman G, Grygier J, Morton D, Schultz G, Staschus K, Stedinger J (1995) Socrates – A system for scheduling hydroelectric generation under uncertainty, Ann Oper Res 59:99–133

Labadie JW (2004) Optimal operation of multireservoir systems: State-of-the-art review, J Water Resour Plann Manag 130:93–111

Latorre JM, Cerisola S, Ramos A, Bellido R, Perea A (2007a) Creation of hydroelectric system scheduling by simulation. In: Qudrat-Ullah H, Spector JM, Davidsen P (eds) Complex decision making: Theory and practice. Springer, Heidelberg, pp. 83–96

Latorre JM, Cerisola S, Ramos A (2007b) Clustering algorithms for scenario tree generation: Application to natural hydro inflows. Eur J Oper Res 181(3):1339–1353

Law AM, Kelton WD (2000) Simulation modeling and analysis, 3rd edn. McGraw-Hill, OH

Padhy NP (2003) Unit commitment problem under deregulated environment – A review. IEEE Power Engineering Society General Meeting 1–4, Conference Proceedings: 1088–1094

Pereira MVF, Pinto LMVG (1991) Multistage stochastic optimization applied to energy planning. Math Program 52:359–375

Red Electrica de Espana (2008) The Spanish Electricity System. Preliminary Report. http://www.ree.es

Rivier M, Ventosa M, Ramos A, Martinez-Corcoles F, Chiarri A (2001) A generation operation planning model in deregulated electricity markets based on the complementarity problem. In: Ferris MC, Mangasarian OL, Pang J-S (eds) Complementarity: Applications, algorithms and extensions. Kluwer, Dordrecht, pp. 273–295

Roman J, Allan RN (1994) Reliability assessment of hydro-thermal composite systems by means of stochastic simulation techniques. Reliab Eng Syst Saf 46(1):33–47

Seifi A, Hipel KW (2001) Interior-point method for reservoir operation with stochastic inflows. J Water Resour Plann Manag 127:48–57

Stein-Erik F, Trine Krogh K (2008) Short-term hydropower production planning by stochastic programming. Comput Oper Res 35:2656–2671

Turgeon A, Charbonneau R (1998) An aggregation-disaggregation approach to long-term reservoir management. Water Resour Res 34(12):3585–3594

Williams HP (1999) Model building in mathematical programming, 4th edn. Wiley, New York

Wirfs-Brock R, Wilkerson B, Wiener L (1990) Designing object-oriented software. Prentice Hall, NJ

Wood AJ, Wollenberg BF (1996) Power generation, operation, and control, 2nd edn. Wiley, New York

Yao J, Oren SS, Adler I (2007) Two-settlement electricity markets with price caps and cournot generation firms. Eur J Oper Res 181(3):1279–1296

A Partitioning Method that Generates Interpretable Prices for Integer Programming Problems

Mette Bjørndal and Kurt Jörnsten

Abstract Benders' partitioning method is a classical method for solving mixed integer programming problems. The basic idea is to partition the problem by dividing the variables into complicating variables, normally the variables that are constrained to be integer valued, and easy variables, normally the continuous variables. By fixing the complicating variables, a convex sub-problem is generated. Solving this convex sub-problem and its dual generates cutting planes that are used to create a master problem, which when solved generates new values for the complicating variables, which have a potential to give a better solution. In this work, we assume that the optimal solution is given, and we present a way in which the partitioning idea can be used to generate a valid inequality that supports the optimal solution. By adding some of the continuous variables to the complicating variables, we generate a valid inequality, which is a supporting hyperplane to the convex hull of the mixed integer program. The relaxed original programming problem with this supporting valid inequality added will produce interpretable prices for the original mixed integer programming problem. The method developed can be used to generate economically interpretable prices for markets with non-convexities. This is an important issue in many of the deregulated electricity markets in which the non-convexities comes from large start up costs or block bids. The generated prices are, in the case when the sub-problem generated by fixing the integer variables to their optimal values has the integrality property, also supported by nonlinear price functions that are the basis for integer programming duality.

Keywords Bender's partitioning method · Indivisibilities · Integer programming · Non-convexities · Pricing

---

M. Bjørndal (✉)
Department of Finance and Management Science, Norwegian School of Economics and Business Administration (NHH), Helleveien 30, N-5045 Bergen, Norway
e-mail: mette.bjorndal@nhh.no

S. Rebennack et al. (eds.), *Handbook of Power Systems II*, Energy Systems,
DOI 10.1007/978-3-642-12686-4_12,

1 Introduction

One of the main reasons for the success of optimization formulations in economics is the existence of shadow prices, dual variables, which can be given economically meaningful interpretations. The dual variables are used to analyze the economic model under study, and to find out if a certain solution is optimal or can be further improved. The dual variables are also used in comparative static analyses, etc. As many economic situations involve non-convexities in the form of economies of scale, start up costs and indivisibilities, there have been several attempts to generate interpretable prices also for non-convex optimization problems. In this paper, we suggest a new way to generate prices for non-convex optimization models. It should be noted that the prices suggested are not intended as a means for finding the optimal solutions of the production problems, but rather as a means for communicating the optimality of the equilibrium to the various market participants. An excellent article on the relations between microeconomics and mathematical programming has been written by Scarf, Scarf (1990).

In the article "The Allocation of Resources in the Presence of Indivisibilities", Scarf (1994) states that, "The major problem presented to economic theory in the presence of indivisibilities in production is the impossibility of detecting optimality at the level of the firm, or for the economy as a whole, using the criterion of profitability based on competitive prices." Scarf illustrates the importance of the existence of competitive prices and their use in the economic evaluation of alternatives by considering a hypothetical economic situation which is in equilibrium in the purest Walrasian sense. It is assumed that the production possibility set exhibits constant returns to scale so that there is a profit of zero at the equilibrium prices. Each consumer evaluates personal income at these prices, and market demand functions are obtained by the aggregation of individual utility maximizing demands. As the system is assumed to be in equilibrium, the market is clearing, and thus, supply equals demand for each of the goods and services in the economy. Because of technological advance, a new manufacturing technology has been presented. This activity also possesses constant returns to scale. The question is if this new activity is to be used at a positive level or not? In this setting the answer to the question is simple and straightforward. If the new activity is profitable at the old equilibrium prices, then there is a way to use the activity at a positive level, so that with suitable income redistributions, the welfare of every member of society will increase. Also, if the new activity makes a negative profit at the old equilibrium prices, then there is no way in which it can be used to improve the utility of all consumers, even allowing the most extraordinary scheme for income redistribution. This shows the strength of the pricing test in evaluating alternatives. However, the existence of the pricing test relies on the assumptions made, that is the production possibility set displays constant or decreasing returns to scale. When we have increasing returns to scale, the pricing test for optimality might fail. It is easy to construct examples showing this, for instance by introducing activities with start up costs. In the failure of a pricing test, Scarf introduces as an alternative a quantity test for optimality. Although elegant and based on a theory that has lead to the development of algorithms for parametric integer programming problems, we find it hard to believe that the quantity test suggested by Scarf will have the same impact in economic theory as the pricing test has had in the case where non-convexities do not exist.

Over time, a number of suggestions have been made to address the problem of finding prices for problems with non-convexities, especially for the case where the non-convexities are modelled using discrete variables. The aim has been to find dual prices and interpretations of dual prices in integer programming problems and mixed integer programming problems. The seminal work of Gomory and Baumol (1960) addresses this issue. The ideas in the Gomory and Baumol's work were later used to create a duality theory for integer programming problems, and Wolsey (1981) gives a good description of this theory, showing that, in the pure integer programming case, we need to expand our view of prices to price functions to achieve interpretable and computable duals. However, these dual price functions are rather complex, and other researchers have suggested approximate alternatives. Alternative prices and price interpretations for integer programming problems have been proposed by Alcaly and Klevorik (1966) and Williams (1996). None of these suggestions have so far, to our knowledge, been used successfully to analyze equilibrium prices in markets with non-convexities.

More recently, O'Neill et al. (2005) presented a method for calculating discriminatory prices, IP-prices, based on a reformulation of the non-convex optimization problem. The method is aimed at generating market clearing prices and assumes that the optimal production plan is known. The reason for this renewed interest in interpretable prices in markets with non-convexities is that most deregulated electricity markets have inherent non-convexities in the form of start-up costs or the allowance of block bidding. The papers by Gribik et al. (2007), Bjørndal and Jørnsten (2004, 2008) and Muratore (2008) are all focusing on interpretable prices in electricity markets.

In this paper, we show that a related idea, which we call modified IP-prices, has the properties that we are seeking. The modified IP-prices are in fact equivalent to the coefficients that are generated from the Benders' sub-problem when the complicating variables are held fixed at their optimal values. These prices are derived using a minor modification of the idea of O'Neill et al., and are based on the generation of a valid inequality that supports the optimal solution and can be viewed as a mixed integer programming version of the separating hyperplane that supports the linear price system in the convex case. The modified IP-prices generated are, for the pure integer case, based on integer programming duality theory, and it can be shown that in this case, there exist a non-linear non-discriminatory price function that supports the modified IP-prices. In the mixed integer programming case, further research is needed to find the corresponding non-discriminatory price function that supports the modified IP-prices. However, the modified IP- prices yield a non-linear price structure consisting of linear commodity prices and a fixed uplift fee.

2 Benders' Partitioning Method

Benders' decomposition, or Benders' partitioning method Benders (1962), was developed in the 1960s as a computational method to calculate an optimal solution for mathematical programming problems in which the variables can be partitioned into complicating and non-complicating variables. In its original version, the method was developed as a means of solving linear mixed integer programming problems. However, the method has been used also to generate computational methods for general mathematical programming problems, where some of the variables make the optimization problem non-convex, whereas the problem is a standard convex optimization problem when these complicating variables are held fixed. The theoretical basis of the method relies on duality theory for convex optimization problems. For the complicating variables held fixed, the remaining problem, Benders' subproblem, is a convex optimization problem. Solving this problem and its dual gives us information in the form of cutting planes that involve a constant and a functional expression containing the complicating variables only.

We present the Benders' decomposition method for a problem of the form

$$\begin{aligned} & Min\ c^T x + f^T y \\ \text{s.t.}\ \ & Ax + Fy \geq b \\ & x \geq 0 \\ & y \in S = \{y \,|Cy \geq d,\ y \geq 0\ and\ y \in Z^+\} \end{aligned}$$

where Z^+ is the set of non-negative integers. Let R be the set of feasible variable values for y, that is

$$\mathrm{R} = \{y \,|\, \text{there exists}\, x \geq 0 \,\text{such that}\, Ax \geq b - Fy\ ,\ y \in S\}$$

By the use of Farkas' lemma, the set R can be rewritten as

$$\mathrm{R} = \{y\, |(b - Fy\,)^T u_i^r \leq 0,\, i = 1, \ldots\ldots, n_r,\, y \in S\}$$

where n_r denotes the number of extreme rays in the cone $\mathrm{C} = \left\{u|A^T\ u \leq 0, u \geq 0\right\}$.

It is clear that if the set R is empty, then the original problem is infeasible. Assume now that we choose a solution $y \in R$. We can now rewrite the original problem into

$$Min_{y \in R}\ f(y) + Min \left\{c^T\ x\,|Ax \geq b - Fy,\ x \geq 0\right\}$$

Using linear programming duality theory, the inner minimization problem can be rewritten as

$$Max\{(b - Fy)^T \,\Big|\, A^T u \leq c,\ u \geq 0\}$$

The polyhedron P $= \{u|A^T u \geq c, u \geq 0\}$ is independent of y and is according to the assumption that $y \in R$, nonempty. The dual optimum is then reached at an extreme point of P. Let u_i^p denote the extreme points of P, where i $= 1, \ldots, n_p$, and n_p denotes the number of extreme points of P.

Using this fact, the original optimization problem can be rewritten as

$$Min_{y \in R} \{f^T y + Max_{1 \leq i \leq n_p} (b - Fy)^T u_i^p\}$$

The solution of the complete master problem gives us the optimal solution and the corresponding optimal objective function value. Benders' partitioning method, when used to generate the optimal solution to the original problem, is an iterative solution method that starts off with a suggested solution for the complicating variables, evaluates this solution and uses the information from the sub-problem and its dual to construct a relaxation of the full master problem which, when it is solved, gives a new suggestion for the complicating variables, which have the potential to give a better objective function value than the current best.

3 Creating Interpretable Equilibrium Prices from Partitioning

O'Neill et al. (2005) presented a novel approach to generate interpretable equilibrium prices in markets with non-convexities. The basic idea is that, if we know the optimal solution, we can rewrite the original optimization problem by adding a set of constraints that stipulate that the integer variables should be held at their optimal values. This modified problem, which is convex, can then be solved, and the dual variables are interpretable as equilibrium prices. The dual variables for the demand constraints give the commodity prices, and the dual variables associated with the added set of constraints yield the necessary uplift fee.

O'Neill et al.'s modified optimization problem is

$$\begin{aligned} & Min\ c^T x + f^T y \\ \text{s.t.}\quad & Ax + Fy \geq b \\ & x \geq 0 \\ & y = y^* \\ & y \in S \end{aligned}$$

where y^* consist of the optimal values for the complicating variables. Let π be the dual variables to the appended equality constraints. It is easily seen that the dual variables associated with the added equality constraints are in fact the dual variables generated from the Benders' sub-problem.

However the inequality $\pi^T y \geq \pi^T y^*$ is not necessarily a valid inequality. This is not surprising, since the Benders' objective function cut, derived from the suggested solution y^*, yields a direction for potential improvement, and was not intended to be used as a cutting plane inequality. In our case, however, we know that y^* is the optimal solution, and that no better solution exists. Hence, we derive a valid inequality that supports the optimal solution, and when added to the original problem, gives us the optimal objective function value when the original problem is solved as a linear programming problem. Since the valid inequality $\pi^T y \geq \pi^T y^*$ does not necessarily possess this property, the prices and uplift charges generated by this procedure can be very volatile to small changes in demand requirements. This is illustrated in the examples in Sect. 5.

4 Modified IP-prices

The modified IP-prices are based on the connection between the IP-prices and the dual information generated when Benders' decomposition is used to solve the resource allocation problem. Benders' decomposition is explained above as a computational method, which is often used to solve models in which a certain set of variables, the complicating variables, are fixed at certain values, and the remaining problem and its dual is solved to get bounds for the optimal value of the problem, and also generates information on how to fix the complicating variables to obtain a new solution with a potentially better objective function value. When Benders' decomposition is used to solve a nonlinear integer programming problem as the problem studied in this paper, the problem is partitioned into solving a sequence of "easy" convex optimization problems, where the complicating integer variables are held fixed, and their duals. These problems, often called the Benders' sub-problems, are generating lower bounds on the optimal objective function value and yield information that is added to the Benders' master problem, a problem involving only the complicating variables. The information generated takes the form of cutting planes.

Here, we are not interested in using Benders' decomposition to solve the optimization problem, as we assume that we already know the optimal solution. However, viewing the reformulation used by O'Neill et al. as solving a Benders' sub-problem, in which the complicating integer variables are held fixed to their optimal values, reveals useful information concerning the IP-prices generated in the reformulation.

In linear and convex programming, the existence of a linear price vector is based on the use of the separating hyperplane Theorem. Based on the convexity assumption, the equilibrium prices are the dual variables or Lagrange multipliers for the market clearing constraints. In the non-convex case, it is well known that not every efficient output can be achieved by simple centralized pricing decisions or by linear competitive market prices. However, there exists an equivalent to the separating hyperplane in the form of a separating valid inequality, that is an inequality that supports the optimal solution, and does not cut off any other feasible solutions. If a supporting valid inequality can be found, this inequality can be appended to the original problem, and when this problem is solved with the integrality requirements relaxed, the objective function value is equal to the optimal objective function value. The dual variables to this relaxed problem are then the equilibrium prices we are searching for.

For some problems, the coefficients in the Benders' cut, derived when the integer variables are held fixed to their optimal values, are in fact coefficients from a separating hyperplane for the mixed integer programming problem studied. For other problems they are not. This means that if we are looking at a class of problems, for which a supporting valid inequality does only include integer variables, the IP-prices generated will yield a supporting valid inequality in the sense that the inequality supports the optimal solution and is a separating hyperplane valid inequality that does not cut off any other feasible solution. In other problems, we need to regard other variables as "price" complicating, that is determine a subset of the integer variables and continuous variables that, when held fixed, yields dual variables that are coefficients in a supporting separating valid inequality.

This means that when using the partitioning idea in Benders' decomposition in order to calculate the optimal solution, it is clear that only the integer variables are complicating, where complicating is interpreted as complicating from a computational point of view. However, when our aim is to generate interpretable prices, another set of variables should be regarded as complicating. We need to find out those variables that should be held fixed at their optimal values, in order for the Benders' cutting plane derived by solving the Benders' sub-problem and its dual, to be a valid inequality that supports the optimal solution. When these variables are held fixed at their optimal values, this cutting plane will, when added to the problem, generate commodity prices and uplift charges that are compatible with market clearing prices, and the right hand side of the added valid inequality will give us the necessary uplift charge. In this way, we have generated a nonlinear price system consisting of linear commodity prices and a fixed uplift fee. We call the prices derived in this way the modified IP-prices. In the examples presented in Sect. 5, we show how these prices can be calculated.

It is clear that both the Walrasian interpretation that supports O'Neill et al.'s market clearing contracts and the partial equilibrium model that is used by Hogan and Ring (2003) can be reformulated for our modified IP-prices. The reformulated problem we need to solve in order to generate the modified IP-prices is

$$
\begin{aligned}
&Min\ c^T x + f^T y \\
&\text{s.t.}\quad Ax + Fy \geq b \\
&\qquad x \geq 0 \\
&\qquad y_i = y_i^* \quad i \in I \\
&\qquad x_k = x_k^* \quad k \in K \\
&\qquad y \in S
\end{aligned}
$$

where the sets I and K denote subsets of the set of all integer and continuous variables in the original problem, respectively.

How then should the variables to be held fixed be selected? The answer is found in the difference between, on the one hand, the structure of the linear programming solution of the original problem, and on the other hand, the structure of the optimal solution. Variables that appear in the optimal integer solution and not in the linear programming solution have to be forced into the solution. The examples in the next section illustrate this technique.

The price determining problem can then be stated as

$$\begin{aligned}
&Min\ c^T x + f^T y \\
&\text{s.t.}\quad Ax + Fy \geq b \\
&\qquad x \geq 0 \\
&\qquad Cy \geq d \\
&\qquad \sum_{i\in I} \pi_i y_i + \sum_{k\in K} \lambda_k x_k \geq \sum_{i\in I} \pi_i y_i^* + \sum_{k\in K} \lambda_k x_k^* \\
&\qquad y \geq 0
\end{aligned}$$

where π, λ are the dual variables of the equality constraints in the reformulated problem over.

When the appended inequality is a supporting valid inequality, the optimal dual variable for this inequality will be one, and the right hand side value will be the necessary uplift charge that is needed to support the prices that are given by the other optimal dual variables generated when the above problem is solved.

5 Examples

5.1 Example 1

We first use an example from Hogan and Ring (2003) to illustrate the generation of the modified IP prices that are supported by a nonlinear price function. The example consists of three technologies, Smokestack, High Tech and Med Tech, with the following production characteristics:

	Smokestack		High tech		Med tech
Capacity	16		7		6
Minimum output	0		0		2
Construction cost	53		30		0
Marginal cost	3		2		7
Average cost	6	0.3125	6	0.2857	7
Maximum number	6		5		5

We can formulate Hogan and Ring's allocation problem as a mixed integer programming problem as follows. Denote this problem P.

$$\begin{aligned}
&Minimize\ 53z_1 + 30z_2 + 0z_3 + 3q_1 + 2q_2 + 7q_3 \\
&\text{s.t.} \\
&\qquad q_1 + q_2 + q_3 = D \\
&\qquad 16z_1 - q_1 \geq 0 \\
&\qquad 7z_2 - q_2 \geq 0 \\
&\qquad 6z_3 - q_3 \geq 0 \\
&\qquad -2z_3 + q_3 \geq 0
\end{aligned}$$

$$
\begin{aligned}
-z_1 &\geq -6 \\
-z_2 &\geq -5 \\
-z_3 &\geq -5 \\
q_1, q_2, q_3 &\geq 0 \\
z_1, z_2, z_3 &\geq 0, \text{ and integer}
\end{aligned}
$$

D denotes the demand, and the construction variables for Smokestack, High Tech, and Med Tech are z_1, z_2 and z_3, respectively, while q_1, q_2 and q_3 denote the level of production using the Smokestack, High Tech, and Med Tech technologies, respectively. Note that, for fixed integer values of z_1, z_2 and z_3, the remaining problem in the continuous variables has the integrality property. Hence, the allocation problem is in fact a pure integer programming problem. The reason for this is the special form of the constraint matrix for this example.

The optimal solutions for demand levels 55 and 56 are given in the table below:

Demand	Smoke-stack		High tech		Med tech		Total cost
	Number	Output	Number	Output	Number	Output	
55	3	48	1	7	0	0	347
56	2	32	3	21	1	3	355

O'Neill et al.'s reformulation of problem P is

$$
\textit{Minimize } 53z_1 + 30z_2 + 0z_3 + 3q_1 + 2q_2 + 7q_3
$$

s.t.

$$
\begin{aligned}
q_1 + q_2 + q_3 &= D \\
16z_1 - q_1 &\geq 0 \\
7z_2 - q_2 &\geq 0 \\
6z_3 - q_3 &\geq 0 \\
-2z_3 + q_3 &\geq 0 \\
-z_1 &\geq -6 \\
-z_2 &\geq -5 \\
-z_3 &\geq -5 \\
z_1 &= z_1^* \\
z_2 &= z_2^* \\
z_3 &= z_3^* \\
q_1, q_2, q_3 &\geq 0 \\
z_1, z_2, z_3 &\geq 0, \text{ and integer}
\end{aligned}
$$

where z_1^*, z_2^* and z_3^* represent the optimal integer solution for the specified demand.

For demand equal to 55, the IP-prices generated are $p = 3$ for the demand constraint and the dual variables for the equality constraints are 53, 23 and 0, respectively. Here 53 and 23 are coefficients in a supporting valid inequality, namely the inequality $53z_1 + 23z_2 + 4q_3 \geq 182$. However, for demand equal to 56, the IP-prices that are generated are $p = 7$ for the demand constraint and -11, -5 and 0 for the other three equality constraints. It is obvious that these numbers cannot be part of a supporting valid inequality.

For the modified IP-prices, the reformulated problem to be solved is

$$\begin{aligned}
&\textit{Minimize } 53z_1 + 30z_2 + 0z_3 + 3q_1 + 2q_2 + 7q_3 \\
&\text{s.t.} \\
&\qquad q_1 + q_2 + q_3 = D \\
&\qquad 16z_1 - q_1 \geq 0 \\
&\qquad 7z_2 - q_2 \geq 0 \\
&\qquad 6z_3 - q_3 \geq 0 \\
&\qquad -2z_3 + q_3 \geq 0 \\
&\qquad -z_1 \geq -6 \\
&\qquad -z_2 \geq -5 \\
&\qquad -z_3 \geq -5 \\
&\qquad z_1 = z_1^* \\
&\qquad z_2 = z_2^* \\
&\qquad q_3 = q_3^* \\
&\qquad q_1, q_2, q_3 \geq 0 \\
&\qquad z_1, z_2, z_3 \geq 0, \text{ and integer}
\end{aligned}$$

With this definition of complicating variables, z_1, z_2 and q_3, the dual variables generated are $p = 3$ and the prices 53, 23 and 4 for the other equality constraints. Hence, both for demands equal to 55 and 56, the supporting valid inequalities $53z_1 + 23z_2 + 4q_3 \geq 182$ and $53z_1 + 23z_2 + 4q_3 \geq 187$ are generated, respectively. The right hand side values 182 and 187 are the respective uplift charges.

5.2 *Example 2*

A second example is taken from a situation with hockey stick bidding in an electricity market with generators having start up costs. The example includes four generators A, B, C and D, each having a limited capacity, an energy price and a start up cost. The data for the four units are as follows

	Capacity	Energy price	Start up cost
Unit A	45	10	0
Unit B	45	20	0
Unit C	10	100	20
Unit D	80	30	2,000

The integer programming formulation for this problem is

$$
\begin{aligned}
Min\ & 20z_3 + 2000z_4 + 10q_1 + 20q_2 + 100q_3 + 30q_4 \\
s.t\ & q_1 + q_2 + q_3 + q_4 \geq 100 \\
& -q_1 + 45z_1 \geq 0 \\
& -q_2 + 45z_2 \geq 0 \\
& -q_3 + 10z_3 \geq 0 \\
& -q_4 + 89z_4 \geq 0 \\
& -z_1 \geq -1 \\
& -z_2 \geq -1 \\
& -z_3 \geq -1 \\
& -z_4 \geq -1 \\
& z_1, z_2, z_3, z_4, q_1, q_2, q_3, q_4 \geq 0 \\
& z_1, z_2, z_3, z_4 \text{ integer}
\end{aligned}
$$

The optimal solution is to use units A and B to their full capacity and committing unit C to produce 10 units. The total cost for this production is 2,370.

Regarding the integer, restricted variables as complicating variables yields the commodity price $p = 100$, that is the committed unit with the highest energy cost sets the market price. This means that the hockey stick bidder sets the market price. The shadow prices for the equality constraints are $-4,050$, $-3,600$, 20 and $-3,600$, which are not coefficients in a supporting valid inequality.

If we instead regard the variables q_3 and z_4 as complicating variables, the commodity price generated is $p = 20$, and the two equality constraints have dual variables equal to 80 and 2,000, corresponding to coefficients in the supporting valid inequality $80q_3 + 2000z_4 \geq 80$ or $q_3 + 25z_4 \geq 10$. This example illustrates that what should be regarded as complicating variables from a computational point of view and from a pricing point of view might be very different.

5.3 Example 3

A third example is a capacitated plant location problem, with three potential facilities and two customers. The mathematical programming formulation is

$$
\begin{aligned}
Min\ & 10y_1 + 8y_2 + 12y_3 + 5x_{11} + 6x_{12} + 6x_{21} + 7x_{22} + 7x_{31} + 3x_{32} \\
s.t.\ & x_{11} + x_{21} + x_{31} = 10 \\
& x_{12} + x_{22} + x_{32} = 2 \\
& -x_{11} - x_{12} + 8y_1 \geq 0 \\
& -x_{21} - x_{22} + 7y_2 \geq 0 \\
& -x_{31} - x_{32} + 6y_3 \geq 0 \\
& x_{ij} \geq 0 \quad i = 1, 2, 3\ j = 1, 2 \\
& y_i \in [0, 1] \quad i = 1, 2, 3
\end{aligned}
$$

The optimal solution to the capacitated facility location problem has objective function value 82. Plants 1 and 3 are to be opened, eight units of goods should be transported from plant 1 to customer 1, two units of goods should be transported from plant 2 to customer 1 and finally two units of goods are to go from plant 3 to customer 2.

Adding constraints $y_1 = 1$, $y_3 = 1$, $x_{31} = 2$ gives dual variables to these equality constraints of 2, 12, and 1, respectively. Based on this result, the price supporting valid inequality

$$2y_1 + 8y_2 + 12y_3 + 3x_{12} + 4x_{22} + 1x_{31} \geq 16$$

can be generated. If the linear relaxation of the original plant location problem is solved with this valid inequality appended, the prices of goods at customers 1 and 2 will be 6 and 3, respectively, and the uplift charge needed is 16, that is the right hand side of the added valid inequality, which in the optimal LP relaxation has an optimal dual price of 1.

Note, however, that when the constraints $y_1 = 1$, $y_3 = 1$, $x_{31} = 2$ are added to the original problem, an alternative dual solution to the equality constraints exists, with dual variables 10, 12, and 2. Using these dual variables leads to a price supporting valid inequality, which reads

$$10y_1 + 8y_2 + 12y_3 + 3x_{12} + 1x_{21} + 4x_{22} + 2x_{31} \geq 26$$

If this inequality is added to the original problem and its LP relaxation is solved, the price for goods at customer 1 is 5 and at customer 2 is 3. Note that with these lower commodity prices the uplift charge is substantially higher. Following Hogan and Ring (2003), it is obvious that a price structure with a lower uplift charge is better, and should be preferred in cases like the one in this example, where degeneracy leads to alternative price structures.

5.4 Example 4

The final example is a modification of the first one, in which the two units Smokestack and High Tech are located in node 1 of a three node meshed electricity network, and where the Med Tech unit is located in node 2. All the demand is located at node 3, and there is a capacity limit of 15 between node 1 and 2. The mathematical programming formulation of this problem is

$$\textit{Minimize } 53z_1 + 30z_2 + 0z_3 + 3q_1 + 2q_2 + 7q_3$$

s.t.

$$q_1 + q_2 - f_{12} - f_{13} = 0$$
$$q_3 + f_{12} - f_{23} = 0$$

$$
\begin{aligned}
&f_{13} + f_{23} = 56 \\
&f_{12} - f_{13} + f_{23} = 0 \\
&16z_1 - q_1 \geq 0 \\
&7z_2 - q_2 \geq 0 \\
&6z_3 - q_3 \geq 0 \\
&-2z_3 + q_3 \geq 0 \\
&-z_1 \geq -6 \\
&-z_2 \geq -5 \\
&-z_3 \geq -5 \\
&f_{12} \leq 15 \\
&-f_{12} \leq 15 \\
&q_1, q_2, q_3 \geq 0 \\
&z_1, z_2, z_3 \geq 0, \text{ and integer}
\end{aligned}
$$

The optimal solution has objective function value 358, and $z_1 = 1,\ z_2 = 5,\ z_3 = 1$ The production from the different producers in the optimal solution is 15.5, 35 and 5.5 respectively, and the flow on the links are as follows: link 1–3 flow is at the capacity limit of 35.3, link 2–3 flow at 20.5 and finally link 1–2 flow at 15. Solving the problem with the constraints $z_1 = 1,\ z_2 = 5,\ f_{12} = 15$ yields dual variables for these constraints, which have the values 53, 23 and −6, respectively. Apart from this, the prices in the three nodes, the dual variables of flow constraints, are 3, 7 and 5.

Using this information, the valid inequality is $53z_1 + 23z_2 - 6f_{12} \geq 78$. When appended to the original problem and solving the LP relaxation, it yields the node prices 3, 7, and 5, and the dual price for the appended inequality is equal to 1. It is notable that the coefficient −6 is the negative of the shadow price on the constrained link 1–2. The uplift charge required to support these commodity prices are as low as eight, since the congestion fee from the constrained link, which is six times $15 = 90$, is used to make the required uplift charge lower.

Note also that in this example the continuous variables are not integer-valued in the optimal solution; hence, we do not know if there exists a nonlinear price function that supports this price structure.

6 Conclusions and Issues for Future Research

In this paper, we have shown that it is possible to construct modified IP-prices as a nonlinear price structure with affine commodity prices, supplemented by a fixed uplift charge. The prices are derived by the use of knowledge of the optimal solution to generate a reformulation that generates coefficients in a supporting valid inequality to the resource allocation optimization problem. When this valid inequality is appended to the original resource allocation problem, and the integer constraints are relaxed, the resulting optimal dual variables are the modified IP-prices, and on the right hand side the uplift charge needed to support these commodity prices.

For resource allocation including demand constraints, it is possible to interpret the modified IP-prices as prices derived from a partial equilibrium model.

How to choose the variables that should be held fixed when deriving the modified IP-prices, in order to get the price structure with the minimum uplift charge, needs to be further investigated. Another interesting research question is to determine how the bidding format and a contract mechanism should be constructed in a market in which modified IP-prices are used. Other research questions that should be addressed are how elastic demand can be incorporated in markets with non-convexities. If and how the uplift charges should be collected among the customers, and how this cost allocation should influence the design of market clearing contracts.

References

Alcaly RE, Klevorik AV (1966) A note on dual prices of integer programs. Econometrica 34: 206–214

Benders JF (1962) Partitioning procedures for solving mixed variables programming problems, Numer Math 4:238–252

Bjørndal M, Jørnsten K (2004) Allocation of resources in the presence of indivisibilities: scarfs problem revisited. SNF Working Paper

BjørndalM, Jørnsten K (2008) Equilibrium prices supported by dual price functions in markets with non-convexities. Eur J Oper Res 190(3):768–789

Gribik PR, Hogan WW Pope SL (2007) Market-clearing electricity prices and energy uplift. Working paper Harvard Enelectricity Policy Group, 31 Dec 2007

Gomory RE, Baumol WJ (1960) Integer programming and pricing. Econometrica, 28:521–550

Hogan WW, Ring BJ (2003) On minimum-uplift pricing for electricity markets. WP, Harvard University

Muratore G (2008) Equilibria in markets with non-convexities and a solution to the missing money phenomenon in energy markets. CORE Discussion paper 2008/6 Louvan la Neuve Belgium

O'Neill RP, Sotkiewicz PM, Hobbs BF, Rothkopf MH, Stewart WR (2005) Equilibrium in markets with nonconvexities. Eur J Oper Res 164(1):768–789

Scarf HE (1990) Mathematical programming and economic theory. Oper Res 38(3):377–385

Scarf HE (1994) The allocation of resources in the presence of indivisibilities. J Econ Perspect 8(4):111–128

Williams HP (1996) Duality in mathematics and linear and integer programming. J Optim Theor Appl 90(2):257–278

Wolsey LA (1981) Integer programming duality: price functions and sensitivity analysis. Math Program 20:173–195

Yan JH, Stern GA (2002) Simultaneous optimal auction and unit commitment for deregulated electricity markets. Electricity J 15(9):72–80.

An Optimization-Based Conjectured Response Approach to Medium-term Electricity Markets Simulation

Julián Barquín, Javier Reneses, Efraim Centeno, Pablo Dueñas, Félix Fernández, and Miguel Vázquez

Abstract Medium-term generation planning may be advantageously modeled through market equilibrium representation. There exist several methods to define and solve this kind of equilibrium. We focus on a particular technique based on conjectural variations. It is built on the idea that the equilibrium is equivalent to the solution of a quadratic minimization problem. We also show that this technique is suitable for complex system representation, including stochastic risk factors (i.e., hydro inflows) and network effects. We also elaborate on the use of the computed results for short-term operation.

Keywords Medium-term market simulation · Conjectured responses · Stochasticity · Network influence on equilibrium · Short and medium-term coordination

1 Introduction

Electricity deregulation has led to a considerable effort in analyzing the behavior of power markets. In particular, a wide range of research is devoted to the analysis of market behavior, which assumes a scope short enough to make unnecessary the study of the investment problem. Even with this simplification, the representation of markets by means of a single model is an extremely difficult task. Therefore, it is common practice to pay attention to only the most relevant phenomena involved in the situation under study. This point of view naturally leads to different modeling strategies depending on the horizon considered. In the short term, the most critical effects concern operational issues. The short-term behavior of the market will be highly influenced by the technical operation of power plants, as well as by short-term uncertainty. However, the modeling of strategic interaction between market

J. Barquín (✉)
Institute for Research in Technology (IIT), Advanced Technical Engineering School (ICAI), Pontifical Comillas University, Alberto Aguilera 23, 28015 Madrid, Spain
e-mail: julian.barquin@iit.upcomillas.es

S. Rebennack et al. (eds.), *Handbook of Power Systems II*, Energy Systems,
DOI 10.1007/978-3-642-12686-4_13,

players may be simplified if one assumes that strategic effects can be obtained by statistical estimation. The main assumption behind this approach is that, in the short term, the fundamental drivers that determine power prices do not change, and thus the complete formulation of the model of strategic interactions may not be required. This approach does not necessarily disregard market power effects in the short term, but it assumes that they can be estimated through recent past market data. A considerable amount of research has been devoted to the characterization and estimation of this short-term market representation, which can be classified under two broad headers. On the one hand, one can find models on the basis of describing the market using an aggregated representation, which is defined by the market price. Time series models, for example, Hamilton (1994) or Harvey (1989), and their financial versions, see Eydeland and Wolyniec (2003) for a recent review, belong to this category. Another useful alternative is based on considering competitors' decisions simply as another uncertain variable that the generator faces when optimizing its operation. The competitors' behavior is thus represented by the residual. Under this statement, the problem consists of an agent that is dealing with a residual demand and wishes to maximize its profit. A review can be found in Baillo et al. (2006).

In longer terms, however, the role of strategic issues becomes more important, because the behavior of market players may change, and consequently should be anticipated. We will refer to this time scope (typically up to 3 years) as medium-term, leaving the definition "long-term" for investment models. A wide range of models have been proposed for analyzing the interaction of competing power producers who price strategically. For instance, emission allowances management, fuel purchases, hydro resources allocation, and forward contracting are typically medium-term decisions, and consequently they will be affected by the strategic interaction between power producers. In addition to their applications as tools to aid the decision-making processes of generation companies, these models are useful for gaining insights into firms' strategic behavior as they allow for comparisons between different market scenarios, and they are therefore suitable for regulatory purposes as well (e.g., impacts of mergers or different market designs). In this context, Barquín (2006), Kiss et al. (2006), or Neuhoff (2003) are examples of regulatory applications.

Therefore, in the medium-term context, the modeling of market players' strategic behavior should be a central issue. When modeling electricity markets, three dimensions should be taken into account. First, the model for market equilibrium should represent the behavior of market players. In addition, an appropriate modeling of system constraints is required, so that it allows for the specification of the power producers' costs. Finally, the algorithms for finding the solution of the model must play a central role when developing the model, so that it allows for the resolution of realistic cases.

One popular alternative for describing power markets consists in using one-stage games to represent oligopolistic competition. The rationale behind this is to find an equilibrium point of the game to describe the market behavior; that is, a set of prices, generator outputs, consumptions, and other relevant numerical quantities, which no market agent can modify unilaterally without a decrease in its profits (Nash 1950).

Although alternative approaches have been proposed, such as agent-based models (Otero-Novas et al. 2000), we prefer to concentrate our review on game-theoretic models. Although the former simulation models may be very useful, especially for scenario generation, we think game-theoretic models are more able to respond to changes in the market structure and more appropriate for gaining insights into the market behavior.

1.1 Strategic Behavior of Market Players

The set of strategies that market agents can play defines the possible agents' behaviors (the actual behavior is determined by the outcome of the game). In fact, from Cournot (1838) and its criticism (Bertrand 1883), economists debate the merits of choosing prices or quantities as strategic variables. Price competition seems to be discarded in power markets, as capacity constraints play an important role in electricity markets, but a great number of models based on a quantity-game have been proposed to describe power-market equilibria, (Scott and Read 1996; Borenstein and Bushnell 1999; Hobbs 2001). These models are built on the idea that market players choose their quantities to maximize their profits, that is, the classic Cournot competition (Daughety 1988). Although their specification is not particularly challenging, they benefit from the fact that system constraints are relatively easy to represent. In addition, Cournot models make possible the representation of the worst possible case for oligopoly in the market, which may be of interest in some regulatory studies.

In power markets, a more realistic set of strategies would consist of a set of price–quantity pairs. Actually, in most pool-based markets, power producers bid a curve made up of the prices for each output. This is the idea behind the supply function equilibrium (Kemplerer and Meyer 1989). Examples of the application of these models to electricity markets can be found in Green and Newbery (1992) and Rudkevich (2005). However, their use in realistic settings is often difficult because simple operational constraints may pose problems in the definition of the supply function. For instance, in Baldick et al. (2004), it is shown that the relatively simple problem of specifying a linear supply function is not a closed question when upper capacity limits on power plants are considered. Nevertheless, there have also been recent promising advances (Anderson and Xu 2006).

Therefore, a compromise, the conjectured-supply-function approach (Vives 1999), has attracted considerable attention. The central idea of the methodology is to restrict the admissible set of supply functions to a much smaller parameterized set of curves. Very often it is assumed that market players bid linear supply functions with fixed and known slopes and decide on the intercept of the curve. There are several ways to define the slope of the conjectured supply function, which include the conjecture of the production response of competitors to market price variation (Day et al. 2002), the residual demand elasticity curve of each agent (García-Alcalde et al. 2002), and the total production variation of competitors with respect to the agent's production (Song et al. 2003). In this context, Bunn (2003) provides a wide range of techniques that can be used to obtain the conjectures. One of the key advantages of this methodology is that it can describe a wide range of market behaviors, from perfect to Cournot competition: zero-price response is, by definition, perfect competition; Cournot competition is represented by equal price response for every market player, given by the slope of the demand curve, see for instance Day et al. (2002). In addition, it is important to note that initially the game was proposed to avoid the estimation of a market state typical of the short-term models. However, when using the conjectured-response approach, the fundamental model is highly sensitive to the value of the conjecture, and so conjectured-supply-function models imply the estimation of the behavior of market players, placing them closer to short-term models.

1.2 Representation of the Power System

In addition, as electricity cannot be stored, operational constraints are of paramount importance even in medium-term analysis, and so there is a need for an adequate representation of technical characteristics of the system. The detail in the system constraints and in the complex operational characteristics of power plants may play a considerable role. In fact, the cost curve of power producers is typically made up of the costs of each power plant in their portfolio and of their maximum and minimum output, which results in a step-wise curve. Modeling, for example, a supply function equilibrium including these constraints is a very difficult task. An important instance of the operational constraints relevance is the representation of the power network. A popular approach to model oligopolistic markets is to disregard the transmission network. However, in some situations this approach may fail to represent adequately the strategic interaction between market players, mainly because the existence of the network can add opportunities for the exercising of market power. For example, congestion in one or more lines may isolate certain subsets of nodes so that supply to those nodes is effectively restricted to a relatively small subset of producers. Then, there will be increased market power, since it is relatively easy for these producers to increase prices in the isolated area by withholding power, irrespective of the competitiveness of the whole system. Furthermore, system variables assumed as input data when defining the game are often subject to uncertainty. In this context, market agents play a static game, and each of them faces a multiperiod optimization subject to uncertainty. Hence, the uncertainty of input data may be represented by means of a scenario tree to describe non-anticipatory decisions of market players.

The previous considerations are strongly linked to the problem of developing algorithms to find the equilibrium. Ideally, a detailed representation of the system would be needed to correctly represent the equilibrium of the game. However, it may result in computationally unaffordable problems. Therefore, the requisite level of simplification is ultimately determined by the algorithms available. Several methodologies have been proposed, but most of them share the characteristic of benefiting from the statement of complementarity conditions. Bushnell (1998) found the solution by means of heuristic methods. The mixed or linear complementarity problem was used by Hobbs (2001) and Rivier et al. (1999). Mixed complementary problem (MCP) has a structure that makes possible the formulation of equilibrium problems, including a higher level of complexity (e.g., network and production constraints or different competitor's behavior). Nevertheless, formulating the problem in this framework does not guarantee either the existence or the uniqueness of the solution. Commercial solvers exist (such as PATH or MILES) that allow this problem to be addressed, but they have some limitations: the size of the problem is restricted and, if a solution is not reached, no conclusions can be drawn regarding its existence. Moreover, if a solution is found, it may not be the only one. A particular situation arises when demand is considered to be affine, cost functions are linear or quadratic, and all the constraints considered are linear. In this case, MCP becomes a linear complementary problem (LCP) (Bushnell 2003; Day et al. 2002; García-Alcalde et al. 2002), and the existence and uniqueness of the solution can be guaranteed in most practical situations (Cottle et al. 1992). Closely related to MCP is the variational inequality (VI) problem (e.g. Yuan and Smeers 1999). MCP is a special case of this wider framework, which enables a more natural formulation of market equilibrium conditions (Daxhelet and Smeers 2001).

1.3 Dynamic Effects in Power Markets

Although we are essentially reviewing static games, there are several examples of dynamic games that are worthy of comment. For example, when dealing with power networks, several authors have proposed to describe the game as one of two stages: market players decide on their output in the first stage and the system operator clears the market in the second. This representation can be found, for instance, in Cardell et al. (1997). Regarding the algorithms developed to solve such equilibria, the equilibrium problem with equilibrium constraints (EPEC) is the most used algorithm to solve these two-stage games in power markets. For example, Hobbs et al. (2000) and Ehrenmann and Neuhoff (2003) use an EPEC formulation to solve the equilibrium, which takes into account the power network.

The strategic use of forward contracting is a typical application of dynamic games techniques. Actually, from Allaz and Villa (1993), there is an increasing body of literature that analyzes the use of forward positions as strategic variables. Yao et al. (2004) propose a similar model, but include a very detailed representation of the power system. In fact, they modeled a two-stage game using an EPEC formulation. However, they are being contested by some recent research. For instance, Liski and Montero (2006) and Le Coq (2004), among others, study the effect of collusive behavior in repeated games, while Zhang and Zwart (2006) is concerned with the effects of reputation.

1.4 The Proposed Methodology

The rest of the chapter is devoted to the description of an alternative methodology of dealing with equilibrium problems. The methodology is built on the idea that, under some reasonable conditions, such as monotonous increasing cost function, the solution of a conjectured-supply-function game is equivalent to the solution of a quadratic minimization problem. The approach will be applied, in the rest of the chapter, to different models of the power system, including the representation of the power network and the consideration of uncertainty. This methodology contributes some important advantages. First, the equilibrium existence is clearly stated, and it can be computed with very efficient optimization techniques. Second, the formulation has a structure that strongly resembles classical optimization of hydrothermal coordination, and thus, it allows for the inclusion of linear technical constraints in a direct way. Third, by using commercial optimization tools, large-size problems can solved. This last point makes possible a more detailed representation of the power system.

The chapter continues with the introduction of a simplified deterministic case and the equivalent minimization problem. After that, the model is generalized to the multiperiod setting. The transmission network is included in the subsequent section, and the uncertainty is then considered in the market model. Finally, there is a description of how the additional information of the medium-term model can become signals for short-term operation. Most of the results described herein have been previously published in the literature by the authors.

2 Market Equilibrium Under Deterministic Conditions: The Single Period Case

This section starts by describing the market model in a simplified version, which is restricted to a single unitary time period and disregards technical constraints and forward contracting. In following sections, the model is generalized to consider a multiperiod setting, in which hydro operation and technical constraints are described. Finally, the effects of the transmission network in the strategic interaction of market agents are included in the model.

The formulation can be interpreted as a conjectured-price-response approach that represents the strategic behavior of a number of firms $u = 1, 2, \ldots, U$ competing in an oligopolistic market. The variation of the clearing price λ with respect to each generation-company production P_u is assumed to be known. Under the logical assumption that this variation is nonpositive, the non-negative parameter θ_u is defined as

$$\theta_u = -\frac{\partial \lambda}{\partial P_u} \geq 0 \qquad (1)$$

Disregarding contracts, the profit B_u that generation company u obtains when remunerated at the marginal price is

$$B_u = \lambda \cdot P_u - C_u(P_u) \tag{2}$$

The first term $\lambda \cdot P_u$ represents the revenue function for firm u, and the second $C_u(P_u)$ its cost function. The equilibrium is obtained by expressing the first-order profit-maximization condition[1] for each generation company

$$\frac{\partial B_u}{\partial P_u} = 0 = \lambda + P_u \cdot \frac{\partial \lambda}{\partial P_u} - \frac{\partial C_u(P_u)}{\partial P_u} \tag{3}$$

The conjectured response of the clearing price with respect to each generation company's production leads to

$$\lambda = \frac{\partial C_u(P_u)}{\partial P_u} + P_u \cdot \theta_u, \forall u \tag{4}$$

To obtain the equilibrium conditions, it is necessary to define the equations that determine the market price, and consequently to take into account the market clearing process (the system operator, who clears the market, is the remaining player of the game). Thus, a generic market clearing process would imply the maximization of the demand utility subject to system constraints. Provided that D^0 is the constant term of the demand function, and α_0 represents the demand slope, the utility function of the demand $U(D)$ is defined as

$$U(D) = \int_0^D \lambda(D) \cdot dD = \frac{1}{\alpha_0} \cdot \left(D \cdot D^0 - \frac{D^2}{2} \right) \tag{5}$$

The only system constraint in this case is the power balance equation. Therefore, the system operator behavior is defined by the problem

$$\begin{aligned} &\min_D -U(D) \\ &\text{s.t. } \sum_{u=1}^{U} P_u - D = 0 : \lambda \end{aligned} \tag{6}$$

Furthermore, the corresponding first optimality conditions of the above problem, that is, the market-clearing conditions, are

$$D = D^0 - \alpha_0 \cdot \lambda \tag{7}$$

[1] This is a consequence of Nash equilibrium. None of the utilities could modify its production without a decrease in its profits. Therefore, the first derivative is null at the point of maximum profit.

$$\sum_{u=1}^{U} P_u - D = 0 \tag{8}$$

Therefore, equilibrium of the game is defined by (4), (7), and (8). This equilibrium point can be alternatively computed by solving an equivalent optimization problem:

$$\begin{aligned} \min_{P_u, D} \; & \sum_{u=1}^{U} \left[\bar{C}_u\left(P_u\right)\right] - U\left(D\right) \\ \text{s.t.} \; & \sum_{u=1}^{U} P_u - D = 0 \quad : \quad \lambda \end{aligned} \tag{9}$$

where $\bar{C}_u\left(P_u\right)$ denotes a term called effective cost function[2]

$$\bar{C}_u\left(P_u\right) = C_u\left(P_u\right) + \frac{P_u^2 \cdot \theta_u}{2} \tag{10}$$

Under the hypothesis of continuous and convex cost functions, and single node market clearing as well as non-negative θ_u, it can be proved that this optimization problem (9) is equivalent to the market equilibrium problem defined by (4), (7), and (8). Note that the dual variable λ of the power-balance constraint is the system's marginal price that clears the market.

Furthermore, the model can be generalized to include forward contracts signed by the utilities (Barquín et al. 2004). Two kinds of contracts are considered:

1. *Forward contracts*. In these contracts the utility agrees to generate a certain power G_u^k at price λ_u^{Gk}.
2. *Contracts for differences*. In these financial instruments, the utility contracts a power H_u^l and it is paid the difference between the clearing price λ and an agreed one λ_u^{Hl}.

When physical contracts are being considered, generated power P_u is different from the tenders in the spot market P_u^S. Specifically,

$$P_u = P_u^S + \sum_k G_u^k \tag{11}$$

The utility profit is

$$B_u = \lambda \cdot P_u^S - C_u\left(P_u\right) + \sum_k \lambda_u^{Gk} \cdot G_u^k + \sum_l \left(\lambda_u^{Hl} - \lambda\right) H_u^l \tag{12}$$

[2] It is important to highlight that the effective cost function derivative is the effective marginal cost function $\overline{MC_u}\left(P_u\right) = MC_u\left(P_u\right) + \theta_u \cdot P_u$. Therefore, the solution of the minimization problem, where the effective cost functions sum is minimized, implies that utilities' effective marginal costs are equal to system marginal price.

The previous equation can be written as

$$B_u = \lambda \cdot P_u - C_u(P_u) + \sum_k \left(\lambda_u^{Gk} - \lambda\right) G_u^k + \sum_l \left(\lambda_u^{Hl} - \lambda\right) H_u^l \quad (13)$$

Profit maximization leads to the equilibrium equations

$$\lambda = \frac{\partial C_u(P_u)}{\partial P_u} + \theta_u \left(P_u - \sum_l H_u^l - \sum_k G_u^k \right), \forall u \quad (14)$$

These equations happen to be the optimality conditions of problem (9) if effective cost is defined as

$$\bar{C}_u(P_u) = C_u(P_u) + \frac{\theta_u}{2} \left(P_u - \sum_l H_u^l - \sum_k G_u^k \right)^2 \quad (15)$$

Finally, some common market situations are represented considering that demand is inelastic, that is, a constant known value. For example, it is usual to consider that the competitors' reaction comes before the demand's reaction, and that consequently the demand elasticity should be disregarded. In this situation, market equilibrium is obtained as the solution of an optimization problem, which does not include the utility function of the demand

$$\begin{aligned} \min_{P_u} \quad & \sum_{u=1}^{U} \bar{C}_u(P_u) \\ \text{s.t.} \quad & \sum_{u=1}^{U} P_u - D^0 = 0 : \lambda \end{aligned} \quad (16)$$

3 Conjectural Price Response's Specification

A key issue when adjusting the model is the adequate estimation of conjectured-price responses. Several papers have dealt with this substantial topic. The existing methodologies can be grouped into three categories: implicit methods, bid-based methods, and costs-based methods.

- Implicit methods take into account actual and past prices and companies' production to estimate the value of conjectures according to (4). Two applications of this methodology can be found in García-Alcalde et al. (2002) and López de Haro et al. (2007).
- Bid-based methods use historical bids made by companies to carry out the estimation. Three different applications can be found in Bunn (2003), where it is shown how conjectures can be estimated through different approximations of bid

functions combined with clustering analysis, time series analysis, or input–output hidden Markov models.
- Finally, cost-based methods are useful when dealing with new markets, or markets in which no information is provided from the auction process (Perry 1982).

There are, both from the theoretical and the practical points of view, important drawbacks to all these approaches. The first two (implicit and bid-based methods) are essentially historical approaches that ultimately rely on the assumption that the future will resemble the past. In other words, no dramatic changes in the economic or regulatory conditions of the system are to be expected in the study horizon. However, they should not be applied if these conditions are not met, for instance, when analyzing the impact of new significant generation capacity additions or that of sufficiently significant regulatory changes like large Virtual Power Plants auctions. On the other hand, they can be recommended for medium-term studies.

The last approach (cost-based methods) can be thought of as a way to determine the conjecture endogenously. These methods can conceivably overcome the drawbacks described in the previous paragraph, but we are hampered for our ignorance of some basic issues regarding the nature of oligopolistic competition. Actually, conjectural variations or responses techniques have been often looked upon with suspicion by a number of economists, as it can be argued that, being able to explain almost anything, they are not useful for predicting anything (Figuières et al. 2004). Our approach here is a pragmatic one. We assume that for medium-term studies the "historical methods" can provide sensible figures, and that for longer-term or regulatory studies the cost based techniques can provide plausible working guides. For instance, when analyzing competition enhancing reforms for the Spanish market in the Spanish White Paper on Electricity, analysts relied on a supply function equilibrium approach to obtain rough estimates of the effectiveness of some proposed methods (Pérez-Arriaga et al. 2005; Barquin et al. 2007). However, we think that it can be fairly stated that currently results from all these approaches are to be taken as suggestive of possible outcomes rather than as reliable quantitative forecasts.

As we move towards multiperiod, networked systems, the number of conjectural parameters are greater, and so, as just stated, they are of concern. Although these are important issues that merit close attention, they will not be pursued further on.

A closely related issue concerns the forward position specification. The assumption implicitly made to obtain (14) is that the effective production subject to oligopolistic behavior is not the actual output, but the output sold in the spot market. Although there is a considerable body of literature that addresses the issue of how forward equilibrium positions are determined, we have adopted the conclusion in Allaz and Villa (1993), which stated that the production previously contracted is not considered by power producers when deciding their output in the spot market. In other words, the contracted production is not an incentive to raise the spot price. Therefore, the production involved in the oligopolistic term of the costs is the output minus the contracted production. The conjectured price response can be estimated from historical data or predicted from the knowledge of the relevant cost functions and economic and regulatory constraints. There are advantages and drawbacks similar to those stated earlier. In the case of using the implicit method, it could be advisable to jointly estimate conjectures and forward positions. Cournot studies typically assume away forward positions and consider conjectured price responses much more elastic than those implied by short-term demand elasticity.

4 Market Equilibrium Under Deterministic Conditions: The Multiperiod Case

Production-cost functions are usually multivariable functions that relate different periods. The structure proposed in this model makes possible the introduction of additional variables and constraints to represent the cost functions and the associated technical constraints within the same optimization problem in a natural way. The characteristics of these cost functions are very dependent on the particular system to be represented. It is very common, in medium-term horizon studies, to represent cost as linear or quadratic cost functions. In any of these situations, the formulation yields a quadratic programming (QP) problem. This schema closely resembles classical optimization medium-term operation models and maintains its main advantageous features: there is a clear model structure (extended objective function and technical constraints), reasonable computing time, and dual information is easily obtained.

As an example of the proposed methodology, a hydrothermal power system constituted by a set of generation companies will be represented with the objective of obtaining a QP problem that could also be achieved with different cost function representations.

For the time representation, the model scope is divided into periods, weeks, or months, denoted as p. Each period is divided into subperiods, denoted as s. A subperiod consists of several load levels, denoted as b. Each load level b of a subperiod s and a period p is characterized by its duration l_{psb}.

The optimization model should yield as optimality conditions a generalization of equilibrium (14), namely

$$\lambda_{psb} = \left.\frac{\partial C_u(P_u)}{\partial P_u}\right|_{psb} + \theta_{upsb}\left(P_{upsb} - G_{upsb} - H_{upsb}\right), \ \forall u, p, s, b \tag{17}$$

G_{upsb} and H_{upsb} are the total physical and financial contracted amounts, respectively. Additional decision variables are required to represent the power system, which consists of a set of thermal and hydro units' productions. All of them have to be replicated for each period p, subperiod s, and load level b.

t_{jpsb} Power generation of thermal unit j (MW)

h_{mpsb} Power generation of hydro unit m (MW)

b_{mpsb} Power consumption of pumped-hydro unit m (MW)

r_{mp} Energy reservoir level of hydro unit m at the end of period p. The initial value, as well as the value for the last period, is considered to be known (MWh)

s_{mp} Energy spillage of hydro unit m at period p (MWh)

Generated power for each company does not require a specific variable, as will be shown later; it may be computed as a linear combination of other decision variables.

Some new parameters should also be considered to make possible the introduction of power system's technical characteristics. Each thermal unit j is defined by using its

$\bar{t}_j$ Maximum power generation (MW)
δ_{jp} Variable cost in period p (e/MWh)
o_j Owner generation company

Hydro units are represented as a single group with pumping capability. Every hydro unit m has an associated reservoir and may have pumping capability. Run-off-the-river production is considered separately for each utility. Hydro units' parameters include

$\bar{h}_{mp}$ Maximum power generation in period p (MW)
o_m Owner generation company
$\bar{b}_m$ Maximum pumping power consumption (MW)
ρ_m Performance of pumping (p.u.)
$\bar{r}_{mp}$ Maximum energy reservoir storage at the end of period p (MWh)
$\underline{r}_{mp}$ Minimum energy reservoir storage at the end of period p (MWh)
f_{upsb} Run-off-the-river hydro energy for the whole company u at load level b of subperiod s and period p (MW)
I_{mp} Hydro inflows, except run-off-the-river, in period p (MWh)
r_{m0} Initial energy reservoir level (MWh)
r_{mp^*} Final energy reservoir level in the last period p^* (MWh)

Each utility u may have signed bilateral (physical) contracts or financial contracts for each load level b of subperiod s and period p

G_{upsb} Amount of physical contracts (MW)
λ^G_{upsb} Price of physical contracts (€/MWh)
H_{upsb} Amount of financial contracts (MW)
λ^H_{upsb} Price of financial contracts (€/MWh)

From the previous definitions, generated power for utility u at load level b of subperiod s and period p is obtained as a linear combinations of decision variables and parameters

$$P_{upsb} = \sum_{j|o_j=u} t_{jpsb} + \sum_{m|o_m=u} \left(h_{mpsb} - b_{mpsb}\right) + f_{upsb} \tag{18}$$

Definition of cost functions requires the addition of the following technical constraints: bounds for the decision variables, power balances, and energy balances. The decision variable bounds correspond to the maximum power generation of thermal unit j

$$t_{jpsb} \leq \bar{t}_j \tag{19}$$

The maximum power generation of hydro unit m is

$$h_{mpsb} \leq \bar{h}_{mp} \tag{20}$$

The maximum pumping power consumption of hydro unit m is

$$b_{mpsb} \leq \bar{b}_{m} \tag{21}$$

And, finally, the minimum- and maximum-level energy reservoir storage of hydro unit m is

$$\underline{r}_{mp} \leq r_{mp} \leq \bar{r}_{mp} \tag{22}$$

Power balance for each load level b of subperiod s and period p is shown next. The dual variable of each one of these constraints will be referred to as π_{psb}, which is necessary to compute the system's marginal price:

$$\sum_{j} t_{jpsb} + \sum_{m} h_{mpsb} + \sum_{u} f_{upsb} - D_{psb} - \sum_{m} b_{mpsb} = 0 \tag{23}$$

Energy balance for each load level b of subperiod s and period p, and hydro unit m is

$$r_{mp} - r_{m,p-1} = \sum_{s}\sum_{b} l_{psb}\left(h_{mpsb} - b_{mpsb}\right) + I_{mp} - s_{mp} \tag{24}$$

The cost function can be now expressed for each company as a linear function of decision variables and can be directly included in the objective function without modifying the QP structure:[3]

$$C_{upsb} = \sum_{j|o_j=u} \left(\delta_{jp} \cdot t_{jpsb}\right) \tag{25}$$

The objective function includes total system operation costs, the quadratic term of the effective cost, and demand utility function:

$$\min \sum_{p}\sum_{s}\sum_{b} l_{psb} \cdot \sum_{u=1}^{U} \left(C_{upsb} + \frac{\left(P_{upsb} - H_{upsb} - G_{upsb}\right)^2 \cdot \theta_{upsb}}{2} \right) - U\left(D\right) \tag{26}$$

where the demand utility function $U(D)$ is computed as

$$U\left(D\right) = \sum_{p}\sum_{s}\sum_{b} l_{psb} \cdot \frac{1}{\alpha_{0psb}} \left(D_{psb} \cdot D^0_{psb} - \frac{D^2_{psb}}{2} \right) \tag{27}$$

[3] The model can easily be extended in order to consider start-up and shut-down costs.

As mentioned, the final problem has a QP structure. Quadratic terms are included only in objective function since all restrictions are linear. Extremely efficient algorithms and solvers are available for this particular schema.

One of the main features of the proposed approach is that marginal price is directly obtained from the results of the optimization problem. Price is computed from the dual variable of the power balance constraint (23) as

$$\lambda_{psb} = \frac{\pi_{psb}}{l_{psb}} \tag{28}$$

In addition, the profit of each utility u is obtained as its income minus its costs:

$$B_u = \sum_p \sum_s \sum_b l_{psb} \cdot \Big[\lambda_{psb} \cdot P_{upsb} - C_{upsb} + \left(\lambda^G_{upsb} - \lambda_{psb} \right) \cdot G_{upsb} + \left(\lambda^H_{upsb} - \lambda_{psb} \right) \cdot H_{upsb} \Big] \tag{29}$$

5 Equilibrium in Power Networks

When firms decide their output, they must anticipate which lines will be congested after the market clearing, in order to anticipate how their production will affect prices. One way of tackling this problem is within the leader–follower framework. The problem can be stated as a two-stage game, where the firms first allocate their output and submit their bids to the central auctioneer (leaders) and then the central auctioneer clears the market, given the bids of the firms (follower). Hereinafter, we will assume a central auctioneer as a representation of an efficient market-clearing process, in spite of the fact that the clearing mechanism may not include a central auction, as in the case of transmission auctions. Unfortunately, when the transmission network is taken into account by means of a DC load flow, a pure strategies equilibrium for such a game may not exist or may be not unique (Ehrenmann 2004; Daxhelet and Smeers 2001), since optimality conditions of the auctioneer problem considered as constraints of the firms' profit maximization result in a nonconvex feasible region. An alternative proposal to cope with the problem is to assume some conjectured reaction of the firms to the decisions of the transmission system operator. This facilitates the convexity of the feasible region, while still playing a quantity game against other firms when deciding their output. This is the case of Metzler et al. (2003) and Hobbs and Rijkers (2004), which assume the reactions of the firms to the decisions of the transmission system operator to be exogenous parameters, regardless of the distribution of the flows in the network: in Hobbs and Rijkers (2004) the conjecture is no reaction, that is, firms are price-takers with respect to transmission prices.

In principle, including transmission constraints in the above model may be dealt with like any other technical constraint. However, the power network adds the spatial dimension to the problem, and consequently the strategic interaction between power producers is extended to the new dimension. The main problem when taking into account the power network is that price sensitivities, modeled above as conjectured variations, are not independent of the set of congested lines of the system. Therefore, power producers must anticipate the result of the market clearing to anticipate how their decisions will affect prices. We describe in this section a way to obtain the market equilibrium avoiding the statement of a two-stage game between power producers and the system operators. The rationale here is to model the decision-making process of the firms as a search for consistent assumptions: firms assume that the central auctioneer will decide that a certain set of transmission lines will be congested and, on that assumption, decide their output. If the market clearing considering these bids results in the assumed set of congested lines, then the assumption is consistent and the solution is a Nash equilibrium. The firms' assumption about congested lines gives their price response, so that any iteration can be thought of as their conjectured-transmission-price response (Metzler et al. 2003; Hobbs and Rijkers 2004). However, the iterative procedure may be viewed as a way of selecting the firms' reactions to the power-network constraints that is consistent with the central auctioneer behavior, and thus of foreseeing the auctioneer's reactions to the firms' bids.

The statement of the multiperiod problem with network constraints is close to the model described in previous sections. However, we choose to describe a simplified single-period model with the aim of highlighting network-related effects.[4] In the same spirit, we do not take into account any forward contract. Nevertheless, both inter-temporal constraints and forward positions can be included as in the sections above. The transmission network has a set of nodes $M = \{1, 2, \dots, N\}$. Market agents will be characterized by their output decisions at each node of the network, denoted by P_u^i. Consequently, there is, in general, a different price at each node λ_i. Thus, firms decide by solving the following program:

$$B_u = \sum_{i \in M} \lambda_i P_u^i - C_u^i \left(P_u^i \right) \tag{30}$$

The above expression is just the multi-node generalization of the single period model, where the profit of a certain firm is the sum of its profit over every node. Again, the set of solutions to the problems for each firm characterizes the market equilibrium:

$$\lambda_i + \left[\sum_{j \in M} \frac{\partial \lambda_j}{\partial P_u^i} P_u^j - \frac{\partial C_u^i \left(P_u^i \right)}{\partial P_u^i} \right] = 0 \tag{31}$$

[4] However, interaction of inter-temporal and network constraints can lead to other concerns, as in Johnsen (2001) and Skaar and Sorgard (2006) in the case of hydro management in a system with network constraints.

Therefore, the conjectured price-response is

$$\theta_u^{ij} = -\frac{\partial \lambda_j}{\partial P_u^i} \tag{32}$$

Under reasonable assumptions, it can be shown that $\theta_u^{ij} = \theta_u^{ji}$ (Barquín 2008). Following the same rationale as in the single node case, we can define an effective cost function, given by the expression

$$\bar{C}_u\left(P_u^i\right) = C_u^i\left(P_u^i\right) + \frac{1}{2}\sum_j P_u^i \theta_u^{ij} P_u^j \tag{33}$$

Note that symmetry of conjectured responses θ_u^{ij} is required in order that optimality conditions be the same as equilibrium conditions.

After defining the effective cost function for market players, following the reasoning of previous sections implies defining the market-clearing conditions. In the single-period, single-node case, market clearing was defined by the demand curve and the power balance equation. When the transmission network is considered, demands and power flows at each node are determined to maximize the utility of demand. That is, the market clearing problem is the solution of an optimal power flow solved by the system operator. Then

$$\begin{aligned} \min_{D,f,\theta} \; & -\sum_i U\left(D_i\right) \\ \text{s.t.} \; & D_i + \sum_j m_{ij} f_j = \sum_u P_u^i \\ & f_j = y_j\left(\theta_i - \theta_k\right) \\ & -f_j^{\max} \le f_j \le f_j^{\max} \end{aligned} \tag{34}$$

The first constraint is the power balance equation: f_j is the flow through the line j. m_{ij} is 1 if the line j is leaving the node i, -1 if the line is arriving at the node i, and 0 otherwise. The second constraint represents the DC power flow equations: y_j is the admittance of the line j, and $\theta_i - \theta_k$ is the difference of voltage phases between the nodes of the line j. The last constraint is the maximum flow through transmission lines. The full equivalent optimization problem is

$$\begin{aligned} \min_{P_{ii} D,f,\theta} \; & \sum_{u,i}\left[\bar{C}_u\left(P_u^i\right)\right] - \sum_i U\left(D_i\right) \\ \text{s.t.} \; & D_i + \sum_j m_{ij} f_j = \sum_u P_u^i \; : \lambda_i \\ & f_j = y_j\left(\theta_i - \theta_k\right) \\ & f_j^{\max} \le f_j \le f_j^{\max} \end{aligned} \tag{35}$$

It is important to notice that the above problem represents the market-clearing process. Therefore, one may argue that many power markets have different mechanisms, and that the problem above represents the particular case of nodal-pricing.

However, it should be understood as a description of any efficient mechanism to clear the market. For example, the result of a market splitting design would be the same as the solution to the above problem.

The previous problem allows for the calculation of the conjectured-response equilibrium, as long as the conjectured price variations θ_u^{ij} are known. In principle, it is possible, as in the single-node model, to consider the variations as known data. However, when the power network is considered, this task is more delicate. The variations represent, by definition, the response of every player in the market. But the transmission constraints complicate the kind of reactions that can be found. Consider, for example, a two-node grid. If the unique transmission line is binding, changes in the production of units located at one of the nodes can be compensated only by units at the same node. Thus, the owner of the unit is competing only with firms owning units at the same node. In contrast, if the line is below its limits, the system may be considered to be a single-node network. In such a case, the previous unit is competing with the whole system. Therefore, in general, the conjectured responses depend on whether the line is binding or not, and hence the maximum flow constraint of the line may be used strategically. Therefore, it is necessary, in principle, to define a set of conjectures θ_u^{ij} for each possible set of constrained lines (hereinafter, we will refer to this set as network state). To take into account the strategic use of congestions, and thus to consider rational agents, one faces the problem of anticipating the changes in price responses with respect to the network state. In other words, rational agents would anticipate the auctioneer's decisions. The problem resulting from such a statement, in addition to being computationally expensive, may have infinite equilibria, or may have none.

As an alternative, we propose an algorithm that is designed to search for consistent price responses as an approximation to the equilibrium of the two-stage game. In the two-node example, consistent price responses imply that the producer is not willing to congest the line, as if he were, and in the next iteration he would produce so that the line is binding. Therefore, although at each step of the algorithm, power producers do not take into account congestions as strategic variables, the complete algorithm describes the market players' selection of the most lucrative set of congested lines. The iterative algorithm that we explain below can be thought of as a way of calculating endogenously the state-dependent responses.

In any case, in the market clearing problem, generated energy P_u^i is no longer a decision variable but an input data. Therefore, the network state as well as the relevant set of conjectured price responses (see below) can be computed as a function of the generated energy. So we state the condition that, as power producers must assume the network state to make their decisions, the network state assumed when deciding the output of each unit must be the same as the network state obtained after the market has been cleared. This is the rationale for the iterative algorithm, which works as follows. First, a network state is assumed. Then, with the corresponding conjectures θ_u^{ij} the optimization problem is solved, resulting in a new network state. If it coincides with the network state previously assumed, then the algorithm ends; if not, the profit-maximization problem is solved again with the new conjectures and the whole procedure iterates again.

The algorithm loops until either convergence is obtained or a limit cycle is detected. The algorithm convergence necessarily implies that a consistent Nash equilibrium has been computed, because price sensitivities assumed by the oligopolistic agents when bidding are the same as those resulting after the market has been cleared. However, the convergence of the algorithm is not assured for a number of reasons: the game may have no pure strategies equilibrium, and so the algorithm will not converge; in addition, the procedure does not consider points where any line is just changing from congested to non-congested, or vice versa, since market players, in each of the iterations, do not take into account that the lines may change their state, for example, from congested to non-congested, as a result of the production decisions. However, in such cases, multiple Nash equilibria (even a continuum of equilibria) may exist (Ehrenmann 2004). Further discussion can be found in Barquín and Vázquez (2008).

In addition, the model can also be written by using power distribution factors. However, we do think that one definite advantage of directly stating a DC load flow representation is that it facilitates, from the numerical point of view, the study of more complex power systems, because the admittance matrix used in the DC load flow formulation is a sparse one.

Even if the market clearing problem makes it possible to relate the production decisions to the network status, the problem of defining a complete set of conjectures θ_u^{ij} is far from having been solved. In particular, it is necessary to define which nodes react to output decisions at a certain node, and this problem is not obvious in complex networks. We will next show how to calculate conjectured responses based on the following Cournot-like assumption: the production decision at a certain node is not affected by the competitors' decisions at the rest of the nodes. However, more general conjectures can be considered. For instance, in Barquin (2008) the case of more general supply functions is analyzed. One of the most important results of all these analysis is a qualitative one: conjectured-price responses θ_u^{ij} are found to be symmetric $\left(\theta_u^{ij} = \theta_u^{ji}\right)$ under general circumstances. To obtain explicit values, the techniques considered in Sect. 3 can be used, although attention must be paid to the network status.

The Cournot-like case is a particularly important instance, as it is assumed in a number of studies with regulatory aims. So let us consider a net additional megawatt at a certain node (in general due to the combination of decisions of every producer located at the node). In this case, from the point of view of the auctioneer, there is an additional power injection in that bus. Therefore, given the new generation, the system operator will react to maximize the total consumer value by means of redistributing power flows and demands. More formally, consider a net additional megawatt at the ith node δP^i. The system operator will decide on the incremental variables δD_i, δf_j, and $\delta \theta_i$ to reoptimize the load flow. In addition, the flow through unconstrained lines may change, while the flow in constrained lines must remain constant, equal to its limit value. Therefore, the system operator solves the following program:

$$\begin{aligned} \min_{D,f,\theta} \quad & -\sum_i U\left(D_i + \delta D_i\right) \\ \text{s.t. } \delta D_i + \sum_j m_{ij}\delta f_j &= \delta P^i \\ \delta f_j &= y_{ij}\left(\delta\theta_i - \delta\theta_k\right) \\ \delta f_{cong} &= 0 \end{aligned} \tag{36}$$

The problem is the incremental version of the program previously described. The last constraint represents the fact that after a small change the network state will not change. This is not a harmless assumption, as the oligopolistic game outcome can result in lines being "critically congested." In these lines the flow is at its maximum, but arbitrarily small changes in the system conditions can result in the line being either congested or not congested (Cardell et al. 1997; Barquín 2006).Therefore, the solution to this problem provides δD_i for every node. Moreover, by solving the problem consecutively for an increment of production δP^i at each of the nodes, it is possible to obtain the derivatives $\frac{\partial D_j}{\partial P^i}$. The last step in calculating the conjectured response is to state the relationship between the previous derivative and the response. To do so, taking into account the definition of the demand curve given in the single-node case, we have that the conjectural response is

$$\frac{\partial \lambda_j}{\partial P_u^i} = -\frac{1}{\alpha_j}\frac{\partial D_j}{\partial P_u^i} \tag{37}$$

Thus, it is enough to obtain the derivative $\frac{\partial D_j}{\partial P_u^i}$. On the other hand, we have

$$\frac{\partial \lambda_j}{\partial P_u^i} = -\frac{1}{\alpha_j}\frac{\partial P^i}{\partial P_u^i}\frac{\partial D_j}{\partial P^i} \tag{38}$$

Finally, armed with the definition of the conjectured responses of the single-node case, we have that

$$\frac{\partial \lambda_j}{\partial P_u^i} = \theta_u^{ij}\frac{\partial D_j}{\partial P^i} \tag{39}$$

where θ_u^i is the variation as defined in the previous sections, that is, disregarding the effects of the rest of the nodes. Examination of the optimality conditions of the incremental program allows us to prove that $\theta_u^{ij} = \theta_u^{ji}$.

6 Consideration of Uncertainty: Stochastic Model

Some studies have dealt with market equilibrium under uncertainty. In Murto (2003), equilibrium with uncertain demand is analyzed and solved within the framework of game theory. Results are obtained analytically, which provides insight into market behavior but is restricted to two periods. The study in Kelman et al. (2001) addresses the representation of static market equilibrium where market players face a multiperiod optimization, and expands the concept of stochastic dynamic programming (SDP) with the objective of assessing market power in hydrothermal systems where inflows are subject to uncertainty. Nash–Cournot equilibrium is solved for each period for different reservoir levels, and a set of future-benefit functions that replace the classical single future-cost function is computed. Convexity of these future-benefit functions is not guaranteed, nor is the convergence of this SDP algorithm. The structure of the optimization problem is suitable for stochastic representations of the operation of hydrothermal generation systems and the consideration of uncertainty (Morton 1993; Gorenstin et al. 1992; Pereira and Pinto 1991). When it comes to medium-term generation operation, several major sources of uncertainty can be identified: hydro inflows, fuel prices, emissions costs, system demand, generating units' failures, and competitors' behavior. Any of these factors can be considered as stochastic with the presented method, which is based on a scenario-tree representation (Dupacova et al. 2000). The proposed model allows for the computation of non-anticipatory decisions for systems with numerous generation units, including a detailed description of the technical characteristics of generating plants that determine the shape of agents' cost functions. Moreover, scenario trees can be extended to a large number of periods. These two characteristics add special interest to the model from a practical point of view. In addition, we do not represent forward contracting to simplify the description of the model.

An extension of the deterministic market representation is required to consider uncertainty for some of the parameters included in it.[5] Scenario analysis arises as the first possibility. The number of alternative scenarios that can be considered has to be reduced due to the size of the problem. This approach consists of first finding market equilibrium separately for each alternative value of the uncertain parameters and then performing a simultaneous analysis of the results obtained, computing, for example, an average value. This method lacks robustness for short-term operation: independent equilibrium computation leads to different solutions for the first periods when a single decision would be desirable. Moreover, the reduced number of scenarios makes the analysis insufficiently accurate if treated as a Monte Carlo simulation.

A second and more sophisticated alternative for taking into account uncertainty, especially when the size of the problem prevents an extensive Monte Carlo analysis, is using a scenario tree, which allows the inclusion of stochastic variables and provides a single decision in the short-term. Figure 1 shows an example of the structure of a scenario tree. Each period p has been assigned a consecutive number, as well as each branch k within each period.

The set $B(p)$ comprises those branches that are defined for each period. The correspondence of precedence $a(p,k)$ establishes a relationship between the branch k of period p with the previous one. These two elements precisely define the tree structure. For example, in Fig. 1, $a(p_3, k_2) = k_1$ and $B(p_3) = \{k_1, k_2, k_3, k_4, k_5\}$.

[5] No transmission network is considered in this section.

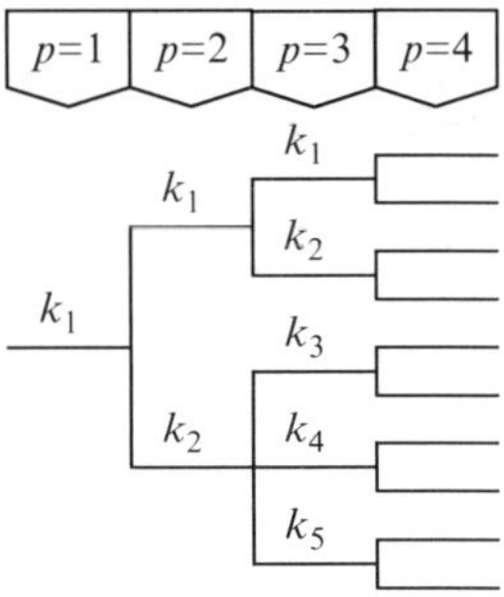

Fig. 1 Example structure of a scenario tree

Each branch is assigned a probability w_{pk}. This assignment must satisfy two coherence conditions. First, the probability of the single branch in the first period must be one, and second, the subsequent branches of a single branch must add to the probability of the previous branch:

$$\sum_{k^*|a(p,k^*)=k} w_{pk^*} = w_{p-1,k} \; \forall p > 1, \; \forall k \in B(p-1) \tag{40}$$

The overall profit function that will be considered in the stochastic model is the expected profit, which will be maximized for every generation company u in the model:

$$B_u = \sum_p \sum_{k \in B(p)} \sum_s \sum_b \left[l_{psb} \cdot w_{pksb} \cdot \left[\lambda_{pksb} \cdot P_{upksb} - C_{upksb}\left(P_{upksb}\right)\right]\right] \tag{41}$$

This extension of profit definition includes different prices and companies' productions for each branch, which allows the inclusion of uncertainty in parameters that affect costs, such as fuel costs, hydro inflows, or the availability of units. Note that the cost function for every branch depends on the production levels in other branches, possibly including the present and all the previous ones. So, it is possible to accommodate hydro-management and other inter-temporal policies. On the other hand, costs of a given company are not dependent on other companies' productions. This excludes issues like joint management of common hydrobasins. Companies are linked only in the electricity market.

Some extensions are required to establish a stochastic formulation of the previous deterministic market equilibrium over a scenario tree structure. First, the conjectured responses of the clearing price with respect to each generation company production are different in each branch of the tree and represent different behaviors of the companies under different circumstances:

$$\theta_{upksb} = -\frac{\partial \lambda_{pksb}}{\partial P_{upksb}} \tag{42}$$

Additionally, demand is also different for each branch, which makes possible the introduction of uncertainty in its value:

$$D_{pksb} = D^0_{pksb} - \alpha_{0pksb} \cdot \lambda_{pksb} \tag{43}$$

Furthermore, system balance constraint must be satisfied for every branch:

$$\sum_{u=1}^{U} P_{upksb} - D_{pksb} = 0 \tag{44}$$

Finally, cost functions $C_{upksb}(P_{upksb})$ are also dependent on the branch. Maximization of the new profit expression for each generation company leads to

$$l_{psb} \cdot w_{pksb} \cdot \lambda_{pksb} = \frac{\partial \sum\limits_{upksb} C_{upksb}\left(P_{upksb}\right)}{\partial P_{upksb}} + l_{psb} \cdot w_{pksb} \cdot P_{upksb} \cdot \theta_{upksb} \tag{45}$$

Equations (43), (44), and (45) define the stochastic market equilibrium. The newly defined equilibrium is obtained from the solution of the following optimization problem:

$$\begin{aligned} \min_{P_{upksb}, D_{pksb}} \quad & \sum_{p} \sum_{k \in B(k)} \sum_{s} \sum_{b} \left[l_{psb} \cdot w_{pksb} \cdot \left[\sum_{u=1}^{U} \left[\bar{C}_{upksb}\left(P_{upksb}\right) \right] - U\left(D_{pksb}\right) \right] \right] \\ \text{s.t.} \quad & \sum_{u=1}^{U} P_{upksb} - D_{pksb} = 0 \qquad : \eta_{pksb} \end{aligned} \tag{46}$$

In this expression, effective cost functions $\bar{C}_{upksb}\left(P_{upksb}\right)$ have been introduced:

$$\bar{C}_{upksb}\left(P_{upksb}\right) = C_{upksb}\left(P_{upksb}\right) + \frac{P^2_{upksb} \cdot \theta_{upksb}}{2} \tag{47}$$

Utility function $U(D_{pksb})$ maintains its definition (5) in each branch

$$U\left(D_{pksb}\right) = \int_0^{D_{pksb}} \lambda_{pksb}\left(D_{pksb}\right) \cdot dD_{pksb} = \frac{1}{\alpha_{0pksb}} \cdot \left(D_{pksb} \cdot D^0_{pksb} - \frac{D^2_{pksb}}{2} \right) \tag{48}$$

The system's marginal clearing price λ_{pksb} for each period p, branch k, subperiod s, and load level b can be computed from the Lagrange multiplier of the constraint as follows:

$$\lambda_{pksb} = \frac{\eta_{pksb}}{l_{psb} \cdot w_{pksb}} \tag{49}$$

For the sake of simplicity, the complete formulation of the stochastic optimization problem is not shown. Nevertheless, the equivalent minimization problem for the deterministic case is easily extended to obtain an equivalent problem for the stochastic one (Centeno et al. 2007). In this study, the above modeling was applied to a medium-term system with significant hydro resources, which were the main uncertainty source and also the most significant reason for inter-temporal constraints. Natural final conditions of the reservoirs' levels were also available: those required at the end of the dry season. Therefore, there was a natural horizon to be analyzed.

7 Medium-Term Signals for Short-Term Operation

The operation of power generation systems has traditionally been organized following a hierarchical structure. In the new liberalized framework, this hierarchy is maintained. Planning decisions belong to a long-, medium-, and short-term level according to their horizon of influence (Pereira and Pinto 1983). Typically, the long-term decision level considers more than 3 years of operation, the medium-term level encompasses from a few months up to 2 years, and the short-term level includes at most the following week. The detail with which the power system and the time intervals are represented diminishes as the time horizon of interest increases.

Longer-term decision levels yield resource allocation requirements that must be incorporated into shorter-term decision levels. This coordination between different decision levels is particularly important to guarantee that certain aspects of the operation that arise in the medium-term level are explicitly taken into account. Therefore, when a constraint is defined over a long- or medium-term scope, it is necessary to apply a methodology, so that it will be considered properly in the short-term operation planning.

Traditional short-term operation-planning tools such as unit-commitment or economic-dispatch models include guidelines to direct their results towards the objectives previously identified by the medium-term models. It is important to highlight that the models adopted in practice to represent medium- and short-term operation differ significantly. Usually, medium-term models are equilibrium problems with linear constraints that represent oligolopolistic markets. However, short-term models represent in detail the problem for a company facing a residual demand function (Baillo et al. 2006). There are two main reasons for not adopting a market-equilibrium model in the short term: the model would become computationally unaffordable, and it is very rare for a company to own detailed data about its competitors' generation units.[6]

When addressing the coordination between its medium-term planning and its short-term operation, the company faces two kinds of situations.

[6] A company can estimate medium-term model data from its knowledge of its own units. Nevertheless, short-term model data include real-time characteristics, which can differ significantly between different units.

- The allocation of limited-energy resources throughout the whole medium-term horizon, for example, hydro resources or maximum production levels for thermal units with limited emission allowances.
- The allocation of obligatory-use resources throughout the whole medium-term horizon, such as minimum production levels for technical, economic, or strategic reasons. Two examples of this situation are a minimum-fuel-consumption requirement due to a take-or-pay contract or a minimum market share for the company to maintain its market position.

Three different approaches for coordinating medium- and short-term models are proposed: the primal-information approach, the dual-information approach, and the marginal resource-valuation function.

7.1 Primal-Information Approach

This has been mostly used in practice to send signals from medium-term generation-planning models to short-term operation models.

Once the market equilibrium is computed with the medium-term model, a resource production level will be obtained for each time period considered in the medium-term scope, either for limited-energy resources or for obligatory-use resources. The strategy followed by the primal-information approach is to strictly impose these production levels of resources, that is, primal signals, as constraints on the short-term operation of the company.

The main advantage of the primal coordination is that it is easily obtained from the medium-term models and is not difficult to implement in the short-term models. Furthermore, the signals provided with this approach are very easy to understand, since they are mere short-term-scope production levels of the resources. Finally, the primal-information approach ensures that the medium-term objective will be satisfied.

However, this methodology has important drawbacks. The main one is the lack of flexibility in the decisions that the company can make in the short term: the resource usage levels provided by the medium-term model may not be optimal, mainly due to the time aggregations in the models and to the presence of uncertainty. In addition, the results obtained with the medium-term model are based on a forecast of the future market conditions that, in the end, may not arise. Finally, the medium-term model deals with aggregated load levels that may distort short-term results.

Hence, when facing the short-term operation, the company may find a market situation considerably different from the one forecasted, and the primal signals may not be consistent or economically efficient.

7.2 Dual-Information Approach

The dual-information approach is based on valuing the company's resources in the medium-term horizon. The proposed medium-term model makes possible the computing of marginal valuations of the resources, either limited-energy resources or obligatory-use resources (Reneses et al. 2004). These valuations are given by the dual variables of the corresponding constraints. For a limited-energy resource, the valuation is the dual variable of the maximum-medium-term-production constraint. For an obligatory-use resource, the valuation is the dual variable of the minimum-medium-term-production constraint.

Once the medium-term planning provides these valuations, that is, dual signals, they can be incorporated into the short-term operation. The short-term model may incorporate the explicit valuation of the company's resources into its objective function, depending on the kind of resource. For a limited-energy resource, its use will be penalized. For an obligatory-use resource, it will be a bonus for its use.

Equation (50) illustrates an example of an objective function. It considers hydro production for a generating unit h valuated through the dual variable μ, and annual minimum production for a thermal unit t valuated through the dual variable π. Both of them are obtained from the medium-term model constraints:

$$B_u = \sum_n \left[\lambda_n \left(P_{un} \right) \cdot P_{un} - C_{un} \left(\dot{P}_{un} \right) + \mu \cdot \sum_{h \in u} P_{hn} + \pi \cdot \sum_{t \in u} P_{tn} \right] \tag{50}$$

where B_u is the profit of the company u and $C_{un}(P_{un})$ is the cost function of the company's generating portfolio in hour n.

Note that μ is nonpositive and π is non-negative. Thus, there is a penalty for the use of hydro production of unit h and a bonus for the use of thermal production of unit t. As a consequence of this, the sign of dual variables directly takes into account the penalty or bonus for different resources.

An interesting aspect of the dual-information approach is the interpretation of the dual signals. They should not be considered as a reflection of how much it *costs* the company to use the resource in the short-term scope, but rather of how much it *costs* in the rest of the medium-term scope.

The main disadvantage of dual coordination is the lack of robustness. A small change in the medium-term valuation can lead to important changes in the short-term operation. Another problem of the dual approach arises when the actual market conditions are considerably different from the forecasts used in the medium-term planning. In this situation, the valuation provided by the medium-term model may be incorrect. One way to avoid these undesirable effects is to combine the primal and dual approaches. The valuation of the resources is included, but the deviation from the medium-term results is limited to a selected range.

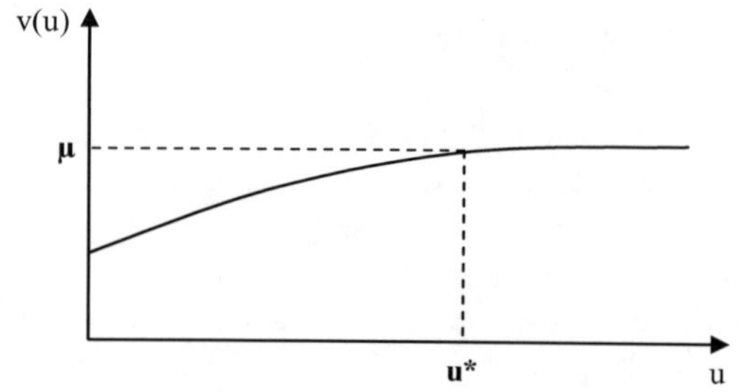

Fig. 2 Marginal valuation function for a limited-energy resource

7.3 Marginal Resource-Valuation Functions

A marginal resource-valuation function is a continuous valuation of a resource, either a limited-energy resource or an obligatory-use resource, for a range of operating points that the company could face. The valuation function is an extension of the dual-information approach, which provides only one point of the function. An example of the absolute value, because it is nonpositive, of a marginal valuation function $v(u)$ for a limited-energy resource u is shown in Fig. 2.

The independent variable of the function is the total use of the resource in the short-term scope. μ is the absolute value of the valuation for the resource used in the dual-information approach, and u^* is the resource level used in the primal-information approach.

Note that the absolute value of the function is nondecreasing, because it will be usually higher if the use of the resource is higher in the short-term operation, and, consequently, it will be lower in the rest of the medium-term scope. However, the marginal valuation for an obligatory-use resource is a nonincreasing function.

The coordination based on marginal-resource-valuation functions improves the dual-information approach, because it eliminates its two main disadvantages: on the one hand, the valuation function provides robustness, since significant changes in short-term operation are discouraged; on the other hand, the function provides information for all the values of the independent variable, even for those far from the medium-term forecast.

The marginal valuation functions are incorporated into the short-term operation through new terms in the objective function of the short-term model. Each term corresponds to the total valuation $V(u)$ of a resource, computed as the integral of the marginal valuation function:

$$V(u) = \int_0^u v(r)\,dr \tag{51}$$

The main drawback of this method is its practical implementation. The computation of the function may be difficult with the medium-term model, although it allows thc obtaining of different points of the function through different executions of the model. The computation of this set of points leads to thc incorporation of (51) as a piecewise-linear function (see Reneses et al. 2006 for a practical application).

Finally, it should be pointed out that marginal resource-valuation functions are actually multidimensional, that is, the valuation of every resource depends not only on its use, but also on the use of the rest of the resources. Hence, marginal valuation functions are really an approximation of these multidimensional functions. The computation and implementation of multidimensional valuation functions is an immediate extension of the proposed methodology. Nevertheless, the computer time requirements could not justify their use. An alternative is making use of approximate techniques that are well known in cost-minimization frameworks. For example, dynamic programming by successive approximations is used in Giles and Wunderlich (1981) in the long-term operation of a large multi-reservoir system.

8 Conclusion

Electricity market's equilibrium computing remains a fascinating area, with a large number of topics open to research. Some of these topics address concerns of interest in fundamental economics, such as the modeling of oligopolistic behavior. In fact, there is not a wide accepted economic theory for this kind of markets, as opposed to the perfect competition and monopoly situations. However, research in electricity markets has contributed some interesting insights in this regard.

The aim of this paper has been more modest: to contribute to the modeling of well-known oligopolistic competition paradigms (namely, Cournot and conjectural variations competition) for huge systems, including all the complexities that have traditionally been considered, especially medium-term uncertainties and network effects. Many topics remain open in both areas, such as risk aversion modeling or complex strategies based on network effects.

Acknowledgements We thank an anonymous reviewer for their helpful remarks.

References

Allaz B, Villa JL (1993) Cournot competition, forward markets and efficiency. J Econ Theory 59:1–16

Anderson EJ, Xu H (2006) Optimal supply functions in electricity markets with option contracts and non smooth costs. Math Methods Oper Rcs 63:387–411

Baillo A, Cerisola S, Fernandez-Lopez JM, Bellido R (2006) Strategic bidding in electricity spot markets under uncertainty: a roadmap. In: Power Engineering Society General Meeting

Baldick R, Grant R, Kahn E (2004) Theory and application of linear supply function equilibrium in electricity markets. J Regul Econ 25(2):143–167

Barquín J, Centeno E, Reneses J (2004) Medium-term generation programming in competitive environments: a new optimization approach for market equilibrium computing. Proc Inst Electr Eng Gener Transm Distrib 151(1):119–126

Barquin J (2006) Critically-congested Nash–Cournot equilibria in electricity networks. In: Proceedings Of the 9th Probabilistic Methods Applied to Power Systems Conference, Stockholm, Sweden, 2006

Barquín J, Rodilla P, Batlle C, Vázquez C, Rivier M (2007) Modeling market power: the Spanish White Paper approach. In: Power Systems Modelling, Athens, Greece, 2007

Barquín J, Vázquez M (2008) Cournot equilibrium in power networks: an optimization approach with price response computation. IEEE Trans Power Syst 23(2):317–26

Barquín J (2008) Symmetry properties of conjectural price responses. In: IEEE General Meeting 2008, Pittsburgh, Pennsylvania, 2008

Bertrand J (1883) Théorie Mathématique de la Richesse Sociale. J Savants 499–508

Borenstein S, Bushnell J (1999) An empirical analysis of the potential for market power in California's electricity industry. J Ind Econ 47:285–323

Bushnell J (1998) Water and power: hydroelectric resources in the era of competition in the Western US. In: Power Conf. Electr. Restruct., Univ. Calif. Berkeley Energy Inst., 1998

Bushnell J (2003) A mixed complementarity model of hydrothermal electricity competition in the Western United States. Oper Res 51(1):80–93

Bunn DW (2003) Modelling prices in competitive electricity markets. Wiley, Chichester, UK

Cardell JB, Hitt CC, Hogan WW (1997) Market power and strategic interaction in electricity networks. Resour Energy Econ 19:109–137

Centeno E, Reneses J, Barquín J (2007) Strategic analysis of electricity markets under uncertainty: a conjectured-price-response approach. IEEE Trans Power Syst 22(1):423–432

Cottle RW, Pang JS, Stone RE (1992) The linear complementarity problem. Academic, San Diego, CA

Cournot A (1838) Recherches sur les principies mathématiques de la théorie des richesses. Paris, Hachette. English Translation: Researches into the mathematical principles of the theory of wealth (1897)

Daughety AF (1988) Cournot oligopoly: Characterization and applications. Cambridge University Press, Cambridge, UK.

Daxhelet O, Smeers Y (2001) Inequality models of restructured electricity systems. In Ferris MC, Mangarasian OL, Pang JS (ed) Complementarity: applications, algorithms and extensions. Kluwer, Norwell, MA

Day CJ, Hobbs BF, Pang JS (2002) Oligopolistic competition in power networks: a conjectured supply function approach. IEEE Trans Power Syst 17(3):597–607

Dupacova J, Consigli G, Wallave W (2000) Scenarios for multistage stochastic programs. Ann Oper Res 100:25–53

Ehrenmann A (2004) Manifolds of multi-leader Cournot equlibria. Oper Res Lett 32(2):121–125

Ehrenmann A, Neuhoff K (2003) A comparison of electricity market designs in networks. CMI. Working Paper 31

Eydeland A, Wolyniec K (2003) Energy and power risk management: new developments in modeling, pricing, and hedging. Wiley

Figuières C, Jean-Marie A, Querou N, Tidball M (2004) Theory of conjectural variations, World Scientific

García-Alcalde A, Ventosa M, Rivier M, Ramos A, Relaño G (2002) Fitting electricity market models. A conjectural variations approach. 14th PSCC, Power System Computation Conference, Seville, Spain, 2002

Giles JE, Wunderlich WO (1981) Weekly multipurpose planning model for TVA reservoir system. Proc ASCE J Water Resour Plan Manag Div 107:495–511

Gorenstin BG, Campodonico NM, Costa JP, Pereira MVF (1992) Stochastic optimization of a hydro-thermal system including network constraints. IEEE Trans Power Syst 7(2):791–797

Green RJ, Newbery DM (1992) Competition in the British electric spot market. J Polit Econ 100:929–53

Hamilton JD (1994) Time series analysis. Princeton University Press, Princeton, NJ
Harvey AC (1989) Forecasting, structural time series and the Kalman filter. Cambridge University Press, Cambridge
Hobbs BF, Rijkers FAM (2004) Strategic generation with conjectured transmission price responses in a mixed transmission pricing system I: formulation and II: application. IEEE Trans Power Syst 19(2):707–717, 872–879
Hobbs BF (2001) Linear complementarity models of Nash–Cournot competition in bilateral and POOLCO power markets. IEEE Trans Power Syst 16(2):194–202
Hobbs BF, Metzler CB, Pang JS (2000) Strategic gaming analysis for electric power systems an MPEC approach. IEEE Trans Power Syst 15(2):638–644
Johnsen TA (2001) Hydropower generation and storage, transmission contraints and market power. Utilities Policy 10(2):63–73
Kelman R, Barroso LA, Pereira MVF (2001) Market power assessment and mitigation in hydrothermal systems. IEEE Trans Power Syst 16(3):354–359
Kemplerer PD, Meyer MA (1989) Supply function equilibria in oligopoly under uncertainty. Econométrica 57(6):1243–1277
Kiss A, Barquín J, Vázquez M (2006) Can closer integration mitigate market power? A numerical modelling exercise. In: Towards more integration of central and eastern energy markets. Ed. REKK. Budapest, Hungary
Liski M, Montero JP (2006) Forward Trading and Collusion in Oligopoly. J Econ Theory 131: 212–230
Le Coq C (2004) Long-term supply contracts and collusion in the electricity market. Stockholm School of Economics. Working Paper Series in Economics and Finance 552
López de Haro S, Sánchez P, de la Hoz JE, Fernández J (2007) Estimating conjectural variations for electricity market models. Eur J Oper Res 181(3):1322–1338
Metzler C, Hobbs BF, Pang JS (2003) Nash–Cournot equilibria in power markets on a linearized DC network with arbitrage: formulations and properties. Netw Spat Econ 3(2):123–150
Morton DP (1993) Algorithmic advances in multi-stage stochastic programming. PhD Diss. Stanford University, Stanford, CA
Murto P (2003) On investment, uncertainty, and strategic interaction with applications in energy markets. PhD Diss., Helsinki University of Technology
Nash JF (1950) Equilibrium points in N-person games. Proc Nat Acad Sci 36:48–49
Neuhoff K (2003) Integrating transmission and energy markets mitigates market power. Cambridge Working Papers in Economics 0310, Faculty of Economics, University of Cambridge
Otero-Novas I, Meseguer C, Batlle C, Alba JJ (2000) Simulation model for a competitive generation market. IEEE Trans Power Syst 15(1):250–256
Pereira MVF, Pinto MVG (1983) Application techniques to the mid- and short-term scheduling of hydrothermal systems. IEEE Trans Power Syst 102(11):3611–3618
Pereira MVF, Pinto MVG (1991) Multi-stage stochastic optimization applied to energy planning. Math Program 52:359–375
Pérez-Arriaga JI, Rivier M, Batlle C, Vázquez C, Rodilla P (2005) White paper on the reform of the regulatory framework of Spain's electricity generation. available at http://www.mityc.es/NR/rdonlyres/5FB9B792–1D76–4FB3B11EB4B3E49A63C7/0/LibroBlanco.pdf (in Spanish).
Perry MK (1982) Oligolpoly and consistent conjectural variations. Bell J Econ 13(1):197–205
Reneses J, Centeno E, Barquín J (2004) Medium-term marginal costs in competitive power markets. Proc Inst Electr Eng Gener Transm Distrib 151(5):604–610
Reneses J, Centeno E, Barquín J (2006) Coordination between medium-term generation planning and short-term operation in electricity markets. IEEE Trans Power Syst 21(1):43–52
Rivier M, Ventosa M, Ramos A (1999) A generation operation planning model in deregulated electricity markets based on the complementary problem. ICCP, International Conference on Complementary Problems, Madison, WI
Rudkevich A (2005) On the supply function equilibrium and its application in electricity markets. Decis Support Syst 40(3–4):409–25

Scott TJ, Read EG (1996) Modeling hydro reservoir operation in a deregulated electricity market. Int Trans Oper Res 3:243–253

Skaar J, Sorgard L (2006) Temporary Bottlenecks, Hydropower and Acquisitions. Scand J Econ 108(3):481–497

Song Y, Ni Y, Wen F, Hou Z, Wu FF (2003) Conjectural variation based bidding strategy in spot markets: fundamentals and comparison with classical game theoretical bidding strategies. Electr Power Syst Res 67:45–51

Vives X (1999) Oligopoly pricing: old ideas and new tools. MIT, Cambridge, MA

Yao J, Oren SS, Adler I (2004) Computing Cournot equilibria in two settlement electricity markets with transmission constraints. 37th International Conference on System Science

Yuan WJ, Smeers Y (1999) Spatial oligopolistic electricity models with Cournot generators and regulated transmission prices. Oper Res 47(1):102–12

Zhang Y, Zwart G (2006) Market power mitigation contracts and contracts duration. Tilburg University (TILEC). mimeo

Part IV
Risk Management

A Multi-stage Stochastic Programming Model for Managing Risk-optimal Electricity Portfolios

Ronald Hochreiter and David Wozabal

Abstract We present a multi-stage decision model, which serves as a building block for solving various electricity portfolio management problems. The basic setup consists of a portfolio optimization model for a large energy consumer, which has to decide about its mid-term electricity portfolio composition. The given stochastic demand may be fulfilled by buying energy on the spot or futures market, by signing a supply contract, or by producing electricity in a small plant. We formulate the problem in a dynamic risk management-based stochastic optimization framework, whose flexibility allows for extensive parameter studies and comparative analysis of different types of supply contracts. A number of application examples is presented to outline the possibilities of using the basic multi-stage stochastic programming model to address a range of issues related to the design of optimal policies. Apart from the question of an optimal energy policy mix for a large energy consumer, we investigate the pricing problem for flexible supply contracts from the perspective of an energy trader, demonstrating the wide applicability of the framework.

Keywords Electricity portfolio management · Multi-stage decision optimization · Real option pricing · Risk management · Stochastic programming

1 Introduction

In this chapter we present a versatile multi-stage stochastic optimization model for calculating optimal electricity portfolios. The model optimizes the decision of a large energy consumer facing the problem of determining its optimal electricity mix consisting of different kinds of supply contracts for electric energy. We assume that

R. Hochreiter (✉)
Department of Finance, Accounting and Statistics, WU Vienna University of Economics and Business, Augasse 2-6, 1090 Vienna, Austria
e-mail: ronald.hochreiter@wu.ac.at

S. Rebennack et al. (eds.), *Handbook of Power Systems II*, Energy Systems,
DOI 10.1007/978-3-642-12686-4_14,

the economic agent under consideration has a diverse set of possibilities to obtain electric energy. Thereby we consider three main sources: products that are traded at the EEX (both spot and futures), producing energy in a (small) plant, and to buy a swing option from an energy trader. The decision that has to be taken is the optimal mix of all these instruments to obtain an energy portfolio given some risk constraints, highlighting the importance of risk management. Stochasticity enters the model both via stochastic demand of the consumer, as well as the stochastic spot price for energy at the EEX. The decision that has to be taken at the root stage of the time horizon under consideration is the design of the supply contract bought from an energy trader and the amount of futures (peak and base) that will be bought. Implicit in this decision is that the rest of the stochastic demand has to be fulfilled by own production or via the spot market. The formulation is that the stochastic program takes these possibilities into account and provides an optimal decision given the possible future situations. We refer to Ruszczyński and Shapiro (2003) for an overview of stochastic programming and to Wallace and Ziemba (2005) for a comprehensive overview of stochastic programming applications. In summary, consider the general formulation of a multistage stochastic optimization program, that is,

$$
\begin{aligned}
\text{minimize } & \mathbf{F}\big(f(x(\xi),\xi)\big)\\
\text{subject to } & (x(\xi),\xi) \in \mathcal{X}\\
& x \in \mathcal{N},
\end{aligned}
$$

where $\xi = (\xi_1, \ldots, \xi_T)$ denotes a multi-dimensional stochastic process describing the future uncertainty. The constraint-set $\mathcal{X}$ contains feasible solutions (x, ξ) and the (non-anticipativity) set $\mathcal{N}$ of functions $\xi \mapsto x$ is necessary to ensure that the decisions x_t are only based on realizations up to stage t, that is, $(\xi_0, \ldots, \xi_t)$. $f(x(\xi), \xi)$ denotes the objective function. An important feature for practical applicability of stochastic programming in general is that contemporary risk management techniques can be integrated naturally by choosing the appropriate risk functional as the respective probability functional $\mathbf{F}$. For more details see, for example, Pflug and Römisch (2007), McNeil et al. (2005), and especially Eichhorn and Römisch (2008) in the context of stochastic programming. From the stochastic programming point of view, a vast range of similar applications in the area of optimal energy and electricity portfolios have been presented, see, for example, Eichhorn et al. (2005), Sen et al. (2006), Schultz et al. (2003), and Hochreiter et al. (2006) and references therein.

However, the main feature of the approach presented in this chapter is that while stochastic programming is often neglected for practical application due to its inherent complexity and implementation overhead, we are trying to present a balanced trade-off by adding a realistically comprehensive set of real-world constraints, while still maintaining reasonable numerical properties. Most importantly, we model a mid-term planning horizon of half a year. Every day in the planning horizon is modeled as one stage in the stochastic program where a decision can be taken. Hence, we consider a stochastic optimization problem with 184 stages, that is, one stage per day. The natural intra-day model would necessitate to describe the energy consumption by 24 different values. This would lead to 24 decision variables in every stage for every possible source of energy. Since this approach would lead to models that are way too large, we simplify the situation by introducing six 4 h blocks and therefore reduce the complexity of the model by 75%. We will try to focus on each step that is necessary to build and solve the optimization model. The general workflow looks like this:

1. Model the uncertain spot price process
2. Generate possible scenarios for the spot price movement and the demand by simulating disturbance in the respective models
3. Construct a scenario tree from the fan of simulated trajectories
4. Choose an appropriate risk measure
5. Model decision problem as multi-stage stochastic optimization program
6. Solve the optimization problem

Each of the above workflow steps is highly non-trivial and deserves in depth discussion. Furthermore, there exist various techniques to achieve the goals outlined in every single step. Aim of this chapter is not to identify the best possible methods for each of the tasks or to discuss the problems arising in great detail, but rather to give an overview of one particular way to solve the problems and show how methods to solve 1–6 can be made to fit together in a stochastic programming workflow.

This chapter is organized as follows: In Sect. 2 we explain how we model uncertainty in our stochastic model, describe the stochastic variables, and the way to construct a joint spot/demand scenario tree, which is necessary to solve the deterministic equivalent formulation of the respective stochastic program with a numerical solver. In Sect. 3 we introduce the main energy mix model, as well as a model for pricing supply contracts, which is obtained by slight rearrangement of the objective function and the constraints. Section 4 summarizes numerical results, while Sect. 5 concludes the chapter.

2 Modeling Uncertainty in Electricity Markets

Statistical modeling energy spot prices has attracted considerable interest in the statistical community (see, e.g. de Jong and Huisman (2002); Escribano et al. (2002); Lucia and Schwartz (2002)).

The aim of this section is to briefly sketch a model for the spot price of energy of the European Energy Exchange (EEX). The model should be sufficiently sophisticated to explain most of the variation in the spot prices, while at the same time be sufficiently simple to use it for the generation of scenarios. These scenarios will be used to generate trees on which the stochastic programs will be formulated. In particular, the requirement to simulate possible future price trajectories from the model restricts us from including explanatory variables which themselves are random factors.

2.1 Econometric Model

Our main focus in this section is to build an econometric model that is capable of capturing the main features of the spot price process and from which we are able to simulate possible future spot price scenarios. The second point is crucial, since it shifts the focus from in-sample accuracy as a measure of quality to the quality (and credibility) of the simulated prices.

The spot market data of the EEX in the past years possessed five distinctive features (see, e.g. Lucia and Schwartz (2002)):

1. Mean spot prices follow a pattern, which depends mainly on the combination of the day of the week and the hour of the day
2. Once in a while extreme price events occur that can lift the price to multiples of its "normal" values (these excursions of the price typically last only 1–2 h)
3. Prices vary with the season (summer/winter and transition time)
4. Prices are lower on public holidays
5. All patterns that can be observed in energy spot prices (including to some degree those described in 1–3) are unstable in the long run

See Fig. 1 for a plot of typical 20 days of EEX spot price that illustrates points 1 and 2 above.

Of course also other factors seem to influence the spot price for energy (or at least exhibit positive correlation). Among these are the price for fossil fuels, temperature and data on sky cover. However 1, 2 and 4 from above seem to explain most of the variance in the spot prices (at least for short time horizons). Furthermore, these

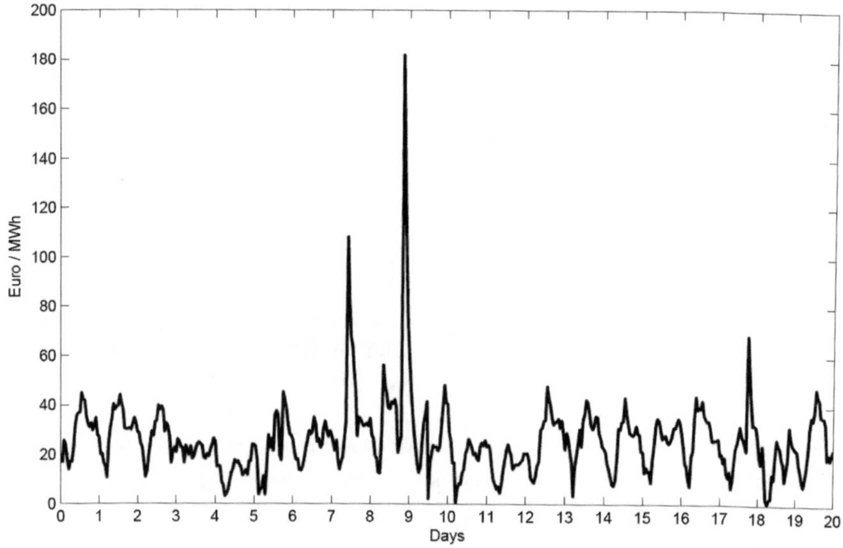

Fig. 1 Twenty days of spot prices at the EEX including two price spikes of moderate intensity

factors are non-stochastic and can therefore easily be incorporated into a simulation framework (as opposed to, e.g. the prices of various fossil fuels).

The spiky behavior of the spot prices cannot really be forecasted or explained in an econometric model. Spikes occur due to the non-storability of electricity and the inelasticity of demand. The demand in a specific hour has to be met exactly in that hour. Only few consumers of energy have the possibility to delay their demand (for a reasonable price), and even fewer are as "energy aware" to actually consider this option. Coupled with the fact that the supply curve for electricity is very steep at the high end (because of high marginal costs of suppliers who normally cover unusually high demands), spikes occur at certain hours of the day when there is an demand/supply mismatch. Typically these hours are characterized by high demand like the hours around noon and in the late afternoon and early evening.

Since modeling all the above would lead to too complicated models, we restrict ourselves to handling spikes as part of the residuals of the model for the spot price and try to capture the nature of the residual distribution by statistical methods that permit us to simulate the residuals to our models and therefore also the spikes that might occur.

We therefore focus on modeling correct average prices for every hour of the day on every day of the week in every season of the year. Additionally we want to capture the effect of temperatures on the price, since it seems to be the only relevant factor (in terms of explanatory power), except the ones already mentioned worthwhile to be included in the model.

The outlined model is estimated using the least squares technique, where the variables indicating hour, day and season are modeled as dummy variables. Since it is unlikely that temperature has the same impact on demand for energy and therefore on the price of energy at different times of the day, we measure this effect separately for six 4 h blocks of the day, that is there is one regressor measuring the impact of temperatures in the hours 0–3, one for 4–7 and so on.

As already mentioned we model the regressors related to time (hour, day and season) as dummy variables (see Escribano et al. (2002) for a similar approach). This yields a group of $24 \times 7 \times 3 = 504$ dummy variables rendering the model computationally intractable. For this reason we try to reduce the number of regressors by a clustering approach. Since the coefficient of a dummy variable will (roughly) be the mean of the data points that it corresponds to, a feasible way to reduce the number of regressors is to compare the means of data corresponding to the different hours on the different days in the different seasons and club two regressors if the means are only insignificantly different. For example, if it turns out that the mean of the prices in summer on Sundays at midnight is statistically insignificantly different from the mean of the prices in winter at the same day and time, it does not make sense to distinguish these two possibilities in the model.

We test for difference in means using the non-parametric Kruskal–Wallis test (see, e.g. Hollander and Wolfe (1973)), which is based on rank orders instead of values. The choice of a non-parametric test is motivated by the fact that the observed market data is far from being normally distributed and therefore the classical tests for equal mean would fail.

With the described clustering procedure we were able to significantly reduce the number of regressors without sacrificing much of the accuracy in predicting the expected (mean) price in the respective hours.

2.2 Spot Price Simulation

For the simulation we group the residuals of the model in the six 4 h blocks mentioned above and fit a stable distribution to every blocks. The class of stable distributions is a parametric family of distributions that also contains extremely heavy tailed distributions, which in the case of electricity prices is needed to incorporate the possibility for occasional spikes. For an in depth description of heavy tailed distributions, their properties and how to fit them, see Rachev and Mittnik (2000). In Bernhardt et al. (2008) and Weron (2006), stable distributions are used to model the spikes that occur in electricity spot markets.

Using the estimates of the parameters of the stable distribution, we simulate residuals and thereby generate possible future price trajectories for the spot price on the EEX.

2.3 Demand Simulation

We use the demand of a medium size municipal utility (located in Upper Austria) and base our model on one yearly consumption pattern. Like with energy prices, the demand is estimated as a linear model with similar regressors as in the case of the spot prices. However, the demand is much less complicated than the spot price process. In particular, the demand series lacks the erratic jump behavior the spot price process exhibits. Correspondingly, the distribution of the residuals does not have that heavy tails and can therefore be modeled as a simple t-distribution. In Fig. 2 typical 2 weeks of energy demand of the considered municipal utility are plotted. The pattern is indeed very similar to the regular pattern the spot market prices follows over the weekly and daily cycles and can be explained with a high accuracy with the modeling approach outlined in the previous section.

2.4 Temperature Simulation

Scenarios for future temperatures in hour i are obtained by simulating from the model $t_i = m_i + y_i$, where m_i is the mean temperature in the month which the hour i belongs to and y_i follows a autoregressive model of order 1, that is $y_i = \rho y_{i-1} + \varepsilon_i$, where the coefficient ρ is estimated from the historical temperature data and ε_i are the independent standard normal innovations of the process.

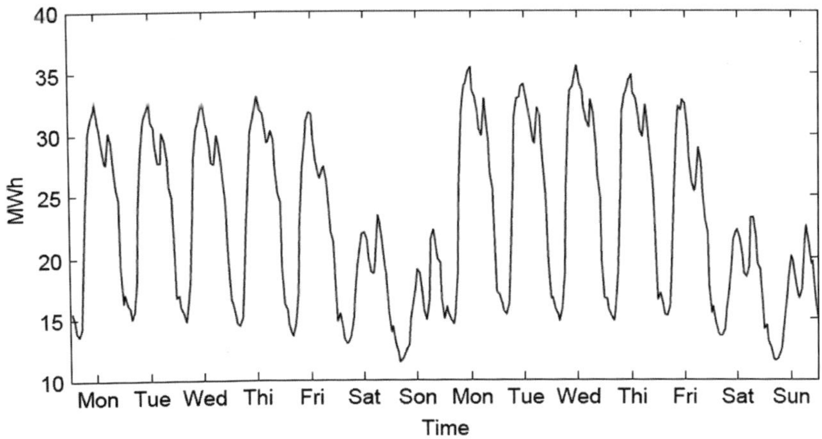

Fig. 2 Typical 2 weeks of the observed demand pattern of the considered municipal utility

See Király and Jánosi (2002) and references therein for a justification of this simple model for temperatures and possible extensions.

2.5 *Scenario Tree Generation*

We are trying to cope with a medium-term electricity portfolio optimization problem. In contrast to low-stage financial long-term planning models, we are facing daily electricity portfolio changes and even hourly price differences. This leads to a highly multi-stage decision horizon. In this regard, it is important to keep the trade-off between level of model detail (variables and constraints) for the underlying uncertainty (dimension of underlying scenario tree) and solvability of the problem in mind. As we are faced with a day-ahead hourly spot price market at the EEX, we can easily reduce the number of stages by designing the uncertainty such that instead of modeling prices and demands as uni-variate time series (i.e., 8,760 stages per year), we define one stage per day, that is a 24-dimensional time series with 365/366 stages per year. This is already a massive reduction. To simplify the model even further, we suggest to consider hour-blocks. Thereby, we group the uncertainty in 6×4 h blocks, that is we are dealing with a six-dimensional instead of a 24-dimensional process.

We sample both possible future price and demand scenarios separately. Since we assume independence in the error terms of the two models we can pairwise merge the scenarios for demand and price development. The resulting scenarios are merged into a tree structure, see Fig. 3.

We aim at calculating an optimal scenario tree given the input simulation paths and apply methods based on theoretical stability considerations, see Rachev and Römisch (2002) and Römisch (2003). The optimal tree needs to be constructed such

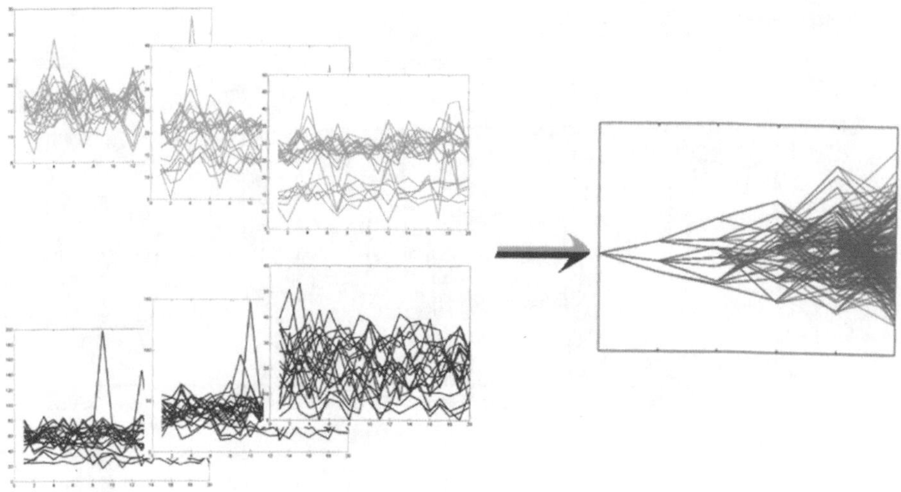

Fig. 3 Spot price and demand sampled independently and directly combined in a scenario tree

that distances (probability metrics) are minimized and a tree structure is created. Given a stochastic process ξ and a multi-stage stochastic program

$$\min\{f(x) = \mathbb{F}(x, \xi) : x \in \mathcal{X}\},$$

we are then able to find an optimal discrete approximation $\hat{\xi}$, such that we can numerically solve the approximated stochastic program

$$\min\{\hat{f}(x) = \mathbb{F}(x, \hat{\xi}) : x \in \mathcal{X}\}.$$

The approximation of the stochastic process ξ by $\hat{\xi}$ translates into an approximation of the objective value of the corresponding problems, that is

$$\sup_{x \in \mathcal{X}} |\hat{f}(x) - f(x)|$$

remains "small" (see Rachev and Römisch (2002); Römisch (2003); Pflug (2001) for further details) and thereby the two problems lead to similar objective values.

Still, even if the theoretical correct probability metric is chosen, we need to choose the appropriate heuristic to approximate the chosen distance. This is a numerically challenging task, and different heuristics are often combined and applied, for different approaches see, for example Heitsch and Römisch (2003) and Dupačová et al. (2003). For this specific application we used the multi-dimensional facility location problem formulation for multi-stage scenario generation, which is shown in Pflug (2001) and implemented for practical use in Hochreiter and Pflug (2007), see also Fig. 4. For highly-multi stage applications, we applied the forward

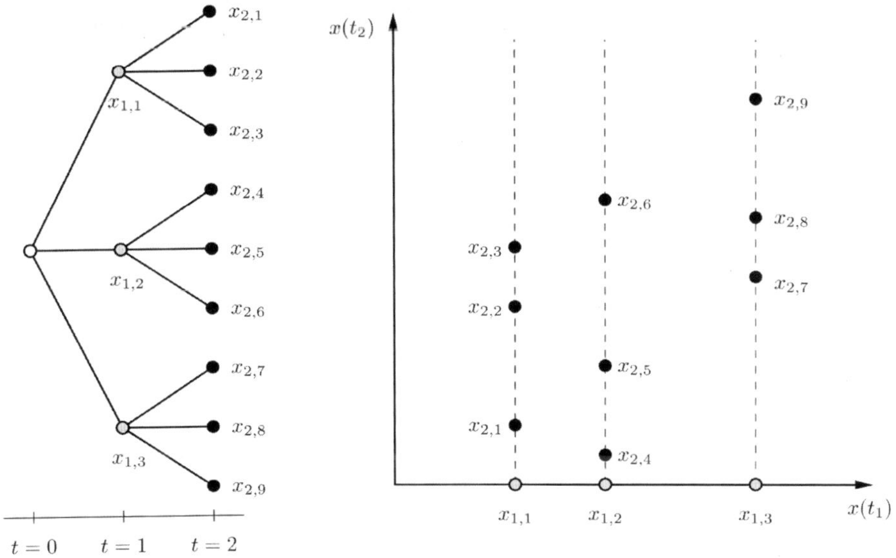

Fig. 4 Multi-stage scenario generation and multi-dimensional facility location

clustering approach using iterative K-Means-style approximations. Further details about the scenario trees are discussed in Sect. 4.

3 Optimization Models

In this section the stochastic optimization model is outlined. With this model it is possible to analyze a broad range of interesting questions. Apart from the problems treated in this paper the framework could, for example, be used to price a production unit in a real option framework. We will demonstrate the flexibility of the approach by discussing two models in detail, that is the problem of finding an optimal energy mix in Sect. 3.2 and the pricing of supply contracts in Sect. 3.3.

For all optimization models, we employ the terminal average value-at-risk (AVaR, also called conditional value at risk or Expected Shortfall) risk measure, mainly because a well-known linear programming reformulation exists, see Rockafellar and Uryasev (2000), and the risk measure is widely accepted for practical use. While we consider only the terminal version of the risk measure, a linear combination of multiple AVaRs in more stages is straightforward.

3.1 Types of Contracts

Liberalized energy markets offer both chances and risks to energy consumers. Especially energy consumers whose demand for energy is big enough to justify a rather involved energy mix nowadays have many options to satisfy their demand for power.

On the one hand, energy can be bought in advance on a yearly, quarterly or monthly basis; on the other hand, energy can be bought on a day ahead basis on the spot market. Additionally, it is always possible to take the service of an energy trader and buy a supply contract for energy. To satisfy the demand for energy in periods of unexpectedly high demand or prices, it might also be advantageous to obtain energy by own production in a small plant. Of course, generally it will be an optimal solution to rely on a mix of the options, since the different contract types have different (expected) prices and are subject to different types of risk. We consider two types of risk here.

1. We refer to the risk of having to pay a higher price for energy as expected as the *price risk*. The price risk is present in scenarios where a considerable portion of energy is bought on the spot market. The spot market for power is extremely volatile and shows highly fluctuating prices (see, e.g. Weron (2006)). Additionally, to short-term fluctuations due to demand supply mismatch, the average price of energy changes yearly because of changing conditions in the energy market.
2. The risk of contracting too much or too little energy at a too high price is referred to as the *contract risk*. The contract risk mainly stems from the inability to accurately predict the demand for energy in the planning period.

The actual strategy will depend – besides risk preferences – on how well the consumer knows her future demand. For example, a plant with a fixed production plan for a whole year may know its energy consumption very well and buy a suitable contract that supplies exactly the required amount with little freedom to deviate. Such a contract eliminates most of the risks and can probably be acquired at a reasonable price. There might be consumers with a somewhat erratic consumption that cannot be forecast very well. We assume that the demand is inelastic with respect to prices, that is demand does not depend on energy prices. This assumption seems reasonable for most of the energy consumers. The different types of contracts we consider in this chapter are described below. Table 1 summarizes the types of risk exposures that originate from different ways of obtaining electric energy.

3.1.1 Supply Contracts

By far the most common way of buying electricity is by buying a supply contract from an energy trader or producer. Details of the contract may vary from a flexible contract that is designed to cover the whole demand to a fixed delivery scheme.

Table 1 Different sources of energy and the associated risks

Risk Type	Contract Type			
	Spot Market	Futures	Power Plant	Supply Contract
Contract Risk	None	High	Mild	Medium – High
Price Risk	High	None	None	None

Different options may have different prices per MWh depending on the conditions that apply. A supply contract shifts the market *price risks* in the energy portfolio to the party that provides the contract. On the other hand, most of the common contracts require the buyer to know its energy demand and thereby induce *contract risk* as described before. The contract type that we consider here is a so-called swing option contract that leaves some flexibility of how much energy to actually buy. The overall quantity that can be bought out of the contract is limited by a fixed quantity C_h that can be decided for every hour block separately. Additionally, the contract sets certain limits on how the available energy might be used over the planning horizon. We consider the following restrictions for every day n

$$\gamma^l C_h \le c_{n,h} \le \gamma^u C_h, \quad \forall h : 1 \le h \le 6, \tag{1}$$

that is the consumption in every stage n and in every hour h is bounded by fractions of the overall contracted energy for that particular hour. We also limit the overall energy that can be used till a certain stage n to lie in a certain predefined cone that also is dependent on the contract volume. The restrictions are modeled as

$$n\delta^l C_h \le \sum_{i=1}^{n} c_{i,h} \le n\delta^u C_h, \quad \forall h : 1 \le h \le 6, \tag{2}$$

where δ^l and δ^u are the upper and the lower slope of the limiting lines of the cone.

Furthermore, we consider a differential constraint

$$|c_{n,h} - c_{n-1,h}| \le \eta_h, \quad \forall h : 1 \le h \le 6, \tag{3}$$

that is for a given hour the demand in consecutive stages can differ only by η_h. Figure 5 illustrates the constraints that a consumption pattern has to fulfill to comply with the rules of the swing option contract.

3.1.2 Futures

Futures are much less risky than buying on the spot market since the *price risk* disappears. However, the disadvantage of futures is that they are very coarse instruments available only in blocks for either base or peak hours. Therefore, a consumer who buys futures will inevitably end up with either too much or too little energy in certain periods of the planning horizon. These positions have to be evened up by one of the other instruments. In the case where there is an excess of energy from a future contract, the energy that cannot be used is simply lost (we assume that the selling of the excess energy on the spot market would lead to administrative costs that are too high to make this option attractive). Since in our model futures are only bought at the beginning of the planning horizon (where prices are known), we do not model the future price process.

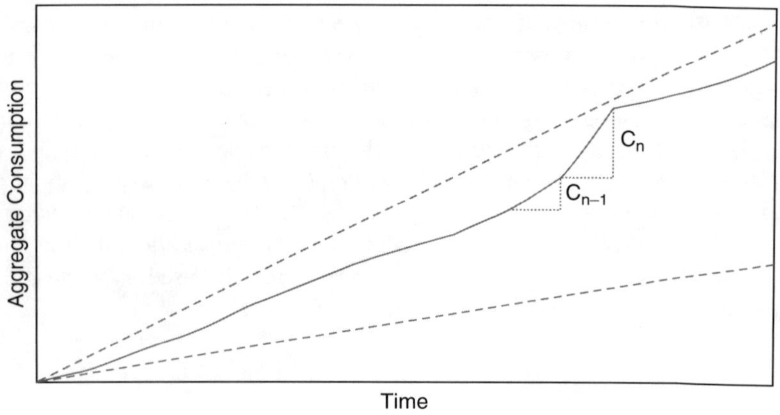

Fig. 5 Plot of the aggregate consumption with the conical bounds as *dashed lines*. Consecutive consumptions c_{n-1} and c_n have to fulfill the differential constraint (3) and every c_n is bounded by constraint (1)

3.1.3 Spot Market

Buying electricity on the spot market on a day ahead basis may lead to lower prices in the mean than buying energy in the form of a contract. Furthermore, the spot market offers full flexibility and the consumer is not bound to any commitments made in earlier stages; therefore, there is no contract risk when buying on the spot market. On the other hand, it results in taking the full *price risk*. We denote the amount of energy bought in stage n for hour h as $s_{n,h}$.

3.1.4 Production

The production from plants that belong to the consumer can typically not cover all the demand for energy and therefore in most cases has the character of a real option, that is the option to produce in an own plant if the market price is too high and thereby limiting the price paid per MWh. The price of the energy produced in the own plant can be assumed to be lower than the peak prices on the spot market but higher than the average prices. The power plant can be used to lower or even eliminate price risk connected to sudden peaks in energy prices, whereas unless it covers all the demand for energy, it can only dampen the price risk connected to changes in the average price levels. We denote the amount of energy produced in stage n for hour h as $p_{n,h}$. Furthermore, we assume that the capacities of the plant are limit by a maximum production P per hour block, that is

$$p_{n,h} \leq \pi.$$

To model a power plant we introduce the constraint that the production in the plant cannot change too much from one hour block to the next (see below). This restriction is realistic for most of the plants and introduces a risk similar to the *contract risk*, that is the problem that earlier decisions influence the profit and loss distribution of the present stage.

To realistically model a power plant, the decision to switch the plant on and off would have to be modeled by integer variables, rendering the considered problems linear mixed integer problems. Since we want to stick with problems that can be solved efficiently with interior point methods, we do not model the power plant in this way.

3.2 *Finding an Optimal Energy Mix*

In our multi-stage stochastic programming setting, let $t = 0, \ldots, T$ denote the stages, $\mathcal{N}$ be the node set of the scenario tree, $\mathcal{S}$ the set of scenarios, $\mathcal{N}^t$ all nodes of stage t, $\mathcal{N}_s$ all the nodes corresponding to scenario s and $S(n)$ the stage of node n. Furthermore, let P_s be the probability of the scenario s. The formulation below already considers the special hour-block structure. The energy portfolio in every hour block h and in every node of the tree n consists of the contracted volume $c_{n,h}$, energy bought via future contracts $f_{n,h}$, energy bought on the spot market $s_{n,h}$ and optionally the energy produced in the plant $p_{n,h}$.

3.2.1 Objective Function

The expenditures in scenario s (over the whole time horizon) is given by

$$e_s = \sum_{n \in \mathcal{N}_s, h=1,\ldots,6} e_{n,h},$$

where $e_{n,h}$ is the expenditure in the node n in hour block h given by

$$e_{n,h} = c_{n,h}C + f_{n,h}F_{n,h} + s_{n,h}S_{n,h} + p_{n,h}P,$$

with C, $F_{n,h}$, $S_{n,h}$, P the prices of one MWh of energy from the supply contract, the future contract, the spot market (stochastic variable) and the power plant, respectively. Our aim is to minimize

$$\sum_{s \in \mathcal{S}} P_s e_s + \kappa AVaR_\alpha \left(\sum_{s \in \mathcal{S}} P_s e_s \right),$$

that is to perform a bi-criteria optimization, which keeps expected costs and the AVaR of expected costs low. The parameter κ can be used to control risk-aversion.

3.2.2 Demand

The stochastic demand $d_{n,h}$has to be met at every node n of the tree and every hour h, that is

$$c_{n,h} + f_{n,h} + s_{n,h} + p_{n,h} \geq d_{n,h}, \quad \forall n \in \mathcal{N}, \ \forall h = 1, \ldots, 6.$$

3.2.3 Supply Contract

As discussed earlier, to meet the conditions of the supply contract the following has to be fulfilled. First, the amount, which is bought every hour is restrained by the constant γ^l and γ^u, and the overall contracted volume C_h for that hour block h (over the whole planning horizon) in the following way:

$$\gamma^l C_h \leq c_{n,h} \leq \gamma^u C_h, \quad \forall n \in \mathcal{N}, \quad h = 1, \ldots, 6.$$

Furthermore, the amount of energy bought in every scenario has to be within the cone described in (3), that is

$$\delta^l t C_h \leq \sum_{n \in \mathcal{N}_s : S(n) \leq t} c_{n,h} \leq \delta^u t C_h, \quad \forall s \in \mathcal{S}, \ \forall t = 1, \ldots, T, \ \forall h = 1, \ldots, 6.$$

Furthermore, from (3) we have

$$|c_{n,h} - c_{n,h-1}| \leq \eta_h, \forall h = 2, \ldots, 5, \quad \forall n \in \mathcal{N}$$

and

$$|c_{n,1} - c_{\text{pred}(n),6}| \leq \eta_h, \quad \forall n \in \mathcal{N}.$$

3.2.4 Stylized Power Plant

To model a stylized power plant, we define that its production $p_{n,h}$ is limited by a maximum amount π, that is

$$p_{n,h} \leq \pi, \quad \forall n \in \mathcal{N}, \quad h = 1, \ldots, 6,$$

and the production of the power plant cannot change more than a parameter β MWh per hour block, that is

$$|p_{n,h} - p_{n,h-1}| \leq \beta, \forall h = 2, \ldots, 5, \quad \forall n \in \mathcal{N}$$

and

$$|p_{n,1} - p_{\text{pred}(n),6}| \leq \beta, \quad \forall n \in \mathcal{N},$$

where $\text{pred}(n)$ denotes the predecessor node of the node n.

3.2.5 Putting It All Together

The final optimization model, which can be used to calculate an optimal electricity mix, can be written as the following linear program:

$$
\begin{array}{llll}
\min \sum_{s\in\mathcal{S}} P_s e_s + \kappa\left(q_\alpha + \sum_{s\in\mathcal{S}} \frac{P_s z_s}{1-\alpha}\right) & & & \\
\text{s.t.}\quad d_{n,h} \le c_{n,h} + f_{n,h} + s_{n,h} + p_{n,h}, & & & \forall n \in \mathcal{N},\ \forall h : 1 \le h \le 6 \\
\quad c_{n,h}C + f_{n,h}F_{n,h} + s_{n,h}S_{n,h} + p_{n,h}P & = & e_{n,h}, & \forall n \in \mathcal{N},\ \forall h : 1 \le h \le 6 \\
\gamma^l C_h \le c_{n,h} & \le & \gamma^u C_h, & \forall n \in \mathcal{N},\ \forall h : 1 \le h \le 6 \\
\delta^l t C_h \le \sum_{n\in\mathcal{N}_s : S(n)\le t, h=1,\dots,6} c_{n,h} & \le & \delta^u t C_h, & \forall s \in \mathcal{S},\ \forall t : t = 1,\dots,T \\
|c_{n,h} - c_{n,h-1}| & \le & \eta_h, & \forall n \in \mathcal{N},\ \forall h : 2 \le h \le 6 \\
|c_{n,1} - c_{\text{pred}(n),6}| & \le & \eta_1, & \forall n \in \mathcal{N} \\
p_{n,h} & \le & \pi, & \forall n \in \mathcal{N},\ \forall h : 1 \le h \le 6 \\
|p_{n,h} - p_{n,h-1}| & \le & \beta, & \forall n \in \mathcal{N},\ \forall h : 2 \le h \le 6 \\
|p_{n,1} - p_{\text{pred}(n),6}| & \le & \beta, & \forall n \in \mathcal{N} \\
\sum_{n\in\mathcal{N}_s} e_{n,h} & = & e_s, & \forall s \in \mathcal{S},\ \forall h : h = 1,\dots,6 \\
e_s - q_\alpha & \le & z_s, & \forall s \in \mathcal{S}.
\end{array}
$$

In the above program all variables are non-negative. The two interesting decisions taken by the above program are the amount of futures to be bought and the amount of energy which is obtained from the swing option contract.[1] Both decisions have to be taken at the beginning of the planning horizon. The decisions have to perform reasonably well in all possible scenarios for demand and energy prices and should enable the customer to purchase energy at acceptable rates.

3.3 Pricing Supply Contracts

To present another interesting application of the proposed framework, we discuss a method of valuing electricity supply contracts. Contrary to the above model we now optimize the decision of an energy broker, who sells the right to consume energy at a given price and in a certain period to a big energy consumer. We assume that the energy broker owns no power production facilities and therefore has to buy the energy to fulfill her contracts in the form of power futures or from the spot market. See also the recent work in Haarbrücker and Kuhn (2009) and Pflug and Broussev (2009), both dealing with similar problems. Our approach differs from the existing approaches in the sense that we assume that the energy consumers demand for energy on a given day does not depend on the spot market prices of that day. We further assume that the consumer has no means to sell surplus energy on the market.

[1] Note that the amount of futures $f_{n,h}$ in node n and hour block h cannot be made independently of the other periods, since future contracts are block contracts. The contracts are available with different maturities and effect hour blocks differently depending on whether the corresponding hours are base or peak hours. For the sake of clarity of exposition, we do not explicitly write all the constraints concerning $f_{n,h}$. Of course, in the implementation of the models these straightforward but tedious constraints have to be included.

These assumptions seem perfectly natural for the vast majority of the industrial energy consumers.

While other approaches assume that the consumer sells the energy drawn from the contract on the open market and therefore are resulting in unrealistically high prices, we apply a principle from classic option pricing theory: the value of the contract is the expected amount of money of hedging the costs that arise from the contract. However, in our case – unlike in classical option pricing theory – it is of course not possible to find risk neutral probability measures and analytical solutions. Therefore, we also have to explicitly care for the *risk dimension* of the contract.

The supply contract in this model is of a little simpler structure than that in the previous model.[2] In particular, we consider a supply contract where the supply in every node n and every hour h is bounded by v_h for $h = 1, \dots, 6$. However, since we also assume that the contract is the only source of energy for the consumer and the demand is stochastic, we have to deal also with the case that $d_{n,h} > v_h$ in some of the hours. If this happens the consumer is charged a fixed higher price X for the extra amount of energy

$$x_{n,h} = [d_{n,h} - v_h]^+$$

that he consumes above the allowed quantity v_h.

The aim is to find the minimal price C (per MWh) for which energy can be sold to the customer, while at the same time keeping the risk from the contract (again for the sake of simplicity modeled by terminal AVaR) bounded by a constant μ. The seller of the contract is assumed to be a pure energy trader, that is she does not own a production plant. Therefore, all the required energy has to be bought either in form of futures or on the spot market, that is

$$f_{n,h} + s_{n,h} \geq d_{n,h}.$$

The expenditure in every node is therefore given by

$$e_{n,h} = f_{n,h} F_{n,h} + s_{n,h} S_{n,h} - d_{n,h} C - x_{n,h} X.$$

Finally, the full model written as linear program reads

$$\begin{aligned}
\min\ & C + \textstyle\sum_{n \in \mathcal{N}, h=1,\dots,6} x_{n,h} X' & & \\
\text{s.t.}\ & d_{n,h} \leq f_{n,h} + s_{n,h}, & & \forall n \in \mathcal{N}, \forall h : 1 \leq h \leq 6 \\
& f_{n,h} F_{n,h} + s_{n,h} S_{n,h} - d_{n,h} C - x_{n,h} X & = e_{n,h}, & \forall n \in \mathcal{N}, \forall h : 1 \leq h \leq 6 \\
& d_{n,h} - v_h & \leq x_{n,h}, & \forall n \in \mathcal{N}, \forall h : 1 \leq h \leq 6 \\
& \textstyle\sum_{n \in \mathcal{N}_s, h=1,\dots,6} e_{n,h} & = e_s, & \forall s \in \mathcal{S} \\
& e_s - q_\alpha & \leq z_s, & \forall s \in \mathcal{S} \\
& \left(q_\alpha + \textstyle\sum_{s \in \mathcal{S}} \frac{P_s z_s}{1-\alpha}\right) & \leq \mu. &
\end{aligned}$$

[2] It would, however, be no problem to generalize to the supply contract used in Sect. 3.2. The only difficulty lies in defining various penalties for the violations of the contract restrictions (1), (2) and (3) (one per restriction) and to cope with the more complicated notation that such a model would make necessary.

Besides the minimal price, the model yields an optimal hedge for the contract in terms of EEX futures. Note that the penalty term $\sum_{n\in\mathcal{N},h=1,\ldots,6} x_{n,h} X'$ (X' being a number bigger than X) is necessary to make sure that the solver indeed chooses the variables $x_{n,h}$ as the excess consumptions in the respective nodes and hour-blocks. While the model looks slightly different from the energy mix model presented in Sect. 3.2, the main ingredients stay the same. This demonstrates the wide applicability of the presented approach.

4 Numerical Results

4.1 Implementation

To obtain numerical results for the presented models, the following implementation setup has been used: The general workflow engine, as well as the simulation and estimation have been developed in MatLab R2007a. The scenario generator has been developed in C++. The multi-stage models have been modeled using AMPL. While different models require slight changes in the model, a model generator has been developed using the scripting language Python. This model generator creates the respective AMPL model, which is needed for a specific optimization run. For solving the deterministic equivalent formulations the MOSEK solver (Version 4) has been used. The parameters of the applied IPM solver had to be tweaked to get good run-time behavior. The computer on which the experiments have been conducted had a Intel(R) Pentium(R) 4 CPU 3.00 GHz, 4 GB RAM running under a Linux OS (Debian Stable, Etch).

Because of the long running time of the interior point solver, alternative solution methods, for example Lagrange or Nested-Benders Decomposition, could be considered, especially as the underlying models are linear, see for example Ruszczyński (2003) and the references therein.

4.2 Estimation, Simulation and Scenario Tree Generation

For all optimization models presented in this chapter, the planning horizon starts from 01.07.2007 and ends at 31.12.2007. For our econometric models we use historic spot prices, and the futures prices for energy in the planning horizon are given in Table 2. The data was obtained from the EEX on 15.03.2007. The econometric models from Sect. 2 were estimated using EEX spot price data and temperature data from 01.01.2006 to 28.02.2007. Using the estimates we calibrate the model in such a way that the mean prices correspond to the future prices given in Table 2. The calibrated model is subsequently used to create 2,000 trajectories

Table 2 Peak and base future prices for the planning horizon

Period	Base future	Peak future
July 2007	41.88	63.42
August 2007	38.78	62.11
September 2007	41.81	63.86
October–December 2007	48.2	70.1

Futures for the first 3 months are available separately, while for the last 3 month only a 3-month block future was available on 15.03.2007

Table 3 Size of LPs and solver run-time

Power plant	No	No	Yes
Contract gamma	No	Yes	Yes
Constraints	541,710	809,982	1,346,524
Variables	805,687	805,687	1,342,230
Nonzeros	2,036,883	3,860,207	31,689,184
Solution time (sec)	450–700	1,000–5,000	13,000–19,500

Entries in row power plant indicate whether the model includes a power plant or not. Similarly the column contract gamma indicates whether constraint (1) – which drastically increases the solution time – is imposed or not

by simulating from the stable distributions[3] fitted to the residuals of the model. Using these scenario paths for the spot price and equally many scenarios for the demand, we construct a scenario tree with 184 stages, 44,712 nodes, 426 scenarios and 12 dimensions (6 spot price, 6 demand). The construction of the tree takes approximately 2 h. The same tree has been used for all calculations described below.

We are using realistically large as the scenario tree, the deterministic equivalent formulations of the multi-stage stochastic optimization problems, which tend to grow with the size of the scenario tree, are huge. Table 3 shows the number of constraints and variables for different models described in this chapter, as well as the run-time for solving one instance of the respective model using the software and hardware infrastructure described above.

4.3 Results

The flexibility of the model allows for a plethora of different studies. Hence, only a small subset of possible evaluations is presented below. The AVaR risk parameters have been fixed to $\alpha = 0.85$ and $\kappa = 1$, although it would definitely be worthwhile to study different levels of risk aversion to obtain an efficient frontier.

Let us first examine the total contracted amount of electricity per hour block as well as the total cost distribution when we calculate optimal energy mix portfolios

[3] The freely available software *stable.exe* (see Nolan (2004)) was used for parameter estimation and simulation.

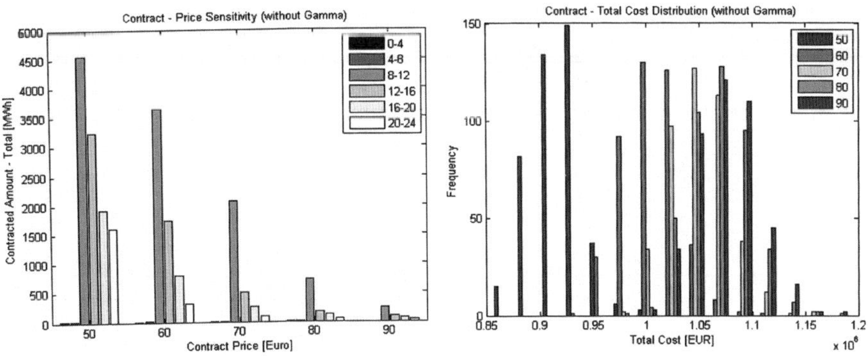

Fig. 6 Contracted amount and total cost – without γ, different contract prices

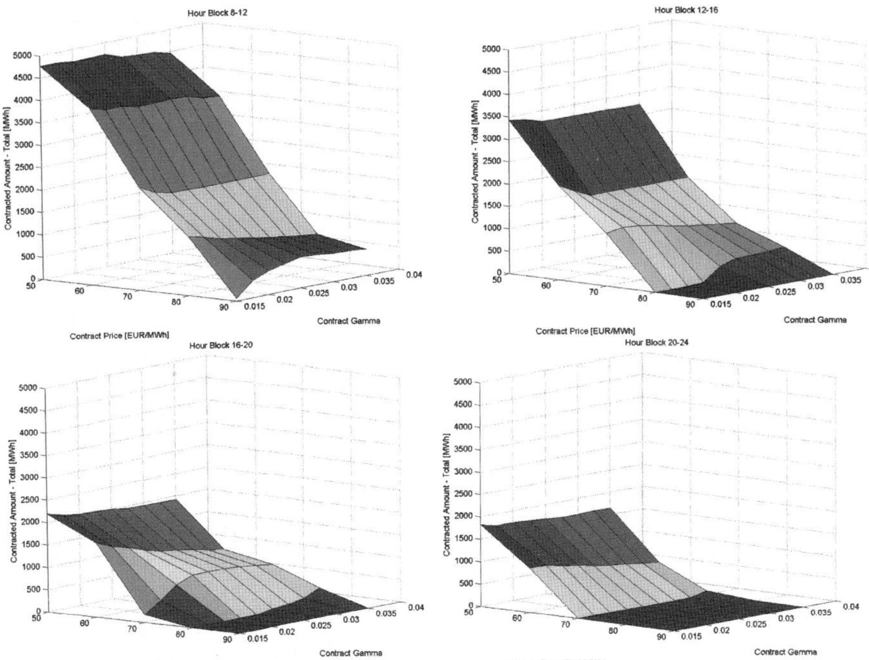

Fig. 7 Contracted amount – different contract prices and different levels of γ

in Fig. 6. For these graphs, the constraints (1) have not been considered. The results of adding (1) on the total contracted amount of electricity for selected hour blocks is shown in Fig. 7. Figure 8 depicts the total cost distribution in those cases. We can observe that the cost-distributions are heavy-tailed, confirming the explicit use of risk management techniques.

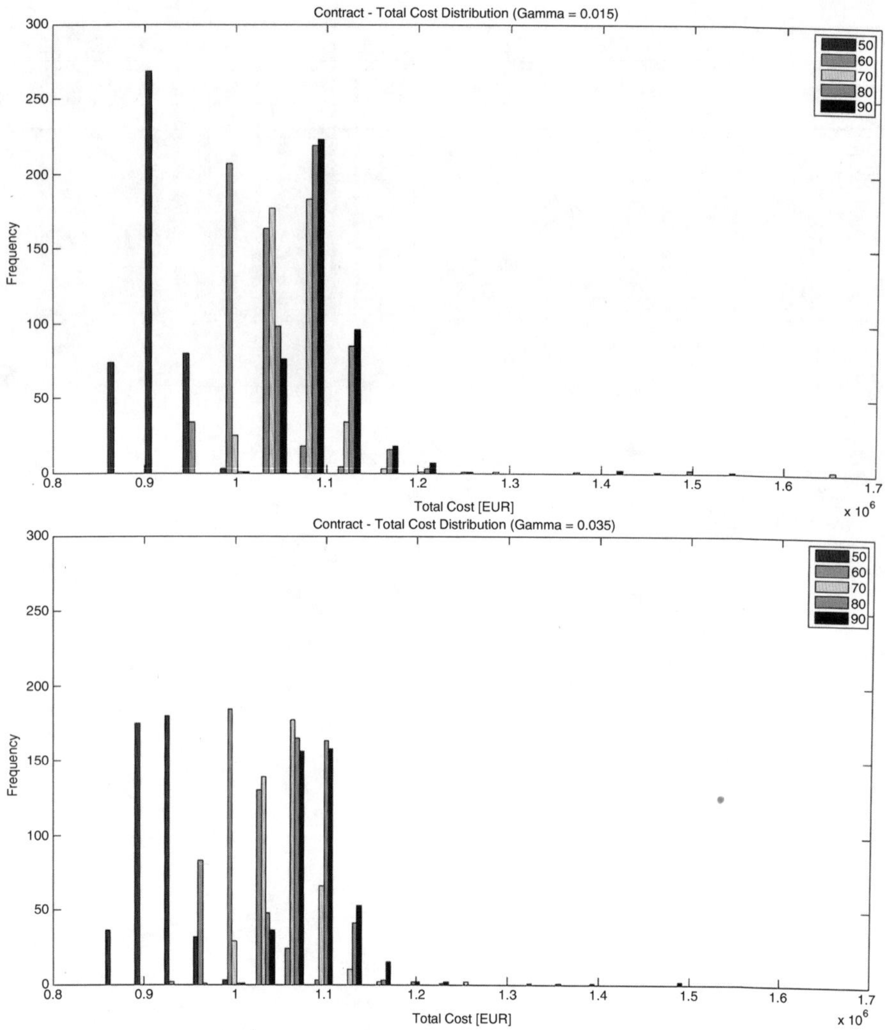

Fig. 8 Total cost distribution – different contract prices and different levels of γ

Finally, we examine the formulation to price supply contracts by iterating over the parameters μ and α. Their impact on the optimal supply contract price is shown in Fig. 9.

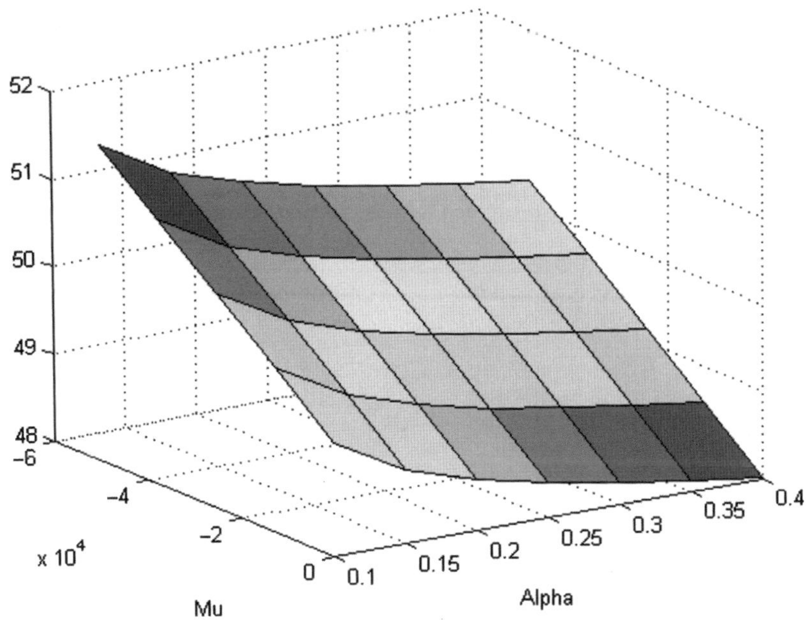

Fig. 9 Supply contract pricing for different levels of μ and α

5 Conclusion

In this chapter we presented a versatile multi-stage stochastic programming model for solving various problems of large energy consumers and producers. Two explicit models, the construction of an optimal energy mix portfolio, as well as the optimal pricing of supply contracts, are shown. The initial setup effort for large-scale multi-stage stochastic programs is huge, as the whole process from data preparation, estimation, simulation, scenario tree generation, optimization modeling and solution has to be tackled. This is the main reason why this valuable modeling technique is often neglected for practical purposes. However, once the whole workflow has been created, many applications and real-world extensions are possible. The aim of this paper is to provide a general outline for implementing the whole process to successfully model and solve large-scale stochastic programs for various energy applications.

References

Bernhardt C, Klüppelberg C, Meyer-Brandis T (2008) Estimating high quantiles for electricity prices by stable linear models. J Energ Market 1(1):3–19

de Jong CM, Huisman R (2002) Option formulas for mean-reverting power prices with spikes. Research paper, Erasmus Research Institute of Management (ERIM), October 2002

Dupačová J, Gröwe-Kuska N, Römisch W (2003) Scenario reduction in stochastic programming. An approach using probability metrics. Math Program A 95(3):493–511

Eichhorn A, Römisch W (2008) Stability of multistage stochastic programs incorporating polyhedral risk measures. Optimization 57(2):295–318

Eichhorn A, Römisch W, Wegner I (2005) Mean-risk optimization of electricity portfolios using multiperiod polyhedral risk measures. IEEE Proceedings - St. Petersburg Power Tech, 2005

Escribano L, Pena JI, Villaplana P (2002) Modeling electricity prices: International evidence. Economics working papers, Universidad Carlos III, Departamento de Economa, 2002

Haarbrücker G, Kuhn D (2009) Valuation of electricity swing options by multistage stochastic programming. Automatica 45(4):889–899

Heitsch H, Römisch W (2003) Scenario reduction algorithms in stochastic programming. Computat Optim Appl 24(2–3):187–206

Hochreiter R, Pflug GC (2007) Financial scenario generation for stochastic multi-stage decision processes as facility location problems. Ann Oper Res 152:257–272

Hochreiter R, Pflug GC, Wozabal D (2006) Multi-stage stochastic electricity portfolio optimization in liberalized energy markets. In System modeling and optimization, vol. 199 of IFIP Int. Fed. Inf. Process. Springer, New York, pp. 219–226

Hollander M, Wolfe A (1973) Nonparametric statistical inference. Wiley, New York

Király A, Jánosi IM (2002) Stochastic modeling of daily temperature fluctuations. Phys. Rev. E 65(5):051102

Lucia JJ, Schwartz ES (2002) Electricity prices and power derivatives: evidence from the Nordic Power Exchange. Rev Derivatives Res 5(1):5–50

McNeil AJ, Frey R, Embrechts P (2005) Quantitative risk management. Princeton Series in Finance. Princeton University Press, NJ

Nolan J (2004) stable.exe. A program to fit and simulate stable laws. http://academic2.american.edu/~jpnolan/stable/stable.html

Pflug GC (2001) Scenario tree generation for multiperiod financial optimization by optimal discretization. Math Program B 89(2):251–271

Pflug GC, Broussev N (2009) Electricity swing options: behavioral models and pricing. Eur J Oper Res 197(3):1041–1050

Pflug GC, Römisch W (2007) Modeling, measuring and managing risk. World Scientific, Singapore

Rachev ST, Mittnik S (2000) Stable paretian models in finance. Wiley, New York

Rachev ST, Römisch W (2002) Quantitative stability in stochastic programming: the method of probability metrics. Math Oper Res 27(4):792–818

Rockafellar RT, Uryasev S (2000) Optimization of conditional value-at-risk. J Risk 2(3):21–41

Römisch W (2003) Stability of stochastic programming problems. In Stochastic programming, vol. 10 of Handbooks Oper. Res. Management Sci. Elsevier, Amsterdam, pp. 483–554

Ruszczyński A (2003) Decomposition methods. In: Ruszczyński A, Shapiro A (eds) Stochastic programming. Handbooks in operations research and management science, vol. 10. Elsevier, Amsterdam, pp. 141–211

Ruszczyński A, Shapiro A (eds) (2003) Stochastic programming. Handbooks in operations research and management science, vol. 10. Elsevier, Amsterdam

Schultz R, Nowak MP, Nürnberg R, Römisch W, Westphalen M (2003) Stochastic programming for power production and trading under uncertainty. In: Mathematics – Key technology for the future. Springer, Heidelberg, pp. 623–636

Sen S, Yu L, Genc T (2006) A stochastic programming approach to power portfolio optimization. Oper Res 54(1):55–72

Wallace SW, Ziemba WT (eds) (2005) Applications of stochastic programming. MPS/SIAM Series on Optimization, vol. 5. Society for Industrial and Applied Mathematics (SIAM), 2005

Weron R (2006) Modeling and forecasting electricity loads and prices: A statistical approach. Wiley, Chichester

Stochastic Optimization of Electricity Portfolios: Scenario Tree Modeling and Risk Management

Andreas Eichhorn, Holger Heitsch, and Werner Römisch

Abstract We present recent developments in the field of stochastic programming with regard to application in power management. In particular, we discuss issues of scenario tree modeling, that is, appropriate discrete approximations of the underlying stochastic parameters. Moreover, we suggest risk avoidance strategies via the incorporation of so-called polyhedral risk functionals into stochastic programs. This approach, motivated through tractability of the resulting problems, is a constructive framework providing particular flexibility with respect to the dynamic aspects of risk.

Keywords Electricity · Energy trading · Multiperiod risk · Polyhedral risk functionals · Power portfolio · Risk management · Scenario reduction · Scenario tree approximation · Stochastic programming

1 Introduction

In medium term planning of electricity production and trading, one typically faces uncertain parameters (such as energy demands and market prices in the future) that can be described reasonably by stochastic processes in discrete time. When time passes, additional information about the uncertain parameters may arrive (e.g., actual energy demands may be observed). Planning decisions can be made at different time stages based on the information available by then and on probabilistic information about the future (non-anticipativity), respectively. In terms of optimization, this situation is modeled by the framework of *multistage stochastic programming*; cf. Sect. 2. This framework allows to anticipate this dynamic decision structure appropriately. We refer to Cabero et al. (2005); Fleten and Kristoffersen (2008); Fleten et al. (2002); Gröwe-Kuska et al. (2002); Krasenbrink (2002); Li et al.

W. Römisch (✉)
Humboldt University, 10099 Berlin, Germany
e-mail: romisch@math.hu-berlin.de

S. Rebennack et al. (eds.), *Handbook of Power Systems II*, Energy Systems, 405
DOI 10.1007/978-3-642-12686-4_15, © Springer-Verlag Berlin Heidelberg 2010

(2007); Pereira and Pinto (1991); Sen et al. (2006); Takriti et al. (2000); Wu et al. (2007) for exemplary case studies of stochastic programming in power planning. For a broad overview on stochastic programming models in energy we refer to Wallace and Fleten (2003).

However, a stochastic program incorporating a (discrete-time) stochastic process having infinite support (think of probability distributions with densities such as normal distributions) is an *infinite dimensional* optimization problem. For such problems a solution can hardly be found in practice. On the other hand, an a priori limitation to stochastic processes having finite support (think of discrete probability distributions) would not be appropriate to many applications (including power planning). Therefore, for practical problem solving, approximation schemes are required such that general given stochastic processes are replaced by discrete ones with finite support (scenario trees) in such a way that the solutions of a stochastic program incorporating the discrete process are somehow close to the (unknown) solutions of the same program incorporating the original process. Such *scenario tree approximation* schemes will be one major topic in this chapter. Within the methods (Heitsch and Römisch 2009; Heitsch 2007; Heitsch and Römisch 2005) to be presented, the closeness of the solutions will be ensured by means of suitable stability theorems for stochastic programs (Römisch 2003; Heitsch et al. 2006).

The second major topic of this chapter will be the incorporation of risk management into power production planning and trading based on stochastic programming. In energy risk management, which is typically carried out ex post in practice, that is, after power production planning, derivative products such as futures or options are traded to hedge a given production plan. However, decisions about buying and selling derivative products can also be made at different time stages, that is, the dynamics of the decisions process here is of the same type as in production and (physical) power trading. Moreover, risk management and stochastic optimization rest upon the same type of stochastic framework. Hence, it is suggestive to integrate these two decision processes, that is, to carry out simultaneously production planning, power trading, and trading of derivative products. For example, in Blaesig (2007); Blaesig and Haubrich (2005) it has been demonstrated that such an integrated approach based on stochastic programming (electricity portfolio optimization) yields additional overall efficiency.

If risk avoidance is an objective of a stochastic optimization model, risk has to be quantified in a definite way. To this end, a suitable risk functional has to be chosen according to the economic requirements of a given application model. While in short-term optimization, simple *risk functionals* (risk measures) such as expected utility or average-value-at-risk might be appropriate, the dynamic nature of risk has to be taken into account if medium or long-term time horizons are considered. In this case, intermediate cash flows as well as the partial information that is revealed gradually at different time stages may have a significant impact on the risk. Therefore, *multiperiod* risk functionals are required (Artzner et al. 2007; Pflug and Römisch 2007). Another important aspect of choosing a risk functional for use in a stochastic programming model is a technical one: How much does a certain risk functional complicate the numerical resolution of a stochastic program? We argue that *polyhedral risk functionals* are a favorable choice with respect to the tractability of stochastic programs (Eichhorn and Römisch 2005). Also the stability theorems known for stochastic programs without any risk functional remain valid (Eichhorn 2007; Eichhorn and Römisch 2008) and, hence, there is a justification for scenario tree approximation schemes. In addition, the class of polyhedral risk functionals provides flexibility, particularly in the multiperiod situation.

This paper is organized as follows: after brief reviews on multistage stochastic programming in Sect. 2, we present scenario tree approximation algorithms in Sect. 3. After that, in Sect. 4, we discuss risk functionals with regard to their employment in electricity portfolio optimization. In particular, our concept of polyhedral risk functionals is presented in Sect. 4.2. Finally, we illustrate the effect of different polyhedral risk functionals with optimal cash flow curves from a medium-term portfolio optimization model for a small power utility featuring a combined heat and power plant (CHP).

2 Multistage Stochastic Programming

For a broad presentation of stochastic programming we refer to Ruszczyński and Shapiro (2003) and Kall and Mayer (2005). Let the time stages of the planning horizon be denoted by $t = 1, \ldots, T$ and let, for each of these time steps, a d-dimensional random vector ξ_t be given. This random vector represents the uncertain planning parameters that become known at stage t, for example, electricity demands, market prices, inflows, or wind power. We assume that ξ_1 is known from the beginning, that is, a fixed vector in $\mathbb{R}^d$. For $\xi_2, \ldots, \xi_T$, one may require the existence of certain statistical moments. The collection $\xi := (\xi_1, \ldots, \xi_T)$ can be understood as multivariate discrete time stochastic process. Based on these notations, a multistage stochastic program can be written as

$$\min_{x_1,\ldots,x_T} \left\{ \mathbb{F}(z_1, \ldots, z_T) \,\middle|\, \begin{array}{l} z_t := \sum_{s=1}^{t} b_s(\xi_s) \cdot x_s, \\ x_t = x_t(\xi_1, \ldots, \xi_t), \quad x_t \in X_t, \qquad (t = 1, \ldots, T) \\ \sum_{s=0}^{t-1} A_{t,s}(\xi_t) x_{t-s} = h_t(\xi_t) \end{array} \right\} \quad (1)$$

where x_t is the decision vector for time stage t. The latter may depend and may only depend on the data observed until time t (non-anticipativity), that is, on $\xi_1, \ldots, \xi_t$, respectively. In particular, the components of x_1 are *here and now* decisions since x_1 may only depend on ξ_1, which was assumed to be deterministic. The decisions are subject to constraints: each x_t has to be chosen within a given set X_t. Typically, each X_t is a polyhedron or even a box, potentially further constrained by integer requirements. Moreover, there are dynamic constraints involving matrices $A_{t,s}$ and right-hand sides h_t, which may depend on ξ_t in an affine linear way. For the objective, we introduce wealth values z_t (accumulated revenues) for each time stage defined by a scalar product of x_t and (negative) cost coefficients b_t. The latter may also depend on ξ_t in an affine linear way. Hence, each z_t is a random variable $(t = 2, \dots, T)$.

The objective functional $\mathbb{F}$ maps the entire stochastic wealth process (cash flow) to a single real number. The classical choice in stochastic optimization is the *expected value* $\mathbb{E}$ (mean) of the overall revenue z_T, that is,

$$\mathbb{F}(z_1, \dots, z_T) = -\mathbb{E}[z_T],$$

which is a *linear* functional. Linearity is a favorable property with respect to theoretical analysis as well as to the numerical resolution of problem (1). However, if risk is a relevant issue in the planning process, then some sort of nonlinearity is required in the objective (or, alternatively, in the constraints). In this presentation, we will discuss *mean-risk* objectives of the form

$$\mathbb{F}(z_1, \dots, z_T) = \gamma \cdot \rho(z_{t_1}, \dots, z_{t_J}) - (1 - \gamma) \cdot \mathbb{E}[z_T],$$

with $\gamma \in [0, 1]$ and ρ being a *multiperiod* risk functional applied to selected time steps $1 < t_1 < \cdots < t_J = T$, allowing for dynamic perspectives to risk.

Though the framework (1) considers the dynamics of the decision process, typically only the first stage solution x_1 is used in practice since it is scenario independent, while x_t is scenario dependent for $t \geq 2$. When the second time stage $t = 2$ is reached in reality, one may solve a new problem instance of (1) such that the time stages are shifted one step ahead (rolling horizon). However, x_1 is a good decision in the sense that it anticipates future decisions and uncertainty.

3 Scenario Tree Approximation

If the stochastic input process ξ has infinite support (infinitely many scenarios), the stochastic program (1) is an infinite dimensional optimization problem. For such problems a solution can hardly be found in practice. Therefore, ξ has to be approximated by another process having finite support (Heitsch and Römisch 2009, 2005). Such an approximation must exhibit tree structure to reflect the monotone information structure of ξ. It is desirable that *scenario tree* approximation schemes rely on approximation or stability results for (1) (cf., e.g., Heitsch et al. (2006); Eichhorn and Römisch (2008); Mirkov and Pflug (2007); Pennanen (2005)) that guarantee that the results of the approximate optimization problem are related to the (unknown) results of the original problem.

The recent stability result in Heitsch et al. (2006) reveals that the multistage stochastic program (1) essentially depends on the probability distribution of the stochastic input process and on the implied information structure. While the probability information is based on the characteristics of the individual scenarios and their probabilities, the information structure says something about the availability of information at different time stages within the optimization horizon. The scenario tree construction approach to be presented next consists of both approximation of the probability information and recovering the information structure (Heitsch (2007)).

Presently, there exist several approaches to generate scenario trees for multi-stage stochastic programs (see Dupačová et al. (2000) for a survey). They are based on several different principles. We mention here (1) bound-based constructions (Casey and Sen 2005; Edirisinghe 1999; Frauendorfer 1996; Kuhn 2005), (2) Monte Carlo-based schemes (Chiralaksanakul and Morton 2004; Shapiro 2003, 2007) or Quasi-Monte Carlo-based methods (Pennanen 2005, 2009), (3) (EVPI-based) sampling within decomposition schemes (Corvera Poiré 1995; Dempster 2004; Infanger 1994), (4) the target/moment-matching principle (Høyland et al. 2003; Høyland and Wallace 2001; Kaut and Wallace 2007), and (5) probability metric-based approximations (Gröwe-Kuska et al. 2003; Heitsch and Römisch 2009; Hochreiter 2005; Hochreiter and Pflug 2007; Pflug 2001).

We propose a technique that belongs to the group (5) and is based on probability distances that are associated with the stability of the underlying stochastic program. Input of the method consists of a finite number of scenarios that are provided by the user and, say, are obtained from historical data by data analysis and resampling techniques or from statistical models calibrated to the relevant historical data. Sampling from historical time series or from statistical models (e.g., time series or regression models) is the most popular method for generating data scenarios. Statistical models for the data processes entering power operation and planning models have been proposed, for example, in Burger et al. (2004); Clewlow and Strickland (2000); Eichhorn et al. (2005); Hochreiter et al. (2006); Schmöller (2005); Schmöller et al. (2003); Sen et al. (2006); Weron et al. (2004).

The actual scenario tree construction method starts with a finite set of typically individual scenarios, where we assume that these scenarios serve as approximation for the original probability information. Although such individual scenarios are convenient to represent a very good approximation of the underlying probability distribution, the approximation with respect to the information structure could be poor. In particular, if sampling is performed from nondiscrete random variables (e.g., random variables having a density function such as normal distributions), the information structure gets lost in general. But, fortunately, it can be reconstructed approximately by applying techniques of *optimal scenario reduction* successively.

3.1 Scenario Reduction

The basis of our scenario tree generation methods is the reduction of scenarios modeling the stochastic data process in stochastic programs. We briefly describe this universal and general concept developed in Dupačová et al. (2003) and Heitsch and Römisch (2003). More recently, it was improved in Heitsch and Römisch (2007) and extended to mixed-integer models in Henrion et al. (2008). It was originally intended for nondynamic (two-stage) stochastic programs and, hence, does not take into account the information structure when applied in a multistage framework. There are no special requirements on the stochastic data processes (e.g., on the dependence structure or the dimension of the process) or on the structure of the scenarios (e.g., tree-structured or not).

Scenario reduction may be desirable in some situations when the underlying optimization models already happen to be large scale and the incorporation of a large number of scenarios leads to huge programs and, hence, to high computation times. The idea of the scenario reduction framework in Dupačová et al. (2003) and Heitsch and Römisch (2003) is to compute the (nearly) best approximation of the underlying discrete probability distribution by a measure with smaller support in terms of a probability metric that is associated to the stochastic program in a natural way by stability theory (Römisch 2003; Heitsch et al. 2006). Here, with regard to problem (1), the *norm* $\|\cdot\|_r$ will be used defined by

$$\|\xi\|_r := \left(\sum_{t=1}^{T} \mathbb{E}[|\xi_t|^r]\right)^{\frac{1}{r}} \tag{2}$$

for a random vector $\xi = (\xi_1, \ldots, \xi_T)$, where $\mathbb{E}[\cdot]$ denotes expectation and $|\cdot|$ denotes some norm in $\mathbb{R}^d$. We aim at finding some $\hat{\xi}$ such that the distance $\|\xi - \hat{\xi}\|_r$ is small. The role of the parameter $r \geq 1$ is to ensure that the stochastic program (1) is well defined, provided that $\|\xi\|_r < \infty$. The choice of r depends on the existing moments of the stochastic input process ξ coming across and on whether ξ enters the right-hand side h_t and/or the costs b_t and/or the (technology) matrices $A_{t,s}$. Typical choices are $r = 1$ if either right-hand sides or costs are random and $r = 2$ if both right-hand sides and costs are random. For further details we refer to Heitsch and Römisch (2009).

The scenario reduction aims at reducing the number of scenarios in an optimal way. If $\xi = (\xi_1, \ldots, \xi_T)$ is a given random vector with finite support, that is, represented by the scenarios $\xi^i = (\xi_1^i, \ldots, \xi_T^i)$ and probabilities p_i, $i = 1, \ldots, N$, then ones may be interested in deleting of a certain number of scenarios for computational reasons. So the main issue here is to find a suitable index subset $J \subset \{1, \ldots, N\}$. Moreover, if J is given, the question arises, what is the best approximation $\tilde{\xi} = (\tilde{\xi}_1, \ldots, \tilde{\xi}_T)$ of ξ supported only by the scenarios $\xi^j = (\xi_1^j, \ldots, \xi_T^j)$, $j \notin J$. The answer to the latter question, however, can be given directly: in our notation using (2) the problem reads

$$\min\left\{\sum_{t=1}^{T}\sum_{i=1}^{N} p_i|\xi_t^i - \tilde{\xi}_t^i|^r \;\middle|\; (\tilde{\xi}_1^i, \ldots, \tilde{\xi}_T^i) \in \{(\xi_1^j, \ldots, \xi_T^j)\}_{j \notin J}\right\} \tag{3}$$

and if we define a mapping $j(i)$ such that

$$j(i) \in \arg\min_{j \notin J} \sum_{t=1}^{T} |\xi_t^i - \xi_t^j|^r, \qquad i \in J,$$

the minimum of (3) is attained for scenarios

$$(\hat{\xi}_1^i, \ldots, \hat{\xi}_T^i) = \begin{cases} (\xi_1^{j(i)}, \ldots, \xi_T^{j(i)}), & \text{if } i \in J, \\ (\xi_1^i, \ldots, \xi_T^i), & \text{if } i \notin J. \end{cases} \tag{4}$$

Hence, the best approximation of ξ is obtained for the random vector $\hat{\xi}$ supported by the scenarios $\hat{\xi}^j = (\hat{\xi}_1^j, \ldots, \hat{\xi}_T^j)$ and probabilities q_j, $j \notin J$, where we have

$$\|\xi - \hat{\xi}\|_r^r = \sum_{i \in J} p_i \min_{j \notin J} \sum_{t=1}^{T} |\xi_t^i - \xi_t^j|^r , \tag{5}$$

$$q_j = p_j + \sum_{\substack{i \in J \\ j(i)=j}} p_i \, . \tag{6}$$

In other words, the *redistribution rule* (6) consists in assigning the new probability to a preserved scenario to be equal to the sum of its former probability and of all probabilities of deleted scenarios that are closest to it.

More complicated is the actual problem of optimal scenario reduction, that is, finding an optimal choice for the index set J with, say, prescribed cardinality. This problem represents a metric k-median problem, which is known to be NP-hard; hence, (polynomial-time) approximation algorithms and heuristics become important. Simple heuristics may be derived from formula (5) for the approximation error. The results are two heuristic algorithms to compute nearly optimal index sets J with given cardinality n.

Algorithm 3.1 (*Forward selection*)

[*Initialization*]
Set $J := \{1, \ldots, N\}$.

[*Index Selection*]
Determine an index $l \in J$ such that

$$l \in \arg\min_{u \in J} \sum_{k \in J \setminus \{u\}} p_k \min_{j \notin J \setminus \{u\}} \sum_{t=1}^{T} |\xi_t^k - \xi_t^j|^r$$

and set $J := J \setminus \{l\}$. If the cardinality of J equals n go to the termination step. Otherwise, continue with a further index selection step.

[*Termination*]
Determine scenarios according to (4) and apply the redistribution rule (6) for the final index set J.

Algorithm 3.2 (*Backward reduction*)

[*Initialization*]
Set $J := \emptyset$.

[*Index Selection*]
Determine an index $u \notin J$ such that

$$u \in \arg\min_{l \notin J} \sum_{k \in J \cup \{l\}} p_k \min_{j \notin J \cup \{u\}} \sum_{t=1}^{T} |\xi_t^k - \xi_t^j|^r$$

and set $J := J \cup \{l\}$. If the cardinality of J equals n, go to the termination step. Otherwise, continue with a further index selection step.

[*Termination*]
Determine scenarios according to (4) and apply the redistribution rule (6) for the final index set J.

3.2 Scenario Tree Construction

Now we turn to the scenario tree construction, where we assume to have a sufficient large set of original or sample scenarios available. Let the (individual) scenarios and probabilities be denoted again by $\xi^i = (\xi_1^i, \ldots, \xi_T^i)$ and p_i, $i = 1, \ldots, N$, respectively, and we assume that $\xi_1^1 = \xi_1^2 = \ldots = \xi_1^N =: \xi_1^*$ (deterministic first stage). The random process with scenarios ξ^i and probabilities p_i, $i = 1, \ldots, N$, is denoted by ξ.

The idea of our tree construction method is to apply the above scenario reduction techniques successively in a specific way. In fact, by the approach of a *recursive scenario reduction* for increasing and decreasing time, respectively, both a forward and backward in time performing method can be derived.

The recursive scenario reduction acts as recovering the original information structure approximately. In the next two subsections we present a detailed description for two variants of our method, the forward and the backward approach. In the following let $I := \{1, \ldots, N\}$.

3.2.1 Forward Tree Construction

The forward tree construction is based on recursive scenario reduction applied to time horizons $\{1, \ldots, t\}$ with successively increasing time parameter t. It successively computes partitions of I of the form

$$\mathcal{C}_t := \{C_t^1, \ldots, C_t^{k_t}\}, \qquad k_t \in \mathbb{N},$$

such that for every t the partitions satisfy the conditions

$$C_t^k \cap C_t^{k'} = \emptyset \quad \text{for} \quad k \neq k', \quad \text{and} \quad \bigcup_{k=1}^{k_t} C_t^k = I.$$

The elements of a partition $\mathcal{C}_t$ are called (scenario) clusters. The following forward algorithm allows to generate different scenario tree processes depending on the parameter settings for the reductions in each step.

Algorithm 3.3 (*Forward construction*)

[*Initialization*]
Define $\mathcal{C}_1 = \{I\}$ and set $t := 2$.

[*Cluster computation*]
Let be $\mathcal{C}_{t-1} = \{C_{t-1}^1, \ldots, C_{t-1}^{k_{t-1}}\}$. For every $k \in \{1, \ldots, k_{t-1}\}$, subject the scenario subsets $\{\xi_t^i\}_{i \in C_{t-1}^k}$ to a scenario reduction with respect to the tth components only. This yields disjoint subsets of remaining and deleted scenarios I_t^k and J_t^k, respectively. Next, obtain the mappings $j_t^k : J_t^k \to I_t^k$ such that

$$j_t^k(i) \in \arg\min_{j \in I_t^k} |\xi_t^i - \xi_t^j|, \quad i \in J_t^k,$$

according to the reduction procedure (cf. Sect. 3.1). Finally, define an overall mapping $\alpha_t : I \to I$ by

$$\alpha_t(i) = \begin{cases} j_t^k(i), & i \in J_t^k \text{ for some } k = 1, \ldots, k_{t-1}, \\ i, & \text{otherwise.} \end{cases} \tag{7}$$

A new partition at t is defined now by

$$\mathcal{C}_t := \left\{ \alpha_t^{-1}(i) \,\middle|\, i \in I_t^k,\ k = 1, \ldots, k_{t-1} \right\},$$

which is in fact a refinement of the partition $\mathcal{C}_{t-1}$. If $t < T$ set $t := t + 1$ and continue with a further cluster computation step; otherwise, go to the termination step.

[*Termination*]
According to the partition set $\mathcal{C}_T$ and the mappings (7), define a scenario tree process ξ_{tr} supported by the scenarios

$$\xi_{\text{tr}}^k = \left(\xi_1^*, \xi_2^{\alpha_2(i)}, \ldots, \xi_t^{\alpha_t(i)}, \ldots, \xi_T^{\alpha_T(i)}\right) \quad \text{for any} \quad i \in C_T^k$$

and probabilities $q_k := \sum_{i \in C_T^k} p_i$, for each $k = 1, \ldots, k_T$ (Fig. 1).

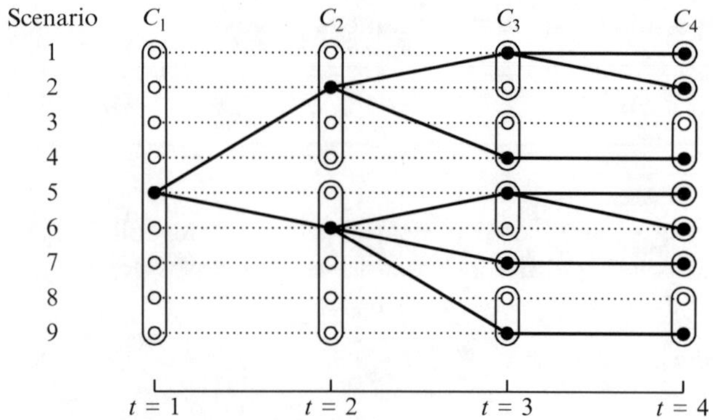

Scenario	α_2	α_3	α_4
1	2	1	1
2	2	1	2
3	2	4	4
4	2	4	4
5	6	5	5
6	6	5	6
7	6	7	7
8	6	9	9
9	6	9	9

Fig. 1 Illustration of the clustering by the forward scenario tree construction algorithm 3.3 (*left*) and the mappings α_t (*right*) for an example

We want to conclude this subsection with two remarks regarding Algorithm 3.3. First, both heuristic algorithms from Sect. 3.1 may be used to compute the scenario reduction within the cluster computation step. Second, according to (5) the error of the cluster computation step t is

$$\mathrm{err}_t := \sum_{k=1}^{k_{t-1}} \sum_{i \in J_t^k} p_i \min_{j \in I_t^k} |\xi_t^i - \xi_t^j|^r .$$

Furthermore, as shown in Heitsch (2007, Proposition 6.6), the estimate

$$\|\xi - \xi_{\mathrm{tr}}\|_r \leq \left(\sum_{t=2}^{T} \mathrm{err}_t \right)^{\frac{1}{r}}$$

holds for the total approximation error. The latter estimate allows to control the construction process by prescribing tolerances for err_t for every $t = 2, \ldots, T$.

3.2.2 Backward Tree Construction

The idea of the backward scenario tree construction consists in recursive scenario reduction on $\{1, \ldots, t\}$ for decreasing t, $t = T, \ldots, 2$. That results in a chain of index sets

$$I_1 := \{i_*\} \subseteq I_2 \subseteq \cdots \subseteq I_{t-1} \subseteq I_t \subseteq \cdots \subseteq I_T \subseteq I = \{1, \ldots, N\}$$

representing an increasing number of scenario realizations over the time horizon. The following backward algorithm is the counterpart of the forward Algorithm 3.3

and allows again to generate different scenario tree processes depending on the parameters for the reduction steps.

Algorithm 3.4 (*Backward construction*)

[*Initialization*]
Define $I_{T+1} := \{1, \ldots, N\}$ and $p^i_{T+1} := p_i$ for all $i \in I_{T+1}$. Further, let α_{T+1} be the identity on I_{T+1} and set $t := T$.

[*Reduction*]
Subject the scenario subset $\{(\xi^i_1, \ldots, \xi^i_t)\}_{i \in I_{t+1}}$ with probabilities p^i_{t+1} $(i \in I_{t+1})$ to a scenario reduction, which results in an index set I_t of remaining scenarios with $I_t \subseteq I_{t+1}$. Let be $J_t := I_{t+1} \setminus I_t$. According to the reduction procedure (cf. Sect. 3.1) obtain a mapping $j_t : J_t \to I_t$ such that

$$j_t(i) \in \arg\min_{j \in I_t} \sum_{k=1}^{t} |\xi^i_k - \xi^j_k|^r , \quad i \in J_t .$$

Define a mapping $\alpha_t : I \to I_t$ by

$$\alpha_t(i) = \begin{cases} j_t(\alpha_{t+1}(i)), & \alpha_{t+1}(i) \in J_t , \\ \alpha_{t+1}(i), & \text{otherwise}, \end{cases} \tag{8}$$

for all $i \in I$. Finally, set probabilities with respect to the redistribution (6), that is,

$$p^j_t := p^j_{t+1} + \sum_{\substack{i \in J_t \\ j_t(i)=j}} p^i_{t+1}.$$

If $t > 2$ set $t := t - 1$ and continue with performing a further reduction step; otherwise, go to the termination step.

[*Termination*]
According to the obtained index set I_T and the mappings (8), define a scenario tree process ξ_{tr} supported by the scenarios

$$\xi^i_{\text{tr}} = \left(\xi^*_1, \xi^{\alpha_2(i)}_2, \ldots, \xi^{\alpha_t(i)}_t, \ldots, \xi^{\alpha_T(i)}_T\right)$$

and probabilities $q_i := p^i_T$, for all $i \in I_T$ (Fig. 2).

We note again that the specific scenario reduction can be performed with both heuristic algorithms of Sect. 3.1. A similar estimate for the total approximation error $\|\xi - \xi_{\text{tr}}\|_r$ holds as for the forward variant. For details we refer to Heitsch and Römisch (2009, Sect. 4.1). Finally, we mention that all algorithms discussed in this section are implemented and available in GAMS-SCENRED (see www.gams.com).

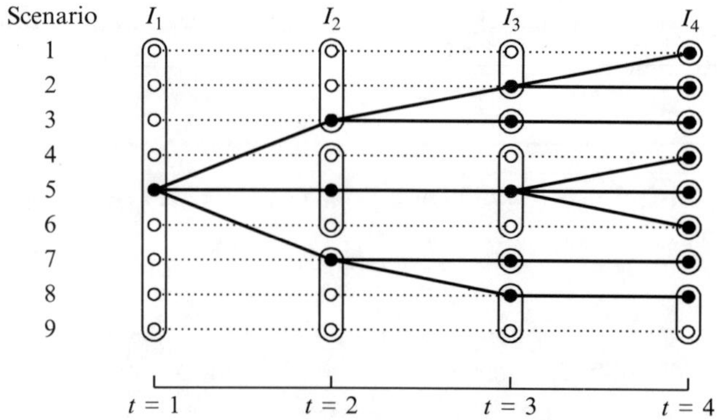

Scenario	α_2	α_3	α_4
1	3	2	1
2	3	2	2
3	3	3	3
4	5	5	4
5	5	5	5
6	5	5	6
7	7	7	7
8	7	8	8
9	7	8	8

Fig. 2 Illustration of the recursive backward scenario tree construction Algorithm 3.4 (*left*) and the mappings α_t (*right*) for an example. Note that the backward construction yields a clustering similar to the forward variant. *Black circles* correspond to scenarios contained in the index sets I_t

4 Risk Avoidance via Risk Functionals

Risk avoidance requirements in optimization are typically achieved by the employment of a certain *risk functional*. Alternatively, *risk probabilistic constraints* or *risk stochastic dominance constraints* with respect to a given acceptable strategy may be incorporated, that is, (1) may adopt constraints of the form

$$\mathbb{P}(z_T \leq z_{\text{ref}}) \geq \alpha \qquad \text{or} \qquad z_T \preceq z_{\text{ref}},$$

with (high) probability $\alpha \in (0, 1]$ and some acceptable reference level z_{ref} or some acceptable reference distribution z_{ref} and a suitable stochastic ordering relation "$\preceq$". For the relevant background of probabilistic constraints we refer to the survey by Prékopa (2003) and to Henrion (2006, 2007). For a systematic introduction into stochastic order relations we refer to Müller and Stoyan (2002) and for recent work on incorporating stochastic dominance constraints into optimization models to Dentcheva and Ruszczyński (2003) and Dentcheva and Ruszczyński (2006).

In this section, we focus on risk functionals ρ with regard to their utilization in the objective $\mathbb{F}$ of (1) as suggested, for example, in Ruszczyński and Shapiro (2006); cf. Sect. 2. Clearly, the choice of ρ is a very critical issue. On the one hand, the output of a stochastic program is highly sensitive to this choice. One is interested in a functional that makes sense from an economic point of view for a given situation. On the other hand, the choice of the risk functional has a significant impact on the numerical tractability of (1) (where ξ may be approximated by a finite scenario tree according to Sect. 3). Note that reasonable risk functionals are never linear (like the expectation functional), but some of them may be reformulated as infimal value of a linear stochastic program (see Sect. 4.2).

4.1 Axiomatic Frameworks for Risk Functionals

Basically, a risk functional in a probabilistic framework ought to measure the danger of ending up at low wealth in the future and/or the degree of uncertainty one is faced with in this regard. However, the question what is a good or what is the best risk functional from the viewpoint of economic reasoning cannot be answered in general. The answer depends strongly on the application context. However, various axioms have been postulated by various authors in the last decade, which can be interpreted as minimum requirements.

A distinction can be drawn between single-period risk functionals evaluating a stochastic wealth value z_T at one single point in time T and multiperiod risk functionals evaluating ones wealth at different time stages, say, $t_1 < t_2 \cdots < t_J$. The latter are typically required for medium- or long-term models. Of course, from a technical point of view single-period risk measurement can be understood as a special case of multiperiod risk measurement. However, with regard to single-period risk functionals, there is a relatively high degree of agreement about their preferable properties (Artzner et al. 1999; Föllmer and Schied 2004; Pflug and Römisch 2007), while the multiperiod case raises a lot more questions. In the following we pass directly to multiperiod risk measurement having single-period risk measurement as a special case in mind.

Let a certain linear space $\mathcal{Z}$ of discrete-time random processes be given. A random process $z \in \mathcal{Z}$ is basically a collection of random variables $z = (z_{t_1}, \ldots, z_{t_J})$ representing wealth at different time stages. The realization of z_{t_j} is completely known at time t_j, respectively. Moreover, at time stage t_j one may have more information about $(z_{t_{j+1}}, \ldots, z_{t_J})$ than before (at earlier time stages $t_1, \ldots, t_{j-1}$). Therefore, a multiperiod risk functional may also take into account *conditional distributions* with respect to some *underlying information structure*. In the context of the multistage stochastic program (1), the underlying information structure is given in a natural way through the stochastic input process $\xi = (\xi_1, \ldots, \xi_T)$. Namely, it holds that $z_{t_j} = z_{t_j}(\xi_1, \ldots, \xi_{t_j})$, that is, z is *adapted* to ξ. In particular, if ξ is discrete, that is, if ξ is given by a finite scenario tree as in Sect. 3, then also z is discrete, that is, z is given by the values $z^i_{t_j}$ $(j = 1, \ldots, J, i = 1, \ldots, N)$ on the scenario tree. However, we will consider general (not necessarily discretely distributed) random processes here and we also write $z^i_{t_j}$ for a realization (outcome) of random variable z_{t_j} even if the number of scenarios (possible outcomes) is infinite.

From a formal point of view, a risk functional ρ is just a mapping

$$z = (z_{t_1}, \ldots, z_{t_J}) \in \mathcal{Z} \quad \mapsto \quad \rho(z) \in \mathbb{R},$$

i.e., a real number is assigned to each random wealth process from $\mathcal{Z}$. One may require the existence of certain statistical moments for the random variables z_{t_j} $(j = 1, \ldots, J)$, that is, $\mathbb{E}[|z_{t_j}|^p] < \infty$ for some $p \geq 1$. The J time steps are denoted by $t_1, \ldots, t_J$ to indicate that, with regard to problem (1), they may be only a subset of the time steps $t = 1, \ldots, T$ of the underlying information structure. We assume $1 < t_1 < \cdots < t_J = T$ and set $t_0 = 1$ for convenience. The special case of *single-period* risk functionals occurs if only one time step is taken into account ($J = 1, t_J = T$).

Now, a high number $\rho(z)$ should indicate a high risk of ending up at low wealth values z_{t_j}, and a low (negative) number $\rho(z)$ indicates a small risk. In Artzner et al. (2007) the number $\rho(z)$ is interpreted as the *minimal amount* μ *of additionally required risk-free capital* such that the process $z_{t_1} + \mu, \dots, z_{t_J} + \mu$ is acceptable. Such and other intuitions have been formalized by various authors in terms of axioms. As a start, we cite the first two axioms from Artzner et al. (2007), in addition to convexity as the third axiom. A functional ρ is called a multiperiod *convex (capital) risk functional* if the following properties hold for all stochastic wealth processes $z = (z_{t_1}, \dots, z_{t_J})$ and $\tilde{z} = (\tilde{z}_{t_1}, \dots, \tilde{z}_{t_J})$ in $\mathcal{Z}$ and for all (non-random) real numbers μ:

- *Monotonicity*: If $z_{t_j} \leq \tilde{z}_{t_j}$ in any case for $j = 1, \dots, J$, then it holds that $\rho(z) \geq \rho(\tilde{z})$.
- *Cash invariance*: It holds that $\rho(z_{t_1} + \mu, \dots, z_{t_J} + \mu) = \rho(z_{t_1}, \dots, z_{t_J}) - \mu$.
- *Convexity*: If $0 \leq \mu \leq 1$ it holds that $\rho(\mu z + (1-\mu)\tilde{z}) \leq \mu\rho(z) + (1-\mu)\rho(\tilde{z})$.

The formulation "$z_{t_j} \leq \tilde{z}_{t_j}$ in any case" means that in each scenario i it holds that $z^i_{t_j} \leq \tilde{z}^i_{t_j}$. The convexity property is motivated by the idea that *diversification* might decrease risk but does never increase it. Sometimes the following property is also required for all $z \in \mathcal{Z}$:

- *Positive homogeneity*: For each $\mu \geq 0$ it holds that $\rho(\mu z) = \mu\rho(z)$.

Note that, for the single-period case $J = 1$, the first three properties coincide with the classical axioms from Artzner et al. (1999); Föllmer and Schied (2002); Frittelli and Rosazza Gianin (2002). A positively homogeneous convex risk functional is called *coherent* in Artzner et al. (1999) and Artzner et al. (2007). We note, however, that other authors do not require positive homogeneity, but claim that risk should rather grow overproportionally, that is, $\rho(\mu z) > \mu\rho(z)$ for $\mu > 1$; cf. Frittelli and Scandolo (2005) and Föllmer and Schied (2004). Clearly, the negative expectation functional $-\mathbb{E}$ is a (single-period) coherent risk functional, whereas the α-value-at-risk given by $\mathrm{VaR}_\alpha(z) = -\inf\{\mu \in \mathbb{R} : \mathbb{P}(z \leq \mu) > \alpha\}$ is not as it is not convex (Artzner et al. 1999).

For the multiperiod case ($J > 1$) the three above axioms are only a basis admitting many degrees of freedom. There are several aspects of risk that could be measured. First of all, one may want to measure the chance of ending up at very low values $z^i_{t_j}$ at each time since very low values can mean bankruptcy (liquidity considerations). In addition, one may want to measure the degree of uncertainty one is faced with at each time step; cf. Fig. 3 (left). A situation where, at some time t_k, one can be sure about the future development of ones wealth z_{t_j} ($j > k$) that may be preferred to a situation continuing uncertainty. For example, low values z_{t_j} may be tolerable if one can be sure that later the wealth is higher again. Hence, one may want to take into account not only the marginal distributions of $z_{t_1}, \dots, z_{t_J}$ but also their chronological order, their interdependence, and the underlying information structure. Therefore, a multiperiod risk functional may also take

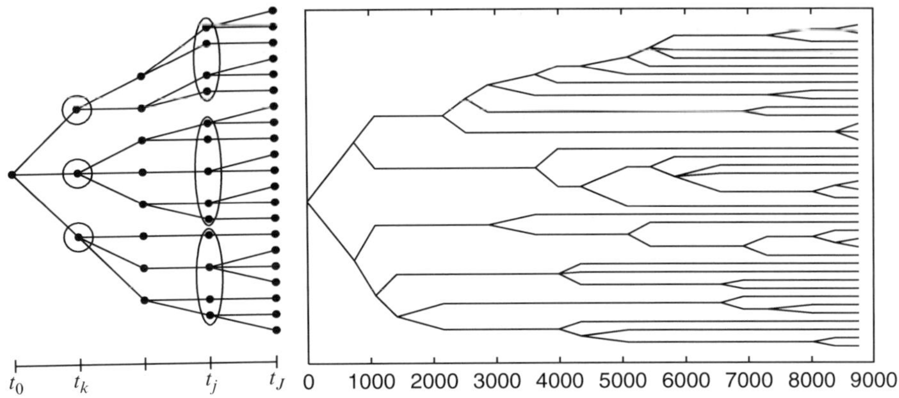

Fig. 3 *Left*: Illustration of the (discretized) information structure of a stochastic wealth process $z_{t_1}, \ldots, z_{t_J}$. At each time stage t_k and in each scenario one can look at subsequent time steps $t_j > t_k$ and consider the discrete (sub-) distribution of z_{t_j} seen from this node. *Right*: Branching structure of an exemplary scenario tree with 40 scenarios, $T = 8,760$ time steps, and approximately $150,000$ nodes used for the simulations in Sect. 4.3. There is a node at each time step for each scenario

into account the conditional distributions of z_{t_j}, given the information $\xi_1, \ldots, \xi_s$ with $s = 1, \ldots, t_j - 1$ $(j = 1, \ldots, J)$; cf. Fig. 3 (left). Clearly, there are quite a lot of those conditional distributions and the question arises which ones are relevant and how to weight them reasonably.

The above axioms leave all these questions open. In our opinion, general answers cannot be given, and the requirements depend strongly on the application context, for example, on the time horizon, on the size and capital reserves of the respective company, on the broadness of the model, etc. Some stronger versions of cash invariance (translation equivariance) have been suggested, for example, in Frittelli and Scandolo (2005) and Pflug and Römisch (2007), tailored to certain situations. However, the framework of polyhedral risk functionals in the next section is particularly flexible with respect to the dynamic aspects.

4.2 Polyhedral Risk Functionals

The basic motivation for polyhedral risk functionals is a technical, but important one. Consider the optimization problem (1). It is basically linear or mixed-integer linear if the objective functional is linear, that is, $\mathbb{F} = -\mathbb{E}$. In this case it is well tractable by various solution and decomposition methods. However, if $\mathbb{F}$ incorporates a risk functional ρ, it is no longer linear since risk functionals are essentially nonlinear by nature. Decomposition structures may get lost and solution methods may take much longer or may even fail. To avoid the worst possible situation one should choose ρ to be at least convex (Ruszczyński and Shapiro 2006). Then (1) is at least a convex problem (except possible integer constraints contained in X_t); hence, any local optimum is always the global one. As discussed earlier, convexity is in accordance with economic considerations and axiomatic frameworks.

Now, the framework of polyhedral risk functionals (Eichhorn and Römisch 2005; Eichhorn 2007) goes one step beyond convexity: polyhedral risk functionals maintain linearity structures even though they are nonlinear functionals. Namely, a polyhedral risk functional ρ is given by

$$\rho(z) = \inf\left\{ \mathbb{E}\left[\sum_{j=0}^{J} c_j \cdot y_j\right] \;\middle|\; \begin{array}{l} y_j = y_j(\xi_1,\dots,\xi_{t_j}) \in Y_j, \\ \sum_{k=0}^{j} V_{j,k} y_{j-k} = r_j \ (j=0,\dots,J), \\ \sum_{k=0}^{j} w_{j,k} \cdot y_{j-k} = z_{t_j} \ (j=1,\dots,J) \end{array} \right\}, \tag{9}$$

where $z = (z_{t_1},\dots,z_{t_J})$ denotes a stochastic wealth process being nonanticipative with respect to ξ, that is, $z_t = z_t(\xi_1,\dots,\xi_t)$. The notation $\inf\{\,.\,\}$ refers to the infimum. The definition includes fixed polyhedral cones Y_j (e.g., $\mathbb{R}_+ \times \cdots \times \mathbb{R}_+$) in some Euclidean spaces $\mathbb{R}^{k_j}$, fixed vectors c_j, r_j $w_{j,k}$, and matrices $V_{j,k}$, which have to be chosen appropriately. We will give examples for these parameters below. However, functionals ρ defined by (9) are always convex (Eichhorn and Römisch 2005; Eichhorn 2007).

Observe that problem (9) is more or less of the form (1), that is, the risk of a stochastic wealth process z is given by the optimal value of a stochastic program. Moreover, if (9) is inserted into the objective of (1) (i.e., $\mathbb{F} = \rho$), one is faced with two nested minimizations which, of course, can be carried out jointly. This yields the equivalent optimization problem

$$\min\left\{ \mathbb{E}\left[\sum_{j=0}^{J} c_j \cdot y_j\right] \;\middle|\; \begin{array}{l} x_t = x_t(\xi_1,\dots,\xi_t) \in X_t, \ \sum_{s=0}^{t-1} A_{t,s}(\xi_t) x_{t-s} = h_t(\xi_t) \\ (t = 1,\dots,T), \\ y_j = y_j(\xi_1,\dots,\xi_{t_j}) \in Y_j, \ \sum_{k=0}^{j} V_{j,k} y_{j-k} = r_j, \\ \sum_{k=0}^{j} w_{j,k} \cdot y_{j-k} = \sum_{s=1}^{t_j} b_s(\xi_s) \cdot x_s \ (j = 1,\dots,J) \end{array} \right\},$$

which is a stochastic program of the form (1) with *linear* objective. In other words, the nonlinearity of the risk functional ρ is transformed into additional variables and additional linear constraints in (1). This means that decomposition schemes and solution algorithms known for linear or mixed-integer linear stochastic programs can also be used for (1) with $\mathbb{F} = \rho$. In particular, as discussed in Eichhorn and Römisch (2005, Sect. 4.2), dual decomposition schemes (like scenario and geographical decomposition) carry over to the situation with $\mathbb{F} = \rho$. However, the dual problem in Lagrangian relaxation of coupling constraints (also called geographical or component decomposition) contains polyhedral constraints originating from the dual representation of ρ.

Furthermore, the linear combination of two polyhedral risk functionals is again a polyhedral risk functional (cf. Eichhorn (2007, Sect. 3.2.4)). In particular, the case

$$\mathbb{F}(z) = \gamma\rho(z) + \sum_{k=1}^{J} \mu_k \mathbb{E}\left[z_{t_k}\right],$$

with a polyhedral risk functional ρ (with parameters c_j, $w_{j,k}$ etc.), and real numbers γ and μ_k, $k = 1, \ldots, J$, can be fully reduced to the case ρ by setting

$$\hat{c}_j := \gamma c_j + \sum_{k=j}^{J} \mu_k w_{k,k-j} \quad (j = 0, \ldots, J)$$

for the vectors in the objective function of the representation (9) of $\mathbb{F}$ and letting all remaining parameters of ρ unchanged.

Another important advantage of polyhedral risk functionals is that they also behave favorable to stability with respect to (finite) approximations of the stochastic input process ξ (Eichhorn and Römisch 2008). Hence, there is a justification for the employment of the scenario tree approximation schemes from Sect. 3.

It remains to discuss the issue of choosing the parameters c_j, h_j, $w_{j,k}$, $V_{j,k}$, Y_j in (9) such that the resulting functional ρ is indeed a reasonable risk functional satisfying, for example, the axioms presented in the previous section. To this end, several criteria for these axioms have been deduced in Eichhorn and Römisch (2005) and Eichhorn (2007) involving duality theory from convex analysis. However, here we restrict the presentation to examples.

First, we consider the case $J = 1$, that is, single-period risk functionals evaluating only the distribution of the final value z_T (total revenue). The starting point of the concept of polyhedral risk functionals was the well-known risk functional *average-value-at-risk* AVaR_α at some probability level $\alpha \in (0, 1)$. It is also known as conditional-value-at-risk (cf. Rockafellar and Uryasev (2002)), but as suggested in Föllmer and Schied (2004) we prefer the name average-value-at-risk according to its definition

$$\text{AVaR}_\alpha(z) := \tfrac{1}{\alpha} \int_0^\alpha \text{VaR}_\beta(z) d\beta$$

as an average of value-at-risks and avoid any conflict with the use of conditional distributions within VaR and AVaR (see Pflug and Römisch (2007) for such constructions). The average-value-at-risk is a (single-period) coherent risk functional which is broadly accepted. $\text{AVaR}_\alpha(z_T)$ can be interpreted as the mean (expectation) of the α-tail distribution of z_T, that is, the mean of the distribution of z_T below the α-quantile of z_T. It has been observed in Rockafellar and Uryasev (2002) that AVaR_α can be represented by

$$\begin{aligned}\text{AVaR}_\alpha(z_T) &= \inf_{y_0 \in \mathbb{R}} \left\{ y_0 + \tfrac{1}{\alpha}\mathbb{E}[(y_0 + z_T)^-] \right\} \\ &= \inf \left\{ y_0 + \tfrac{1}{\alpha}\mathbb{E}[y_{1,2}] \;\middle|\; \begin{array}{l} y_0 \in \mathbb{R}, \\ y_1 = y_1(\xi_1, \ldots, \xi_T) \in \mathbb{R}^2_+, \\ y_0 + z_T = y_{1,1} - y_{1,2} \end{array} \right\},\end{aligned}$$

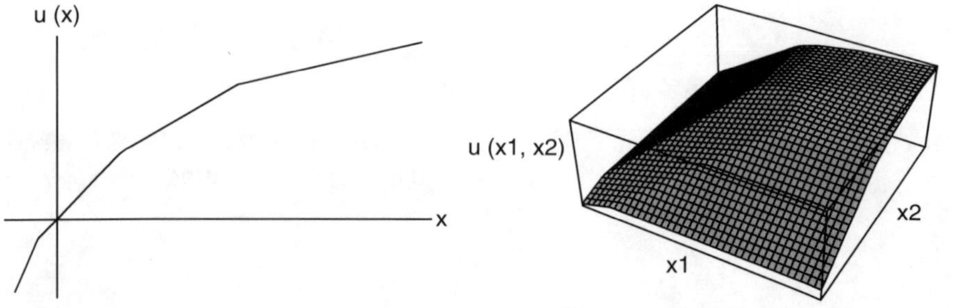

Fig. 4 Monotone and piecewise linear concave utility functions, single-period (*left*) and two-period ($J = 2$) (*right*)

where $(\,.\,)^-$ denotes the negative part of a real number, that is, $a^- = \max\{0, -a\}$ for $a \in \mathbb{R}$. The second representation is deduced from the first one by introducing stochastic variables y_1 for the positive and the negative part of $y_0 + z_T$. Hence, AVaR_α is of the form (9) with $J = 1$, $c_0 = 1$, $c_1 = (0, \frac{1}{\alpha})$, $w_{1,0} = (1, -1)$, $w_{1,1} = -1$, $Y_0 = \mathbb{R}$, $Y_1 = \mathbb{R}^2_+ = \mathbb{R}_+ \times \mathbb{R}_+$, and $h_0 = h_1 = V_{0,0} = V_{1,0} = V_{1,1} = 0$. Thus, it is a (single-period) polyhedral risk functional.

Another single-period example for a polyhedral risk functional (satisfying monotonicity and convexity) is expected utility, that is, $\rho_u(z_T) := -\mathbb{E}[u(z_T)]$ with a nondecreasing concave utility function $u : \mathbb{R} \to \mathbb{R}$; cf. Föllmer and Schied (2004). Typically, nonlinear functions such as $u(x) = 1 - e^{-\beta x}$ with some fixed $\beta > 0$ are used. Of course, in such cases ρ_u is not a polyhedral risk functional. However, in situations where the domain of z_T can be bounded a priori, it makes sense to use piecewise linear functions for u (see Fig. 4, left). Then, according to the infimum representation of piecewise linear convex functions (Rockafellar 1970, Corollary 19.1.2), it holds that

$$\rho_u(z_T) = \inf \left\{ \mathbb{E}\left[c \cdot y_1\right] \,\middle|\, \begin{array}{l} y_1 = y_1(\xi_1, \ldots, \xi_T) \in \mathbb{R}^{n+2}_+, \\ w \cdot y_1 = z_T, \; \sum_{i=1}^n y_{1,i} = 1 \end{array} \right\},$$

where n is the number of cusps of u, $w_1, \ldots, w_n$ are the x-coordinates of the cusps, and $c_i = -u(w_i)$ $(i = 1, .., n)$. Thus, ρ_u is a polyhedral risk functional. This approach can also be generalized to the multiperiod situation in an obvious way by specifying a (concave) utility function $u : \mathbb{R}^J \to \mathbb{R}$ (see Fig. 4, right). However, specifying an adequate utility function may be difficult in practice, in particular in the multiperiod case. Furthermore, expected utility is not cash invariant (cf. Sect. 4.1), neither in the single-period nor in the multiperiod case. Therefore, we will focus on generalizations of AVaR_α to the multiperiod case.

In the multiperiod case $J > 1$, the framework of polyhedral risk functionals allows to model different perspectives to the relations between different time stages. In Eichhorn and Römisch (2005), Eichhorn and Römisch (2006), Eichhorn (2007), and Pflug and Römisch (2007), several examples extending AVaR_α to the multiperiod situation in different ways have been constructed via a bottom-up approach

Table 1 Representation (9) of $\mathrm{AVaR}_\alpha(\frac{1}{J}\sum_{j=1}^{J} z_{t_j})$ (above) and $\mathrm{AVaR}_\alpha(\min\{z_{t_1},\dots,z_{t_J}\})$ (below)

$\mathrm{AVaR}_\alpha(z_0)$	Polyhedral representation (9)
$z_0 = \frac{1}{J}\sum_{j=1}^{J} z_{t_j}$	$\inf\left\{ \frac{1}{J}\left(y_0 + \sum_{j=1}^{J} \frac{1}{\alpha}\mathbb{E}\left[y_{j,2}\right]\right) \middle\vert \begin{array}{l} y_0 \in \mathbb{R},\ y_j = y_j(\xi_1,\dots,\xi_{t_j}) \\ \in \mathbb{R}\times\mathbb{R}_+ \ (j=1,\dots,J-1), \\ y_J = y_J(\xi_1,\dots,\xi_{t_J}) \in \mathbb{R}_+\times\mathbb{R}_+ \\ y_{j,1} - y_{j,2} = z_{t_j} + y_{j-1,1} \\ (j=1,\dots,J) \end{array} \right\}$
$z_0 = \min\{z_{t_1},\dots,z_{t_J}\}$	$\inf\left\{ y_0 + \frac{1}{\alpha}\mathbb{E}\left[y_{J,2}\right] \middle\vert \begin{array}{l} y_0 \in \mathbb{R},\ y_j = y_j(\xi_1,\dots,\xi_{t_j}) \in \mathbb{R}_+\times\mathbb{R}_+\times\mathbb{R}_+ \\ (j=1,\dots,J), \\ y_{1,2} - y_{1,3} = 0,\ y_{j,2} - y_{j,3} - y_{j-1,2} = 0 \\ (j=2,\dots,J), \\ y_{j,1} - y_{j,2} - y_0 = z_{t_j} \ (j=1,\dots,J) \end{array} \right\}$

using duality theory from convex analysis. Here, we restrict the presentation to the most obvious extensions that can be written in the form $\mathrm{AVaR}_\alpha(z_0)$ with a suitable *mixture* z_0 of $(z_{t_1},\dots,z_{t_J})$. We consider

$$z_0 = \frac{1}{J}\sum_{j=1}^{J} z_{t_j} \qquad \text{or} \qquad z_0 = \min\{z_{t_1},\dots,z_{t_J}\}$$

where "$\sum$" and "min" are understood scenariowise, that is, $z_0^i = \frac{1}{J}\sum_{j=1}^{J} z_{t_j}^i$, respectively, $z_0^i = \min\{z_{t_1}^i,\dots,z_{t_J}^i\}$ for each scenario i. Hence, in both cases the risk functional $\mathrm{AVaR}_\alpha(z_0)$ depends on the multivariate distribution of $(z_{t_1},\dots,z_{t_J})$.

As shown in Table 1, both $\mathrm{AVaR}_\alpha(\frac{1}{J}\sum_{j=1}^{J} z_{t_j})$ and $\mathrm{AVaR}_\alpha(\min\{z_{t_1},\dots,z_{t_J}\})$ can be written in the form (9), that is, they are indeed *multiperiod polyhedral risk functionals*. Moreover, they are multiperiod coherent risk functionals in the sense of Sect. 4.1. Clearly, the latter of the two functionals is the most reasonable multiperiod extension of AVaR with regard to liquidity considerations, since AVaR is applied to the respectively lowest wealth values in each scenario; this worst case approach has also been suggested in Artzner et al. (2007, Sect. 4).

4.3 Illustrative Simulation Results

Finally, we illustrate the effects of different polyhedral risk functionals by presenting some optimal wealth processes from an electricity portfolio optimization model (Eichhorn et al. 2005; Eichhorn and Römisch 2006). This model is of the form (1); it considers the 1 year planning problem of a *municipal power utility*, that is, a price-taking retailer serving heat and power demands of a certain region; see Fig. 5. It is assumed that the utility features a combined heat and power (CHP) plant that can serve the heat demand completely but the power demand only in part. In addition, the utility can buy power at the day-ahead spot market of some power exchange,

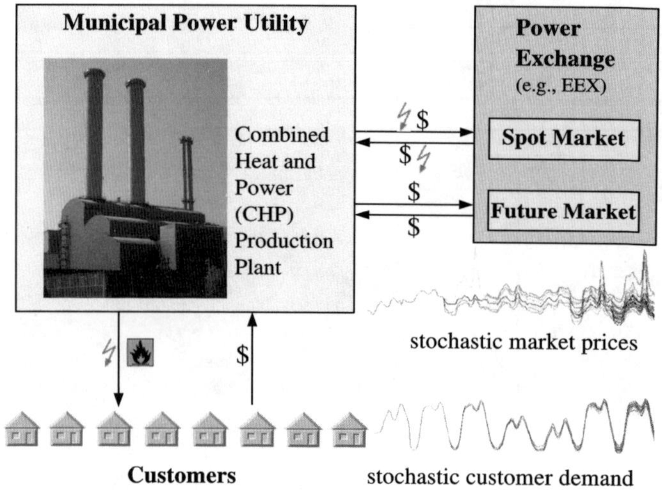

Fig. 5 Schematic diagram for a power production planning and trading model under demand and price uncertainty (portfolio optimization)

for example, the European Energy Exchange (EEX). Moreover, the utility can trade monthly (purely financial) futures (e.g., Phelix futures at EEX).

The objective $\mathbb{F}$ of this model is a mean-risk objective as discussed in Sect. 1 incorporating a polyhedral risk functional ρ and the expected total revenue $\mathbb{E}[z_T]$; the weighting factor is set to $\gamma = 0.9$. Time horizon is 1 year in hourly discretization, that is, $T = 8760$. Time series models for the uncertain input data (demands and prices) have been set up (see Eichhorn et al. (2005) for details) and approximated according to Sect. 3 by a finite scenario tree consisting of 40 scenarios; see Fig. 3 (right). The scenario tree has been obtained by the forward construction procedure of Algorithm 3.3. It represents the uncertainty well enough on the one hand, and, on the other hand, the moderate size of the tree guarantees computational tractability. For the risk time steps t_j, we use 11 PM at the last trading day of each week ($j = 1, \ldots, J = 52$). Note that, because of the limited number of branches in the tree, a finer resolution for the risk time steps does not make sense here. The resulting optimization problem is very large-scale; however, it is numerically tractable due to the favorable nature of polyhedral risk functionals. In particular, since we modeled the CHP plant without integer variables, it is a linear program (LP) that could be solved by ILOG CPLEX in about 1 h.

In Fig. 6 (as well as in Fig. 7) the optimal cash flows are displayed, that is, the wealth values z_t for each time step $t = 1, \ldots, T$ and each scenario, obtained from optimization runs with different mean-risk objectives. The price parameters have been set such that the effects of the risk functionals may be observed well, although these settings yield negative cash values. These families of curves differ in shape due to different policies of future trading induced by the different risk functionals; see Fig. 8. Setting $\gamma = 0$ (no risk functional at all) yields high spread for z_T and there is

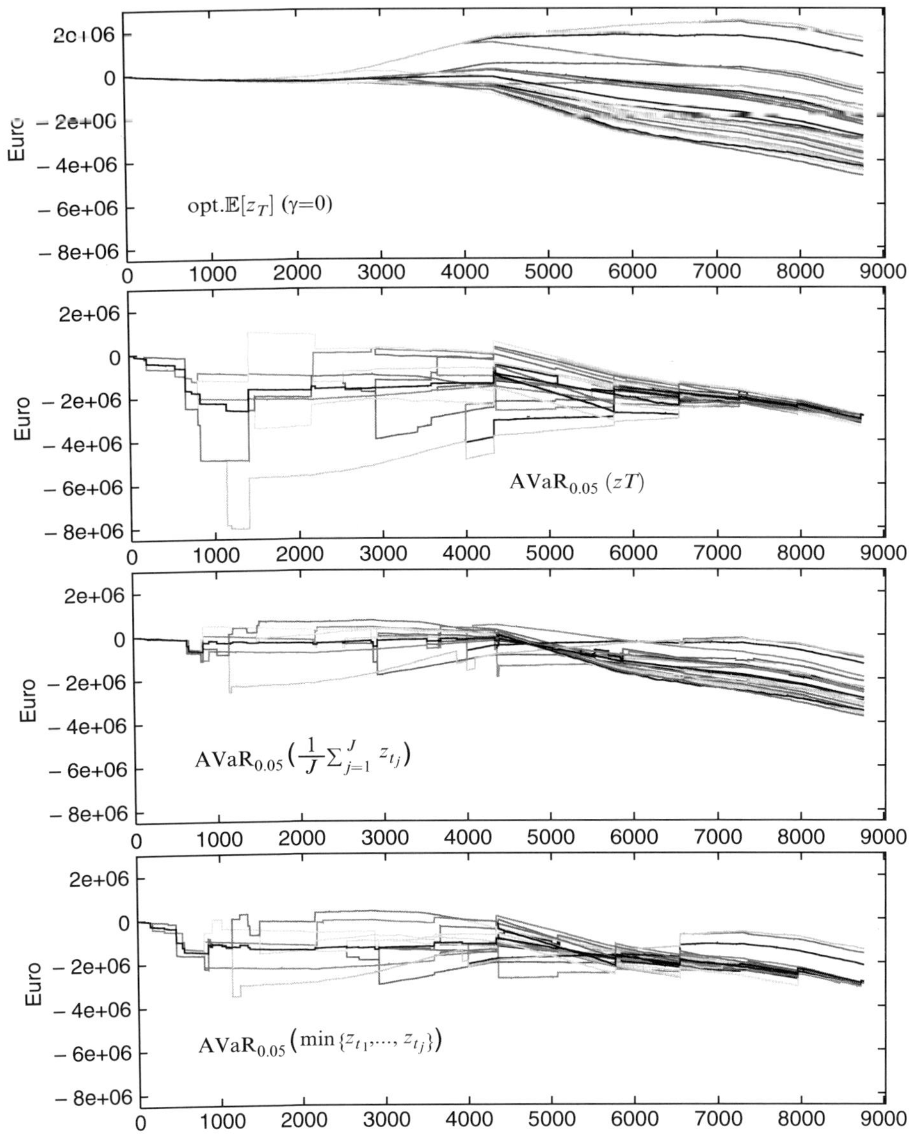

Fig. 6 Optimal cash values z_t (wealth) over time ($t = 1, \ldots, T$) with respect to different risk functionals. Each *curve* in a graph represents one of the 40 scenarios. The expected value of the cash flows is displayed in *black*

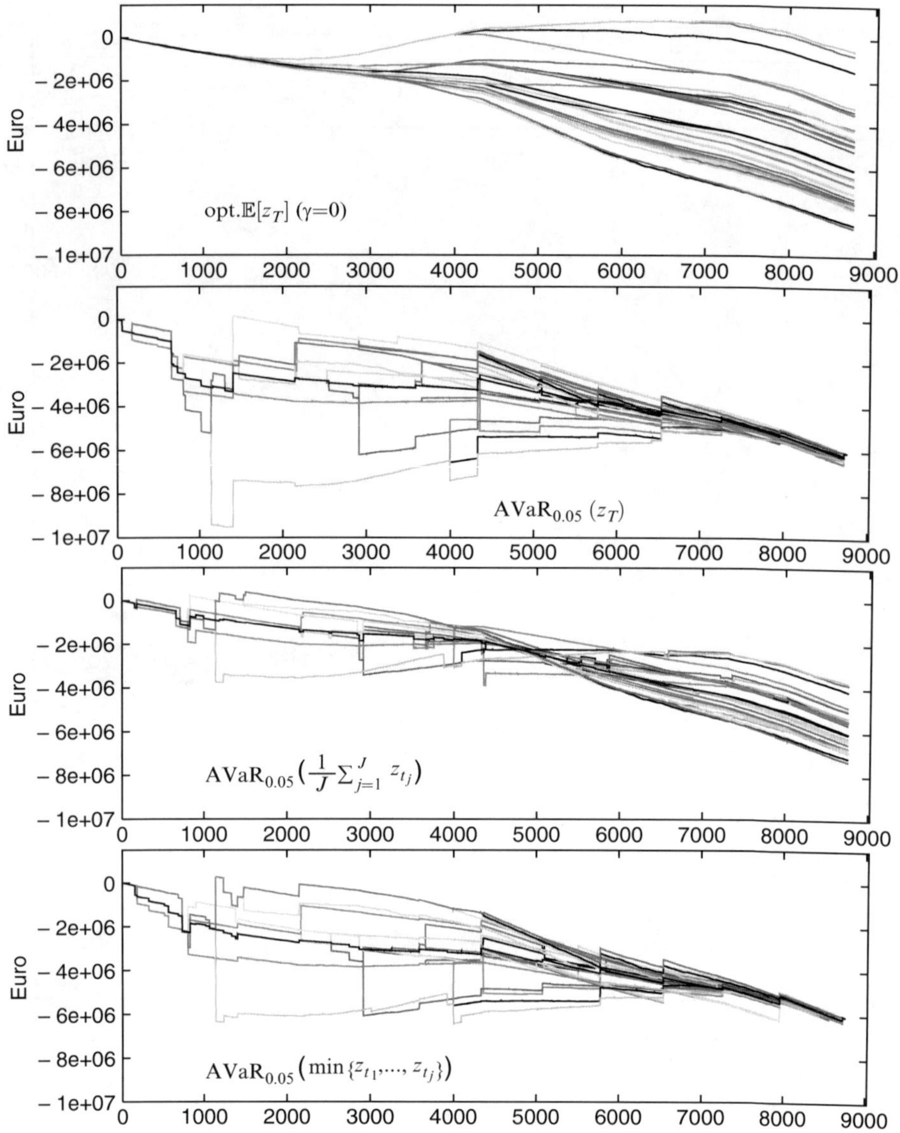

Fig. 7 Optimal cash values z_t (wealth) over time ($t = 1, \ldots, T$) with respect to different risk functionals. For these calculations, slightly higher fuel cost parameters have been used such that the graphs demonstrate the nature of the risk functionals best

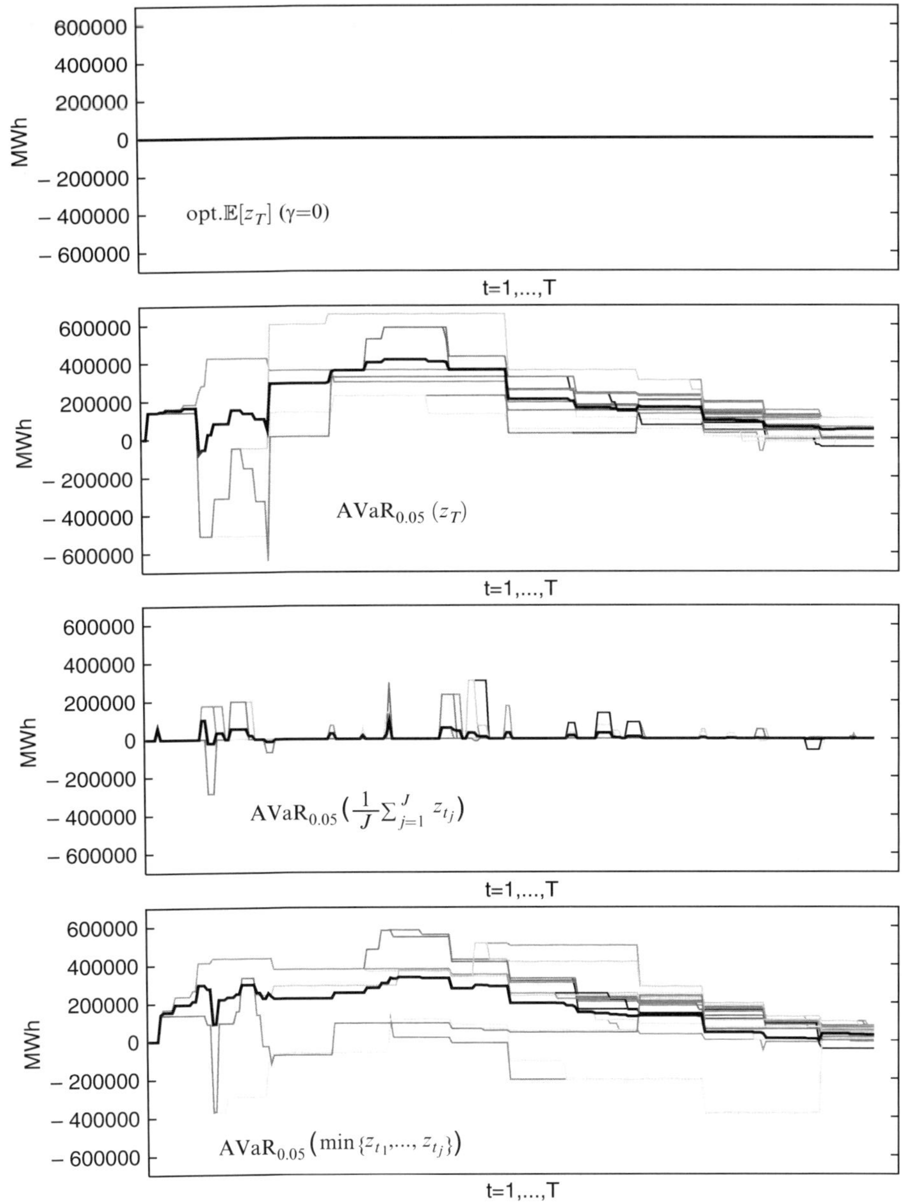

Fig. 8 Optimal future stock over time with respect to different polyhedral risk functionals. The expected value of the future stock over time is displayed in *black*

no future trading at all (since we worked with fair future prices). Using $\mathrm{AVaR}_\alpha(z_T)$ ($\gamma = 0.9$) yields low spread for z_T but low values and high spread at $t < T$. This shows that, for the situation here, single-period risk functionals are not appropriate. The employment of multiperiod polyhedral risk functionals yields spread that is better distributed over time. However, the way how this is achieved is different: The functional $\mathrm{AVaR}_\alpha(\min\{z_{t_1}, \ldots, z_{t_J}\})$ aims at finding a level y_0 as high as possible such that the curves rarely fall below that level, while $\mathrm{AVaR}_\alpha(\frac{1}{J}\sum_{j=1}^{J} z_{t_j})$ aims at equal spread at all times. In the latter case, futures are held only for short time periods, while in the other cases futures are held for longer. Finally, we note that the effects of the risk functionals cost only less than 1% of the expected overall revenue $\mathbb{E}[z_T]$.

5 Conclusions

Multistage stochastic programming models are discussed as mathematical tools for dealing with uncertain (future) parameters in electricity portfolio and risk management. Since statistical information on the parameters (like demands, spot prices, inflows, or wind speed) is often available, stochastic models may be set up for them so that scenarios of the future uncertainty are made available. To model the information flow over time, the scenarios need to be tree-structured. For this reason a general methodology is presented in Sect. 3, which allows to generate scenario trees out of the given set of scenarios. The general method is based on stability argument for multistage stochastic programs and does not require further knowledge on the underlying multivariate probability distribution. The method is flexible and allows to generate scenario trees whose size enables a good approximation of the underlying probability distribution, on the one hand, and allows for reasonable running times of the optimization software, on the other hand. Implementations of these scenario tree generation algorithms are available in GAMS-SCENRED.

A second issue discussed in the paper is risk management via the incorporation of risk functionals into the objective. This allows maximizing expected revenue and minimizing risk simultaneously. Since risk functionals are nonlinear by definition, a natural requirement consists in preserving computational tractability of the (mixed-integer) optimization models and, hence, in reasonable running times of the software. Therefore, a class of risk functionals is presented in Sect. 4.2 that allow a formulation as linear (stochastic) program. Hence, if the risk functional (measure) belongs to this class, the resulting optimization model does not contain additional nonlinearities. If the expected revenue maximization model is (mixed-integer) linear, the linearity is preserved. A few examples of such *polyhedral* risk functionals are provided for multiperiod situations, that is, if the risk evolves over time and requires to rely on multivariate probability distributions. The simulation study in Sect. 4.3 for the electricity portfolio management of a price-taking retailer provides some insight into the risk minimization process by trading electricity derivatives.

It turns out that the risk can be reduced considerably for less than 1% of the expected overall revenue.

Acknowledgements This work was supported by the DFG Research Center MATHEON *Mathematics for Key Technologies* in Berlin (http://www.matheon.de) and by the German Ministry of Education and Research (BMBF) under the grant 03SF0312E. The authors thank the anonymous referees for their helpful comments.

References

Artzner P, Delbaen F, Eber JM, Heath D (1999) Coherent measures of risk. Math Financ 9:203–228
Artzner P, Delbaen F, Eber JM, Heath D, Ku H (2007) Coherent multiperiod risk adjusted values and Bellman's principle. Ann Oper Res 152:5–22
Blaesig B (2007) Risikomanagement in der Stromerzeugungs- und Handelsplanung, Aachener Beiträge zur Energieversorgung, vol. 113. Klinkenberg, Aachen, Germany, PhD Thesis
Blaesig B, Haubrich HJ (2005) Methods of risk management in the generation and trading planning. In: IEEE St. Petersburg PowerTech Proceedings
Burger M, Klar B, Müller A, Schindlmayr G (2004) A spot market model for pricing derivatives in electricity markets. Quant Financ 4:109–122
Cabero J, Baillo A, Cerisola S, Ventosa M, Garcia-Alcalde A, Beran F, Relano G (2005) A medium-term integrated risk management model for a hydrothermal generation company. IEEE Trans Power Syst 20:1379–1388
Casey MS, Sen S (2005) The scenario generation algorithm for multistage stochastic linear programming. Math Oper Res 30:615–631
Chiralaksanakul A, Morton DP (2004) Assessing policy quality in multi-stage stochastic programming. Preprint, Stochastic Programming E-Print Series, www.speps.org
Clewlow L, Strickland C (2000) Energy derivatives: pricing and risk management. Lacima, London
Corvera Poiré X (1995) Model generation and sampling algorithms for dynamic stochastic programming. PhD thesis, Department of Mathematics, University of Essex, UK
Dempster MAH (2004) Sequential importance sampling algorithms for dynamic stochastic programming. Zapiski Nauchnykh Seminarov POMI 312:94–129
Dentcheva D, Ruszczyński A (2003) Optimization with stochastic dominance constraints. SIAM J Optim 14:548–566
Dentcheva D, Ruszczyński A (2006) Portfolio optimization with stochastic dominance constraints. J Bank Finance 30:433–451
Dupačová J, Consigli G, Wallace SW (2000) Scenarios for multistage stochastic programs. Ann Oper Res 100:25–53
Dupačová J, Gröwe-Kuska N, Römisch W (2003) Scenario reduction in stochastic programming: An approach using probability metrics. Math Program 95:493–511
Edirisinghe NCP (1999) Bound-based approximations in multistage stochastic programming: Nonanticipativity aggregation. Ann Oper Res 85:103–127
Eichhorn A (2007) Stochastic programming recourse models: Approximation, risk aversion, applications in energy. PhD thesis, Department of Mathematics, Humboldt University, Berlin
Eichhorn A, Römisch W (2005) Polyhedral risk measures in stochastic programming. SIAM J Optim 16:69–95
Eichhorn A, Römisch W (2006) Mean-risk optimization models for electricity portfolio management. In: Proceedings of the 9th International Conference on Probabilistic Methods Applied to Power Systems (PMAPS), Stockholm, Sweden
Eichhorn A, Römisch W (2008) Stability of multistage stochastic programs incorporating polyhedral risk measures. Optimization 57:295–318

Eichhorn A, Römisch W, Wegner I (2005) Mean-risk optimization of electricity portfolios using multiperiod polyhedral risk measures. In: IEEE St. Petersburg PowerTech Proceedings

Fleten SE, Kristoffersen TK (2008) Short-term hydropower production planning by stochastic programming. Comput Oper Res 35:2656–2671

Fleten SE, Wallace SW, Ziemba WT (2002) Hedging electricity portfolios via stochastic programming. In: Greengard C, Ruszczyński A (eds) Decision making under uncertainty: energy and power. IMA volumes in mathematics and its applications, vol. 128. Springer, New York Greengard and Ruszczyński (2002), pp. 71–93

Föllmer H, Schied A (2002) Convex measures of risk and trading constraints. Finance Stochast 6:429–447

Föllmer H, Schied A (2004) Stochastic finance. An introduction in discrete time. De Gruyter Studies in Mathematics, vol. 27, 2nd edn. Walter de Gruyter, Berlin

Frauendorfer K (1996) Barycentric scenario trees in convex multistage stochastic programming. Math Program 75:277–293

Frittelli M, Rosazza Gianin E (2002) Putting order in risk measures. J Bank Finance 26:1473–1486

Frittelli M, Scandolo G (2005) Risk measures and capital requirements for processes. Math Finance 16:589–612

Greengard C, Ruszczyński A (eds) (2002) Decision making under uncertainty: energy and power. IMA Volumes in Mathematics and its Applications, vol. 128. Springer, New York

Gröwe-Kuska N, Kiwiel KC, Nowak MP, Römisch W, Wegner I (2002) Power management in a hydro-thermal system under uncertainty by Lagrangian relaxation. In: Greengard C, Ruszczyński A (eds) Decision making under uncertainty: energy and power. IMA Volumes in Mathematics and its Applications, vol. 128. Springer, New York , pp. 39–70

Gröwe-Kuska N, Heitsch H, Römisch W (2003) Scenario reduction and scenario tree construction for power management problems. In: Borghetti A, Nucci CA, Paolone M (eds). IEEE Bologna PowerTech Proceedings

Heitsch H (2007) Stabilität und Approximation stochastischer Optimierungsprobleme. PhD thesis, Department of Mathematics, Humboldt University, Berlin

Heitsch H, Römisch W (2003) Scenario reduction algorithms in stochastic programming. Comput Optim Appl 24:187–206

Heitsch H, Römisch W (2005) Generation of multivariate scenario trees to model stochasticity in power management. In: IEEE St. Petersburg PowerTech Proceedings

Heitsch H, Römisch W (2007) A note on scenario reduction for two-stage stochastic programs. Oper Res Lett 35:731–738

Heitsch H, Römisch W (2009) Scenario tree modeling for multistage stochastic programs. Math Program 118:371–406

Heitsch H, Römisch W, Strugarek C (2006) Stability of multistage stochastic programs. SIAM J Optim 17:511–525

Henrion R (2006) Some remarks on value-at-risk optimization. Manag Sci Eng Manag 1:111–118

Henrion R (2007) Structural properties of linear probabilistic constraints. Optimization 56: 425–440

Henrion R, Küchler C, Römisch W (2008) Discrepancy distances and scenario reduction in two-stage stochastic mixed-integer programming. J Ind Manag Optim 4:363–384

Hochreiter R (2005) Computational optimal management decisions – the case of stochastic programming for financial management. PhD thesis, University of Vienna, Austria

Hochreiter R, Pflug GC (2007) Financial scenario generation for stochastic multi-stage decision processes as facility location problems. Ann Oper Res 152:257–272

Hochreiter R, Pflug GC, Wozabal D (2006) Multi-stage stochastic electricity portfolio optimization in liberalized energy markets. In: System modeling and optimization. IFIP International Federation for Information Processing, Springer, Boston, MA, USA, pp. 219–226

Høyland K, Wallace SW (2001) Generating scenario trees for multi-stage decision problems. Manag Sci 47:295–307

Høyland K, Kaut M, Wallace SW (2003) A heuristic for moment-matching scenario generation. Comput Optim Appl 24:169–185

Infanger G (1994) Planning under uncertainty – solving large-scale stochastic linear programs. Scientific Press Series, Boyd & Fraser, Danvers, MA, USA

Kall P, Mayer J (2005) Stochastic linear programming. Springer, Berlin Heidelberg New York

Kaut M, Wallace SW (2007) Evaluation of scenario generation methods for stochastic programming. Pacific J Optim 3:257–271

Krasenbrink B (2002) Integrierte Jahresplanung von Stromerzeugung und -handel, Aachener Beiträge zur Energieversorgung, vol. 81. Klinkenberg, Aachen, Germany, PhD Thesis

Kuhn D (2005) Generalized bounds for convex multistage stochastic programs. Lecture Notes in Economics and Mathematical Systems, vol. 548. Springer, Berlin Heidelberg New York

Li T, Shahidehpour M, Li Z (2007) Risk-constrained bidding strategy with stochastic unit commitment. IEEE Trans Power Syst 22:449–458

Mirkov R, Pflug GC (2007) Tree approximations of dynamic stochastic programs. SIAM J Optim 18:1082–1105

Müller A, Stoyan D (2002) Comparison methods for stochastic models and risks. Wiley, Chichester, UK

Pennanen T (2005) Epi-convergent discretizations of multistage stochastic programs. Math Oper Res 30:245–256

Pennanen T (2009) Epi-convergent discretizations of multistage stochastic programs via integration quadratures. Math Program 116:461–479

Pereira MVF, Pinto LMVG (1991) Multi-stage stochastic optimization applied to energy planning. Math Program 52:359–375

Pflug GC (2001) Scenario tree generation for multiperiod financial optimization by optimal discretization. Math Program 89:251–271

Pflug GC, Römisch W (2007) Modeling, measuring, and managing risk. World Scientific, Singapore

Prékopa A (2003) Probabilistic programming. In: Ruszczyński A, Shapiro A (eds) (2003) Stochastic programming. Handbooks in operations research and management science, vol. 10, 1st edn. Elsevier, Amsterdam. chap 5, pp. 267–351

Rockafellar RT (1970) Convex analysis. Princeton University Press, Princeton, NJ, USA

Rockafellar RT, Uryasev S (2002) Conditional value-at-risk for general loss distributions. J Bank Finance 26:1443–1471

Römisch W (2003) Stability of stochastic programming problems. In: Ruszczyński A, Shapiro A (eds) (2003) Stochastic programming. Handbooks in operations research and management science, vol. 10, 1st edn. Elsevier, Amsterdam. chap 8, pp. 483–554

Ruszczyński A, Shapiro A (eds) (2003) Stochastic programming. Handbooks in operations research and management science, vol. 10, 1st edn. Elsevier, Amsterdam

Ruszczyński A, Shapiro A (2006) Optimization of convex risk functions. Math Oper Res 31: 433–452

Schmöller HK (2005) Modellierung von Unsicherheiten bei der mittelfristigen Stromerzeugungs- und Handelsplanung, Aachener Beiträge zur Energieversorgung, vol. 103. Klinkenberg, Aachen, Germany, PhD Thesis

Schmöller HK, Hartmann T, Kruck I, Haubrich HJ (2003) Modeling power price uncertainty for midterm generation planning. In: Borghetti A, Nucci CA, Paolone M (eds) IEEE Bologna PowerTech Proceedings

Sen S, Yu L, Genc T (2006) A stochastic programming approach to power portfolio optimization. Oper Res 54:55–72

Shapiro A (2003) Inference of statistical bounds for multistage stochastic programming problems. Math Methods Oper Res 58:57–68

Shapiro A (2007) Stochastic programming approach to optimization under uncertainty. Math Program 112:183–220

Takriti S, Krasenbrink B, Wu LSY (2000) Incorporating fuel constraints and electricity spot prices into the stochastic unit commitment problem. Oper Res 48:268–280

Wallace SW, Fleten SE (2003) Stochastic programming models in energy. In: Ruszczyński A, Shapiro A (eds) (2003) Stochastic programming. Handbooks in operations research and management science, vol. 10, 1st edn. Elsevier, Amsterdam. chap 10, pp. 637–677
Weron R, Bierbrauer M, Trück S (2004) Modeling electricity prices: jump diffusion and regime switching. Physica A 336:39–48
Wu L, Shahidehpour M, Li T (2007) Stochastic security-constrained unit commitment. IEEE Trans Power Syst 22:800–811

Taking Risk into Account in Electricity Portfolio Management

Laetitia Andrieu, Michel De Lara, and Babacar Seck

Abstract We provide an economic interpretation with utility functions of the practice consisting in incorporating risk measures as constraints in a classic expected return maximization problem. We also establish a dynamic programming equation. Inspired by this economic approach, we compare two ways to incorporate risk (Conditional Value-at-Risk, CVaR) in generation planning in electrical industry: either as constraints or making use of utility functions deduced from the risk constraints.

Keywords Risk management · Risk measures · Stochastic optimization · Utility functions

1 Introduction

Liberalization of energy markets displays new issues for electrical companies, which now have to master both traditional problems (such as optimization of electrical generation) and emerging problems (such as integration of financial markets and risk management). The question of how to take risk into account has been studied since long in mathematical and economic literature (among many, we just point out two references: Gollier (2001) and Pflug and Römisch (2007)). Let us briefly describe two classical approaches to deal with risk in a decision problem. On the one hand, we may maximize the expected return of a portfolio under explicit risk constraints, such as variance or CVaR; we shall coin this practice as belonging to the engineers' or practitioners' world. On the other hand, we may maximize the expected utility of the return of a portfolio by using a utility function that captures more or less risk aversion (in the so-called expected utility theory, or more general functionals else); this is the world of economists. In this work, we focus on the

L. Andrieu (✉)
EDF R&D OSIRIS, 1 avenue du Général de Gaulle 92140 Clamart, France
e-mail: laetitia.andrieu@edf.fr

S. Rebennack et al. (eds.), *Handbook of Power Systems II*, Energy Systems,
DOI 10.1007/978-3-642-12686-4_16,

links between this two approaches and provide applications to electricity portfolio management.

The paper is organized as follows. In Sect. 2, we specify a wide class of risk measures and we give an equivalence between a profit maximization problem under risk constraints (belonging to the specified class of risk measures) and a maximin problem involving an infinite number of utility functions. We generalize the above static approach to the dynamic case in Sect. 3; we also exhibit a dynamic programming equation to theoretically solve a class of optimization problems under risk constraints. In Sect. 4, we present numerical experiments comparing different ways to take risk into account in practical multistage stochastic optimization problems, such as a generation planning in electrical industry.

2 From Risk Constraints to Utility Functions in Stochastic Optimization

Details of what follows may be found in Andrieu et al. (2008).

2.1 The Infimum of Expectations Class of Risk Measures

Let a probability space $(\Omega, \mathcal{F}, \mathbb{P})$ be given with $\mathbb{P}$ as a probability measure on a σ-field $\mathcal{F}$ of events on Ω. We are thus in a *risk* decision context. The expectation of a random variable on $(\Omega, \mathcal{F}, \mathbb{P})$ will be denoted by $\mathbb{E}$. We introduce a class of risk measures, which covers many of the usual risk measures (variance, CVaR, optimized certainty equivalent, etc.) and which will be proven adequate for optimization problems.

Let a function $\rho : \mathbb{R} \times \mathbb{R} \to \mathbb{R}$ be given. Let $L_\rho(\Omega, \mathcal{F}, \mathbb{P})$ be a set of random variables X defined on $(\Omega, \mathcal{F}, \mathbb{P})$ such that $\rho(X, \eta)$ is measurable and integrable for all η and such that the following risk measure

$$R_\rho(X) = \inf_{\eta \in \mathbb{R}} \mathbb{E}\big[\rho(X, \eta)\big], \tag{1}$$

is well defined and finite ($R_\rho(X) > -\infty$).

We shall say that the risk measure formulation $(\rho, L_\rho(\Omega, \mathcal{F}, \mathbb{P}))$ of R_ρ in (1) is *nice* – more briefly that R_ρ is *nice* – if the function $\rho : \mathbb{R} \times \mathbb{R} \to \mathbb{R}_+$ and the set $L_\rho(\Omega, \mathcal{F}, \mathbb{P})$ satisfy the following properties (introduced mainly for technical reasons):

H1. The function $\eta \in \mathbb{R} \mapsto \rho(x, \eta)$ is convex, for all $x \in \mathbb{R}$,
H2. For all $X \in L_\rho(\Omega, \mathcal{F}, \mathbb{P})$, $\eta \in \mathbb{R} \mapsto \mathbb{E}\big[\rho(X, \eta)\big]$ is continuous and goes to infinity at infinity ($\lim_{|\eta| \to +\infty} \mathbb{E}\big[\rho(X, \eta)\big] = +\infty$).

Such a definition of infimum of expectations class of nice risk measures extends, in a sense, the notion of the optimized certainty equivalent introduced in Ben-Tal and Teboulle (1986) (see also Ben-Tal and Teboulle (2007)). This latter is grounded in economics. for a concave and nondecreasing $U : \mathbb{R} \to \mathbb{R}$, the optimized certainty equivalent is $\mathcal{S}_U(X) = \sup_{\nu \in \mathbb{R}} \left(\nu + \mathbb{E}\left[U(X-\nu)\right]\right)$, and the associated risk measure is $R_\rho(X) := -\mathcal{S}_U(-X) = \inf_{\eta \in \mathbb{R}} \left(\eta - \mathbb{E}\left[U(\eta - X)\right]\right)$. With specific requirements on U, we can obtain a nice risk measure.

Other examples belonging to the infimum of expectations class of risk measures are the following. In Markowitz (1952), Markowitz uses variance as a risk measure. A well known formula for the variance is $\mathsf{var}\left[X\right] = \inf_{\eta \in \mathbb{R}} \mathbb{E}\left[(X-\eta)^2\right]$. With $\rho_{\mathsf{var}}(x,\eta) := (x-\eta)^2$ and $L_\rho(\Omega, \mathcal{F}, \mathbb{P}) = L^2(\Omega, \mathcal{F}, \mathbb{P})$, $\mathsf{var} = R_{\rho_{\mathsf{var}}}$ is nice. In Ogryczak and Ruszczynski (1999), the weighted mean deviation from a quantile is given by $\mathsf{WMd}_p(X) = \inf_{\eta \in \mathbb{R}} \mathbb{E}\Big[\max\left\{p(X-\eta), (1-p)(\eta - X)\right\}\Big]$.

We shall particularly concentrate on the CVaR. The risk measure CVaR at confidence level[1] $p \in]0,1[$ is given by the following formula to be found in Rockafellar and Uryasev (2000):

$$\mathsf{CVaR}_p(X) = \inf_{\eta \in \mathbb{R}} \left(\eta + \frac{1}{1-p} \mathbb{E}\left[\max\{0, X-\eta\}\right]\right).$$

The function

$$\rho_{\mathsf{CVaR}}(x,\eta) := \eta + \frac{1}{1-p} \max\{0, x-\eta\}$$

is convex with respect to η, and taking $L_{\rho_{\mathsf{CVaR}}}(\Omega, \mathcal{F}, \mathbb{P}) = L^1(\Omega, \mathcal{F}, \mathbb{P})$, assumption H2 is satisfied. Thus, $R_{\rho_{\mathsf{CVaR}}}$ is nice.

2.2 *A Maxmin Reformulation of a Profit Maximization Problem Under Risk Constraints*

We now state an equivalence between a profit maximization problem under risk constraints (belonging to the infimum of expectations class of risk measures) and a maximin problem involving an infinite number of utility functions. Let $J : \mathbb{R}^n \times \mathbb{R}^p \to \mathbb{R}$, let $\mathbb{A} \subset \mathbb{R}^n$, and let ξ be a random variable defined on $(\Omega, \mathcal{F}, \mathbb{P})$ with values in $\mathbb{R}^p$. We interpret $J(a,\xi)$ as the random return of a portfolio with respect to a decision variable $a \in \mathbb{A}$, and where the random variable ξ stands for the uncertainties that can affect the return.

Assume that R_ρ is nice, that $J(a,\xi) \in L_\rho(\Omega, \mathcal{F}, \mathbb{P})$ for all $a \in \mathbb{A}$, and that the infimum in (1) is achieved for any loss $X = -J(a,\xi)$ when a varies in $\mathbb{A}$.

[1] In practice, p is rather close to 1 ($p = 0.95$, $p = 0.99$). The value-at-risk $\mathsf{VaR}_p(X)$ is such that $\mathbb{P}(X \leq \mathsf{VaR}_p(X)) = p$. Then $\mathsf{CVaR}_p(X)$ is interpreted as the expected value of X knowing that X exceeds $\mathsf{VaR}_p(X)$.

Then it can be proved (see Andrieu et al. (2008), but this is also a Corollary to Proposition 1) that the maximization under risk constraint[2] problem

$$\sup_{a \in \mathbb{A}} \mathbb{E}\big[J(a,\xi)\big] \tag{2a}$$

$$\text{under constraint} \quad R_\rho\big(-J(a,\xi)\big) \le \gamma \tag{2b}$$

is equivalent to the following maximin problem

$$\sup_{(a,\eta)\in\mathbb{A}\times\mathbb{R}} \inf_{U\in\mathcal{U}} \mathbb{E}\left[U\left(J(a,\xi),\eta\right)\right] . \tag{3}$$

The set $\mathcal{U}$ of functions $\mathbb{R}^2 \to \mathbb{R}$ over which the infimum is taken is

$$\mathcal{U} = \left\{U^{(\lambda)} : \mathbb{R}^2 \to \mathbb{R}\,,\ \lambda \ge 0 \mid U^{(\lambda)}(x,\eta) = x + \lambda\big(-\rho(-x,\eta)+\gamma\big)\right\} . \tag{4}$$

Our approach uses duality theory and Lagrange multipliers. However, it does not focus on the optimal multiplier, and this is how we obtain a family of utility functions and an economic interpretation (though belonging to nonexpected utility theories as in Maccheroni (2002)), and not a single utility function. This differs from the result in Dentcheva and Ruszczynski (2006), where the authors prove, in a way, that utility functions play the role of Lagrange multipliers for second order stochastic dominance constraints. With this, Dentcheva and Ruszczynski (2006) prove the equivalence between portfolio maximization under second order stochastic dominance constraints and expected utility maximization, for one single utility function. However, such utility function is not given a priori and may not be interpreted economically before the decision problem.

2.3 *Conditional Value-at-Risk and Loss Aversion*

Suppose that the risk measure is the CVaR $R_{\rho_{\text{CVaR}}}$. The utility functions associated to this risk measure are

$$U^{(\lambda)}(x,\eta) = x - \frac{\lambda}{1-p}(-x-\eta)_+ - \lambda\eta + \lambda\gamma .$$

Let us consider only the above function as function of x (profit, benefit, etc.). We interpret the $-\eta$ argument as an *anchorage* parameter: for $x \ge -\eta$, $x \mapsto U^{(\lambda)}(x,\eta) = x - \lambda\eta + \lambda\gamma$ has slope 1, while it has slope

[2] The risk constraint is not on the return $J(a,\xi)$, but on the loss $-J(a,\xi)$.

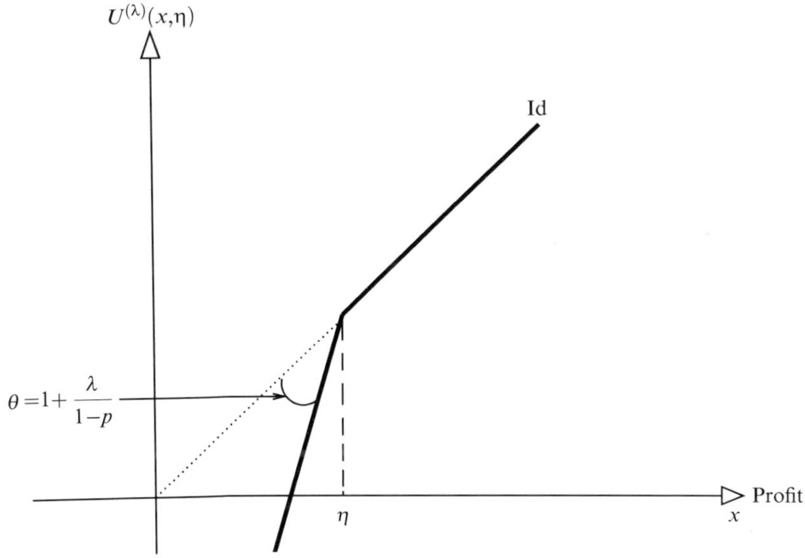

Fig. 1 Utility function attached to CVaR. For $x \geq -\eta$, the slope is one. For $x \leq -\eta$, the slope is $\theta > 1$, which measures a kind of "loss aversion" parameter

$$\theta := 1 + \frac{\lambda}{1-p}$$

for x lower than $-\eta$, as in Fig. 1. We interpret the parameter θ as a *loss aversion* parameter introduced by Kahneman and Tversky (see Kahneman and Tversky (1992)). Indeed, this utility function $x \mapsto U^{(\lambda)}(x,\eta)$ expresses the property that one monetary unit more than the anchorage $-\eta$ gives one unit of utility, while one unit less gives $-\theta$.

3 Dynamic Profit Maximization Under Risk Constraints

Now, we try and generalize the above static approach to the dynamic case. We take static risk measures to formulate risk constraints on each time period; this will prove to be appropriate to obtain a dynamic programming equation. We shall also give an equivalent economic formulation to the original problem. We expect that this new formulation can give ideas to formulate other, approximate, ways to deal with risk. We shall give examples in the numerical applications.

All mappings will be supposed to have proper measurability and integrability assumptions ensuring that mathematical expectations can be performed.

3.1 Problem Statement

Let $t_0 < T$ be two integers, and $(\xi(t))_{t=t_0,\dots,T-1}$ be an $\mathbb{R}^p$-valued stochastic process defined on a probability space $(\Omega, \mathcal{F}, \mathbb{P})$; mathematical expectations are denoted by $\mathbb{E}$. Let a dynamic mapping $F : \mathbb{R}^n \times \mathbb{R}^m \times \mathbb{R}^p \to \mathbb{R}^n$ be given. Define a *feedback*

$$\hat{a}(\cdot) := \big(\hat{a}(t_0, \cdot), \dots, \hat{a}(T-1, \cdot)\big) \tag{5}$$

as a sequence of measurable mappings $\hat{a}(t, \cdot) : x \in \mathbb{R}^n \mapsto \hat{a}(t, x) \in \mathbb{A}$[3]; the set of all feedbacks is $\mathcal{A}$. Given $\hat{a}(\cdot) \in \mathcal{A}$, a state and control stochastic process is generated by $X(t+1) = F\big(X(t), a(t), \xi(t)\big)$, $a(t) = \hat{a}\big(t, X(t)\big)$ and is evaluated by the criterion $\mathbb{E}\Big[\sum_{t=t_0}^{T-1} J\big(X(t), a(t), \xi(t)\big) + K(X(T))\Big]$.

We consider the optimization problem

$$\sup_{\hat{a}(\cdot)} \mathbb{E}\Big[\sum_{t=t_0}^{T-1} J\big(X(t), a(t), \xi(t)\big) + K(X(T))\Big] \tag{6a}$$

subject to dynamic constraints

$$X(t+1) = F\big(X(t), a(t), \xi(t)\big) \tag{6b}$$

$$a(t) = \hat{a}\big(t, X(t)\big) \tag{6c}$$

and risk constraints

$$R_\rho\Big(-J\big(X(t), a(t), \xi(t)\big)\Big) \le \gamma(t), \quad t = t_0, \dots, T-1, \tag{6d}$$

where

- $J : \mathbb{R}^m \times \mathbb{R}^n \times \mathbb{R}^p \to \mathbb{R}$ is the instantaneous return (profit, etc.), and $K : \mathbb{R}^n \to \mathbb{R}$ is the final one
- R_ρ is a risk measure as in (1), together with the level constraint $\gamma \in \mathbb{R}$.

We have taken stationary dynamics, profits, and risk measures for the sake of notational simplicity, but the nonstationary case can be treated in the same way. Thus, discounting might be included.

We consider risk constraints in the intermediary problems, with the idea to control all periods risks. However, both risk constraint for the last period and on the cumulated return can be treated at the price of burdening notations. Indeed, a risk constraint for the last period would notationally differ from the others by the absence of a control since there is no $a(T)$. This also allows to have a risk constraint on

[3] We recall that $\mathbb{A} \subset \mathbb{R}^m$. We could also consider that $\hat{a}(t, x) \in \mathbb{A}(t, x)$, where $\mathbb{A}(t, x) \subset \mathbb{R}^m$. However, we will not do it for notational simplicity.

the final cumulated return $\sum_{t=t_0}^{T-1} J\big(X(t), a(t), \xi(t)\big) + K(X(T))$ by introducing cumulated returns as a new state.

3.2 Economic Formulation

The following Proposition is the dynamic version of the static result presented in Sect. 2. Its proof is given in the Appendix, Sect. 6. We denote $\eta(\cdot) = \big(\eta(t_0), \dots, \eta(T-1)\big) \in \mathbb{R}^{T-t_0}$ and $\lambda(\cdot) = \big(\lambda(t_0), \dots, \lambda(T-1)\big) \in \mathbb{R}^{T-t_0}$.

Proposition 1. *Assume that the risk measure R_ρ in (1) is* nice. *Assume also that the solution of (6) is such that $J\big(X(t), a(t), \xi(t)\big) \in L_\rho(\Omega, \mathcal{F}, \mathbb{P})$ for all $t = t_0, \dots, T-1$, and that the infimum in (1) is achieved for any $X = -J\big(X(t), a(t), \xi(t)\big)$.*

The dynamic maximization problem under risk constraints (6) is equivalent to

$$\sup_{(\hat{a}(\cdot), \eta(\cdot)) \in \mathcal{A} \times \mathbb{R}^{T-t_0}} \inf_{(U_{t_0}, \dots, U_{T-1}) \in \mathcal{U}^{T-t_0}} \mathbb{E}\Big[\sum_{t=t_0}^{T-1} U_t\big(J\big(X(t), a(t), \xi(t)\big), \eta(t)\big) + K(X(T))\Big], \tag{7}$$

where the set of (utility) functions $\mathcal{U}$ is defined by

$$\mathcal{U} := \Big\{U^{(\lambda)} : \mathbb{R}^2 \to \mathbb{R}\,, \quad \lambda \geq 0 \mid U^{(\lambda)}(x, \eta) = x + \lambda\big(-\rho(-x, \eta) + \gamma\big)\Big\}. \tag{8}$$

Notice that the formulation (7) of Problem (6) leads us to introduce a new decision variable η. This decision variable comes from our choice of risk measure given by (1).

3.3 Dynamic Programming Equation

Now, we show how the optimization problem under risk constraints (6) can be theoretically solved by stochastic dynamic programming.

For any feedback $\hat{a}(\cdot) \in \mathcal{A}$, we define the function $\Psi_{\hat{a}(\cdot)} : \mathbb{R}^{T-t_0} \times \mathbb{R}^{T-t_0} \to \mathbb{R}$ by

$$\Psi_{\hat{a}(\cdot)}\big(\lambda(\cdot), \eta(\cdot)\big) := \mathbb{E}\Big[\sum_{t=t_0}^{T-1} \Big(J\big(X(t), a(t), \xi(t)\big) - \lambda(t)\rho\big(-J\big(X(t), a(t), \xi(t)\big), \eta(t)\big) + \lambda(t)\gamma(t)\Big) + K(X(T))\Big]. \tag{9}$$

Proposition 2. *Suppose that the following function $\mathcal{L}$ admits a saddle point on $\mathcal{A} \times \mathbb{R}^{T-t_0}$:*

$$\mathcal{L}(\hat{a}(\cdot), \lambda(\cdot)) := \sup_{\eta(\cdot) \in \mathbb{R}^{T-t_0}} \Psi_{\hat{a}(\cdot)}\big(\lambda(\cdot), \eta(\cdot)\big) . \tag{10}$$

Then, the optimization problem (6) under risk constraints is equivalent to

$$\inf_{\lambda(\cdot) \in \mathbb{R}_+^{T-t_0}} \sup_{\eta(\cdot) \in \mathbb{R}^{T-t_0}} V_{t_0}^{(\lambda(\cdot),\eta(\cdot))}\big(x(t_0)\big) \tag{11}$$

where, for fixed $(\lambda(\cdot), \eta(\cdot)) \in \mathbb{R}_+^{T-t_0} \times \mathbb{R}^{T-t_0}$, the sequence of value functions $V_t^{(\lambda(\cdot),\eta(\cdot))}$ is given by the backward induction

$$\left\{ \begin{aligned} V_t^{(\lambda(\cdot),\eta(\cdot))}(x) = \sup\nolimits_{a \in \mathbb{A}} \mathbb{E}\Big[& J(x, a, \xi(t)) - \lambda(t)\rho(-J(x, a, \xi(t)), \eta(t)) \\ & + \lambda(t)\gamma(t) + V_{t+1}^{(\lambda(\cdot),\eta(\cdot))}\Big(F\big(x, a, \xi(t)\big)\Big)\Big] \end{aligned} \right. \tag{12}$$

starting from $V_T^{(\lambda(\cdot),\eta(\cdot))}(x) = K(x)$.

Supposing solve the family of dynamic programming equations, giving the initial value functions $V_{t_0}^{(\lambda(\cdot),\eta(\cdot))}$, the existence of optimal parameters $\lambda^\star(\cdot)$ and $\eta^\star(\cdot)$ for (11) results from the properties that the function $V_{t_0}^{(\lambda(\cdot),\eta(\cdot))}$ is

- Convex in $\lambda(\cdot)$, as supremum of linear and convex functions of $\lambda(\cdot)$ by induction (12)
- Concave in $\eta(\cdot)$, by definition of a nice risk measure and by induction (12).

4 Numerical Applications

For a low-dimensional dynamic portfolio model with energy assets and financial contracts, we can compute numerically the maximal expected return (benchmark). Then, we incorporate risk (CVaR) either by constraints or with utility functions (a family or a single one). The last two ways are not equivalent to the first profit maximization problem under risk constraints, because we consider restricted classes of utility functions (included in, but smaller than, the whole class of Proposition 1). We shall compare the numerical results given by these three methods with the benchmark, both on the expected return and on the violation of risk constraints criterion.

4.1 Problem Statement

We consider a portfolio consisting of energy assets (hydraulic reservoirs, thermal units) and financial contracts. The numerical results that we present are obtained for a low-dimensional case: one reservoir, one thermal unit, one financial contract (future).

This problem can be written as the canonical problem of Sect. 3.1. The generation planning is designed for the optimization horizon of 18 months and is based on a half-month discretization (hence $t_0 = 0, T = 36$). The state $X(t)$ is made of reservoir level and market prices. The controls $a(t)$ are the generation decisions and the trading activities. The stochastic component $\xi(t)$ includes hydraulic inflows, breakdowns on thermal units, energy demand, and Gaussian factors influencing prices. The dynamics of market prices is given by a Gaussian two factors model, which represents the main characteristics of prices: mean-reverting and short-term and long-term volatilities. A dynamic equation links the level of a reservoir between time periods, depending on the stochastic inflows and on the production during the period. Notice that, since hydraulic production is very cheap but limited, the good use of the hydraulic reservoir is important to improve the global cost of the generation planning and the risk indicators: thus, if the reservoir is empty while the load is high, the generation planning will be very expensive. Thermal and reservoir controls are bounded.

The monetary instantaneous return during each period is denoted by $G(t) = J\big(X(t), a(t), \xi(t)\big)$. It includes costs of production and of trading activities, and also a penalization of the energy balance constraint. Indeed, this is a global constraint stating that the difference between the generation and the demand can be exchanged on markets at each stage. We are interested in maximizing the expected sum $\mathbb{E}[\sum_{t=t_0}^{T-1} G(t)]$. The planning should give the proper controls – here, the generation decisions and the trading activities – to face the variable demand from customers.

4.2 Some Numerical Results

An upper bound is obtained by solving (by dynamic programming) this portfolio optimization problem without risk constraint. Let us denote by $G^*(t)$ the instantaneous optimal return of the problem without risk constraint at time t.

We then implement and compare three ways to take risk into account in this problem.

1. Like for the optimization problem (6), we shall consider one financial risk constraint of CVaR by period, namely $\mathsf{CVaR}_p\big(G(t)\big) \leq \gamma(t)$; this corresponds to the optimization problem (6) under risk constraints; the solution is found by dynamic programming as in Proposition 2.
2. In Proposition 1, we stated an equivalence between a profit maximization problem under risk constraints and a maxmin problem involving a family of utility functions (in other words, an infinite number of utility functions); from a numerical point of view, we shall not consider an infinity number of utility functions but one single utility function by fixing λ for every stage; the optimization problem that we solve is then not equivalent to the profit maximization problem under

risk constraints, since we do not consider an infinity number of utility functions but a single utility function; in this second approach, we therefore shall consider as set of (utility) functions one subfamily – only indexed by η with λ fixed for every stage – of the family $\mathcal{U} = \Big\{U^{(\lambda)} : \mathbb{R}^2 \to \mathbb{R}\ ,\quad \lambda \geq 0 \mid U^{(\lambda)}(x,\eta) = x - \frac{\lambda}{1-p}(x+\eta)_- - \lambda\eta + \lambda\gamma\Big\}$ of utility functions introduced in Sect. 3 and associated to CVaR constraints.

3. An intertemporal expected utility maximization $\sup \mathbb{E}\Big[\sum_{t=t_0}^{T-1} U\big(G(t)\big)\Big]$, with a utility function $U(x) = x + \eta + (1-\theta)(-x-\eta)_+$ inspired by the above family $\mathcal{U}$, where the parameter θ can be interpreted as loss aversion "à la Kahneman and Tversky" (see Kahneman and Tversky (1992)); this formulation is not equivalent to any of the the previous ones.

The first way consists in an optimization problem *with* constraints, while the two alternative ways are without constraints since they follow the expected utility economic approach. This is why we expect these latter ones to be easier to solve. Since the two alternative ways of modelling risk are not equivalent to the original optimization problem with constraints, our aim is to compare them with respect to expected cost and constraints satisfaction.

In these numerical experiments, we focus only on how sensitive the expected returns are to changes in parameters introduced by our specific approaches.

4.2.1 Conditional Value-at-Risk Constraints

We solve the optimization problem with a sequence of $cvar$ constraints, one by period t (we fix $p = 0.95$). To be sure that the constraints are active, we choose the level constraints, for $t = t_0, \ldots, T-1$,

$$\gamma(t) = \gamma = \inf_{s=t_0,\ldots,T-1} CVaR(-G^*(s))\,.$$

The problem is solved by the dynamic programming equation of Proposition 2. The optimal expected return $\mathbb{E}[\sum_{t=t_0}^{T-1} G_1(t)]$ of this problem with risk constraints corresponds to a diminution of 58% of the optimal return $\mathbb{E}[\sum_{t=t_0}^{T-1} G^*(t)]$ of the problem without risk constraint; this strong diminution is obviously related to the choice of low constraint levels.

4.2.2 Sub-Families of Utility Functions Associated to Conditional Value-at-Risk Measure

In this approach, $\lambda(\cdot) \in \mathbb{R}_+^{36}$ is fixed and we solve

$$\sup_{\hat{a}(\cdot)\in\mathcal{A}} \sup_{\eta(\cdot)\in\mathbb{R}^{T-t_0}} \mathbb{E}\Big[\sum_{t=t_0}^{T-1} G(t) + \lambda(t)\Big(-\eta(t) - \frac{1}{1-p}\big(-G(t)-\eta(t)\big)_+ + \gamma\Big)\Big]. \quad (13)$$

The problem is solved by the dynamic programming equation of Proposition 2, where the optimization with respect to $\lambda(\cdot)$ disappears since $\lambda(\cdot) \in \mathbb{R}_+^{36}$ is fixed.

For the sake of simplicity, we choose a stationary $\lambda(\cdot) = (\lambda, \lambda, \ldots, \lambda) \in \mathbb{R}_+^{36}$ and we study how sensitive the expected returns are to changes in λ. We solve this new problem for different selected values of λ in the interval [0.01; 100]. For all chosen values, numerical experiments lead to expected returns $\mathbb{E}[\sum_{t=t_0}^{T-1} G_2(t)]$, which correspond to a diminution between 60 and 65% of the optimal expected return $\mathbb{E}[\sum_{t=t_0}^{T-1} G^*(t)]$ of the problem without risk constraint (to be compared to the 58% diminution in the case of risk constraints). Moreover, the CVaR constraints are violated at all periods, the deviation being of the order of the bound γ (meaning that constraints with a bound of 2γ are almost always satisfied).

4.2.3 Loss Aversion Utility Function

We solve the portfolio optimization problem formulated with a "loss aversion utility function" (see Sect. 2.3):

$$U(x) = x + \eta + (1-\theta)(-x-\eta)_+ \quad \forall x \in \mathbb{R}. \tag{14}$$

In this approach we have to tune two parameters, the loss aversion parameter θ and an auxilliary parameter η used to write the risk measure as an infimum of expectations. We thus solve the problem

$$\sup_{\hat{a}(\cdot)\in\mathcal{A}} \mathbb{E}\Big[\sum_{t=t_0}^{T-1} U\big(G(t)\big)\Big] = \sup_{\hat{a}(\cdot)\in\mathcal{A}} \mathbb{E}\Big[\sum_{t=t_0}^{T-1} G(t)+\eta+(1-\theta)(-G(t)-\eta)_+\Big]. \tag{15}$$

Guided by the fact that, in the CVaR formulation, the optimal value of η corresponds to the VaR, we choose $\eta = \frac{1}{T}\sum_{t=t_0}^{T-1} VaR(-G^*(t))$. For the loss aversion, we fix $\theta = 10$. With these parameters, the optimal expected return $\mathbb{E}[\sum_{t=t_0}^{T-1} G_3(t)]$ of this problem corresponds to a diminution of only 5% of the optimal expected return $\mathbb{E}[\sum_{t=t_0}^{T-1} G^*(t)]$ of the problem without risk constraint, and in this case the constraints are not satisfied for less than 20% of the periods (for $\theta = 4$, constraints are violated for all periods).

4.2.4 Conclusions

In Table 4.2.4, we observe how different ways of taking risk into account lead to different performances, measured by the expected return and by the violation of risk constraints criterion.

Taking risk as constraints in the optimization problem leads to a strong decrease of the optimal return of the portfolio, but with the guarantee that all the constraints are satisfied. With a loss aversion utility function, the optimal return is relatively

Approach	% Optimal return	% Violation of risk constraints on the horizon
Optimization with CVaR constraints	42%	0%
Sub-families of utility functions associated to CVaR	35–40%	100%
Loss aversion utility function	95%	80%

better if the decision maker agrees not to satisfy the risk constraints for some periods.

5 Conclusion

We established an equivalence between a profit maximization dynamic problem under risk constraints (belonging to a specified class of risk measures) and a maximin problem involving an infinite number of utility functions. We then compared for a low-dimensional electricity portfolio model optimization the practice consisting in incorporating risk measures as constraints in a classical expected cost minimization problem and the practice consisting in using utility functions. The class of utility functions chosen is deduced from the risk constraint and inspired by our equivalence result.

Acknowledgements We thank two anonymous Reviewers for their careful reading and their helpful comments.

6 Proofs

6.1 *Proof of Proposition 1*

6.1.1 Dualization of Risk Constraints

Let us introduce a Lagrange multiplier $\lambda(\cdot) := (\lambda(t_0), \dots, \lambda(T-1)) \in \mathbb{R}_+^{T-t_0}$ associated to risk constraints (6d) and denote by $\eta(\cdot) := (\eta(t_0), \dots, \eta(T-1)) \in \mathbb{R}^{T-t_0}$ the auxiliary variable allowing to write the risk measure under the form (1).

The Lagrangian associated to risk constraints (6d) may be written

$$\mathcal{L}(\hat{a}(\cdot), \lambda(\cdot)) = \mathbb{E}\Big[\sum_{t=t_0}^{T-1} J\big(X(t), a(t), \xi(t)\big) + K(X(T))\Big]$$
$$- \sum_{t=t_0}^{T-1} \lambda(t)\Big(R_\rho(-J\big(X(t), a(t), \xi(t)\big) - \gamma(t)\Big)$$

$$
\begin{aligned}
&= \mathbb{E}\Big[\sum_{t=t_0}^{T-1} J\big(X(t), a(t), \xi(t)\big) + K(X(T))\Big] - \\
&\quad \sum_{t=t_0}^{T-1} \lambda(t)\Big(\inf_{\eta(t)} \mathbb{E}\big[\rho\big(-J\big(X(t), a(t), \xi(t)\big), \eta(t)\big)\big] - \gamma(t)\Big) \\
&= \sup_{\eta(\cdot)} \mathbb{E}\Big[\sum_{t=t_0}^{T-1} J\big(X(t), a(t), \xi(t)\big) + K(X(T)) - \\
&\quad \lambda(t)\rho\Big(-J\big(X(t), a(t), \xi(t)\big), \eta(t)\Big) + \lambda(t)\gamma(t)\Big] \\
&\quad \text{(since the } \eta(t) \text{ appear independently in each term of the sum)} \\
&= \sup_{\eta(\cdot)\in\mathbb{R}^{T-t_0}} \Psi_{\hat{a}(\cdot)}\big(\lambda(\cdot), \eta(\cdot)\big) \text{ by (9).}
\end{aligned}
\tag{16}
$$

Problem (6) is equivalent to

$$
\sup_{\hat{a}(\cdot)\in\mathcal{A}} \inf_{\lambda(\cdot)\in\mathbb{R}_+^{T-t_0}} \sup_{\eta(\cdot)\in\mathbb{R}^{T-t_0}} \Psi_{\hat{a}(\cdot)}\big(\lambda(\cdot), \eta(\cdot)\big). \tag{17}
$$

6.1.2 Interverting $\inf_{\lambda(\cdot)\in\mathbb{R}_+^{T-t_0}}$ and $\sup_{\eta(\cdot)\in\mathbb{R}^{T-t_0}}$

Now, we shall show that $\inf_{\lambda(\cdot)\in\mathbb{R}_+^{T-t_0}}$ and $\sup_{\eta(\cdot)\in\mathbb{R}^{T-t_0}}$ can be swapped in problem (17). We distinguish two cases.

In the first case, we use the property that

$$
\inf_{\lambda(\cdot)\in\mathbb{R}_+^{T-t_0}} \sup_{\eta(\cdot)\in\mathbb{R}^{T-t_0}} \Psi_{\hat{a}(\cdot)}\big(\lambda(\cdot), \eta(\cdot)\big) \geq \sup_{\eta(\cdot)\in\mathbb{R}^{T-t_0}} \inf_{\lambda(\cdot)\in\mathbb{R}_+^{T-t_0}} \Psi_{\hat{a}(\cdot)}\big(\lambda(\cdot), \eta(\cdot)\big), \tag{18}
$$

and show that the largest one equals $-\infty$, so that both quantities are equal. This is the result of the following Lemma 1 (we shall not detail its proof, which is rather straightforward).

Lemma 1. *If there exists $t^\star$ such that $\gamma(t^\star) R_\rho(-J(X(t^\star), a(t^\star), \xi(t^\star)))$, then*

$$
\inf_{\lambda(\cdot)\in\mathbb{R}_+^{T-t_0}} \sup_{\eta(\cdot)\in\mathbb{R}^{T-t_0}} \Psi_{\hat{a}(\cdot)}\big(\lambda(\cdot), \eta(\cdot)\big) = \sup_{\eta(\cdot)\in\mathbb{R}^{T-t_0}} \inf_{\lambda(\cdot)\in\mathbb{R}_+^{T-t_0}} \Psi_{\hat{a}(\cdot)}\big(\lambda(\cdot), \eta(\cdot)\big) = -\infty.
$$

In the second case, we use the property that, if the function $\Psi_{\hat{a}(\cdot)}$ admits a saddle point $\big(\overline{\lambda}(\cdot), \overline{\eta}(\cdot)\big)$, that is

$$
\Psi_{\hat{a}(\cdot)}\big(\overline{\lambda}(\cdot), \eta(\cdot)\big) \leq \Psi_{\hat{a}(\cdot)}\big(\overline{\lambda}(\cdot), \overline{\eta}(\cdot)\big) \leq \Psi_{\hat{a}(\cdot)}\big(\lambda(\cdot), \overline{\eta}(\cdot)\big),
$$

then we obtain the reverse inequality of that in (18), hence both quantities are equal.

Lemma 2. *If* $\gamma(t) \geq R_\rho(-J(X(t), a(t), \xi(t)))$ *for* $t = t_0, \ldots, T-1$, *the function* $\Psi_{\hat{a}(\cdot)}$ *defined in (9) admits a saddle point in* $\mathbb{R}_+^{T-t_0} \times \mathbb{R}^{T-t_0}$.

Proof. Let $\eta^\star(\cdot) = (\eta^\star(t_0), \ldots, \eta^\star(T-1))$ be such that (see the assumptions of Proposition 1 ensuring that the infinimum in (1) is achieved)

$$R_\rho\Big(-J\big(X(t), a(t), \xi(t)\big)\Big) = \mathbb{E}\Big[\rho\Big(-J\big(X(t), a(t), \xi(t)\big), \eta^\star(t)\Big)\Big] \text{ for } t = t_0, \ldots, T-1\,.$$

We distinguish two cases. If $\gamma(t) = R_\rho(J(X(t), a(t), \xi(t)))$ for all $t = t_0, \ldots, T-1$, then $(\lambda(\cdot), \eta^\star)$ is a saddle point because, by (9), we have $\Psi_{\hat{a}(\cdot)}(\lambda(\cdot), \eta^\star(\cdot)) = \mathbb{E}\Big[\sum_{t=t_0}^{T-1} J\big(X(t), a(t), \xi(t)\big) + K(X(T))\Big]$.

If there exists $\bar{t}$ such that $\gamma(\bar{t}) > R_\rho\Big(-J(x(\bar{t}), a(\bar{t}), \xi(\bar{t}))\Big)$, we shall check the conditions of Barbu and Precupanu (1986, Chap. 2, Corollary 3.8), which ensure the existence of a saddle point for the function $\Psi_{\hat{a}(\cdot)}(\lambda(\cdot), \eta(\cdot))$ in (9). This latter is

- Linear with respect to $\lambda(\cdot)$ and thus convex in $\lambda(\cdot)$
- Concave with respect to $\eta(\cdot)$, because the function $\eta \mapsto -\rho\Big(-J\big(X(t), a(t), \xi(t)\big), \eta\Big)$ is concave

Now, by assumption on the function ρ and on the set $L_\rho(\Omega, \mathcal{F}, \mathbb{P})$, for all $J\big(X(t), a(t), \xi(t)\big) \in L_\rho(\Omega, \mathcal{F}, \mathbb{P})$, we have that

- $\eta(\cdot) \mapsto \mathbb{E}\Big[\sum_{t=t_0}^{T-1} \Big(\rho\big(-J(X(t), a(t), \xi(t), \eta(t)) - \gamma(t)\big)\Big)\Big]$ is continuous
- $\lim_{\|\eta(\cdot)\| \to +\infty} \mathbb{E}\Big[\sum_{t=t_0}^{T-1} \Big(\rho\big(-J\big(X(t), a(t), \xi(t)\big), \eta(t)\big) - \gamma(t)\Big)\Big] = +\infty$

Thus the function $\Psi_{\hat{a}(\cdot)}$ is convex–concave, l.s.c.–u.s.c, and satisfies $\lim_{\|\eta(\cdot)\| \to +\infty} \Psi_{\hat{a}(\cdot)}(\lambda(\cdot), \eta(\cdot)) = -\infty$. Since $\gamma(\bar{t}) > R_\rho\Big(-J\big(x(\bar{t}), a(\bar{t}), \xi(\bar{t})\big)\Big)$, we obtain that

$$\begin{aligned}\Psi_{\hat{a}(\cdot)}(\lambda(\cdot), \eta^\star(\cdot)) = \mathbb{E}\Big[\sum_{t=t_0}^{T-1} J\big(X(t), a(t), \xi(t)\big)\Big] + \sum_{t=t_0}^{T-1} \lambda(t)\Big(\gamma(t) \\ -R_\rho\Big(-J\big(X(t), a(t), \xi(t)\big)\Big)\Big) + \mathbb{E}\big[K(X(T))\big] \to +\infty,\end{aligned}$$

when $\| \lambda(\cdot) \| \to +\infty$. Hence, the function $\Psi_{\hat{a}(\cdot)}$ admits a saddle point in $\mathbb{R}^{T-t_0}+ \times \mathbb{R}^{T-t_0}$ by Barbu and Precupanu (1986, Chap. 2, Corollary 3.8).

6.1.3 Interpretation

We conclude by noticing that, in the expression (10), we have

$$J(X,a,\xi) - \lambda\rho(-J(X,a,\xi),\eta) + \lambda\gamma = U^{(\lambda)}(J(X,a,\xi),\eta)\ ,$$

where $U^{(\lambda)}(x,\eta) = x + \lambda\big(-\rho(-x,\eta) + \gamma\big)$ is given in (8).

6.2 *Proof of Proposition 2*

Proof. We sketch the proof in three steps.

1. By dualization, we have obtained in (10) the Lagrangian $\mathcal{L}(\hat{a}(\cdot),\lambda(\cdot))$ associated to risk constraints (6d). We have admitted that $\mathcal{L}$ admits a saddle point[4] so that

$$\sup_{\hat{a}(\cdot)\in\mathcal{A}}\ \inf_{\lambda(\cdot)\in\mathbb{R}_+^{T-t_0}} \mathcal{L}(\hat{a}(\cdot),\lambda(\cdot)) = \inf_{\lambda(\cdot)\in\mathbb{R}_+^{T-t_0}}\ \sup_{\hat{a}(\cdot)\in\mathcal{A}} \mathcal{L}(\hat{a}(\cdot),\lambda(\cdot))\ . \tag{19}$$

2. Still because of (10), we notice that

$$\begin{aligned}\sup_{\hat{a}(\cdot)\in\mathcal{A}} \mathcal{L}(\hat{a}(\cdot),\lambda(\cdot)) &= \sup_{\hat{a}(\cdot)\in\mathcal{A}}\ \sup_{\eta(\cdot)\in\mathbb{R}^{T-t_0}} \Psi_{\hat{a}(\cdot)}\big(\lambda(\cdot),\eta(\cdot)\big) \\ &= \sup_{\eta(\cdot)\in\mathbb{R}^{T-t_0}}\ \sup_{\hat{a}(\cdot)\in\mathcal{A}} \Psi_{\hat{a}(\cdot)}\big(\lambda(\cdot),\eta(\cdot)\big)\ .\end{aligned}$$

   There are indeed no difficulties in swapping both sup above.
3. Once $(\lambda(\cdot),\eta(\cdot))$ is fixed, we consider the problem

$$\begin{aligned}\sup_{\hat{a}(\cdot)\in\mathcal{A}} \Psi_{\hat{a}(\cdot)}\big(\lambda(\cdot),\eta(\cdot)\big) = \sup_{\hat{a}(\cdot)\in\mathcal{A}} \mathbb{E}\Big[\sum_{t=t_0}^{T-1} J\big(X(t),a(t),\xi(t)\big) \\ -\lambda(t)\rho(-J\big(X(t),a(t),\xi(t)\big),\eta(t)) + \lambda(t)\gamma(t) + K(X(T))\Big]\ .\end{aligned}$$

   This problem is naturally solved by dynamic programming, introducing the sequence of value functions $V(t)^{(\lambda(\cdot),\eta(\cdot))}$ indexed by the following:

   - The multipliers $\lambda(\cdot)$ associated to the risk constraints
   - The auxiliary variables $\eta(\cdot)$ allowing to write the risk measure under the form (1).

[4] In the proof of Proposition 1, we have shown the existence of a saddle point for $\Psi_{\hat{a}(\cdot)}(\lambda(\cdot),\eta(\cdot))$ defined in (9), and not for $\mathcal{L}(\hat{a}(\cdot),\lambda(\cdot)) = \sup_{\eta(\cdot)} \Psi_{\hat{a}(\cdot)}(\lambda(\cdot),\eta(\cdot))$.

References

Andrieu L, De Lara M, Seck B (2008) Profit maximization under risk constraints, an economic interpretation as maxmin over a family of utility functions, Working paper

Barbu V, Precupanu T (1986) Convexity and optimization in Banach spaces. D. Reidel Publishing Company, Bucarest

Ben-Tal A, Teboulle M (1986) Expected utility, penality functions and duality in stochastic nonlinear programming. Manag Sci 32(11):1445–1446

Ben-Tal A, Teboulle M (2007) An old-new concept of convex risk measures: the optimized certainty equivalent. Math Finance 17(3):449–476

Dentcheva D, Ruszczyński A (2006) Portfolio optimization with stochastic dominance constraints. J Bank Finance 30(2):433–451

Gollier C (2001) The economics of risk and time. MIT, Cambridge

Kahneman D, Tversky A (1992) Advances in prospect theory: cumulative representation of uncertainty. J Risk Uncertainty 5(4):297–323

Maccheroni F (2002) Maxmin under risk. Econ Theor 19:823–831

Markowitz H (1952) Portfolio selection. J Finance 7:77–91

Ogryczak W, Ruszczynski A (1999) From stochastic dominance to mean-risk model: semideviations as risk measures. Eur J Oper Res 116:217–231

Pflug GC, Werner R (2007) Modeling, measuring and managing risk. World Scientific, New Jersey

Rockafellar RT, Uryasev S (2000) Optimization of conditional value-at-risk. J Risk 2:21–41

Aspects of Risk Assessment in Distribution System Asset Management: Case Studies

Simon Blake and Philip Taylor

Abstract The reliability of power systems is a major concern for network designers and operators. For several reasons, the level of electricity distribution network risk, in the UK as elsewhere, is perceived to be increasing in the short term, and further sources of increased risk can be anticipated in the longer term. An asset management approach seeks to optimise the balance between capital expenditure and the risk to customer supplies. To do this effectively, network risk must first be defined and measured. This chapter describes ways of modelling, evaluating and comparing the levels of risk in different parts of the network and under various operating assumptions. It also enables different strategies for mitigating that risk to be quantified, including maintenance policies, replacement and reinforcement construction projects, network reconfiguration and changes to operating practices. In the longer term, it allows evaluation of the impact on network risk of initiatives such as widespread distributed generation, demand side management and smart grids.

Keywords Asset management · Electrical network risk · Electricity distribution case studies · Electricity distribution construction projects · Evaluating risk · Reliability · Risk mitigation · Sub-transmission voltages

1 Introduction

In common with other developed countries in Europe and North America, the annual growth rate of the power distribution network in the UK has decreased from around 10% before 1970 to just over 1% in recent years (Boyle et al. 2003). This decrease has tended to limit the options available for capital investment in the network. Where once asset renewal and reinforcement could be justified as part of system expansion to meet increasing customer demand, now each proposal for capital expenditure is

S. Blake (✉)
Department of Engineering and Computing, Durham University, Durham, UK,
e-mail: s.r.blake@durham.ac.uk

S. Rebennack et al. (eds.), *Handbook of Power Systems II*, Energy Systems,
DOI 10.1007/978-3-642-12686-4_17,

subject to detailed scrutiny and needs to be fully justified as being cost effective. In addition, new environmental, financial and regulatory pressures have each had their impact on what can, or indeed on what must, be done to maintain the integrity and enhance the reliability of the network.

Where in the past low-risk network configurations with high levels of built-in redundancy were preferred, in particular at higher voltages, today's asset management emphasis has meant that it is appropriate to re-examine the balance between system reliability and capital expenditure. Optimum asset utilization may be preferred over comfortable levels of redundancy, in the quest for best value for money in the system as a whole.

However, at the same time, the continuity of supply to customers is more closely monitored than ever before. In the United Kingdom, the system average interruption frequency index (SAIFI) and the system average interruption duration index (SAIDI) are both measured by the distribution companies, as a statutory requirement, and monitored by the industry regulator. Other countries have their own similar monitoring regimes; the UK monitoring system is described in further detail in Sect. 3. SAIFI and SAIDI are two elements to be considered in the definition of network risk, as a prelude to measuring it.

1.1 Short-term Risk

Regardless of the method of measurement, network risk in the United Kingdom is perceived to be increasing. There are a number of reasons for this. In the short term, these reasons include (Northern Electric Distribution Ltd. 2007; Blake et al. 2008) the following:

- Increasing utilisation of distribution networks. The 1% average annual growth rate comprises a wide variation, from former heavy industrial areas with sharply decreasing demand to new towns, industrial estates and regenerated city centres where demand is increasing rapidly. As a result, some parts of the network are under-utilised, while other parts may be overloaded, particularly if there are system outages.
- Severe weather conditions occurring more often (Billinton and Acharya 2006; Kaipa et al. 2007). Flooding in parts of England during summer 2007 affected a number of major supply points (SPs) on the network, causing loss of supply to thousands of customers for several days. These kind of extreme weather events are expected to happen with increasing frequency, possibly as a result of global warming.
- Increasing levels of both accidental and malicious damage. A cable can be damaged unintentionally by a contractor working nearby, and there seems to be an increasing number of contractors as the range of activities increases (e.g. building new roads or installing new communications networks). There has also been a substantial increase in pointless arson attacks and in thefts of cable (even when live), probably as a result of higher scrap copper prices.

- Loss of engineering experience and expertise as older engineers retire. This is a consequence of the demographic profile of European societies in general and of the contraction of employment in the industry over the past 30 years, as it has sought to increase manpower productivity. This could have a particular effect on network versatility and on restoration times.
- Not only the engineering profession, but also the equipment itself is ageing. Much of it was installed in the 1960s and 1970s, and is now reaching the end of its nominal life. Replacement policies have often been cut back, and so the average asset age is increasing.
- The density of recent building underneath overhead power lines is not only increasing some risk factors, but also making it increasingly difficult to service lines and to replace or augment them, particularly given the erosion of statutory access rights. Health and safety regulations, and the sensitivity of consumer equipment, mean that restoration may have to be undertaken more cautiously than that in the past, thus potentially extending outage times.
- Possible growing regulatory focus on levels of network risk. While at the moment the industry regulator focuses mainly on the consequences of risk (such as SAIFI and SAIDI), the regulatory environment is updated every 5 years, and it is considered quite likely that statutory measurement of and reporting on levels of risk may be introduced in the near future. While such regulations would not actually increase the level of risk, they would increase awareness of it.
- Increasing customer requirements. Partly this may be due to an increase in overall consumer awareness and to the unbundling of the UK electricity industry with consequent competition, in particular between suppliers. Also, interruption of supply, even for a very short time, can cause severe and costly disruption to an increasing number of sensitive industrial processes (e.g. electronics manufacture).

1.2 Long-term Risk

In the longer term, a number of additional challenges could be expected to increase demands on the network and, perhaps as a consequence, levels of network risk, although these risks are clearly more conjectural at present (PB Power report 2007). Examples of such future challenges include the following:

- Increased awareness of global warming leading to additional demands on the system and regulatory pressures, for example a requirement to reconfigure, rebuild and operate the network so as to reduce power losses.
- Increased levels of renewable and distributed generation (DG), of varying types and sizes, connected at a wide range of different voltage levels. These would require more frequent and possibly intermittent reverse-direction power flows, putting increased strain on assets that may not have been designed for such loading.

- Demand side management, where customers respond either manually or automatically to overall power demand by selectively shedding load at peak times. This would require more sophisticated control technology, which might itself introduce extra sources of risk.
- The development of Smart Grids and active network management to respond to the variability of all these changes. Again, this would require a significant increase in control technology and non-standard methods of operating the network, both of which could imply increased levels of risk.
- Possible further unbundling and reconfiguration of the industry. Although this has already progressed further in the UK than in some other European countries, further initiatives are likely, including for example an increase in the number of Registered Power Zones and Energy Services Companies, both of which introduce additional complexity into normal network operation.

1.3 Chapter Summary

This chapter is concerned with both short-term and longer-term sources of network risk. Following this introductory Sect. 1, there is a brief review of the relevant research background (Sect. 2), followed by Sect. 3 outlining the structure and operation of the national power system with which the authors are most familiar, that of the UK.

Section 4 describes some of the standard techniques that can be used to analyse networks and compute levels of risk. It defines a methodology based on these techniques, and illustrates that methodology with reference to a case study based on the UK 66 kV system. Section 5 includes a number of further illustrations to demonstrate the versatility of this approach.

2 Research Background

To some extent, Allan's comment from 1979 remains true: 'It is known that the distribution system accounts for about 99% of the consumers' unavailability although in the past, the emphasis has generally been directed towards the generation and transmission sections of the system' (Allan et al. 1979). In Billinton and Allan's seminal treatment, 'Reliability Evaluation of Power Systems', to which the reader is referred for detailed technical treatment of the mathematics underlying the case studies in this chapter, the primary emphasis is on generating capacity and transmission systems (Billinton and Allan 1996). The distribution system is the origin of smaller, and therefore less newsworthy, supply shortfalls – but, as Allan claims, they are relatively far more frequent. Other useful general texts are listed in the 'References' section (Cooper et al. 2005; Grey 1995; Vose 1996).

2.1 Medium and High Voltages

Within the distribution system in particular, many investigative studies concentrate on optimizing the performance of feeders, in terms of cost and reliability, at medium voltages such as 20 or 11 kV (Freeman et al. 2005; Slootweg and Van Oirsouw 2005; Oravasaari et al. 2005). One possible reason for this is that most of the supply loss to consumers occurs at this level, and therefore improvements in medium voltage feeder reliability carry the greatest potential network reliability gains. Another reason is that such feeders are topologically fairly similar to one another and can therefore be represented generically. This makes them easier to model, and possibly therefore a more promising target for research.

At higher voltages, typically in the 30–150 kV range, there tends to be rather less research material. One reason for this lack of research is perhaps the wide range of ownership and operating philosophies from one country to another. In Greece, for example, the 150 kV system that covers the whole of the mainland is owned by one company, but operated by another, as part of the transmission network (Hatziargyriou 2008). In the UK, the 132 kV system is rather more piecemeal. It was originally built and operated as part of the transmission network, but ownership and operating responsibility were separately transferred to the distribution network operators (DNOs).

A second reason for the lack of research at these voltage levels is that there is more variability in network architecture. In Greece, at 150 kV, a ring configuration is generally preferred where the loads justify it (Hatziargyriou 2008). In the UK, at 132 kV, a radial double-circuit arrangement is more normal, although in one region a more expensive but generally more reliable meshed architecture has been preferred (Stowell 1956). Within these general preferences, there is a fairly ad hoc approach, based on local network topology and local load profile. As a result, it is harder to generalise research findings, which perhaps makes research at this level of the network more challenging.

At present, there is a perception within the industry that the increasing levels of risk detailed in Sect. 1 will increase the vulnerability, as well as the strategic importance, of those parts of the network that operate at these subtransmission voltages, and it is with networks operating at these voltages that the present chapter is particularly concerned. However, the techniques developed in this chapter are illustrative of a more general approach, and could be applicable to generation and transmission, as well as to distribution at all voltage levels.

2.2 Single Risk Studies

A number of studies show how network risk can be affected by one particular source of failure. One situation that greatly affects system reliability is severe weather. Events such as flooding or falling trees due to high winds, snow or ice cause line faults and interrupt the supply, increasing both SAIDI and SAIFI. The same weather may also limit the repair crew's ability to reach or work on the fault, thereby further increasing SAIDI. One way of evaluating the likely effects of severe weather is by using Markov modelling techniques. Good and bad weather are then two different states, with probabilities of transition between them and different rates of failure and restoration within them for any given component. A recent Canadian study (Billinton and Acharya 2006) extended this to three different states – normal, adverse and extreme weather – with transitions and rates for each. Cases were examined where, for example, 40% of total failures occurred in bad weather (adverse or extreme), and of these 10% occurred in extreme weather. The expected effects of this on SAIFI and SAIDI were calculated. These figures can then be used to evaluate possible expenditure options. Another study, from Finland, analysed the cost of storm damage, broken down into customer outage costs, customer compensation fees, and fault repair costs (Kaipa et al. 2007).

Other studies look at the effects of asset ageing, typically using the 'bathtub curve' to model failure rates, as shown in Fig. 1. A recent paper uses the Weibull distribution with three parameters to model the onset of stage C when failure rates start to increase as a consequence of ageing (Li et al. 2007). In this case, the location parameter of the distribution corresponds to the component age at the commencement of wear-out, the shape parameter gives the rapidity of onset of this increase in failure rate, and the size parameter measures the factor by which the failure rate eventually increases.

As a final instance of a single risk study, one work in the literature assesses the impact of the increasing penetration of DG (Vovos et al. 2007). There has traditionally been some reluctance from DNOs to pick up the additional risks of connecting DG to their networks, even though they may have a licensed obligation to do so. This has largely been for operational reasons, as well as for the potential to undermine the protective system; DG can control the voltage levels and power factor more problematic, and this has limited the penetration of DG. From a reliability point of view, there are also potential problems as a result of increasing network complexity, but there are also potential advantages from having a greater diversity of infeeds to the distribution system.

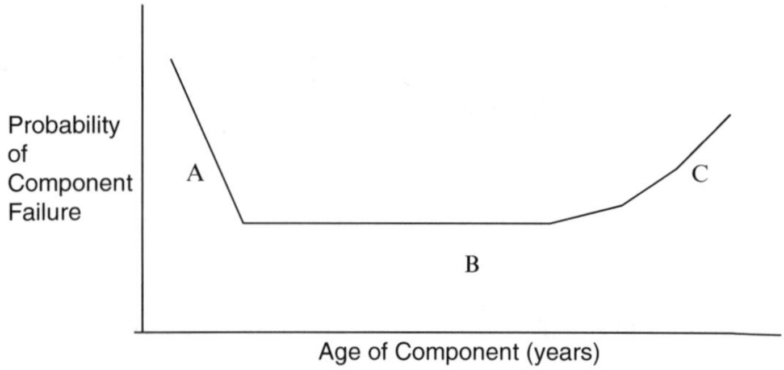

Fig. 1 The bathtub curve

2.3 Asset Management and National Regulation

The growth of an asset management approach within the industry has meant that, instead of simply maximising performance and avoiding risk, utilities increasingly identify prudent performance targets and actively manage risk (Brown and Spare 2004). Asset management is a complex process with this focussed outlook. Norgard et al. list 31 distinct challenges to distribution utilities, grouped into six categories: legal, organizational, technical, environmental, reputation and societal. They state that the electricity distribution industry increasingly regards the concept of risk assessment as an important tool in distribution system asset management (Nordgard et al. 2007).

This chapter concentrates on the UK regulatory environment, which requires compliance with certain technical standards such as P2/6 set out by the regulator office for gas and electricity markets (OFGEM) (ENA Engineering Recommendation P2/6 2006). Other countries have broadly similar regulatory environments, with differences in emphasis. In Canada, for example the power utilities manage both the network assets and the sale of electricity, functions that have been separated in the United Kingdom (Billinton and Zhaoming 2004). In other countries, the regulatory system tends to be less rigorously implemented (Douglas et al. 2005).

Bollen et al. address the limitations of measuring average performance and propose an alternative 'Customer Dissatisfaction Index' (CDI) for use in Sweden (Bollen et al. 2007). They found that, "there seems to be a rather sharp threshold value for customer satisfaction regarding reliability of supply. Complaints started when a customer experienced more than three interruptions per year, or one interruption longer than 8 h." On this basis, the CDI is defined as the proportion of customers who passed that threshold in any year. This could be used in place of or alongside the current measures of SAIFI and SAIDI to trigger rewards and penalties. The study used probability calculations to evaluate the expected impact of using the CDI, and concluded that it could be of use for network planning, detailed feeder design and performance reporting.

In the United Kingdom, with its low demand growth, the emphasis is less on expanding the network than on managing existing assets effectively, using a risk-based approach. However, risk can also be a factor in determining configurations for new networks. A Spanish study considers the balance between the risk (measured as expected energy loss in kilowatt-hour) associated with managing the existing network and total capital cost of network improvements (Ramirez-Rosado and Bernal-Agustin 2001). Optimal solutions can be found, which depend on the relative importance of these two factors.

2.4 Maintenance Issues

When a component fails in service, it can usually be repaired (possibly by the replacement of failed sub-components) and returned to service. This is one aspect of maintenance. The failure rate of the repaired component is usually assumed to be the same after the repair as it was before. This is a compromise between the possibility of an increased rate (a patch is never as reliable as the original) and a decreased rate (the servicing that accompanies the repair improves reliability).

Alternatively, a planned maintenance strategy inspects components at predetermined intervals and carries out sub-component replacement based on their observed condition or on elapsed time. This systematic maintenance may be considered to improve reliability, not least by extending the useful life of the component by deferring the onset of the wear-out phase.

Sometimes the planned maintenance activity is so extensive that it approaches asset replacement in both cost and outage duration. It might then class as a construction project; the boundary between maintenance and construction is not always clear. However, as the asset base ages, projects tend to increase in scope and duration, so that construction becomes an increasing part of the overall maintenance activity and a more significant part of the risk. Network operators generally have standard maintenance policies applied throughout their networks. However, if a region of higher risk is identified, on the basis of failure history or on possible consequences of failure, including the number of customers supplied, that region could be given a more rigorous maintenance regime, including, for example (Bayliss and Hardy 2007) the following:

- More frequent monitoring of asset condition, for example aerial overhead line surveys, switchgear inspection, oil pressure checks on paper-insulated underground cables.
- Testing the operation of rarely used components such as circuit breakers.
- Cleaning or replacing components liable to degradation, such as insulators or oil in transformers.

The higher cost of these measures would ideally be offset by a reduced failure rate. Risk analyses such as those described in this chapter could be applied to determine the least-cost maintenance solutions.

A seminal paper (Endrenyi et al. 2001) published in 2001, as the report of an international task force set up by IEEE in 1995 to investigate maintenance strategies and their effect on reliability, states:

> Maintenance is just one of the tools ensuring satisfactory component and system reliability. Others include increasing system capacity, reinforcing redundancy and employing more reliable components. At a time, however, when these approaches are heavily constrained, electric utilities are forced to get the most out of the devices they already own through more effective operating policies, including improved maintenance programs. In fact, maintenance is becoming an important part of what is often called asset management.

Electric utilities have always relied on maintenance programs to keep their equipment in good working condition for as long as it is feasible. In the past, maintenance routines consisted mostly of pre-defined activities carried out at regular intervals (scheduled maintenance). However, such a maintenance policy may be quite inefficient: it may be too costly (in the long run) and may not extend component lifetime as much as possible. In the last 10 years, many utilities replaced their fixed-interval maintenance schedules with more flexible programs based on an analysis of needs and priorities, or on a study of information obtained through periodic or continuous condition monitoring (predictive maintenance). The benefits of maintenance were also discussed: 'In the case of random failures, the constant failure-rate assumption leads to the result that maintenance cannot produce any improvement, because the chances of a failure occurring during any future time-interval are the same with or without maintenance'.

Maintenance, and more particularly construction outage planning, needs to take account of the increased risk to operations. One such study, based on a Canadian case-study, used a model incorporating time shift simulation, end-of-life failure models and dynamic rating values for cables to calculate risk and to evaluate different strategies, balancing the requirements of both maintenance and operations (Li and Korczynski 2004).

Two papers addressed the issue of the duration of maintenance outages, which can on occasion result in increased customer loss of supply and the triggering of compensation payments. One (Chow et al. 1996) used extensive data analysis to evaluate the effects on time of outage restoration (TOR), of occurrence of the outage (time of day, day of the week, month), of outage consequence (number of phases affected, which protection devices are activated by it), of the weather (which can affect repair access and duration), and of the cause of the outage. The other (Scutariu and Albert 2007) examined in particular the effects of contracting out some or all of the maintenance activity. They examined alternative ways of modelling this and concluded that the TOR could no longer be adequately represented by an exponential distribution.

2.5 Choice of Methodology

For all these studies, a range of methodologies are available and the most appropriate must be chosen. Brown and Hanson, in a seminal paper, compare 'four common methodologies used for distribution reliability assessment' (Brown and Hanson 2001). All four are described in detail, and applied to power systems, in the standard textbooks, for example, Billinton and Allan (1996).

Network Modelling translates a physical network into a reliability network on the basis of series and parallel component connections. The method is straightforward to implement, but cannot easily handle complex switching behaviour and sequential system responses to contingencies. It is better suited for modelling steady-state power flows and nodal voltages than it is to evaluating reliability.

Markov Modelling is a powerful method based on defining a number of discrete system states and also the transition rates between these states. Its main advantage is its general acceptability within the academic community. But it has a number of significant disadvantages. The states are memory-less – transition out of a state cannot depend on how the state was reached. The method predicts only how much time the system spends in each state – for example, 30 h per year with one circuit down – and not the frequency, duration and pattern of those occurrences. And the mathematical and computational complexity of the method becomes unhelpful for more than a small number of states, representing a relatively simple network.

Analytical Simulation models each system contingency, computes its impact and weights that impact based on the expected frequency of the contingency. This approach can accurately model complex system behaviour. It is versatile and can cope with high impact low probability events as well as with more frequent but less extensive outages. However, its variability makes it hard to encode as a standard technique that fits all situations. Every problem is to some extent unique and requires a fresh approach.

Monte-Carlo Simulation is like analytical simulation, but models random contingencies rather than expected contingencies. Parameters are modelled by probability distributions rather than by average or expected values. Its disadvantages include possible computational intensity and a degree of irreproducibility, as multiple analyses on the same system will produce slightly different answers. This variability decreases, although the computational intensity increases, with the number of sample runs. The randomness also means that some rare but important occurrences may be overlooked. The great advantage of Monte Carlo simulation is that it produces a range of possible outcomes, not just the expected value. So the cost of circuit outages can be evaluated for the worst year in 50 or 100 years, as well as the average expected annual cost.

In the case study described in Sect. 4, all four methodologies were applied as appropriate, and in turn, to evaluate the methodologies themselves as much as to solve the case-study problems. But first, this case study, and the other case studies described in Sect. 5, need to be placed in their UK industrial context.

2.6 Underlying Mathematics

Before turning to the industrial context and the case studies, however, the mathematics underlying the four common methodologies introduced in Sect. 2.5 will be described. This section details the mathematical equations used by each of the methodologies in turn.

Network Modelling operates by calculating load flows in a network in a steady state. Digital solutions of power flow problems follow an iterative process by assigning estimated values to unknown busbar voltages and by calculating a new value for each busbar voltage from the values at the other busbars and the real and reactive power specified. A new set of values for the voltage at each busbar is thus obtained, and the iterative process is repeated until the changes at each busbar are less than a specified value.

Radial networks require less sophisticated methods than closed loops. Large complex power systems require sophisticated iterative methods such as Gauss Seidel, Newton Raphson or Fast Decoupled. All these methods are based on busbar admittance equations. The key equations here are

$$P_i = \sum_{k=1}^{k=n} |V_i|V_k| \,(g_{ik} \cos \vartheta_{ik} + b_{ik} \sin \theta_{ik}) \quad \text{and} \tag{1}$$

$$Q_i = \sum_{k=1}^{k=n} |V_i|V_k| \,(g_{ik} \sin \vartheta_{ik} - b_{ik} \cos \theta_{ik}) \tag{2}$$

where Y, the self-admittance of each busbar, is given by Y = G + jB, with the appropriate subscripts in each case. Calculations can be done manually for a small network of, say, three busbars. For larger networks, a number of load flow computer packages are available, such as IPSA and Dig Silent.

Markov Modelling requires the identification of a discrete number of states of a given system, designated $s_1,\ s_2, \ldots s_n$. The probability that the system makes a transition from state m to state k in a certain time interval is given by p_{mk}. The n by n matrix **P** is constructed for all such probabilities p_{mk}. This represents the transition from one set of probable states to another, according to the equation

$$\boldsymbol{b}_{t+1} = \mathbf{P}\,\boldsymbol{b}_t \tag{3}$$

where $\boldsymbol{b}_t$ is the vector of probabilities that the system is in each of the n possible states at time t, and $\boldsymbol{b}_{t+1}$ is the vector as it appears one time unit later. Then, if $\boldsymbol{b}$ is the eigenvector of **P** with unit eigenvalue, such that

$$\boldsymbol{b} = \mathbf{P}\boldsymbol{b} \tag{4}$$

$\boldsymbol{b}$ represents the steady-state condition of the system. Each entry b_k in $\boldsymbol{b}$ is the steady-state probability that the system is in state k. These probabilities can be used to evaluate the cost of operating the system. If a cost c_k is associated with each of the n states, then this is

$$\sum_{1}^{n} b_k c_k \tag{5}$$

Analytical Simulation is a more heuristic approach. In the following case-studies it is used to derive overall risk levels, measured in financial terms (£k for a UK utility). It will be seen that these equations are not complex, but have been derived with the end-user in mind, namely a practising electrical engineer working in the electricity distribution industry.

The customer interruption (CI) component of risk is given by

$$CI = \sum \lambda_z \times D \times C \times \pounds F \tag{6}$$

where each λ_z is the average failure rate for a circuit component (e.g. overhead line, underground cable, transformer, switch), derived from appropriate historical data and modified according to local circumstances. Then $\boldsymbol{\Sigma}\lambda_z$ is the overall failure rate for the whole circuit.

In most circuits at the sub-transmission voltage levels under consideration, there is duplication of supply routes, so that the failure of one circuit will not in general result in the loss of supply to customers. But on occasion, it will, either because the failure is common mode, or because failure in one circuit triggers a consequent failure, for example due to the increased load, in its companion, or because a second circuit fails before the first one can be restored. The variable D represents the probability that this occurs. In the UK, a typical value for D might be 0.20.

The variable C stands for the number of customers affected by loss of supply at the load point under consideration, and $\pounds F$ is the penalty imposed by the regulator per CI.

It should be emphasised that these input variables need not be single values. Where Monte-Carlo simulation is used, any or all of them can be sampled from a probability distribution, and then the output value of CI will itself be a probability distribution. In this case, care must be taken to use a sufficiently large sample (around 50,000 is generally adequate) to represent the full range of possibilities.

The customer minutes lost (CML) component of risk is given by

$$CML = \sum \lambda_z \times D \times C \times \pounds M \times T \tag{7}$$

where $\boldsymbol{\Sigma}\lambda_z$, D and C are as given in (1), $\pounds M$ is the penalty imposed by the regulator per CML, and T is a weighted average of interruption duration given by

$$T = R \times t(S) + (1 - R) \times t(L) \tag{8}$$

In (3), R is the proportion of customers who can be supplied at a lower voltage from an alternative load point following remotely controlled reconfiguration of circuits at that lower voltage. Then $t(S)$ is the short time taken on average to achieve such reconfiguration, while $t(L)$ is the generally longer time taken to restore supply at the sub-transmission voltage. Again, it would be possible to use probability distributions instead of single values for all these variables.

The third component of risk is the incremental asset repair cost, including an allowance for accelerated asset deterioration, CR, given by

$$CR = \sum \lambda_z \times \pounds R \tag{9}$$

where $\pounds R$ is the average (or probability distribution) repair cost for the group of assets. Note that this equation does not include D, as repair costs are incurred whether or not the failure results in customer loss.

The total network risk (TNR) is then derived by summing the three components already calculated

$$TNR = CI + CML + CR \tag{10}$$

Monte Carlo Simulation extends the above analytical simulation methods to cope with a variable input. If any of the input has been in the form of a probability distribution, for example the λ_z applying to overhead lines (to allow for different weather conditions), then $\Sigma\lambda_z$ will also be in the form of a distribution, as will be CI, CML and CR.

In practice, there is uncertainty about almost all the input parameters, and so several of them could be expressed as probability distributions, either using standard probability models (Normal, Weibull, Poisson, Exponential etc.) or by specifying customised histograms. The calculations derived for the analytical simulation are then repeated for a large number of times (50,000 would be typical), each time based on a different sample from each input parameter distribution. A value of the output statistic (TNR in this case) is calculated for each of the 50,000 input scenarios. In practice, one of a number of computer packages can be used to carry out the large number of calculations effortlessly and, for a small enough system, relatively quickly.

The output can be presented as a string of values, or as a histogram of values, or it can be summarised in terms of relevant percentiles. The expected value and the variance can be calculated as in Billinton and Allan (1996) according to

$$E(x) = \frac{1}{N}\sum_{i=1}^{N} x_i \quad \text{and} \tag{11}$$

$$V(x) = \frac{1}{N-1}\sum_{i=1}^{N} (x_i - E(x))^2 \tag{12}$$

One utility with which the authors have worked like their Monte-Carlo results to be given as three numbers: expected value, 90th percentile ('the worst year in a decade') and the 99th percentile ('the worst year in a century').

3 UK Industrial Context

This section details some of the principal features of the UK power network, in particular where they are relevant to the case studies that follow. Other networks, particularly in Europe and North America, are similar in general terms, but differ in detail. For instance, in many other countries (e.g. in Greece), voltages between 100 and 200 kV are part of the national transmission network, rather than belonging to the distribution companies as in the United Kingdom, and this can lead to a different philosophy of operation. Furthermore, in Greece, the operation of this network has been separated from the asset ownership, which can also lead to differing priorities (Hatziargyriou 2008). Differences between national systems lead to differences of emphasis, as shown, for example in the studies from Canada (Billinton and Acharya 2006), Finland (Kaipa et al. 2007), Sweden (Bollen et al. 2007), Spain

(Ramirez-Rosado and Bernal-Agustin 2001) and Romania (Scutariu and Albert 2007). These differences in detail would need to be incorporated in any adaptation of the techniques used in the case studies to network risk in other countries.

The UK power network in Great Britain supplies a peak load of around 50 GW to around 30 million customers. Generation is mainly at coal-fired, gas-fired and some nuclear generating stations, with a small but growing contribution by renewables, principally onshore wind at present. Transmission around the UK is at 400 and at 275 kV, with the principal power flow usually from north to south (Boyle et al. 2003).

European competition policy led to extensive unbundling of the industry around 20 years ago. The main organisations now involved in electricity generation, transmission and distribution include the following:

- A number of generating companies, owning and operating one or more generating stations, ranging in capacity from 2 GW coal-fired down to a single 900 kW wind turbine.
- A single transmission operator, the National Grid, which owns and operates the 400 and 275 kV national networks. It also operates the half-hourly balancing market for planned and unplanned purchases and sales of energy.
- 14 DNOs, each with a regional monopoly, owning and operating the assets for the distribution of energy from 132 kV down to 230 V to individual customers. DNOs do not buy or sell energy.
- Several supply companies of varying size. Each company contracts to purchase energy from generators and sell it to the customers who choose that company, regardless of location. The supply company must pay the Grid and the DNOs an agreed fee for carrying the energy. Supply companies do not own or operate any power assets.
- A wide range of ancillary companies, from major equipment manufacturers to those who provide a specialised service such as metering or consultancy.
- A single quasi-governmental regulator, OFGEM, whose task it is to oversee and regulate not only the energy markets, but also other aspects of the industry, including security of supply.

Although these companies operate separately and independently, there has been a growing trend for holding companies to accumulate a portfolio of subsidiary companies, including perhaps one or more generators, one or more DNOs and one or more supply companies. It is the responsibility of OFGEM to ensure that this does not lead to conflict of interest or to diminished competition.

3.1 Network Design Standards

One of the concerns of the regulator OFGEM is minimising network risk at a reasonable cost. Minimising network risk starts with the initial design and subsequent development of the network topology. The standard here is set out in Engineering

Table 1 P2/6 requirements for each demand group

Class of supply	Range of group demand (GD)	First circuit outage	Second circuit outage
A	Up to 1 MW	In repair time (GD)	NIL
B	Over 1–12 MW	(a) Within 3 h (GD minus 1 MW) (b) In repair time (GD)	NIL
C	Over 12–60 MW	(a) Within 15 min (Smaller of GD minus 12 MW and 2/3 GD) (b) Within 3 h (GD)	NIL
D	Over 60–300 MW	(a) Immediately (GD minus up to 20 MW) (b) Within 3 h (GD)	(c) Within 3 h (for GD greater than 100 MW, smaller of GD minus 100 MW and 1/3 GD) (d) Within time to restore arranged outage (GD)
E	Over 300–1,500 MW	(a) Immediately (GD)	(b) Immediately (All customers at 2/3 GD) (c) Within time to restore arranged outage (GD)
F	Over 1,500 MW	*Separate Regulations*	

Recommendation P2/6 (ENA Engineering Recommendation P2/6 2006), produced by the Energy Networks Association for the guidance of DNOs. P2/6 specifies how long it should take to restore power supplies in the event of a single circuit outage, often referred to as (n – 1), and also after a second circuit outage, that is a fault which occurs in one circuit while another nearby circuit is out for maintenance, often referred to as (n – 2). It does not prescribe a maximum frequency for such outages. These recommendations are detailed in Table 1

It can be seen that the restoration requirement depends on the size of that part of the network which has lost power, as measured by the maximum demand in megawatt of that group of customers. The larger the group, the more onerous are the restoration requirements. In particular, below 60 MW the specified (n-2) requirement is nil. This means that a network requires more redundancy to be built in at higher demand levels, typically in the supply to a Primary substation at 66 or 33 kV, than is required for the local 20 or 11 kV secondary distribution network. This requirement is reflected in differences in design philosophy between HV circuits, which are typically doubled, and MV circuits, which are single circuits (whether three-phase or single phase).

It can also be seen that, as group size increases, the time allowed for restoration decreases. The practical implication is that more automation is required for circuits serving larger groups of customers, and this is reflected in the circuit configurations and their associated assets. However, design standards are not the only way in which OFGEM act to try to minimize network risk.

3.2 Operational Incentives

The performance of each DNO is reviewed annually and compared with previous years', with other DNOs, and with the individual targets set by OFGEM as part of the regulatory framework and guaranteed standards of performance. This performance is publicly reported (Brown and Hanson 2001) under three headings: customer telephone response, CIs and CML. CI is a measurement of SAIFI, defined as the number of interruptions of over 3 min duration per 100 customers during the year. CML is a measurement of SAIDI, defined as the average number of minutes without supply (in excess of 3 min per incident) per customer per year. Some events, for example extreme weather, can be omitted by agreement from this total.

Performance against target is subject to financial rewards and penalties imposed by OFGEM. For example, in 2005/2006, the targets set for one DNO were 74.5 interruptions per 100 customers per year, and an average of 71.4 min lost per customer per year. Actual performance was around 10% better than this target, triggering a reward of £1.98 million. Another DNO was around 10% above target for CI (reward £0.78 million), but 10% below target for CML (penalty £1.21 million).The typical unit cost to a DNO is £6 per customer disconnected, plus a further £6 per hour of disconnection.

There are also additional compensation payments for any customer who has been disconnected for a continuous period of over 18 h. This is currently levied at £50 per customer for the first 18 h, plus £25 for each succeeding completed period of 12 h. This quickly mounts up, and is the reason for DNOs to adopt a target emergency return to service time of 18 h maximum.

3.3 DNO Organisation

Over recent years, with changes of ownership and of regulatory environment, the concept of asset management appears to have taken an increasingly central role within the typical DNO. In one case, this is expressed by their Asset Management Policy (Northern Electric Distribution Ltd. 2007), which is:

> to develop and maintain safe, efficient, co-ordinated and economical electricity distribution systems that sustainably serve the needs of our customers and maximise the long-term return on investment for our owners. We shall comply with all legal, regulatory and environmental requirements placed upon us and will not compromise the safety of our employees, our customers or the public.

Implementation of this policy is the role of the Asset Management function within the company. They have to plan and oversee each asset-enhancing project, including the areas of policy, strategy and planning, design and delivery assurance. This includes the assessment and mitigation of network risk before, during and after the duration of the project.

Other functions within the organisation that are directly concerned with network risk include Engineering Projects, which oversees and possibly carries out the construction project, and Network Control, which operates the network on a day-to-day basis, managing the risk that arises from both planned and unplanned events. The research underpinning this chapter was undertaken in partnership with two of the 14 DNOs operating in the United Kingdom. It included time spent with all three functions to gain familiarity with both the principles and the practice of managing and operating a typical UK distribution network. The case studies that follow are all based on these experiences.

4 Choice of Methodology: A First Case Study

Figure 2 shows a part of a 66 kV network, feeding four primary substations (designated A, B, C and D) from a single SP, which is itself fed by the National Grid. Transformers T3 at A, T1 and T2 at C, and T1 and T2 at D feed single industrial customers. The other six transformers supply load, which is mostly residential.

The overhead lines between the SP and A are around 6 km in length, and are projected to become overloaded due to demand increases within the next 5 years. In addition, they are about 40 years old, in a hilly, coastal, industrial environment, and are in poor condition. It is planned to replace the conductors of both circuits in successive summer seasons (when demand drops to around 70% of the normal winter maximum).

Although the circuits to C and D are electrically independent of those to A and B, they are adjacent geographically. At a point about half way between SP and C, those circuits cross over the lines between A and B, their respective towers being about 100 m apart. This gives the possibility of making a temporary interconnection between the circuits (shown as a dotted line in Fig. 2), as a way of mitigating network risk when one of the circuits from SP to A is out of service for the conductor replacement work, which is scheduled to last for about 4 months.

The questions raised by this case study include the following:

- How great is the present level of network risk?
- How much is risk increased during construction?
- How much is risk reduced once it is completed?
- Can risk during construction be reduced by temporary interconnection?

One of the aims of this case study was to apply each of the four Brown and Hanson methodologies (see Sect. 2.5), in turn, to evaluate their suitability for this kind of network risk analysis.

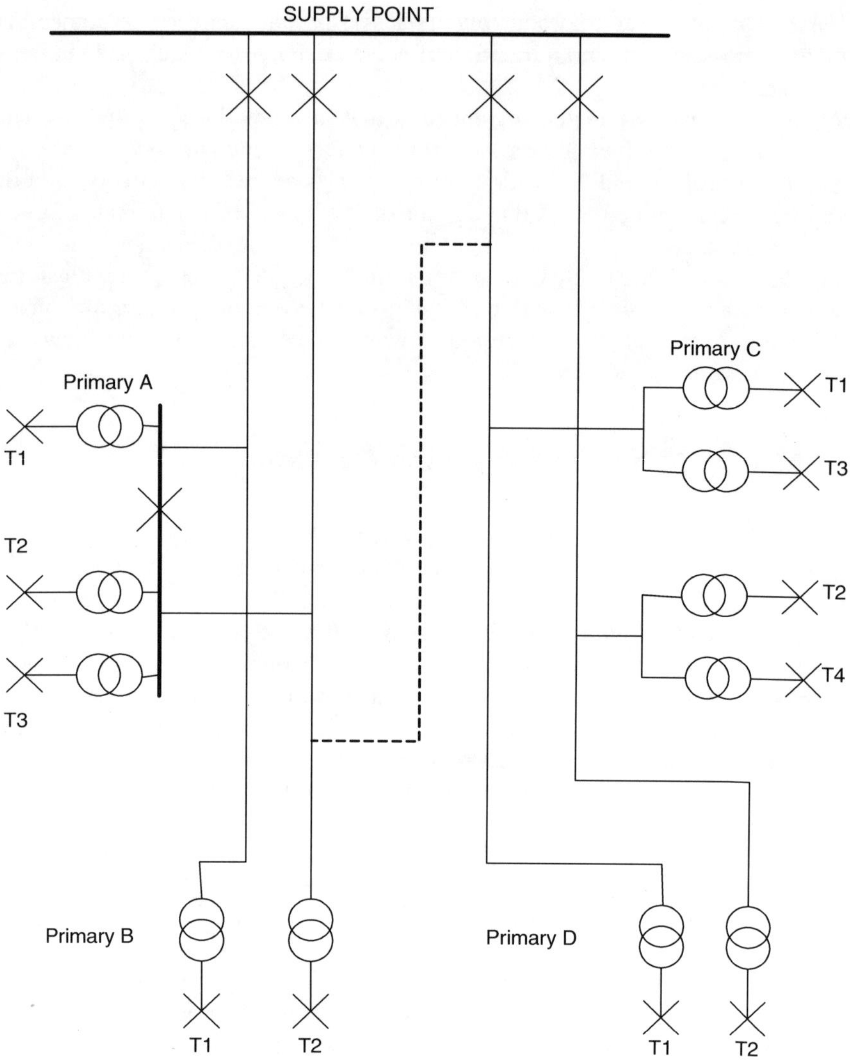

Fig. 2 First case study network configuration

4.1 Network Modelling

Network modelling was used to evaluate voltages and power flows both under normal operation and with one or more circuits open, as would be the case during construction. As expected, in both normal operation and in most of the (n – 1) scenarios, the power flows and voltage drops were within statutory limits. However, at maximum winter loads, some power flows in some (n – 1) conditions exceeded the capability of the circuit, which provided the justification for the re-conductoring project. Because of the network topology, a number of (n – 2) scenarios, that is losing a second circuit while the first is out for maintenance, results in the loss of supply for some customers.

With the temporary interconnection installed between circuits, however, an improved level of reliability was possible in (n-2) scenarios, with lower voltage sags and power flows. The situation was further helped by the possible availability of DG from one of the industrial sites.

4.2 Markov Modelling

Markov modelling was set up with four system states and six transitions between them, as shown in Fig. 3. In this model, P2 is the state in which both circuits are in operation (and also the probability of being in that state). P1 is the (n − 1) state, with a single circuit in operation, and P0 is the (n-2) state, with both circuits out and customers off supply. P0+ designates the state where both circuits are still out, but some customers have been restored by rerouting their supply from other substations.

The solid lines represent transitions by single circuit failure (P2 to P1, and P1 to P0) or by dual circuit failure (P2 to P0). The dotted lines represent successive stages of repair or restoration of supply. National historic data was used to derive failure rates and repair/restoration rates. Constructing the Markov matrix with these values and solving it gives the results shown in Table 2 for normal network operation. Adjustments can be made to the parameter values, and to the number of states and of transitions, to represent other configurations, for example during the construction project, with or without the additional interconnection. In each case, the expected number of hours per year in each defined state can be calculated, and used to derive the expected annual costs (increased penalty or de creased rewards) of CI and of CML.

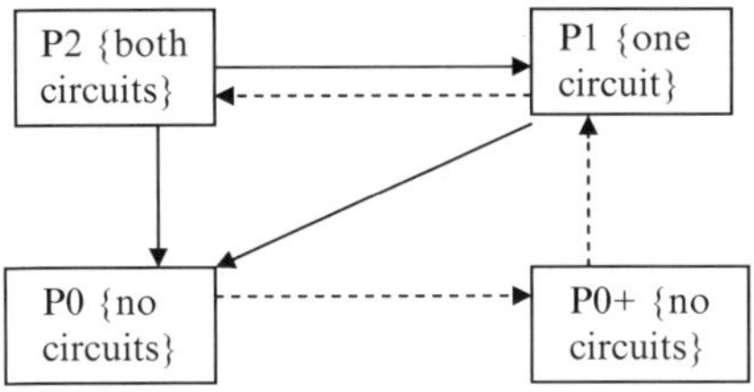

Fig. 3 Markov states and transitions

Table 2 Markov results (normal operation)

State	Probability	Hours/Year
P2	0.9966	8, 730
P1	0.00339	29.7
P0	0.0000137	0.12
P0+	0.0000437	0.38

4.3 Analytical Simulation

Analytical simulation was used in two ways. First, it was applied to the reinforcement required as a result of expected increased loads. The expected load profile over the year, the thermal characteristics of the overhead line in different seasons and the likelihood of (n – 1) where the full load is carried by a single circuit were all input to a calculation, which suggested that excessive conductor temperature leading to an infringement of ground clearances, although possible, was unlikely. In this event, the situation would probably last for a few hours, and there would be the possibility of improving the situation by re-routing a proportion of the load.

Second, analytical simulation was applied to the case for replacing the conductors on the grounds of excessive age and wear. An equation for asset failure rate was derived with the help of historic national data:

$$\mathbf{R} = \mathbf{R_0} + \mathbf{R_A}(\mathbf{1}+\mathbf{p})^{\mathrm{N}}$$

where $\mathbf{R_0}$ is the rate of failure not related to age, **p** is the annual rate of increase in age-related failures, **N** is the age of the asset in years, and $\mathbf{R_A}$ is a constant selected to fit the available data, then **R** gives the overall failure rate. In this case study, values were substituted to give

$$\mathbf{R} = 0.008 + 0.000284(1.05)^N$$

leading to annual expected failure rates for each kilometre of line as shown in Table 3:

These results indicate that failure rates can be expected to increase gradually at first, but more steeply after 40 years and much more steeply after 60 years. If this is so, then conductor replacement becomes progressively more cost-effective as the asset ages. In theory, an optimal age for asset replacement could be calculated. In practice, the cost of increased failure rates (CI, CML and repairs) incurred by not replacing an asset can be compared with the expected capital cost of replacing it to derive a value for the project, in terms of payback time or of discounted cash flow.

Table 3 Predicted failure rates

Age	Rate
0	0.0083
20	0.0088
40	0.0100
60	0.0133
80	0.0221

Table 4 Probabilities of various line temperatures

Year	< 50°C	50–65°C	65–75°C	> 75°C
2008	0.99981	0.00019	0	0
2014	0.99924	0.00068	0.00008	0
2020	0.99823	0.00117	0.00043	0.00017

4.4 Monte-Carlo Simulation

Monte-Carlo simulation was used as an add-on to the Analytical Simulations already described. In the reinforcement case, a run of 100,000 trials was used to investigate how often the conductors would be operating above the design temperature of 50°C. This was run to represent the years from 2008 through to 2020, by which time the maximum level of demand would have increased and the condition of the assets would have further deteriorated due to age, resulting in an increased proportion of time under single circuit operation. Results are shown in Table 4:

The design temperature for these circuits is 50°C, and while permanent damage due to melting of conductor grease or permanent deformation may not occur if the conductor is operated above 50°C for short periods of time, statutory ground clearances might be infringed.

A probability of 0.00017 corresponds to a single 3-h excursion once every 20 years, which might be considered an acceptable risk, particularly as load could be re-routed at the 11 kV level under these circumstances. It seems, therefore, that the case for replacing the conductors from a reinforcement perspective alone before the year 2020 is not strong.

Applying Monte-Carlo simulation to the replacement of ageing assets, inputs that had been single values in the analytical simulation could now be modelled as probability distributions. For example, the time taken to restore customers by rerouting through the 11 kV network was taken to be a rectangular distribution between 20 and 120 min. The number of customers thus restored was modelled as a triangular distribution between 4,000 and 19,000.

The output was then not only expected values, but also the possible range of those values. For example, the expected cost of CML under one particular set of input parameters was £43 k. But the range was from zero (in over 90% of runs) up to £360 k (in the fifth worst year per century) and just over £1,000 k (in the worst year per century). This kind of detailed profile of the risk to which the network is exposed by certain circumstance enables the industrial managers to take informed decisions about the acceptability of risks, and the appropriate level of mitigating measures to take.

4.5 Case Study Conclusions: Measuring Network Risk

As a result of this first case study, it became clear that a definitive measurement of network risk was required. If possible, this should reflect the additional costs of each failure, multiplied by the probability of such failure.

The costing of network risk could include a number of factors. For example, the cost of loss of supply to customers is clearly an ingredient, and one that is included in some studies as part of the overall cost (Kaipa et al. 2007). However, it is difficult to assess this cost accurately and objectively, and it is not a cost directly borne by the DNO. It could be argued that these costs should be reflected in the CI and CML rewards and penalties imposed by OFGEM, and that including them separately is a form of double-counting. It was decided to include the expected costs of CI and of CML in the definitive measurement of network risk, but not to attempt to include any customer costs.

It was also decided to exclude customer compensation payments, although these costs are incurred directly by the DNO. The reason for this is that most such costs arise from faults at lower voltages (20 kV and below), while the present study concentrates primarily at HV (33 kV and above). Awareness of the 18 h cut-off duration for triggering compensation payments tends to dictate restoration strategies, and makes it unlikely that a fault at HV would be allowed to exceed this duration if at all possible. This kind of consideration makes it difficult to model as an essentially random, probabilistic process.

One cost that was included, however, is an estimate of the additional unplanned repair costs incurred by a circuit failure. This is a highly variable quantity, depending as it does on the precise nature of the fault. It is a significant amount, however (a figure of £20 K would be typical), and it arises whether or not the fault causes customer loss of supply.

The definition adopted of network risk, measured in £K, was the sum of three components:

- The expected extra penalty, or reduced reward, for CI
- The expected extra penalty, or reduced reward, for CML
- The expected incremental cost of unplanned repairs

As an illustration of this, the extra risk to supply during a construction project could be evaluated and set against the reduced risk that could be expected as a result of renewed assets with lower failure rates, as shown in Fig. 4.

The dotted line indicates the rate at which network risk would continue to increase as a result of ageing assets if no rebuilding were done. The solid line shows a greatly increased risk during summer 2011 when the first circuit is switched out to allow the re-conductoring (and the aged second circuit takes the full load). The risk is still increased, but less so, in the following summer when the second circuit is switched out for rebuilding and the new circuit takes the full load. The two-circuit risk decreases when one line has been replaced, and decreases further when both lines have been replaced. It then remains at a constant low level for a number of years.

The overall expected value of the project, in terms of reduced risk, can be determined by subtracting the area under the solid line from the area under the dotted line. This process is subject to a number of management decisions – how many years of future to include, whether and how much to discount future cash flows, and whether to use expected values or some kind of worst-year scenario. But the technique of

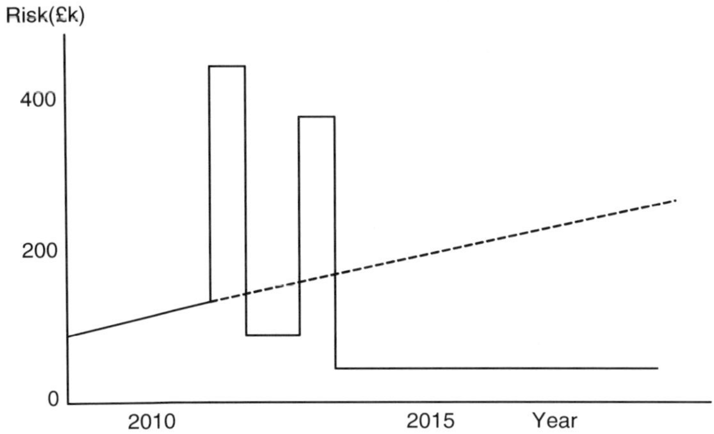

Fig. 4 Profile of risk against time

evaluating risk in probabilistic terms is of value in making investment decisions of this kind.

As a second illustration of this technique, it was used to evaluate the temporary interconnector shown in Fig. 2. The reduction in network risk over the full duration of the construction project as a result of the interconnector was estimated at around £200 K. This could be compared with the capital cost of installing such an interconnector (likely to be considerably less) to help the project managers decide whether such interconnection would be worthwhile.

4.6 Case Study Conclusions: Preferred Methodologies

Perhaps the main result of this first case study was its detailed evaluation of the four methodologies. This has in turn influenced subsequent case studies, as described in Sect. 5. The conclusions are as follows:

Network modelling considers only a single steady state, although multiple runs can be used to investigate discrete changes. There is neither inbuilt time dimension, nor any treatment of probabilities. As such, it is useful mainly as a supplementary tool in network risk studies to establish the acceptable operating window.

Markov modelling is a well-accepted and commonly used tool. It does not incorporate a time dimension, and uses only average or expected values of input and output data. This means that it can only predict average system behaviour. It gives no indication of the variation of that behaviour, for example as regards outage durations. It is heavily mathematical, making it harder for non-mathematical users to understand and computationally demanding. Because of its rigid and prescriptive format, it tends to force case-studies into its own mould, possibly distorting them.

As such, it is judged to be a tool for occasional use within the present set of network risk studies.

Analytical simulation is not a single method, but encompasses a variety of techniques with a common probabilistic approach. It has the versatility to adapt to the range of issues involved in evaluating network risk. Its main drawbacks are its dependency on a wide range of data inputs, not all of which are equally reliable, and its use of average or expected values rather than a probability distribution. Because of its variety, it is harder to automate, and possibly more demanding of the time of skilled network engineers. It is, as Brown and Hanson also concluded, the best overall methodology for these network risk studies (Douglas et al. 2005).

Monte-Carlo simulation uses probability distributions in repeated trials to predict a pattern of behaviour, not just an average. This is especially valuable for sensitivity studies. Its main disadvantage is that modelling a network in this way is complicated and can be computationally demanding. It is a very useful tool for network risk studies, within an overall analytical simulation approach.

5 Further Short-term Case Studies

In the following section, a number of additional case studies are briefly discussed. Each has been included because it illustrates certain features of short-term network risk analysis, typically measured in £K.

5.1 Suburban and Rural Networks

The second case study concerns a complete rebuild, planned for a 66/11 kV primary substation serving a small coastal town and surrounding rural area (see Fig. 5), supplied by a long, double circuit carried on wooden poles at 66 kV. The feeding arrangement was found to pose a relatively high risk even under normal operation, and this risk clearly increased when one circuit was out of service for a number of weeks to enable the rebuilding works to be completed.

One issue considered to be material to this project is the likelihood of double circuit failure. This could be due to the following:

- A single cause such as the collapse of a support carrying the two circuits or the accidental or deliberate cutting of two cables in the same trench.
- An incipient fault that manifests itself when one circuit carries the increase in load imposed on it during a circuit outage.
- Coincidental failure of the second circuit while the first is out of service for planned maintenance, construction, or unplanned failure that has not yet been repaired.

The total probability of these three kinds of double circuit failure could typically be as high as 20% of all failures, and they will generally result in customer loss of supply. This has to be accurately modelled.

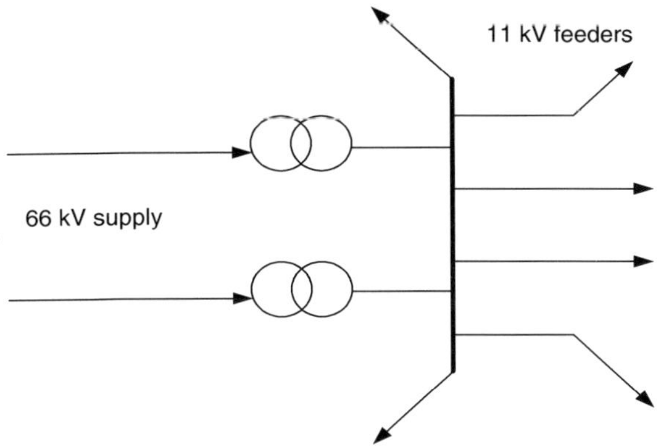

Fig. 5 Rural/suburban primary substation

A second issue is the possibility of re-routing supply from an alternative primary substation through the MV (11 kV) network. There were three such alternative primaries, with direct or indirect interconnection to all six of the 11 kV feeders from the assessed primary.

- The alternative primaries have limited spare capacity, particularly at times of maximum load.
- The interconnecting 11 kV feeders (typically underground cables, with a normally open point which could be closed) have limited capacity.
- Reconfiguration of 11 kV is usually not automated, and so must be done manually, which takes time.
- Two landfill gas generators are connected to the network and could be relied on to some extent.
- The alternative primaries are all themselves supplied from the same 132/66 kV SP, via long rural lines, and so are themselves vulnerable (and no use in the event of a problem at the SP or its incoming circuits).

All these considerations were modelled at an appropriate level of detail, using both analytical and Monte-Carlo simulation. The extra risk to supply during construction could be evaluated and set against the reduced risk that could be expected as a result of renewed assets with lower failure rates, as shown in the first case study (Fig. 4)

It was concluded that the risk, particularly at a time of planned outages during construction, could be reduced by active network management at the 11 kV level, with due regard for voltage levels and current limitations. However, in general, it is hard to quantify both the increase in network risk during the construction project and the possible decrease as a result of project completion. This is primarily because of the large number of available reconfiguration options at the 11 kV level. In the case study, these factors were not separately quantified, mainly because of lack of time. It would be possible to do so, however, by applying analytical

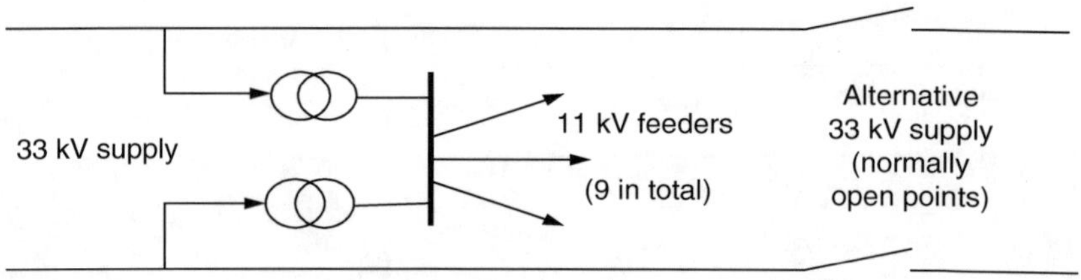

Fig. 6 Central urban primary substation

simulation techniques, possibly incorporating Monte Carlo variability, on an individual feeder-by-feeder basis.

5.2 City Centre Networks

The third case study differs from the second in a number of respects (see Fig. 6):

- The primary substation is fed by underground cables at 33 kV, instead of overhead lines at 66 kV.
- The rebuilding project is for the 11 kV switchboard only (the transformers having been recently replaced).
- The substation is located in a city centre, with considerably more interconnection, making a meshed network, which is, however, operated radially.

This extra interconnection significantly reduces the level of risk before, during and after the reconstruction work. To be more specific, there are four separate 33 kV circuits (two of which are normally open) feeding into the substation from two different directions and two different SPs. Each of the nine 11 kV feeders has at least two interconnections, to different primary substations, typically with loads well below their capacity and fed from different SPs.

Applying similar analysis to this situation, the levels of risk became vanishingly small. In all the scenarios considered, involving the loss of 1, 2 or even 3 independent circuits, there was a range of possible restoration options for all customers, usually in a short time. The rebuilding of the switchboard could be justified in terms of good stewardship of time-expired assets, but would be difficult to justify in terms of quantifiably reduced risk alone.

5.3 Subtransmission at 132 kV

Figure 7 shows the network around a nodal point (NP), which includes a switching station whose aged assets are due for replacement. Load D (at Substation D) and a proportion of load B under normal operating conditions are supplied from the

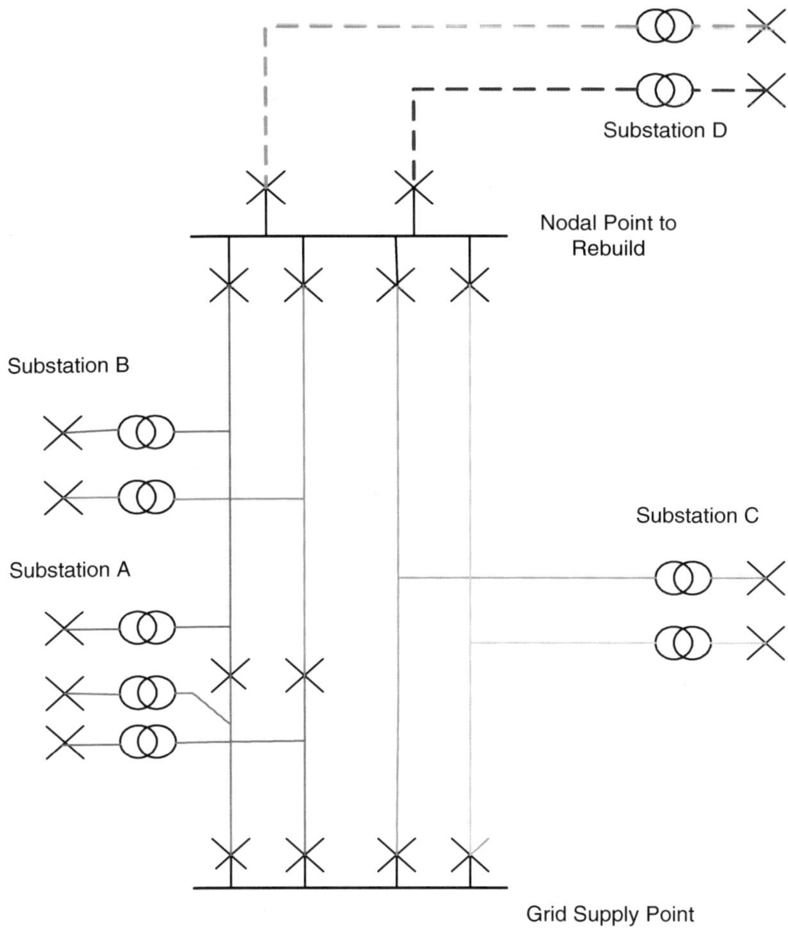

Fig. 7 132 kV configuration

grid supply point (GSP) via NP. Loads A, load C and the remainder of load B are supplied directly from GSP. Double circuits are used throughout this part of the network. Dotted lines indicate underground cable, solid line represents overhead lines and X represents a circuit breaker.

The replacement of assets is a major construction project, expected to last over two summer seasons (when UK demand is lowest) and to cost around £16 M. During construction, one or more circuits will be out of service, and the risk to customer supplies would then be increased. There are a number of options for carrying out the project, such as whether to rebuild in situ or on an adjacent green field site. The in situ solution costs less, but probably involves longer circuit outages and consequently a greater risk.

The rebuilt NP would have an expected lifetime of at least 40 years, and this raises questions as to whether the straightforward replacement of the existing configuration is the best solution to those challenges to network reliability that are likely to arise over the next 40 years. One option concerns a separate 132 kV double circuit, not shown in Fig. 7, which feeds two other loads from a different GSP, but passes within a few metres of NP. It would be relatively straightforward, and incur relatively little extra cost, to connect this double circuit into NP. This would give an extra supply option for all six loads, thereby increasing network flexibility and possibly overall reliability. This option has been included as part of the construction project.

Interconnecting this adjacent double circuit at NP not only gives immediate flexibility benefits, but also makes the network more robust to face the challenges of the next 40 years. However, it needs to be determined whether this extra work would be cost-effective. Indeed, it could be possible to eliminate NP from the network altogether. Figure 8 shows a possible reconfiguration that does this. This would cost significantly less than the replacement of NP, and leave a simpler network, although one with rather fewer options and possibly increased risk.

A model was constructed incorporating probabilities of component failure, and likely consequences of such failure, for various scenarios. The base run of this model evaluated the network risk inherent in the present configuration across all six loads. The total risk came to £67 K for CI, £63 K for CML and £32 K for repairs, a total of £162 K. It is against these levels of risk that different construction and network configuration options should be assessed.

For example, the additional risk incurred due to circuit outages across the whole construction period is an expected £42 K if construction takes place on a green field site. If, however, construction has to take place in situ, then the extended outage periods increases this risk by £66 K, to £108 K. This extra £66 K of risk can be compared with the additional capital cost of purchasing a new site. The risk reduction once new equipment is installed proved harder to assess, as the technology used is still fairly new at this voltage level, and so accurate expected failure rates are not known.

The alternative configuration shown in Fig. 8 was also evaluated using this short-term risk methodology. The expected annual network risk increases by £45 K over the base run. This extra annual cost can be compared with the one-off capital cost of around £8 M to rebuild NP. On this basis, the alternative configuration appears to be a cost-effective option. However, there are other considerations to be taken into account. The first concerns power flows within the revised configuration. A network analysis was carried out to compute steady-state power flows and voltage levels throughout the revised network. This modelling confirmed that, under all $(n - 1)$ outage scenarios, the revised network was not overloaded, and that voltage levels remained within statutory limits.

A second consideration concerns the size of protection zones (PZ) resulting from this configuration. Figure 2 shows that one circuit creates a PZ with switching on four separate sites. This is the upper limit permitted by the UK grid code

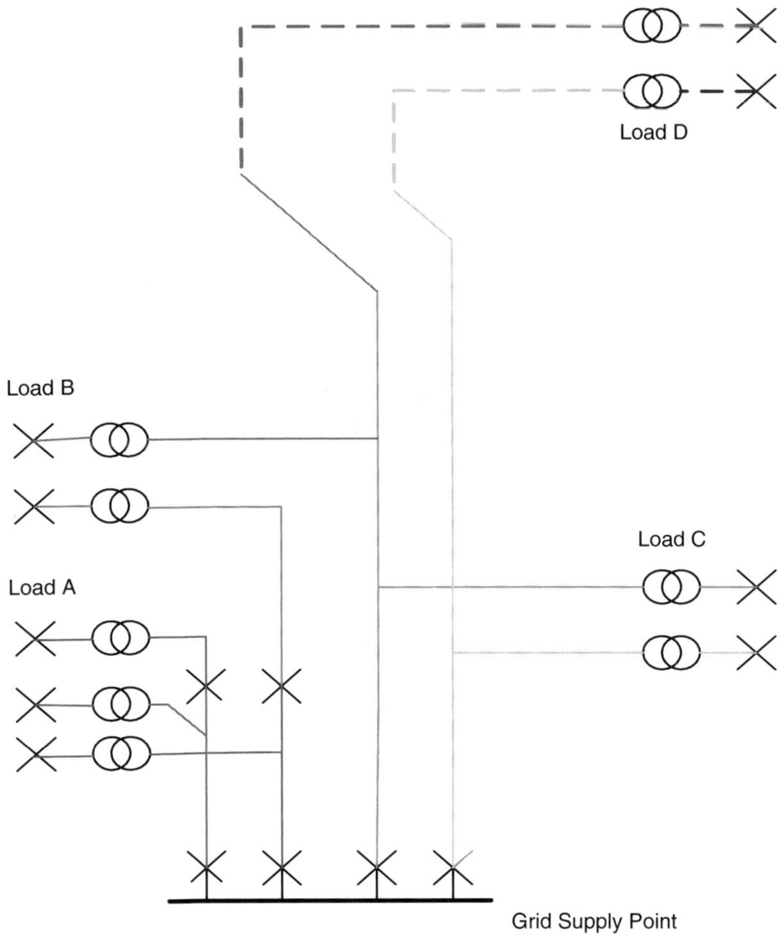

Fig. 8 Alternative configuration without NP

for operating 132 kV circuits, and is above the upper limit of three separate sites currently recommended by the DNO.

A third consideration is the use of average values for model input parameters, such as failure rates and restoration times, which produces the average or expected risk cost per year over a number of years. However, CE Electric UK are also concerned about their liability, an indication of the possible risk cost in a particularly bad year, such as the worst in 10 years or even in 100 years. To address this concern, Monte Carlo simulation was used, using the risk solver package.

In this methodology, the frequency and severity of circuit failures are not fixed at average values, but are sampled from a representative probability distribution. Restoration times are likewise sampled from an underlying distribution. A large number of possible years (50,000 was the number set in the model) were evaluated and ranked. The 90th percentile of this output then gives the extra network risk incurred by configuring the network without NP in the worst year in 10 years. This came to £140 K (as against £45 K for the average year). The worst year in 100 years (the 99th percentile) gave an additional risk of £277 K. These are the figures that the DNO management would consider in deciding whether the additional liability was acceptable when set against possible capital expenditure savings.

6 Conclusions

From this brief overview of network risk quantification and mitigation, a number of significant points have emerged. These include in particular the following:

- The level of risk of customer loss of supply in electricity distribution networks is presently relatively low. However, a number of factors lead to expectations of increased risk in the short term (Sect. 1.1) and still more in the longer term (1.2).
- Research into network risk has tended to concentrate on generation and transmission rather than on distribution. Within distribution networks, the emphasis has tended to be on medium voltages rather than on high voltages (33–132 kV) (2.1).
- There has been a shift in emphasis towards an asset management approach within distribution companies, accompanied by increased levels of regulation. This has had significant impact on the way that network risk is perceived and managed (2.3).
- Maintenance issues can be addressed by risk analysis, and a least-cost solution involving a balance between increased maintenance costs and reduced outage costs can be determined (2.4)
- A number of methodologies are available to analyse network risk (2.5). Of these, the most generally suitable seems to be analytical simulation (4.6), with network modelling, Markov modelling and Monte Carlo simulation providing additional support as appropriate.
- The organisation and regulation of the UK network is reasonably representative of that in other developed countries (3.1–3.3), and is used as the basis for the case studies considered in this chapter.
- Short-term network risk can be measured by the expected annual cost of CI, CML and extra unscheduled repairs (4.5). The units for such measurements are £K. This methodology can be applied to rural and suburban networks (5.1), central urban networks (5.2) and subtransmission systems (5.3).
- The authors have continued to use these techniques to analyse both short-term and long-term network risk issues within the UK network, including applying them to more complex network topologies and evaluating the automation of manually operated components such as switches. Their experience has been that the heuristic approach described in these case studies is particularly well-suited to the problems that arise at sub-transmission voltage levels, because the versatility of the methodology allows for the wide variety of network topologies and operating issues that occur at these voltages.

References

Allan RN, Dialynas EN, Homer IR (1979) Modelling and evaluating the reliability of distribution systems. IEEE Trans Power Apparatus Syst PAS-98(6):2181–2189

Bayliss C, Hardy B (2007) Transmission and Distribution Electrical Engineering, Kindle

Billinton R, Acharya JR (2006) Weather-based distribution system reliability evaluation. IEE Proc Generat Transm Distrib, 153(5):499–506

Billinton R, Allan RN (1996) Reliability evaluation of power systems, 2nd edition. Plenum, New York and London

Billinton R, Zhaoming P (2004) Historic performance-based distribution system risk assessment. IEEE Trans Power Deliv 19(4):1759–1765

Blake SR, Taylor PC, Creighton AM (2008) Distribution network risk analysis, MedPower 2008 conference, Thessaloniki, Greece, November 2008

Bollen M, Holm A, He Y, Owe P (2007) A customer-oriented approach towards reliability indices. CIRED 18th international conference on electricity distribution, Vienna, Austria, May 2007

Boyle G, Everett B, Ramage J (2003) Energy systems and sustainability. Oxford University Press, Oxford

Brown RE, Hanson AP (2001) Impact of two-stage service restoration on distribution reliability. IEEE Trans Power Syst 16(4):624–629

Brown RE, Spare JH (2004) Asset management, risk and distribution system planning. IEEE PES Power System Conference and Exposition, 2004

Chow M Y, Taylor LS, Chow M-S (1996) IEEE Trans Power Deliv 1652–1658

Cooper D, Grey S, Raymond G, Walker P (2005) Managing risk in large projects and complex procurements. Wiley, London

Douglas J, Millard R, Fourie L (2005) Reliability worth for a restructured electricity distribution industry. CIRED 18th international conference on electricity distribution, Turin, Italy, June 2005

ENA Engineering Recommendation P2/6 (2006) "Security of Supply", Energy Network Association Engineering Directorate, London, July 2006, http://ena-eng.org

Endrenyi J, Aboresheid S, Allan RN, Anders GJ Asgarpoor S, Billinton R, Chowdhury N, Dialynas EN, Fipper M, Fletcher RH, Grigg C, McCalley J, Meliopoulos S, Mielnik TC, Nitu P, Reppen ND, Salvaderi L, Schneider A, Singh Ch (2001) The present status of maintenance strategies and the impact of maintenance on reliability. IEEE Trans Power Syst 16(4):638–646

Freeman LA, Van Zandt DT, Powell LJ (2005) Using a probabilistic design process to maximise reliability and minimize cost in urban central business districts. CIRED 18th international conference on electricity distribution, Turin, Italy, June 2005

Grey S (1995) Practical risk assessment for project management. Wiley, London

Hatziargyriou N (chair) (2008) Electricity business in South-Eastern Europe. MedPower 2008 conference, Thessaloniki, Greece, November 2008

Kaipa T, Lassila J, Partanen J, Lohjala J (2007) A cost analysis method for storm caused extensive outages in distribution networks. CIRED 19th international conference on electricity distribution, Vienna, Austria, May 2007

Li W, Korczynski J, (2004) A reliability-based approach to transmission maintenance planning and its application in BC hydro system. IEEE Trans Power Deliv 19(1):303–308

Li W, Zhou J, Lu J, Yan W (2007) A probabilistic analysis approach to making decision on retirement of aged equipment in transmission systems. IEEE Trans Power Deliv 22(3):1891–1896

Nordgard DE, Sand K, Gjerde O, Catrinu MD, Lassila J, Partanen J, Bonnoit S, Aupied J (2007) A risk based approach to distribution system asset management and a survey of perceived risk exposure among distribution companies. CIRED 19th international conference on electricity distribution, Vienna, Austria, May 2007

Northern Electric Distribution Ltd. (2007) Long term development statement, http://ceelectricuk.com/lib

Oravasaari M, Pylvaneinen J, Jarvinen J, Maniken A, Verho P (2005) Evaluation of alternative mv distribution network plans from reliability point of view. CIRED 18th international conference on electricity distribution, Turin, Italy, June 2005

PB Power report (2007) Future Network Architecture, commissioned by BERR, http://www.berr.gov.uk/files/file46168.pdf

Ramirez-Rosado IJ, Bernal-Agustin JL (2001) Reliability and costs optimization for distribution networks expansion using an evolutionary algorithm. IEEE Trans Power Syst 16(1): 111–118

Scutariu M, Albert H (2007) Corrective maintenance timetable in restructured distribution environment. IEEE Trans Power Deliv 22(1):650–657

Slootweg JG, Van Oirsouw PM (2005) Incorporating reliability calculations in routine network planning: theory and practice. CIRED 18th international conference on electricity distribution, Turin, Italy, June 2005

Stowell P (1956) d'E A review of some aspects of electricity distribution. address to Mersey and North Wales Centre of IEE, 1956–7 session

Vose D (1996) Quantitative risk analysis. Wiley, London

Vovos PN, Kiprakis AE, Wallace AR, Harrison GP (2007) Centralized and distributed voltage control: impact on distributed generation penetration. IEEE Trans Power Syst 22(1):476–483

Index

Printed by Printforce, the Netherlands